4th Maritime Safety International Conference (MASTIC) 2024

IOP Conference Series: Earth and Environmental Science
Volume 1423

Bali, Indonesia
25-28 August 2024

ISBN: 979-8-3313-1154-4
ISSN: 1755-1307

Printed from e-media with permission by:

Curran Associates, Inc.
57 Morehouse Lane
Red Hook, NY 12571

Some format issues inherent in the e-media version may also appear in this print version.

Institute of Physics
Dirac House, Temple Back
Bristol BS1 6BE UK

Phone: 44 1 17 929 7481
Fax: 44 1 17 920 0979

techtracking@iop.org

Additional copies of this publication are available from:

Curran Associates, Inc.
57 Morehouse Lane
Red Hook, NY 12571 USA
Phone: 845-758-0400
Fax: 845-758-2633
Email: curran@proceedings.com
Web: www.proceedings.com

TABLE OF CONTENTS

IOP Conf. Series: Earth and Environmental Science **1423** (2024) 011001 doi:10.1088/1755-1315/1423/1/011001

Preface

Welcome to the 4th Maritime Safety International Conference (MASTIC) 2024

The Department of Marine Engineering, Faculty of Marine Technology, Institut Teknologi Sepuluh Nopember (ITS), Surabaya, in collaboration with the Center of Excellence for Maritime Safety and Marine Installation (PUI KEKAL), is proud to present the 4th MASTIC 2024. This year, we are delighted to host the conference started from August 25th to 28th in Prima Plaza Hotel Sanur, Bali, Indonesia.

The conference theme, "*Enhancing Maritime Safety towards Sustainable Marine Technologies and Environmental Protection*", attracted a diverse range of participants, presentations, keynote speeches and workshop session. Four prominent keynote speakers are Ir. Raymond Ivan H.A.S. that talked about the role of government to the sustainable shipping and digitalization, Professor Akira Sou of Kobe University presented about Recent Technology Development for Marine Engines, Dr. Junghan Lee as the CEO of Cluster LNG South Korea that delivered a presentation about LNG Technology in Indonesia and the last speaker, Bani Maulana Mulia, the President Director of PT Samudera Indonesia Tbk who delivers recent developments in Samudera Indonesia that support sustainable shipping and emission reduction.

After being held online twice, this year the conference brings the in-person format that allows more meaningful interactions, networking opportunities, and a deeper engagement for all participants. The conference featured 49 accepted papers that authored by researchers from 6 different countries: Indonesia, Japan, South Korea, Croatia, Montenegro, and Germany. These papers represented the latest advancements in the field of maritime safety and are a valuable contribution to the conference.

We would like to express our sincere gratitude to all the authors who submitted their research, the Technical Program Committee and reviewers for their dedication, and the partnership sponsors, PT Samudera Indonesia Tbk, BP Indonesia, PT Semen Indonesia (Persero) Tbk, PT Terminal Peti Kemas Surabaya, PT Petrokopindo Cipta Selaras, PT Pertamina Nusantara Regas, PT Pelindo Marine Service, and Husky-CNOOC Madura Limited for their generous support.

We are also pleased to announce that the MASTIC 2024 proceedings are published in the IOP Conference Series: Earth and Environmental Science (EES). We hope that this conference proceeding could be a valuable resource for researchers and practitioners in the field of maritime safety and marine engineering.

General Chair of MASTIC 2024
Dr. Fadilla I Prastyasari
Institut Teknologi Sepuluh Nopember, Surabaya, Indonesia
September 2024

IOP Conf. Series: Earth and Environmental Science **1423** (2024) 011001 doi:10.1088/1755-1315/1423/1/011001

List of Conference Committee
The 4th Maritime Safety International Conference (MASTIC) 2024

ADVISORY COMMITTEE

1	Bambang Pramujati, S.T., M.Sc.Eng., Ph.D.	Rector of Institut Teknologi Sepuluh Nopember
2	Dr. Eng. Kriyo Sambodho, ST., M.Eng.	Director of DIKST Institut Teknologi Sepuluh Nopember
3	Dr. Trika Pitana, ST., M.Sc.	Dean of Marine Technology Faculty, Institut Teknologi Sepuluh Nopember
4	Fadillatul Taufany, S.T., Ph.D.	Director of DRPM Institut Teknologi Sepuluh Nopember
5	Beny Cahyono, S.T., M.T., Ph. D	Head of Department of Marine Engineering, Institut Teknologi Sepuluh Nopember
6	D.Eng. Ir. AA Masroeri, M. Eng	Department of Marine Engineering, Institut Teknologi Sepuluh Nopember

TECHNICAL PROGRAM COMMITTEE

1.	Prof. Dr. Ketut Buda Artana	Institut Teknologi Sepuluh Nopember
2.	Prof. Ir. I Ketut Aria PU., M.Sc., Ph.D.	Institut Teknologi Sepuluh Nopember
3.	Prof.Ir. Eko Budi Djatmiko, M.Sc., Ph.D	Institut Teknologi Sepuluh Nopember
4.	Dr. Ir. Wasis Dwi Aryawan, M.Sc.	Institut Teknologi Sepuluh Nopember
5.	Dr.-Ing. Ir. Setyo Nugroho	Institut Teknologi Sepuluh Nopember
6.	Dr. Herman Pratikno, ST., MT.	Institut Teknologi Sepuluh Nopember
7.	Prof. Dr. Takeshi Nakazawa	International Association of Maritime Universities
8.	Prof. Takeshi Shinoda	Kyushu University
9.	Dr.-Ing. Wolfgang Busse	Hochschule Wismar University University
10.	Prof. Koji Gotoh	Kyushu University
11.	Prof. Hirotada Hashimoto	Kobe University
12.	AAB. Dinariyana Dwi P., ST., M.ES., Ph.D.	Institut Teknologi Sepuluh Nopember
13.	Dr. Eng. Rudi Walujo Prastianto, ST., MT.	Institut Teknologi Sepuluh Nopember
14.	Prof. Himsar Ambarita	University of Sumatera Utara
15.	Dr. Fredy Kurniawan	Biro Klasifikasi Indonesia
16.	Dr. Mohammad Zaki	Biro Klasifikasi Indonesia
17.	Dr. Arif Kurniawan	Biro Klasifikasi Indonesia
18.	Dr. M. Badrus Zaman, ST., M.Sc.	Institut Teknologi Sepuluh Nopember
19.	Sutopo Purwono Fitri, S.T., M.Eng., Ph.D.	Institut Teknologi Sepuluh Nopember
20.	Dr. Eng. Yeyes Mulyadi, ST., M.Sc.	Institut Teknologi Sepuluh Nopember
21.	Prof. Dr. I Made Ariana,S.T.,M.Sc	Institut Teknologi Sepuluh Nopember
22.	Dr. Eng. I Putu Sindhu Asmara, S.T., M.T.	Surabaya State Shipping Polytechnic

IOP Conf. Series: Earth and Environmental Science **1423** (2024) 011001 doi:10.1088/1755-1315/1423/1/011001

ORGANIZING COMMITTEE

1	Dr. Fadilla Indrayuni Prastyasari M.Sc.	Department of Marine Engineering, Institut Teknologi Sepuluh Nopember
2	Dr. Dhimas Widhi Handani	Department of Marine Engineering, Institut Teknologi Sepuluh Nopember
3	Dr. Emmy Pratiwi, S.T.	Department of Marine Engineering, Institut Teknologi Sepuluh Nopember
4	Thariq A. Akbar S.T., M.T.	Department of Marine Engineering, Institut Teknologi Sepuluh Nopember
5	Dwi Suasti Antara, M.MT.	Department of Marine Engineering, Institut Teknologi Sepuluh Nopember
6	Hayy Nur Abdillah, S.T., M.T	Surabaya State Shipping Polytechnic
7	Danuja Wijayanto, S.T., B.Eng., M.Sc.	Department of Marine Engineering, Institut Teknologi Sepuluh Nopember
8	Rafi Maulana	Department of Marine Engineering, Institut Teknologi Sepuluh Nopember
9	Dimas Akmal Fakhriyza	Department of Marine Engineering, Institut Teknologi Sepuluh Nopember
10	Galang Jaya Kusuma	Department of Marine Engineering, Institut Teknologi Sepuluh Nopember
11	I Wayan Gede Anantha Pradnya Diva	Department of Marine Engineering, Institut Teknologi Sepuluh Nopember
11	Zahrina Nur Amalina Syamsudin	Department of Marine Engineering, Institut Teknologi Sepuluh Nopember
12	Hera Dian Marsseila	Department of Marine Engineering, Institut Teknologi Sepuluh Nopember
13	Ade Riyan Sukardi	Department of Marine Engineering, Institut Teknologi Sepuluh Nopember
14	Kadek Putra Puja Wirawan	Department of Marine Engineering, Institut Teknologi Sepuluh Nopember
15	Hera Dian Marsseila	Department of Marine Engineering, Institut Teknologi Sepuluh Nopember
16	Michiko Laura	Department of Marine Engineering, Institut Teknologi Sepuluh Nopember

KEYNOTE SPEAKER

1	Prof. Akira Sou	Kobe University, Japan
2	Ir. Raymond Ivan H.	Kementerian Perhubungan RI
3	Dr. Junghan Lee	Cluster LNG, Korea
4	Bani Maulana Mulia	PT. Samudera Indonesia, Tbk

IOP Conf. Series: Earth and Environmental Science **1423** (2024) 011001 doi:10.1088/1755-1315/1423/1/011001

WORKSHOP SPEAKER

1	Dr. Dra. Dian Saptarini, M.Sc	Department of Biology, Institut Teknologi Sepuluh Nopember
2	Wira Sanjaya	Coral Triangle Center Indonesia
3	I Gusti Ngurah Agung Putra Adnyana	Head of Benoa VTS
4	Dr. Eng. Dhimas Widhi Handani	The Center for Excellence of Maritime Safety and Marine Installation (PUI KEKAL)
5	Dr. Eng. Putu Sindhu Asmara	Surabaya State Shipping Polytechnic (PPNS)
6	Prof. I Made Oka Widyantara	Universitas Udayana

REVIEWER

1	A.A. Bagus Dinariyana D. P., S.T., MES., Ph.D	Department of Marine Engineering, Institut Teknologi Sepuluh Nopember
2	Dr. Achmad Baidowi, S.T., M.T.	Department of Marine Engineering, Institut Teknologi Sepuluh Nopember
3	Dr. Ayudhia Pangestu Gusti S.T.	National Research and Innovation Agency (BRIN)
4	Dr. Eddy S. Koenhardono, S.T., M.T.	Department of Marine Engineering, Institut Teknologi Sepuluh Nopember
5	Dr. Emmy Pratiwi S.T.	Department of Marine Engineering, Institut Teknologi Sepuluh Nopember
6	Dr. Eng. Dhimas Widhi Handani, S.T., M.Sc.	Department of Marine Engineering, Institut Teknologi Sepuluh Nopember
7	Dr. Eng. I Gusti Ngurah Sumanta Buana, S.T., M.Eng.	Department of Sea Transportation, Institut Teknologi Sepuluh Nopember
8	Dr. Eng. I Putu Sindhu Asmara, S.T., M.T.	Surabaya State Shipping Polytechnic
9	Dr.Eng. Muhammad Badrus Zaman, S.T., M.T.	Department of Marine Engineering, Institut Teknologi Sepuluh Nopember
10	Dr.Eng. Trika Pitana S.T., M.Sc.	Department of Marine Engineering, Institut Teknologi Sepuluh Nopember
11	Fadilla Indrayuni Prastyasari, S.T., M.Sc. Ph.D.	Department of Marine Engineering, Institut Teknologi Sepuluh Nopember
12	Dr. Hozairi, S. ST., M.T.	Universitas Islam Madura
13	Dr. I Made Ariana, S.T., M.T.	Department of Marine Engineering, Institut Teknologi Sepuluh Nopember
14	Dr. Indra Ranu Kusuma, S.T., M.Sc.	Department of Marine Engineering, Institut Teknologi Sepuluh Nopember
15	Dr. Ir. A. A. Masroeri, M.Eng.	Department of Marine Engineering, Institut Teknologi Sepuluh Nopember
16	Dr. Radityo Anggoro, S.Kom., M.Sc.	Department of Informatics, Institut Teknologi Sepuluh Nopember
17	Juniarko Prananda, S.T., M.T.	Department of Marine Engineering
18	Prof. Dr. Ir. I Ketut Suastika, M.Sc.	Department of Naval Architecture, Institut Teknologi Sepuluh Nopember
19	Prof. Dr. Ketut Buda Artana, S.T., M.Sc.	Department of Marine Engineering, Institut Teknologi Sepuluh Nopember
20	Prof. Ir. I Ketut Aria Pria Utama, M.Sc., Ph.D.	Department of Naval Architecture, Institut Teknologi Sepuluh Nopember

IOP Conf. Series: Earth and Environmental Science **1423** (2024) 011001 doi:10.1088/1755-1315/1423/1/011001

21	Putri Dyah Setyorini S.T., M.T.	Department of Marine Engineering, Institut Teknologi Sepuluh Nopember
22	Putu Dana Karningsih S.T., M.Eng.Sc., Ph.D.	Department of Industrial Engineering, Institut Teknologi Sepuluh Nopember
23	Dr. Taufik Fajar Nugroho, S.T., M.Sc.	Department of Marine Engineering, Institut Teknologi Sepuluh Nopember
24	Yoyok Setyo Hadiwidodo, S.T., M.T., Ph.D.	Department of Ocean Engineering, Institut Teknologi Sepuluh Nopember
25	Dr. Eng. Adi Kurniawan S.T., M.T	Department of Marine Engineering, Institut Teknologi Sepuluh Nopember
26	Dr. I Made Manik Sukanegara Adhita, S.T., M.Sc.	Department of Marine Engineering, Institut Teknologi Sepuluh Nopember
27	Thariq Arafatul Akbar, S.T., B. Eng., M.T.	Department of Marine Engineering, Institut Teknologi Sepuluh Nopember
28	Dr. Adhi Iswantoro, S.T., M.T.	Department of Marine Engineering, Institut Teknologi Sepuluh Nopember
29	Andi Trimulyono, S.T., M.T., Ph.D.	Diponegoro University

IOP Conf. Series: Earth and Environmental Science **1423** (2024) 011002 doi:10.1088/1755-1315/1423/1/011002

Peer review declaration

All conference organisers/editors are required to declare details about their peer review. Therefore, please provide the following information:

- **Type of peer review: Single-blind**
- **Conference submission management system: Easychair**
- **Number of submissions received: 145**
- **Number of submissions sent for review: 137**
- **Number of submissions accepted: 43**
- **Acceptance Rate (Number of Submissions Accepted / Number of Submissions Received X 100): 29.6%**
- **Average number of reviews per paper: 2-3 reviews**
- **Total number of reviewers involved: 47**
- **Any additional info on review process:**

  - All paper has been checked for its similarity index by using IThenticate plagiarism detection software. Similarity index threshold is 20 %.

  - After revision has been submitted by author, the committee recheck the paper to ensure if the review of reviewer has been accommodated in the paper. Further correspondence with author is done if necessary, to improve quality of paper based on contents and format.

- **Contact person for queries: dhimas@ne.its.ac.id**

Please submit this form along with the rest of your files on the submission date written in your publishing agreement.

The information you provide will be published as part of your proceedings.

IOP Conf. Series: Earth and Environmental Science **1423** (2024) 012001 doi:10.1088/1755-1315/1423/1/012001

Evaluation of an automated vessel recognition system: case study in Port of Cristo

Livia Maglić[1*], Lovro Maglić[2], Antonio Blažina[2] and Ivana Ognjanović[3]

[1] Department of Transport and Transportation Technology, University of Rijeka, Faculty of Maritime Studies, Rijeka, Croatia
[2] Center for Marine Technologies, University of Rijeka, Faculty of Maritime Studies, Rijeka, Croatia
[3] University of Donja Gorica, Faculty for Information Systems and Technologies, Podgorica, Montenegro

*E-mail: livia.maglic@pfri.uniri.hr

Abstract. An automated vessel recognition system is a sophisticated technological solution for the automatic detection, identification, and classification of vessels in a maritime environment. This paper presents the evaluation of an automatic vessel recognition system implemented in the Port of Cristo on the island of Mallorca in Spain. The system employs PTZ camera and algorithm for image analysis and object recognition to accurately detect, classify and track different types of small vessels entering or departing from the port. The results revealed that the system successfully detected the event of the arrival and departure of vessels in 95% of the cases. In addition, the system was able to identify the vessel by recognizing the registration number in 75.81% of the cases and classifying the vessel type in 89.07% of the cases. The paper presents the SWOT analysis based on the knowledge and experience gained during the testing period.

1. Introduction

An automated vessel recognition system (AVRS) is a sophisticated technological solution designed to automatically detect, identify, and classify vessels in a maritime environment. Such systems may use various sensors such as radar, sonar, lidar, and optical cameras in combination with advanced algorithms, including machine learning and computer vision. They are designed to perform one or more of the following key functions: detection, identification, tracking, alerts and notifications on specific triggers (usually motion or movement patterns), integration and data management. AVRS play a crucial role in enhancing maritime security, safety, and port efficiency. These systems generally reduce the need for physical monitoring by port employees and provide accurate and real-time information on vessel traffic or berth occupancy in a port, which is essential for applications such as port (berth) management, coastal surveillance, search and rescue operations, pollution investigation and navigation in busy waterways.

The existing Automatic Identification System (AIS) is a VHF (very high frequency) based short-range identification and tracking system for ships. It is developed to automatically provide information about a ship to other ships within its range and appropriately equipped coastal stations (AIS receiver stations) in order to increase safety at sea. There are two classes of AIS systems: Class A and Class B type. According to the International Convention for the Safety of Life at Sea (SOLAS), AIS Class A is mandatory for all ships of 300 gross tonnage and above on

IOP Conf. Series: Earth and Environmental Science **1423** (2024) 012001 doi:10.1088/1755-1315/1423/1/012001

international voyages, for cargo ships of 500 gross tonnage and above on non-international voyages and for all passenger ships regardless of size. This system is mandatory from December 31st 2004 (SOLAS regulation V/19), with well-established infrastructure worldwide, and completely suitable for larger vessels.

On the other hand, more affordable AIS Class B with less stringent requirements has been developed on a voluntary basis for smaller (non-SOLAS) vessels. It is mainly used by pleasure crafts and fishing vessels that want to take advantage of the safety benefits but with limited functionality compared to AIS Class A type. Besides all the good points of the AIS system, there are significant limitations in AIS type B usage, the most important of which are: vessels are not obliged to be equipped with and use AIS B (subject to costal state laws and regulations for certain vessel type and/or size), the possibility of AIS being switched off, occasional position errors, lower update rates, etc. Hence, for smaller vessels, other approaches, like visual recognition and identification systems that do not require any equipment on the vessel itself are extensively investigated and tested. Some researchers have been investigating complementary techniques where they collected maritime traffic data using AIS and optical imagery techniques [1].

Considering port operations, information on vessel traffic and berth occupancy is essential for safe and efficient port management [2]. The increasing demand for efficient, safe, and secure port management, especially in ports with increasing traffic density, has led to the development of automated systems that can enhance existing practices relying solely on staff performance.

Following the widespread use of imaging devices and associated advanced software solutions, camera-based data analysis is becoming the predominant method to sense the navigation environment [3]. In recent years, computer vision and image processing technologies have been widely used in maritime surveillance. Hence, vessel detection and their classification from images in general have become an important research area [4], using optical or thermal sensors, or both [5]. In some ports intelligent marine surveillance systems are used, where camera/image sensors are installed in strategic port areas to automatically detect and identify vessels using automatic object recognition methods [6,7,8,9]. Some tests proved that video recordings from existing monitoring systems for ports, terminals, channels or rivers with different camera characteristics may be used for vessels detection and tracking [10]. This research is aiming to investigate the accuracy, reliability and efficiency of such systems so that end users know their capabilities, drawbacks and limitations.

Figure 1. Port of Cristo, Mallorca, Spain

The contribution of this paper lies primarily in the evaluation of the operational effectiveness of the automated vessel recognition system during the testing period and presenting its drawbacks for future improvements. The test was conducted in the Port of Cristo on Mallorca Island in Spain (Figure 1) as part of the PSAMIDES project (Ports Small and Medium Alliance for Sustainable Development under the Interreg MED Program, co-funded by the European Regional Development Fund). The port is local, accommodating small vessels, primarily private boats and yachts, smaller fishing ships and boats, and passenger ships for excursions. The port accommodates the vessels up to 20 m in length.

2. Testing process

The purpose of the test was to assess the reliability and effectiveness of an AVRS. The system is evaluated considering: a) detection of vessel arrivals and departures, b) recognition and identification of vessels based on specific parameters, c) record of navigation speed in the port area. The main parameters used for recognition and identification are: a vessel's registration number and vessel's type (hull shape) and length.

The testing process took place from 23 June to 1 September 2021. The summer period was chosen due to increased vessel traffic. The increased traffic allowed for a more reliable assessment of the technology's performance.

One PTZ (Pan Tilt Zoom) camera was installed in the port on one side of the port entrance (Figure 2). The PTZ camera provides a wide field of view (60°) and can pan, tilt and zoom to focus on a specific area within the port entrance. In particular, the 30x zoom function enables the recognition of vessel registration numbers, even on moored vessels in the vicinity, up to a distance of approximately 130 meters. A camera can monitor an area of about 600-800 m^2.

Figure 2. Installation of the camera in the Port of Cristo

The camera is strategically placed and oriented almost perpendicular to the entrance, where all arriving and departing vessels are passing by. In that manner, the vessels' profiles can be seen and analyzed, so that the main details are captured. The main details include a registration number or name and key particulars: length and shape. Digital images captured by the PTZ camera are integrated into image analysis software. This software is trained to isolate the vessels from the background and extract the required information. The software training included vessels mainly from 4 to 20 m in length. The tested automated vessel recognition system had to meet the following criteria:

- Installation of a video surveillance camera at the port entrance,
- Conversion of the captured images into information, including registration numbers or vessel names,
- Recognition of new registration numbers or vessel names not existing in the database,
- Recognition of vessel length, shape and speed.

Implementation of the system required careful planning, including finding a strategic point for mounting PTZ cameras and integrating hardware and software components for seamless operation. In addition, ongoing maintenance and updating of the system were deemed essential to ensure optimal performance and accuracy. After completion of initial setup and ensuring the system is operative, for the first day the system was monitored (the computer results and the port entrance simultaneously) to ensure it's proper operation and detection capabilities. After that the system was periodically checked to verify its operation, without deeper analysis, until the end of testing.

3. Results

Arrival and departure detection. The total number of vessels that passed through the port entrance during the testing period was 16,403 (Figure 3). Of these, 8,405 were vessel arrivals and 7,948 were vessel departures, indicating a noticeable imbalance. Monthly this discrepancy averaged approximately 200 passes.

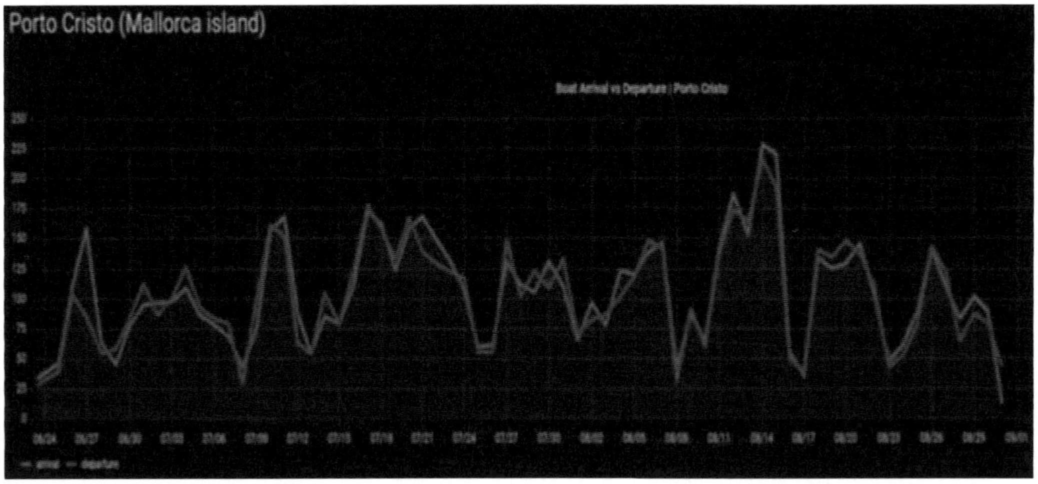

Figure 3. Number of vessels that passed through the port between 23 June and 1 September 2021 (green line - arrivals, red line - departures)

A closer analysis of this discrepancy reveals that the major problem lies in the proximity of passenger ships' berths. As shown in Figure 4, the camera was positioned on the NW entrance to the port to monitor the narrow passage. The berths of the passenger ships are located just next to the camera pole.

The system occasionally counted these ships multiple times per arrival in cases when they significantly changed the direction of movement during docking on berths that are adjacent to the camera's field of view. However, when they would leave the port when the manoeuvre was much simpler, they were counted only once. This resulted in a 5% error in the total count. The simplest solution to issues like this is to position the camera or its orientation in such a way that the manoeuvring areas are not in their field of view.

Figure 4. Camera position at the port entrance with its field of view of 60°

Nevertheless, the system shows a high degree of accuracy in port arrival and departure detection. The accuracy of these detections is maintained whether they are made in daylight or at night.

Vessels recognition. The vessel registration number recognition was accurate in 75.81% of cases. The analysis revealed that the vessels whose registration number could not be recognized were vessels that: a) had a registration number positioned in such a way that it was not visible to the camera, or b) had no registration number at all. The vessel's registration number is not fully standardized in terms of the position where it must be placed, the size of the letters/numbers and the colour to meet the minimum contrast with the colour of the ship's hull or superstructure (Figure 5).

IOP Conf. Series: Earth and Environmental Science **1423** (2024) 012001 doi:10.1088/1755-1315/1423/1/012001

Figure 5. An example of successful detection (registration, time, speed)

Furthermore, the number is sometimes placed only on one side of the vessel. In addition, sea agitation, meteorological visibility, and night and even cases when vessels are crossing next to each other further reduced the number and accuracy of readings. This implies that numerous issues should be considered to improve the recognition of vessels based on their unique registrations.

Speed records. The system records the speed of vessels on arrival or departure. The average speed of vessels during arrival in port was 2.84 knots, while during departure it was 2.48 knots. This information is rather simple, but it is very important for the safety within the port area. For example, the system software could quite easily be programmed to raise an alarm if a particular vessel exceeds a determined maximum speed set by the port authority, e.g. 5 knots. An alarm on the computer can comprise an audio signal, notification and screenshot of the vessel concerned. Upon such alarms port staff could react a) immediately by radio station if necessary, b) later when a detected port user returns to its berth and when a warning and/or penalty may be issued or c) by forwarding the information to another responsible body (e.g. harbormaster's office).

Vessel type classification. Classification of vessel type had an accuracy of 89.07%. In this test, the system was programmed to recognize and classify vessels in one of the following categories: a) boat; b) sailing boat; and c) wood boat - considering small private and fishing boats (Figure 6). The two categories "wood boat" and "boat" are too similar, and the programmed system often did not manage to distinguish these vessels at night and during adverse weather conditions.

It is obvious that such a categorization was not optimal because of its similarity and because the material of the vessels is not easy to recognize visually, i.e. a wooden painted hull looks the same as a GRP (glass reinforced plastic) vessel or even a metal ship (steel or aluminium), especially if observed from a certain distance. Therefore, the categories should be divided

IOP Conf. Series: Earth and Environmental Science **1423** (2024) 012001 doi:10.1088/1755-1315/1423/1/012001

according to the capabilities of the system, i.e. according to the distinguishable shape of the vessel (e.g. motor boat/sailing boat/ship) and/or vessel length. As for the vessel's length, most of them were about 5 meters long (small private boats), and the passenger excursion ships were approximately 20 meters long.

ActivitatPortoCristo

Timestamp	Out (0) / In (1)	Knots	Boat number	Length (m)	Type
2021-06-30 04:40:50	0	3,9	N/A	7,46	Wood boat
2021-06-30 07:12:12	0	4,09	N/A	5,59	Wood boat
2021-06-30 08:57:11	0	1,91	N/A	6,59	Boat
2021-06-30 08:59:01	0	1,32	6te121	9,75	Sailing boat
2021-06-30 08:59:47	0	1,53	whv978	10,25	Sailing boat
2021-06-30 09:05:01	0	1,62	61260310	4,46	Boat
2021-06-30 09:10:35	0	2,97	N/A	6,46	Boat
2021-06-30 09:15:52	0	2,35	N/A	4,44	Boat

Figure 6. Screenshot of results

Considering all, it is concluded that the camera and the system showed good results in detecting arrivals and departures. However, the system was not adequately efficient in vessel registration number recognition and categorization, especially during reduced visibility (at night, during rain, strong wind, mist periods etc.). This indicates that the system is effective at basic vessel detection but its ability to accurately identify and analyze vessel characteristics in challenging situations and environmental conditions is limited.

Furthermore, the system was newly installed and the vessel database was empty (vessels calling port frequently). The system can be improved so that the database contains all the information about vessels that frequently call the port (name, registration number, length, vessel category and other data). In this way, a system could recognize a particular vessel even with imperfect recognition by comparing image data with the database.

4. SWOT analysis

This SWOT analysis provides an overview of the factors that can affect the success and effectiveness of an automated vessel recognition system. It can guide decision-makers in identifying areas where the system can be improved and opportunities to capitalize on while recognizing potential challenges and threats that need to be managed.

STRENGTHS:

Enhanced Security: The system can significantly improve maritime security by identifying and tracking suspicious vessels and helping authorities respond more effectively to potential threats or investigate security incidents.

IOP Conf. Series: Earth and Environmental Science **1423** (2024) 012001 doi:10.1088/1755-1315/1423/1/012001

Enhanced Safety: The system can enhance navigational safety within the port, e.g. in the event of an engine failure, vessel collisions at the port entrance, in marine accident investigations and similar scenarios.

Enhanced port planning and logistics: It can help with statistical analysis of traffic density in the port, number of ports calls in different periods, trends, analysis of traffic structure, planning of logistics processes and improvement of Vessel Traffic Service (VTS) systems in ports where it is applicable.

Efficiency: Automation reduces the need for personal monitoring and can operate 24/7, increasing efficiency and reducing labour costs.

Real-time Data: The system can provide real-time data on vessel traffic to support traffic management and navigational safety.

Accuracy: Advanced sensor technology and machine learning algorithms ensure high accuracy in vessel identification and reduce false alarms and errors.

Scalability: The system can be scaled to monitor and recognize a wide range of vessel types and adapt to different environments and conditions.

Data Integration: Integration with other maritime systems, such as AIS and radar, improves overall situational awareness and decision-making.

Operational Flexibility: The system can be deployed in various maritime applications, including defence, commercial shipping, search and rescue and environmental monitoring.

WEAKNESSES:

Cost: Implementing and maintaining an automated vessel recognition system can be expensive, requiring advanced sensors and computer equipment.

Operational Challenges: As with any automated system, there is a risk of false positive (misidentified vessels) and false negative (unidentified vessels) detections.

Environmental Factors: Adverse weather conditions, poor visibility, and environmental factors such as waves can affect the performance of the system.

Maintenance: It can be a challenge to keep the exposed sensors and equipment in good condition, especially in harsh maritime environments.

OPPORTUNITIES:

Market Expansion: The need for maritime security and surveillance is growing, creating opportunities for companies offering vessel recognition systems.

Technological Advancements: Ongoing advances in sensor technology and machine learning algorithms can enhance the capabilities and reliability of existing systems.

Regulatory Compliance: As maritime regulations continue to evolve, there is an opportunity for the system to adapt and support these requirements.

Customization: Tailoring the system to specific customer needs and applications can open up new markets and revenue streams.

THREATS:

Cybersecurity Risks: Automated systems are vulnerable to cyber-attacks that could compromise data integrity and system security.

Competition: Growing interest in maritime surveillance may lead to competition from other companies developing vessel recognition systems.

Regulatory Hurdles: Evolving and stringent maritime regulations may pose challenges for system providers.

Public Concerns: Privacy concerns may arise, especially if the system detects vessels near coastal areas with residential populations.

Technical Challenges: Environmental conditions and the complexity of the maritime environment may pose technical challenges to system reliability. The problem of human dependency on automated systems is pervasive in various fields. It includes both over-dependence, where people rely excessively on automated systems without questioning, testing or checking them, and under-dependence, where people stick to traditional methods and fail to take advantage of novel systems. To address these issues, it is usually necessary to educate and train the staff and users concerned to understand the benefits and limitations of these novel systems.

Based on the presented SWOT analysis, several strategies may be recommended:

- Numerous ports and marinas have already in operation a certain surveillance system for security reasons. To reduce initial costs and maintenance of separate designated equipment, it is recommended to investigate if one or more existing cameras may be used for monitoring port entrance for AVRS purposes. In many cases adjusting the field of view, or strategically changing location of one or more cameras may cover the port entrance without losing the ability to monitor port infrastructure.
- The detection or recognition may be reduced during night or poor weather conditions. This problem usually is resolved by installing quality cameras or additional sensors (thermal). Other approaches may include: a) reducing the range from camera to the vessel's paths (mounting cameras closer) and b) installing appropriate lights at the port entrance. The latter in general is applicable for smaller ports with narrower port entrance, where light pollution and direction should be considered (not to reduce safety by blinding the ship's captain or skipper).
- Machine learning algorithms must be continuously trained and improved to reduce the number of misidentified and unidentified vessels. Hence, it is important to create a comprehensive training program/module for employees to understand and efficiently use the automated system. Such training program does not need to be extensive and may be incorporated in other training programs related to marina/port safety or security.

Implementing these strategies could effectively leverage the strengths of the automated vessel recognition system to address and mitigate its weaknesses, leading to a more robust, reliable, and cost-effective solution.

5. Conclusion

The case study in the Port of Cristo, Spain, demonstrated the practical application of an automated vessel recognition system. The system showed a rather high level of accuracy in detecting arriving and departing vessels, regardless of whether it was day or night. However, a discrepancy was found in vessel counting, mainly due to vessels manoeuvring near the camera's field of view. This discrepancy can be mitigated by adjusting the camera positioning.

When recognising vessel registration numbers, the system achieved a hit rate of 75.81%. An in-depth analysis revealed that several factors influenced the recognition. These factors included the positioning and visibility of the registration numbers, as well as the effects of varying sea conditions. However, the system's efficiency in accurately recognising vessel numbers proved to be limited, especially in situations with poor visibility.

The system successfully recorded the speed of vessels on arrival and departure. This data is of great value in improving safety in the port by enabling the identification of vessels exceeding the permitted speed.

A hit rate of 89.07% was achieved in the classification of vessel types. Nevertheless, the categorisation showed suboptimal performance, especially in distinguishing between "wood boat" and "boat". It was recommended to further refine the categories based on distinguishable vessel characteristics or length to increase accuracy.

The conclusion from the tests is that the automated vessel recognition system was promising in basic vessel detection, but had its limitations in accurate vessel detection, especially under adverse meteorological conditions. Several recommendations were made to address these limitations:

- Ongoing database maintenance: populate the system's database with information about vessels that frequently call at the port, including their names, registration numbers, lengths and categories. This would improve recognition as the system could compare the incoming data with the database.
- Adjust the camera positioning: Reposition or adjust the camera to avoid counting the same vessels more than once, especially when passenger vessels change direction while mooring in close proximity to the camera's field of view.
- Improve categorisation: Improve the categorisation system to better distinguish vessel types and materials. Consider distinguishable shapes or lengths to improve accuracy.

The results of this test show the potential and limitations of automated vessel recognition systems in monitoring maritime traffic. While they offer benefits such as increased safety and efficiency, it is important to continuously improve and adapt these systems to overcome their limitations, especially in challenging meteorological conditions. These recommendations aim to maximise the effectiveness and accuracy of the system and contribute to safer and more efficient port operations.

It is important to point out the limitation of the test itself. The system was monitored personally for the first day after starting and periodically to verify its operation. For the two months testing period responsible persons to personally monitor concurrently the port entrance and the system recordings were not organized or employed. This control mechanism is especially important during bad weather, precipitations and nights, when camera recordings are limited and difficult to read for later analysis. In that respect there is a lack to verify the ratio of undetected vessels. Similar future tests should include responsible persons for periods (visibility and weather states) when it is known that the cameras could have poor recordings. Furthermore, to improve future analysis of a large scale (considering number of vessel passages) should be planned to divide and categorize the results of detections and recognitions on the time of day (during daylight and night), weather state (e.g., sunny, overcast, rain), sea state (calm and different significant wave conditions), and visibility.

Acknowledgment

The present study was carried out within the framework of the Interreg Mediterranean project "Ports Small and Medium Alliance for Sustainable Development" (PSAMIDES), reference number 5350, co-financed by the European Regional Development Fund.

References

[1] P. D. O'Hara, N. Serra-Sogas, L. McWhinnie, K. Pearce, N. Le Baron, G. O'Hagan, A. Nesdoly, T. Marques, R. Canessa. "Automated identification system for ships data as a proxy for marine vessel related stressors", Science of The Total Environment, Vol. 865, 2023, doi: 10.1016/j.oceaneng.2024.116771.

[2] S. Amarasinghe, N. D. Kodikara and D. Sandaruwan, "Location estimation in a maritime environment using a monocular camera," 14th International Conference on Advances in ICT for Emerging Regions (ICTer), Colombo, Sri Lanka, 2014, pp. 21-28, doi: 10.1109/ICTER.2014.7083874.

[3] J. Qu, Y. Guo, Y. Lu, F. Zhu, Y. Huan and R. W. Liu, "Intelligent maritime surveillance framework driven by fusion of camera-based vessel detection and AIS data," 2022 IEEE 25th International Conference on Intelligent Transportation Systems (ITSC), Macau, China, 2022, pp. 2280-2285, doi: 10.1109/ITSC55140.2022.9921786.

[4] X. Chen, L. Qi, Y. Yang, O. Postolache, Z. Yu, X. Xu, "Port ship detection in complex environments," 2019 International Conference on Sensing and Instrumentation in IoT Era (ISSI), Lisbon, Portugal, 2019, pp. 1-6, doi: 10.1109/ISSI47111.2019.9043694.

[5] C.P. Simonsen, F.M. Thiesson, Ø. Holtskog, R. Gade. Detecting and Locating Boats using a PTZ Camera with Both Optical and Thermal Sensors. 15th International Conference on Computer Vision Theory and Applications, 2020, doi: 10.5220/0008972603950403

[6] Y. Duan, Z. Li, X. Tao, Q. Li, S. Hu and J. Lu, "EEG-based maritime object detection for IoT-driven surveillance systems in smart ocean," in IEEE Internet of Things Journal, vol. 7, no. 10, pp. 9678-9687, Oct. 2020, doi: 10.1109/JIOT.2020.2991025.

[7] M.D. Rodriguez Sullivan, M. Shah. "Visual surveillance in maritime port facilities", Proceedings of SPIE - The International Society for Optical Engineering, 2008, doi: 10.1117/12.777645

[8] M.H. Zwemer, H. G. J. Groot 1, R. G. J. Wijnhoven, E. Bondarev, P. H.N. de With. "Multi-Camera Vessel-Speed Enforcement by Enhancing Detection and Re-Identification Techniques", Sensors, 2021, doi: 10.3390/s21144659.

[9] D. Bloisi, L. Iocchi, M. Fiorini, G. Graziano, "Camera Based Target Recognition for Maritime Awareness", Proceedings of the Fifteen International Conference on Information Fusion (FUSION), 2012.

[10] N. Wawrzyniak, T. Hyla, A. Popik, "Vessel Detection and Tracking Method Based on Video Surveillance", Sensors, Vol. 19, 2019, doi:10.3390/s19235230.

Improving safety by fatigue life assessment of pressure hull considering corrosion factors: a review

Arif Pambudi[1]*, Agoes Santoso[1] and Ahmad Baidowi[1]

[1]Department of Marine Engineering, Faculty of Marine Technology, Institut Teknologi Sepuluh Nopember Surabaya, Keputih, Sukolilo, Surabaya, East Java 60111, Indonesia

*E-mail: 7019231002@student.its.ac.id

Abstract. Research on determining fatigue endurance of submarine pressure-resistant hulls is crucial for determining their remaining operational lifespan. These studies typically focus on the effects of hydrostatic pressure on fatigue under ideal conditions without taking into account other significant factors. However, the actual condition of pressure hulls is influenced by multiple elements, one of which is operational environmental factors such as corrosion. While current research has examined these factors individually, there is a notable absence of a comprehensive analysis that integrates all of them. Corrosion can reduce structural strength due to the resulting plate thinning effect caused by hydrostatic pressure and low cyclic fatigue. This oversight hampers the development of accurate models for predicting fatigue life and compromises the safety and reliability of underwater structures. This paper addresses the research gap by emphasizing corrosion factors in fatigue life assessments using the low-cyclic fatigue method. By incorporating these elements into the assessment methodology, a more comprehensive understanding of pressure hull fatigue can be achieved, leading to more accurate predictions of pressure hull remaining lifespan.

1. Introduction

Corrosion could be broadly termed the damaging process wherein a metal undergoes chemical reactions with its surrounding environment [1]. Corrosion leads to a reduction in material thickness [2] and a loss of strength [3], significantly affecting the structural integrity and performance of components. Besides having significant economic impacts [4], it also reduces the material lifespan[5]. Fatigue, characterized by material exhaustion resulting from repeated loading that cause local plastic deformations, is a critical phenomenon in various materials. In submarines, fatigue life pressure hull calculations have been performed, considering factors such as hydrostatic pressure [6] and residual stress [7], but corrosion factors have not been included. However, corrosion is a natural phenomenon that is highly likely to occur.

Fatigue life calculations for structures during the operational phase usually rely on experimental or partly experimental theoretical models purposed for analysis of the influence of existing effects. Accurate fatigue life prediction is very important and must be considered in mechanical components. Therefore, it is essential to develop a fatigue life method that takes into account the corrosion factor. In this paper, a review of fatigue life calculations of submarine pressurized hulls is proposed, taking into account corrosion in addition to hydrostatic pressure and residual stresses during submersion and immersion periods.

2. Corrosion

Corrosion is a natural phenomenon that gradually deteriorates materials, typically alloys, due to of chemical or electrochemical interactions with their surroundings. It converts refined metals into more chemically stable oxides. There are many explanations for corrosion. In Barbara's [8], it is the degradation of materials due to their interaction with the environment, leading to a gradual deterioration of their physical and chemical properties driven by a decrease in Gibbs energy. Hammonds [9] defines corrosion as material breakdown via environmental interaction, primarily electrochemical. It extends to natural elements like granite. Processes involve electrochemistry and consideration of thermodynamic energy exchange and reaction kinetics at metal-solution interfaces. Lyon [10] specifically refers to the process causing a decline in the performance of a material, ultimately leading to corrosion damage. It can be defined as a physicochemical interaction that substantially degrades the functional properties of either the material itself, the surrounding environment, or both. Yongmei [11] observed that the surface conditions of a specimen significantly influence its endurance limit. From the standpoint of stress intensification, the shape and arrangement of corrosion indentations are crucial factors affecting specimen strength. Koliosa [12] suggests that such deterioration could cause a notable decrease in the ultimate load of a sheet. Consideration of bends due to welds and additional residual stresses increases the impact of oxidation on the strength capability, reducing the load-bearing ability and ductility of the material [13]. Corrosion reduces during coarsening, and strength and stiffness with enhanced strengthening originate from the development of nanoporous structures.

The increased strength of tiny ligaments at the nanoscale due to size-related effects [14], including loss of ductility and strength [15], decreases the bond stress value, and the decrease in the maximum compressive load or bending moment caused by pitting corrosion is less than that affecting tensile strength when considering equivalent thickness [16]. Severely corroded specimens show a significant decrease in mechanical properties, such as yield and ultimate strength [17]. In Wang's view [18], that depth of grooving corrosion has a more significant impact on reducing ultimate strength relative to corrosion width, even as they result in equal volume loss. This type of corrosive deterioration has the potential to reduce ultimate strength by as much as 26%. Zhang [19] observed that as the level of oxidation increased, the edge strength sections notably reduced compared to those flat sections. At a corrosion rate of 36%, the yield strength ratio of the corner sections remained notably high at 0.99, despite the extension being only 2.8%. Oxidation worsened the impact of cold-forming on the strength and flexibility of materials formed at low temperatures. Ye [14] observed that the corrosion-induced strengthening phenomenon was attributed to the creation of nanoporous structures and the naturally elevated strength of nanoscale ligaments due to a by product.Top of Form Bottom of Form Findings from an experiment conducted to explore the theory further validate that the most significant breaches were likely due to the collective impacts of erosion and corrosion on the iron hull exposed to the marine environment.

IOP Conf. Series: Earth and Environmental Science **1423** (2024) 012002 doi:10.1088/1755-1315/1423/1/012002

3. Experimental studies

3.1 Hydrostatic pressure
Higher hydrostatic pressure can indeed intensify corrosion rates through various mechanisms. One significant way in which this occurs is by promoting the dissolution of oxygen in water, leading to increased oxygen concentrations. Oxygen catalyzes many corrosion processes, according to Choi et al. [20] According to Lie and co-workers [21], the presence of oxygen can accelerate corrosion rates, as observed in studies where the corrosion rates first increased rapidly with pressure and then stabilized. Subsequent research by Jie et al [22], investigated two low-alloy steels, demonstrating a degradation in corrosion resistance with increasing hydrostatic pressure, attributed to accelerated anodic reaction rates and notable changes in surface morphology, Yang et al. [23] explored X70 steel corrosion in a controlled deep-sea environment and it was observed that hydrostatic pressure facilitated corrosion processes, including the initiation of pitting and the transport of chloride ions, over the duration of the immersion period.

Hydrostatic pressure significantly influences the corrosion characteristics of materials in deep-sea environments, particularly in the pressure hulls of maritime or underwater structures. Studies have shown that hydrostatic pressure can expedite corrosion processes by influencing the properties of protective films on metal surfaces [24], promoting the onset and progression of pitting corrosion [25], and increasing the dissolution rate within cavities [26]. Moreover, hydrostatic pressure has been observed to expedite corrosion rates and shift general corrosion towards localized oxidation occurring in deep-sea environments [27]. Understanding how hydrostatic pressure influences corrosion is crucial for designing and maintaining pressure hull structures to ensure their structural integrity and longevity in challenging underwater conditions. Research on how different materials corrode under hydrostatic pressure, such as high-strength steel [28], low-alloy steel [29], and nickel alloys [24], offers valuable insights for the development of corrosion-resistant materials and protective coatings for pressure hulls. Overall, these studies collectively highlight the detrimental impact of the influence of hydrostatic pressure on the corrosion behaviour of different materials Top of Form Bottom of Form and provide insights crucial for understanding corrosion mechanisms in high-pressure environments.

3.2 Low Cyclic Fatigue
In deep-sea environments, the interaction between low-cycle fatigue and corrosion in metallic materials is a critical factor affecting the structural integrity of submerged structures. Low-cycle fatigue, which refers to the failure of materials under cyclic loading at stress levels below their ultimate strength, is exacerbated by the corrosive conditions prevalent in deep-sea environments [30]. Corrosion weakens materials by deteriorating their surfaces, making them more susceptible to fatigue failure [31]. Understanding this connection is essential for designing resilient structures capable of withstanding dynamic loads and harsh corrosive elements present in deep-sea environments, ensuring their long-term durability and safety. Research has shown that the impact of loading frequency on the initiation life of corrosion fatigue cracks depends on how the metal interacts with the environment [32]. Additionally, the ability of metallic materials to resist corrosion in deep-sea environments can be affected by factors such as hydrostatic pressure and the concentration of dissolved oxygen [33]. Studies have also highlighted the influence of surface and internal plastic deformations on the corrosion resistance and corrosion fatigue performance of materials such as AISI 316L [34]. Furthermore, the oxidation responses of various materials in deep-sea environments have been extensively studied, including the corrosion behavior of aluminum alloys [35], titanium alloys [36], and nickel-base superalloys [37].

Low cycle fatigue (LCF) can greatly influence corrosion processes in submarines due to the cyclic loading and stress levels experienced by underwater structures. The repetitive stress cycles endured during submarine operations can induce microstructural changes and surface deformations, leading to localized stress concentrations that promote the initiation and propagation of corrosion-induced cracks. The combination of alternating stress and corrosion can result in corrosion fatigue., where the presence of corrosive environments accelerates crack propagation and reduces the fatigue life of materials [38], [39]. Kashani et al, mitigation strategies that consider the effects of plastic buckling, deterioration due to low-cycle fatigue, and corrosion-induced damage on materials can help address the challenges posed by the interaction between LCF and corrosion in submarine structures [40], [41], [42]. Moreover, research on how corrosion affects low-cycle fatigue behavior and the role of corrosion patterns on cyclic response can provide valuable insights into developing effective maintenance and protection strategies for submarines operating in corrosive environments [37], [43]. By taking into account how corrosion affects both the fatigue life and behaviour of corroded materials under cyclic loading, it is possible to enhance the reliability and safety of submarines subjected to dynamic underwater conditions [31], [44].

3.3 Strength

The strength of pressure hulls significantly influences the structural integrity and operational reliability of submarines and underwater vehicles. Understanding the complex interactions between material properties, geometric configurations, and external loading conditions is crucial for accurately predicting collapse pressures and ensuring the safety of these vessels. Strength is often represented by collapse pressure. Collapse pressure refers to the external hydrostatic pressure at which a pressure hull or vessel collapses or undergoes structural failure. It is a critical parameter in assessing the structural integrity and safety of submarine pressure hulls and similar structures designed to withstand high-pressure environments. Therefore, collapse pressure serves as a crucial measure of the strength and structural capacity of a pressure hull.

Corrosion is a significant factor that can significantly impact the structural integrity of pressure hulls in submarines and, furthermore, underwater vehicles. As corrosion gradually eats away at the metal surfaces of the hull, it leads to thinning of the shell material, reducing its overall structural integrity. This decrease in material thickness directly impacts the hull's capability to withstand external pressure loads, such as hydrostatic pressure underwater. Studies like MacKay [45] and [46] have shown that corrosion-induced thinning can introduce substantial stresses and destabilizing bending forces at the corroded site regions, leading to premature yielding and ultimate collapse. Furthermore, the depth and extent of corrosion thinning can exacerbate the reduction in collapse strength, as observed in experiments conducted by [46]. The orientation and alignment of corrosion patches relative to the dominant out-of-circularity mode can influence the severity of the strength reduction [47].

The literature on the strength of pressure hulls offers diverse perspectives and methodologies for assessing collapse pressure and its relationship to hull strength. Smith [48] provides a comprehensive approach to evaluating the strength of uniformly stiffened cylinders, considering interframe collapse, overall collapse, and localized stiffener failure. In contrast, MacKay [49] emphasizes the need for improved analytical predictions of pressure hull strength, promoting the adoption of nonlinear numerical methods to mitigate the inherent conservatism and rigidity in conventional design approaches. Meanwhile, MacKay [47] highlights the sensitivity of collapse to material strength and strain hardening, underscoring the importance of accurate modeling of material behaviour in predicting collapse pressures. MacKay [46] and [45] delve into the impact of corrosion thinning and out-of-circularity imperfections on collapse strength, stressing the necessity of accurately capturing hull shape and material behaviour in numerical models. Graham [50] echoes the importance of detailed structural information for accurate prediction of collapse pressures. McEvely [51] adds another dimension by discussing the prediction of fatigue strength in filament-wound composite cylinders, offering insights into a different aspect of pressure hull performance.

Blachut [52] investigates the collapse performance of bowed-out shells, highlighting the role of imperfections and structural behaviour in determining collapse strength. Hagen [53] proposes a collapse design optimization approach for ring-stiffened cylindrical pressure hulls, highlighting the impact of design factors on collapse performance. Cho [54] contributes to the literature by deriving an empirical equation for forecasting the collapse strength of filament-wound composite cylinders, emphasizing geometric dimensions and orientation angles of layers. Ross [55] explores collapse performance regarding pressure vessels, particularly in relation to design standards like PD 5500. Jasion et al [56] examine the stress and the stability of clothoid shells of revolution, offering insights into about design and behaviour of such structures.

3.4 Thickness

Corrosion is a significant phenomenon that can lead to the thinning of materials due to chemical reactions with the environment, resulting in surface degradation and loss of material thickness, compromising structural integrity and functionality. Wright et all[57] found the degradation caused by corrosion is influenced by factors such as the type of corrosive agent, environmental conditions, and material composition. The degradation influences the rate and extent of corrosion-induced thickness loss in Souza et al[58]. Lou et al studied various mechanisms incorporating uniform corrosion, pitting corrosion, and stress-induced corrosion cracking that contribute to material thinning [59], impacting structural integrity and material functionality [58].

Li et al research has shown that corrosion-induced thickness loss is a common issue across multiple industries [60]. Ding et al, monitoring corrosion is crucial to preventing catastrophic events resulting from material degradation [61]. Studies have highlighted the importance of understanding corrosion behaviours under various conditions, such as exposure to salts or molten chloride salts, to develop effective mitigation strategies [61] [62]. Talakokula et al found the oxidation of steel reinforcement in concrete structures is a significant issue of concern, leading to significant expenditures for inspecting, maintaining, and replacing infrastructure [63]. Techniques such as cyclic potentiodynamic polarization tests have demonstrated changes in pitting corrosion resistance and localized corrosion with thickness reduction [58]. Another paper, Talakokula et al, states that electrochemical methods like impedance spectroscopy and linear polarization are conventionally used for corrosion detection [63]. Yang et al, studies have also highlighted the influence of corrosion on the mechanical characteristics of materials, with observations of thickness and mass loss induced by corrosion damage [64]. The thickness of protective films on materials affects oxygen diffusion rates and corrosion product accumulation, influencing the corrosion process in Li et al [65].

Corrosion not only affects metallic materials but also materials like carbon in fuel cells, where corrosion can impact long-term performance and durability [66], [67] flexibility, longevity, and overall performance in different applications [68]. Jia et al paper that microbiologically influenced corrosion is another area of concern, necessitating the development of current mitigation strategies to combat its effects [69]. Corrosion can occur in diverse environments, such as in concentrated solar power plants, emphasizing the need for effective mitigation strategies to combat material degradation [61], [62]. In submarines, corrosion-induced thinning of submarine hull materials is a critical issue that significantly impacts the structural integrity and performance of underwater vessels. The aggressive marine environment exposes submarine hulls to seawater and various corrosive agents like chlorides [70] and sulfides [71], leading to gradual material thinning and compromising strength and resistance to external pressures such as hydrostatic forces and impacts from marine debris[72].

Studies on the characteristics of materials such as HY 80 steel from submarine hulls following years of operation in severe marine environments by Szturomski & Kiciński, provide insights into the effects of long-term exposure to marine environments on material integrity [73]. The corrosion rate of HY-80, determined using the potentiodynamic method, ranges from 0.17 to 0.18 mm/year, and 0.19 mm/year by the electrochemical method [74]. Helal et al, wrote that numerical analyses of sandwich composite deep submarine pressure hulls offer valuable information on failure criteria and structural considerations in submarine design [75]. Research on the tribocorrosion characteristics of high-strength naval steel in saline solutions highlights about the combined impact of mechanical and corrosive factors on material loss in marine environments in Wu et al [76].

4. Corrosion Examination Sunken Submarine

Researchers conducted corrosion studies on a sunken submarine to obtain accurate data on the rate of submarine hull deterioration due to corrosion. In August 2002, researchers from the Hawaiian Undersea Research Laboratory (HURL) examined a Japanese submarine that had been submerged in Hawaiian waters for approximately 61 years at a depth of 406.7 meters. The submarine's outer shell was constructed from cast iron with a mixture of geothite, magnetite, and sederite minerals. Estimated corrosion rate data of 0.014 mmpy were obtained, and it was estimated that the submarine had lost 11% of its original hull thickness[77].

In 1996, research was conducted on the Resurgam submarine, made from wrought iron, which sank off the north coast of Wales on Tuesday, February 24th, 1880. Experts examined its hull's corrosion. Quantifying the corrosion rate was challenging due to spalling, but an average rate of 0.1 mm per year was calculated where profiles were retained. The methodology involved measuring electrochemical parameters like surface pH and corrosion potential (Ecorr) on-site to evaluate the status of iron artifacts [78]. Another result from an experiment conducted to provide evidence that the H.L. Hunley submarine immersed in 1864 was that the most significant breaches were probably due to the degradation from both erosion and corrosion of the iron hull in the marine setting [79]. An expense for addressing corrosion-related issues, including repairing protective coatings and dealing with metal loss, on a Collins Class Submarine amounts to around $8.9 million per submarine during a Full Cycle Docking (FCD), which occurs roughly every 72 months [80].

5. Fatigue Life Analysis Pressure Hull

The pressure hull is a critical component that protects personnel and equipment in underwater vehicles. Several studies calculate lifetime pressure hulls based on fatigue life pressure using various approaches. Research conducted by Robles et al. considered hydrostatic pressure and residual stress, applying linear elastic fracture mechanics using the Paris law equation. The operational life of the pressure hull was estimated to be 40 years [81]. Nugroho et al. estimated the operational life of submarine pressure hulls made of HY80, considering hydrostatic pressure. The projected operational life obtained was approximately 29 years. This study employed the Palmgren Miner rule method with principal stress data obtained from modeling the pressure hull structure of the submarine using finite element analysis [82]. Hua et al. calculated the service life of manned submersibles simulated to operate at a depth of 4500 meters. The results indicated that the shell could be safely operated up to 4000 times. Fatigue life analysis was conducted using structural reliability theory and finite element analysis with stress interference modeling based on stress concentration [83]. Yu et al. computed the fatigue life of manned submersibles made of titanium alloy TC4, considering the field of temperature and residual stress. The remaining life was estimated to be 49 years under loads ranging from 10 MPa to 20 MPa at a depth of 7000 meters [84]. Robertson analyzed analyzing the stress in submarine pressure hulls subjected to hydrostatic loading using finite element techniques. The accurate stress analysis results led to the precise design use of safety factors of 1.5 or less [85]. Overall, the summary of all research in a table containing methods, considerations, advantages, and disadvantages as presented below.

Tabel 1. Comparative analysis of research fatigue life analysis

No	Methods	Considered	Advantages	Disadvantages	Reff
1	Fracture mechanics that adhere to linear elasticity and employ the Paris equation	Hydrostatic pressure and residual stress of welding process	Evaluation specific to the operational profile	That accuracy depends on estimating residual stress and the submarine's operational profile	[81]
2	Finite element method to model the structure.	Hydrostatic pressure	This method allows quite accurate prediction of the fatigue life of submarine compressed body structures	The calculation only considers the ideal pressure hull condition, smooth without defects	[82]
3	Structural reliability theory and finite element analysis to conduct fatigue analysis.	Hydrostatic pressure and considering stress concentration at the shell's opening	The use of a stress-interference model provides a more accurate assessment of the fatigue reliability	The reliance on theoretical and simulated data might limit the empirical validation of the results. More real-world testing and validation are necessary to confirm the model's accuracy and reliability under operational condition	[83]
4	The equivalent structural stress method is proposed, with techniques for developing a detailed model of the weld area.	Hydrostatic pressure and residual stress of welding factors titanium alloy TC4	The method of equivalent structural stress now better calculates convergence and accurately reflects stress concentration phenomena	Achieving accuracy while thoroughly considering the impact of welding processes and field practices.	[84]
5	Analytical methods of calculation and in the calculation of fatigue loading	Hydrostatic pressure and residual stress welding process	They provide a clear overview of the stress distribution caused by straightforward geometric load distributions like hydrostatic pressure.	Analytical solutions are insufficient for determining stresses at numerous structural points where fatigue cracking may begin.	[85]

From all the research data in the table above, the calculation of hull fatigue life pressure that has been carried out only takes into account hydrostatic pressure effects and residual stresses due to welding, but does not yet consider corrosion factors, and this is a free space for further research. For more details, it can also be seen in the bibliometrics in Figure 1. It can be seen that the average bubble is small, which indicates that very few people have researched it. Fatigue life and fatigue corrosion have not been connected, which suggests that no one has researched this, and this became a gap in the research.

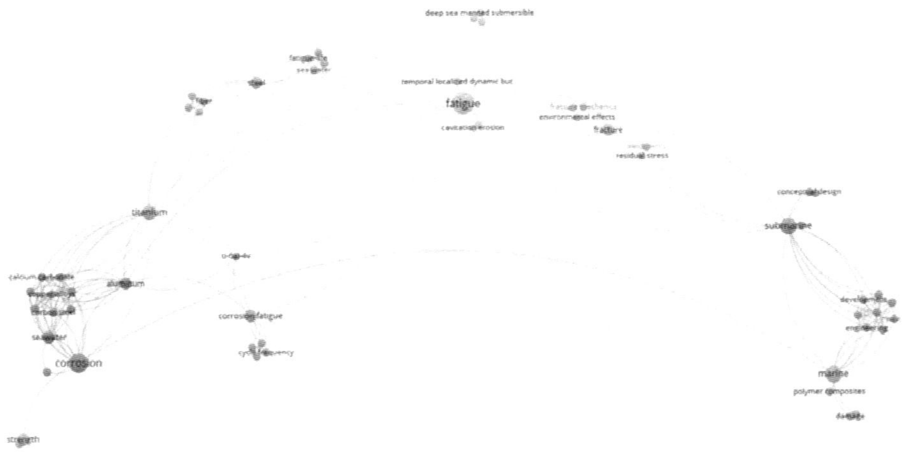

Figure 1. Bibliometrics fatigue life assessment of pressure hull research

6. Conclusion

In submarine operational conditions, the pressure hull is an important component where the crew and equipment live. It is designed to minimize the influence of hydrostatic pressure and cyclic loading on reducing its operational life. With such environmental influences, corrosion is prone to occur, which reduces the strength of the structure and the thickness of the material. The combination of all of the above causes corrosion fatigue, which reduces fatigue life. It is crucial to calculate the service life of a pressure hull by considering corrosion factors other than hydrostatic pressure and residual stress. Overcoming the challenges posed by corrosion caused by low-cycle fatigue to ensure the longevity, reliability, and safety of submarine structures throughout their operational lives became a novelty for the writing of my dissertation.

References

[1] E. McCafferty, Introduction to Corrosion Science. New York, NY: Springer New York, 2010. doi: 10.1007/978-1-4419-0455-3.

[2] W. Li, J. Wang, T. Liu, and M. Luo, "Electromechanical impedance instrumented circular piezoelectric-metal transducer for corrosion monitoring: modeling and validation," Smart Mater. Struct., vol. 29, no. 3, p. 035008, Mar. 2020, doi: 10.1088/1361-665X/ab675c.

[3] A. A. P. Sidharth, "EFFECT OF PITTING CORROSION ON ULTIMATE STRENGTH AND BUCKLING STRENGTH OF PLATES – A REVIEW," 2009. [Online]. Available: https://api.semanticscholar.org/CorpusID:29597716

[4] H. P. Mamgain, K. K. Samanta, R. Brajpuriya, P. R. Pati, R. Gupta, and J. K. Pandey, "Fabrication of Nanostructured Corrosion-Resistant Superhydrophobic Coating on Copper by Electrodeposition: A Comprehensive Critical Review," ECS J. Solid State Sci. Technol., vol. 13, no. 4, p. 043010, Apr. 2024, doi: 10.1149/2162-8777/ad3c25.

[5] X. Sun, O. W. H. Wai, J. Xie, and X. Li, "Biomineralization To Prevent Microbially Induced Corrosion on Concrete for Sustainable Marine Infrastructure," Environ. Sci. Technol., vol. 58, no. 1, pp. 522–533, Jan. 2024, doi: 10.1021/acs.est.3c04680.

[6] C. C.Liang, C. Y.Shu, T. L.Teng, and C. Y Jen, "Estimation of the low cycle fatigue life for a submarine pressure hull," Transactions on Engineering Sciences, vol. 40, p. 13, 2003, doi: 10.2495/FDM030111.

[7] Y. Liu, H. Yi, and L. Chen, "Submarine pressure hull butt weld fatigue life reliability prediction method," Marine Structures, vol. 36, pp. 51–64, Apr. 2014, doi: 10.1016/j.marstruc.2013.12.002.

[8] B. Shaw and R. Kelly, "What is Corrosion?," Electrochem. Soc. Interface, vol. 15, no. 1, pp. 24–26, Mar. 2006, doi: 10.1149/2.F06061IF.

[9] P. Hammonds, "Chapter 4 An Introduction to Corrosion and its Prevention," in Comprehensive Chemical Kinetics, vol. 28, Elsevier, 1989, pp. 233–279. doi: 10.1016/S0069-8040(08)70400-X.

[10] S. Lyon, "Overview of corrosion engineering, science and technology," in Nuclear Corrosion Science and Engineering, Elsevier, 2012, pp. 3–30. doi: 10.1533/9780857095343.1.3.

[11] Z. Yongmei et al., "Fatigue and Corrosion Fatigue of 18Ni Maraging Steel," Archives of Metallurgy and Materials, pp. 321–330, Sep. 2020, doi: 10.24425/amm.2021.134790.

[12] A. Kolios, S. Srikanth, and K. Salonitis, "Numerical Simulation of Material Strength Deterioration due to Pitting Corrosion," Procedia CIRP, vol. 13, pp. 230–236, 2014, doi: 10.1016/j.procir.2014.04.040.

[13] K. M. Kumar, R. N, Y. N, and R. S, "Fatigue Life Estimation of Machine Components," IRJET, vol. 5, no. 6, pp. 1328–33, Jun. 2018.

[14] X. Ye and H. Jin, "Corrosion-Induced Strengthening: Development of High-Strength Nanoporous Metals," Adv Eng Mater, vol. 18, no. 6, pp. 1050–1058, Jun. 2016, doi: 10.1002/adem.201500521.

[15] P. Jurišić and J. Parunov, "Influence of corrosion-related degradation of mechanical properties of shipbuilding steel on collapse strength of plates and stiffened panels," in Towards Green Marine Technology and Transport, C. Soares, R. Dejhalla, and D. Pavletic, Eds., CRC Press, 2015, pp. 427–432. doi: 10.1201/b18855-56.

[16] A. Shetty, K. Venkataramana, and K. S. B. Narayan, "Effect of corrosion on flexural bond strength," J. Electrochem. Sci. Eng., vol. 4, no. 3, pp. 121–134, Sep. 2014, doi: 10.5599/jese.2014.0052.

[17] T. Nakai, H. Matsushita, and N. Yamamoto, "Effect of pitting corrosion on the ultimate strength of steel plates subjected to in-plane compression and bending," J Mar Sci Technol, vol. 11, no. 1, pp. 52–64, Mar. 2006, doi: 10.1007/s00773-005-0203-4.

[18] Y. Wang, J. A. Wharton, and R. A. Shenoi, "Ultimate strength assessment of steel stiffened plate structures with grooving corrosion damage," Engineering Structures, vol. 94, pp. 29–42, Jul. 2015, doi: 10.1016/j.engstruct.2015.02.028.

[19] Z. Zhang, S. Xu, and R. Li, "Comparative investigation of the effect of corrosion on the mechanical properties of different parts of thin-walled steel," Thin-Walled Structures, vol. 146, p. 106450, Jan. 2020, doi: 10.1016/j.tws.2019.106450.

[20] Y.-S. Choi, S. Nesic, and D. Young, "Effect of Impurities on the Corrosion Behavior of CO 2 Transmission Pipeline Steel in Supercritical CO 2 -Water Environments," Environ. Sci. Technol., vol. 44, no. 23, pp. 9233–9238, Dec. 2010, doi: 10.1021/es102578c.

[21] S. Li, K. Zhang, and Q. Wang, "Experimental study on the corrosion of a downhole string under flue gas injection conditions," Energy Science & Engineering, vol. 7, no. 6, pp. 2620–2632, Dec. 2019, doi: 10.1002/ese3.448.

[22] J. LIU, X. LI, and J. WANG, "Effect of hydrostatic pressure on the corrosion behaviors of two low alloy steels," Act Metal Sin, vol. 47, no. 6, pp. 697–705, Dec. 2010, doi: 10.3724/SP.J.1037.2010.00705.

[23] Z. X. Yang, B. Kan, J. X. Li, Y. J. Su, and L. J. Qiao, "Hydrostatic pressure effects on corrosion behavior of X70 pipeline steel in a simulated deep-sea environment," Journal of Electroanalytical Chemistry, vol. 822, pp. 123–133, Aug. 2018, doi: 10.1016/j.jelechem.2018.05.010.

[24] B. Liu, T. Zhang, Y. Shao, G. Meng, and F. Wang, "Effect of hydrostatic pressure on the nature of passive film of pure nickel," Materials & Corrosion, vol. 62, no. 3, pp. 269–274, Mar. 2011, doi: 10.1002/maco.200905372.

[25] Q. Xia, P. Ren, and H. Meng, "Corrosion Resistance of Amorphous-Nanocrystalline Composite Structure Materials," ACS Omega, vol. 8, no. 3, pp. 3348–3353, Jan. 2023, doi: 10.1021/acsomega.2c07073.

[26] Z. Yang, B. Kan, J. Li, L. Qiao, A. A. Volinsky, and Y. Su, "A Statistical Study on the Effect of Hydrostatic Pressure on Metastable Pitting Corrosion of X70 Pipeline Steel," Materials, vol. 10, no. 11, p. 1307, Nov. 2017, doi: 10.3390/ma10111307.

[27] T. Duan et al., "Corrosion behaviour investigation of 460 low alloy steels exposed in the natural deep-sea environment," Corrosion Engineering, Science and Technology, vol. 54, no. 6, pp. 485–492, Aug. 2019, doi: 10.1080/1478422X.2019.1619290.

[28] Y. Yang, T. Zhang, Y. Shao, G. Meng, and F. Wang, "New understanding of the effect of hydrostatic pressure on the corrosion of Ni–Cr–Mo–V high strength steel," Corrosion Science, vol. 73, pp. 250–261, Aug. 2013, doi: 10.1016/j.corsci.2013.04.013.

[29] Y. Wang et al., "Effect of hydrostatic pressure on galvanic corrosion of low-alloy steel in simulated deep-sea environments," Corrosion Engineering, Science and Technology, vol. 57, no. 7, pp. 673–679, Oct. 2022, doi: 10.1080/1478422X.2022.2119006.

[30] P. Traverso and E. Canepa, "A review of studies on corrosion of metals and alloys in deep-sea environment," Ocean Engineering, vol. 87, pp. 10–15, Sep. 2014, doi: 10.1016/j.oceaneng.2014.05.003.

[31] T. Palin-Luc, R. Pérez-Mora, C. Bathias, G. Domínguez, P. C. Paris, and J. L. Arana, "Fatigue crack initiation and growth on a steel in the very high cycle regime with sea water corrosion," Engineering Fracture Mechanics, vol. 77, no. 11, pp. 1953–1962, Jul. 2010, doi: 10.1016/j.engfracmech.2010.02.015.

[32] M. El May, T. Palin-Luc, N. Saintier, and O. Devos, "Effect of corrosion on the high cycle fatigue strength of martensitic stainless steel X12CrNiMoV12-3," International Journal of Fatigue, vol. 47, pp. 330–339, Feb. 2013, doi: 10.1016/j.ijfatigue.2012.09.018.

[33] H. Su, Y. Liang, S. Wei, Y. Wang, B. Wang, and H. Tong, "Couple effect of hydrostatic pressure and dissolved oxygen on corrosion behaviour of low-alloy high strength steel in 3.5 wt-% NaCl solution," Corrosion Engineering, Science and Technology, vol. 54, no. 4, pp. 330–338, May 2019, doi: 10.1080/1478422X.2019.1590959.

[34] A. A. Ahmed, M. Mhaede, M. Wollmann, and L. Wagner, "Effect of surface and bulk plastic deformations on the corrosion resistance and corrosion fatigue performance of AISI 316L," Surface and Coatings Technology, vol. 259, pp. 448–455, Nov. 2014, doi: 10.1016/j.surfcoat.2014.10.052.

[35] E. Canepa, R. Stifanese, L. Merotto, and P. Traverso, "Corrosion behaviour of aluminium alloys in deep-sea environment: A review and the KM3NeT test results," Marine Structures, vol. 59, pp. 271–284, May 2018, doi: 10.1016/j.marstruc.2018.02.006.

[36] S. Chen, L. Qiu, S. Sun, J. Yang, Q. Meng, and W. Yang, "Research Progress on Corrosion of Equipment and Materials in Deep-Sea Environment," Advances in Civil Engineering, vol. 2021, pp. 1–12, Dec. 2021, doi: 10.1155/2021/7803536.

[37] S. L. Li, X. G. Yang, H. Y. Qi, L. Q. Ma, J. S. Yang, and D. Q. Shi, "Experimental investigation and life prediction of hot corrosion pre-exposure on low-cycle fatigue of a directionally solidified nickel-base superalloy," Fatigue Fract Eng Mat Struct, vol. 38, no. 10, pp. 1155–1166, Oct. 2015, doi: 10.1111/ffe.12295.

[38] R. A. Hawileh, J. A. Abdalla, A. Al Tamimi, K. Abdelrahman, and F. Oudah, "Behavior of Corroded Steel Reinforcing Bars Under Monotonic and Cyclic Loadings," Mechanics of Advanced Materials and Structures, vol. 18, no. 3, pp. 218–224, Apr. 2011, doi: 10.1080/15376494.2010.499023.

[39] M. M. Kashani, P. Alagheband, R. Khan, and S. Davis, "Impact of corrosion on low-cycle fatigue degradation of reinforcing bars with the effect of inelastic buckling," International Journal of Fatigue, vol. 77, pp. 174–185, Aug. 2015, doi: 10.1016/j.ijfatigue.2015.03.013.

[40] M. M. Kashani, L. N. Lowes, A. J. Crewe, and N. A. Alexander, "Phenomenological hysteretic model for corroded reinforcing bars including inelastic buckling and low-cycle fatigue degradation," Computers & Structures, vol. 156, pp. 58–71, Aug. 2015, doi: 10.1016/j.compstruc.2015.04.005.

[41] M. M. Kashani, A. K. Barmi, and V. S. Malinova, "Influence of inelastic buckling on low-cycle fatigue degradation of reinforcing bars," Construction and Building Materials, vol. 94, pp. 644–655, Sep. 2015, doi: 10.1016/j.conbuildmat.2015.07.102.

[42] M. M. Kashani, L. N. Lowes, A. J. Crewe, and N. A. Alexander, "Finite element investigation of the influence of corrosion pattern on inelastic buckling and cyclic response of corroded reinforcing bars," Engineering Structures, vol. 75, pp. 113–125, Sep. 2014, doi: 10.1016/j.engstruct.2014.05.026.

[43] G. S. Mahobia et al., "Effect of hot corrosion on low cycle fatigue behavior of superalloy IN718," International Journal of Fatigue, vol. 59, pp. 272–281, Feb. 2014, doi: 10.1016/j.ijfatigue.2013.08.009.

[44] M. C. Vasco, P. Polydoropoulou, A. N. Chamos, and S. G. Pantelakis, "Effect of corrosion and sandblasting on the high cycle fatigue behavior of reinforcing B500C steel bars," Frattura ed Integrità Strutturale, vol. 11, no. 42, pp. 9–22, Sep. 2017, doi: 10.3221/IGF-ESIS.42.02.

[45] J. R. MacKay, M. J. Smith, F. Van Keulen, T. N. Bosman, and N. G. Pegg, "Experimental investigation of the strength and stability of submarine pressure hulls with and without artificial corrosion damage," Marine Structures, vol. 23, no. 3, pp. 339–359, Jul. 2010, doi: 10.1016/j.marstruc.2010.06.001.

[46] J. R. MacKay, "Experimental investigation of the strength of damaged pressure hulls - Phases 5 & 6".

[47] J. R. MacKay and F. V. Keulen, "The Sensitivity of Overall Collapse of Damaged Submarine Pressure Hulls to Material Strength," Journal of Offshore Mechanics and Arctic Engineering, vol. 135, no. 2, p. 021403, May 2013, doi: 10.1115/1.4007051.

[48] N. G. Pegg and D. R. Smith, "PRHDEF - Stress and Stability Analysis of Ring Stiffened Submarine Pressure Hulls".

[49] J. R. MacKay, "Experimental Investigation of the Strength of Damaged Pressure Hulls - Phase 1".

[50] D. Graham, "Predicting the collapse of externally pressurised ring-stiffened cylinders using finite element analysis," Marine Structures, vol. 20, no. 4, pp. 202–217, Oct. 2007, doi: 10.1016/j.marstruc.2007.09.002.

[51] A. Mcevily and M. Endo, "Prediction of the influence of flaws on the fatigue strength under biaxial loading," International Journal of Fatigue, vol. 28, no. 5–6, pp. 504–507, May 2006, doi: 10.1016/j.ijfatigue.2005.07.043.

[52] J. Błachut, "Buckling of externally pressurised barrelled shells: a comparison of experiment and theory," International Journal of Pressure Vessels and Piping, vol. 79, no. 7, pp. 507–517, Jul. 2002, doi: 10.1016/S0308-0161(02)00040-6.

[53] K. Hagen, "Collapse Design Optimisation of Ring-Stiffened Cylindrical Pressure Hulls," University of Stavanger, Norway, 2020. [Online]. Available: https://uis.brage.unit.no/uis-xmlui/bitstream/handle/11250/2678520/Hagen_Kine.pdf?sequence=1

[54] Y. S. Cho, D. H. Oh, and J. K. Paik, "An empirical formula for predicting the collapse strength of composite cylindrical-shell structures under external pressure loads," Ocean Engineering, vol. 172, pp. 191–198, Jan. 2019, doi: 10.1016/j.oceaneng.2018.11.028.

[55] C. T. F. Ross, "Design of submarine pressure hulls to withstand buckling under external hydrostatic pressure," in Pressure Vessels, Elsevier, 2011, pp. 355–360. doi: 10.1533/9780857092496.355.

[56] P. Jasion and K. Magnucki, "Elastic buckling of clothoidal–spherical shells under external pressure – theoretical study," Thin-Walled Structures, vol. 86, pp. 18–23, Jan. 2015, doi: 10.1016/j.tws.2014.10.001.

[57] R. F. Wright, P. Lu, J. Devkota, F. Lu, M. Ziomek-Moroz, and P. R. Ohodnicki, "Corrosion Sensors for Structural Health Monitoring of Oil and Natural Gas Infrastructure: A Review," Sensors, vol. 19, no. 18, p. 3964, Sep. 2019, doi: 10.3390/s19183964.

[58] C. S. Souza, R. B. Soares, R. A. D. De Faria, A. Q. Bracarense, M. Des M. M. R. Castro, and V. De F. C. Lins, "Effect of the prior cold work on the corrosion resistance of lean duplex stainless steel welded by GMAW process," Materials & Corrosion, vol. 69, no. 11, pp. 1548–1559, Nov. 2018, doi: 10.1002/maco.201709950.

[59] W. Luo, T. Liu, W. Li, and M. Luo, "Pitting corrosion prediction based on electromechanical impedance and convolutional neural networks," Structural Health Monitoring, vol. 22, no. 3, pp. 1647–1664, May 2023, doi: 10.1177/14759217221109944.

[60] W. Li, J. Wang, T. Liu, and M. Luo, "Electromechanical impedance instrumented circular piezoelectric-metal transducer for corrosion monitoring: modeling and validation," Smart Mater. Struct., vol. 29, no. 3, p. 035008, Mar. 2020, doi: 10.1088/1361-665X/ab675c.

[61] W. Ding et al., "Molten chloride salts for next generation concentrated solar power plants: Mitigation strategies against corrosion of structural materials," Solar Energy Materials and Solar Cells, vol. 193, pp. 298–313, May 2019, doi: 10.1016/j.solmat.2018.12.020.

[62] W. Ding, A. Bonk, and T. Bauer, "Corrosion behavior of metallic alloys in molten chloride salts for thermal energy storage in concentrated solar power plants: A review," Front. Chem. Sci. Eng., vol. 12, no. 3, pp. 564–576, Sep. 2018, doi: 10.1007/s11705-018-1720-0.

[63] V. Talakokula, S. Bhalla, and A. Gupta, "Corrosion assessment of reinforced concrete structures based on equivalent structural parameters using electro-mechanical impedance technique," Journal of Intelligent Material Systems and Structures, vol. 25, no. 4, pp. 484–500, Mar. 2014, doi: 10.1177/1045389X13498317.

[64] Y. Liu and X. Feng, "Monitoring corrosion-induced thickness loss of stainless steel plates using the electromechanical impedance technique," Meas. Sci. Technol., vol. 32, no. 2, p. 025104, Feb. 2021, doi: 10.1088/1361-6501/abbb67.

[65] W. Li et al., "Numerical simulation of carbon steel atmospheric corrosion under varying electrolyte-film thickness and corrosion product porosity," npj Mater Degrad, vol. 7, no. 1, p. 3, Jan. 2023, doi: 10.1038/s41529-022-00316-3.

[66] P. T. Yu, Z. Liu, and R. Makharia, "Investigation of Carbon Corrosion Behavior and Kinetics in Proton Exchange Membrane Fuel Cell Cathode Electrodes," J. Electrochem. Soc., vol. 160, no. 6, pp. F645–F650, 2013, doi: 10.1149/2.120306jes.

[67] Z. Y. Liu et al., "Transmission Electron Microscopy Observation of Corrosion Behaviors of Platinized Carbon Blacks under Thermal and Electrochemical Conditions," J. Electrochem. Soc., vol. 157, no. 6, p. B906, 2010, doi: 10.1149/1.3391737.

[68] G. Zhang, J. Shi, Y. Zhang, and H. Li, "Life Prediction of Cable Steel Bracket Considering Carbon Steel Corrosion," J. Phys.: Conf. Ser., vol. 2527, no. 1, p. 012048, Jun. 2023, doi: 10.1088/1742-6596/2527/1/012048.

[69] R. Jia, T. Unsal, D. Xu, Y. Lekbach, and T. Gu, "Microbiologically influenced corrosion and current mitigation strategies: A state of the art review," International Biodeterioration & Biodegradation, vol. 137, pp. 42–58, Feb. 2019, doi: 10.1016/j.ibiod.2018.11.007.

[70] P. Yu, J. Liao, Y. Zhou, and S. He, "Analysis of Beibu Gulf tidal zone exposure on properties of carbon steel," Corrosion Engineering, Science and Technology: The International Journal of Corrosion Processes and Corrosion Control, vol. 59, no. 1, pp. 13–26, Feb. 2024, doi: 10.1177/1478422X231218072.

[71] R. E. Melchers, "Microbiological and abiotic processes in modelling longer-term marine corrosion of steel," Bioelectrochemistry, vol. 97, pp. 89–96, Jun. 2014, doi: 10.1016/j.bioelechem.2013.07.002.

[72] M. Shah, M. T. M. Ayob, N. Yaakob, Z. Embong, and N. K. Othman, "Effect of hydrogen sulfide and chloride on the passive film structures of stainless-steel type 316 L," Materialwissenschaft Werkst, vol. 53, no. 2, pp. 201–207, Feb. 2022, doi: 10.1002/mawe.202100212.

[73] S. Bogdan and K. Radosław, "Material Properties of HY 80 Steel after 55 Years of Operation for FEM Applications," Materials, vol. 14, no. 15, p. 4213, Jul. 2021, doi: 10.3390/ma14154213.

[74] H. C. Viljoen, N. Mahomed, L. H. Cupido, and G. P. Mitchell, "Effect of corrosion thinning on depth of operation: Case study of an HY-80 steel submarine pressure hull," Marine Structures, vol. 81, p. 103103, Jan. 2022, doi: 10.1016/j.marstruc.2021.103103.

[75] M. Helal, H. Huang, D. Wang, and E. Fathallah, "Numerical Analysis of Sandwich Composite Deep Submarine Pressure Hull Considering Failure Criteria," JMSE, vol. 7, no. 10, p. 377, Oct. 2019, doi: 10.3390/jmse7100377.

[76] H. Wu et al., "Influences of Load and Microstructure on Tribocorrosion Behaviour of High Strength Hull Steel in Saline Solution," Tribol Lett, vol. 67, no. 4, p. 124, Dec. 2019, doi: 10.1007/s11249-019-1237-9.

[77] B. M. Wilson et al., "Corrosion studies on the USS Arizona with application to a Japanese midget submarine," JOM, vol. 59, no. 10, pp. 14–18, Oct. 2007, doi: 10.1007/s11837-007-0124-2.

[78] D. Gregory, "In situ corrosion studies on the submarine Resurgam : A preliminary assessment of her state of preservation," Conservation and Management of Archaeological Sites, vol. 4, no. 2, pp. 93–100, Jan. 2000, doi: 10.1179/135050300793138363.

[79] M. Jacobsen, V. Y. Blouin, and W. Shirley, "Does Erosion Corrosion Account for Intriguing Damage to the Civil War Submarine H.L. Hunley? 1," mar technol soc j, vol. 46, no. 6, pp. 38–48, Nov. 2012, doi: 10.4031/MTSJ.46.6.2.

[80] R. Dankiw, "Structural corrosion issues on Collins class submarines," Submarine Engineering Structural & Naval Architecture Group, melbourne Australia, modified summary, April2018. [Online]. Available: https://www.corrosion.com.au/structural-corrosion-issues-on-collins-class-submarines-2006/

[81] L. Robles, "A method for the evaluation of the fatigue operational life of submarine pressure hulls," International Journal of Fatigue, vol. 22, no. 1, pp. 41–52, Jan. 2000, doi: 10.1016/S0142-1123(99)00102-4.

[82] Wibowo H Nugroho and Ahmad S. Mujahid, "On the fatigue life prediction of submarine pressure hull due to alternating hydrostatic loads," Majalah Ilmiah Pengkajian Industri, vol. 9, no. 3, pp. 139–146, Sep. 2023, doi: 10.29122/mipi.v9i3.1644.

[83] J. H. LI, D.-Q. BAI, and T.-S. LIU, "Fatigue reliability abalysis of pressure shell structure for manned submersible," Shandong Science, vol. 32, no. 2, Apr. 2019, doi: 10.3976/j.issn.1002-4026.2019.02.006.

[84] C. Yu, Q. Guo, X. Gong, Y. Yang, and J. Zhang, "Fatigue life assessment of pressure hull of deep-sea submergence vehicle," Ocean Engineering, vol. 245, p. 110528, Feb. 2022, doi: 10.1016/j.oceaneng.2022.110528.

[85] I. M. Robertson, "Analytical calculations of fatigue loading of submarine hulls p 0076 DSTO Materials Research Laboratory," Materials Research Labs Ascot Vale (Australia), Victoria Australia, investigation MRL-TR-90-26, DODA-AR-006-326, Nov. 1990. [Online]. Available: https://apps.dtic.mil/sti/citations/tr/ADA232543

IOP Conf. Series: Earth and Environmental Science **1423** (2024) 012003 doi:10.1088/1755-1315/1423/1/012003

Analysis of ship collision accidents in Indonesia using fault tree analysis (FTA) method

M I Firdaus[1*], M B Zaman[1] and R O S Gurning[1]

[1]Department of Marine Engineering, Faculty of Marine Technology, Sepuluh Nopember Institute of Technology, Surabaya, Indonesia

*E-mail: 7019222005@student.its.ac.id

Abstract. As an archipelagic country with emerging economies, maritime transportation has significantly contributed to the Indonesian economy by ensuring the fullness of domestic demand for goods and services, accelerating annually. However, with the increase in sea traffic, the probability of collision accidents may also grow, posing non-negligible risks to individuals and society in various aspects. Considering this condition, this paper assesses and evaluates past maritime collision accidents in Indonesia to observe and analyze the tendency of factors contributing to collision accidents. The analysis was conducted using the Fault Tree Analysis (FTA) method, qualitative and quantitative, on data obtained from the collision investigation report in the last 20 years. The qualitative approach is used to propose the minimal cut set, and the quantitative analysis is conducted to estimate the probability. Based on the obtained probability and the produced cut set diagram, the conclusion states that human-related factors contribute to the top even occurrence at most, implying that the human factor is the most contributing factor to collisions. Those factors include the lack of training, lack of experience, and improper decisions.

1. Introduction

Marine transportation is vital in supporting domestic and international trade, especially for countries with long coastlines and many islands, such as Indonesia, making shipping modes a part of economic growth and human development [1]. Nevertheless, behind the benefits of growth in shipping modes, the increasing congestion on sea shipping routes resulting from the growing number of vessels raises new problems, such as the risk of collision accidents at sea, which is growing and becoming more complex. Zhou et al. in 2023 analyzed the evolutionary pattern of maritime accidents from 2001 to 2020 [2]. It was found that ship collision accident has the highest level of occurrence in the 20-year span, with a relatively large percentage of around 22.1% as shown in Figure 1. In Indonesia, collision accidents have also become a tremendous marine event with severe consequences. In the last 20 years, more than 20 collision accidents have been recorded in Indonesia. Additionally, the collision accident may also trigger a more extensive and more severe accident, such as the sinking accident, which happened to Bhaita Jaya Samudra after the collision with Elisabet [3].

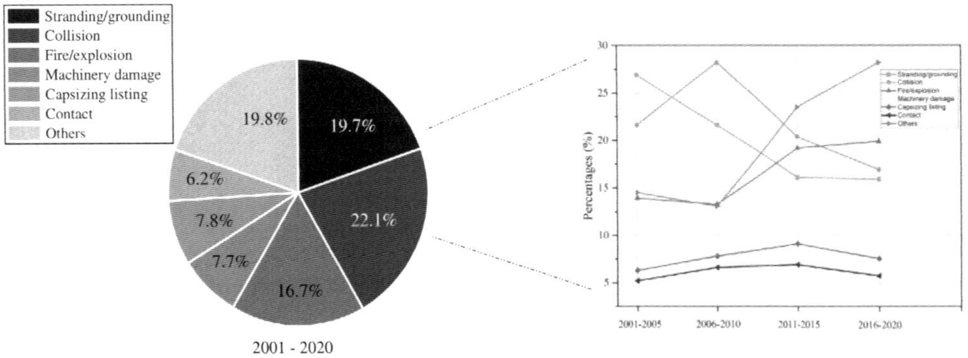

Figure 1. Data on global marine accident reports by accident type (reproduced from [2]).

Although collision accident risk assessments have been conducted, the analyses focus on specific cases that may be limited to certain regions and not representative of other regions with different traffic and geographic characteristics. Therefore, progress in marine transportation safety assessment is still needed. This paper emphasizes the assessment of collision accidents in Indonesia, their causal factors, and their likelihood. The analysis uses Fault Tree Analyses based on ship accident data for the past 20 years. With the data on ship accidents in Indonesia, we can see the similarities and differences in causal factors. This understanding is important for developing effective prevention strategies and improving national shipping safety. Table 1 and Figure 2 show the map of marine accidents in Indonesia.

Table 1. Description of the maritime collision accidents map, from the Republic of Indonesia KNKT investigation report.

Code	Accident	Code	Accident
A	(2003) Uni Chart – Mandiri Nusantara	F	(2011) Marina Nusantara – Pulau Tiga
	(2009) Tanto - Mitra Ocean		(2020) Ocean River 01 – Tugboat
	(2012) Alken Pesat – Alpine	G	(2015) Thorco Cloud – Stolt Commitment
	(2014) Journey – Fatima III – Lambelu		(2019) Antea – Star Centurion
B	(2010) Soechi Chemical XIX - Dian no 1	H	(2016) Victory Prima – Jaya II
	(2013) Lintas Bahari Utama – Lintas Bengkulu		(2017) Kartika Segara – JBB De Rong 19
	(2017) Elisabeth – Bhaita Jaya Samudra	I	(2018) Bunga Melati 79 – Tongkang Golden Way 3310
C	(2010) Bosowa VI - Shinpo 18	J	(2019) Mellinda – Rezeki Penuh 1
	(2022) Gerbang Samudra 2 – Trisila Bhakti II	K	(2019) Prince Soya – Cattleya Express
D	(2010) Indimatam V - Trisal Pratama	L	(2019) Virgo – Windu Karya Dwitya
E	(2011) Gloria Sentosa – Jukung Irpansya	M	(2021) Habco Pioneer – Barokah Jaya

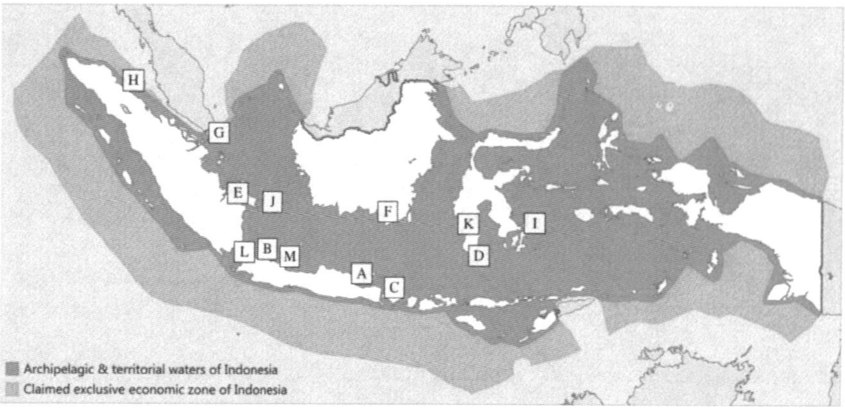

Figure 2. Map of maritime accidents in Indonesia.

2. Literature review

Fault Tree Analysis (FTA) is widely used in the maritime industry to identify potential accidents and analyze factors that may cause system failures [4]. In the marine context, the FTA helps to detect and identify risk factors that could potentially cause system failures on the ship, including damage to equipment, human error, and poor environmental conditions [5]. The FTA also enables comprehensive risk assessment and the development of mitigation strategies to prevent or mitigate the impact of potential accidents [6].

Uğurlu et al in 2015 applied FTA analyses to study collision and grounding data of oil tanker ships from 1998 to 2010, focusing on risk assessment [7]. With the FTA method, the cause factors of the accident are carefully identified, and the early events that triggered the accident are determined with the clinician. Studies show that crashes on tanker ships cause economic losses (81%), pollution (6%), and death or injury (13%). According to the FTA, human error is a significant cause of accidents, such as violations of the Convention on the International Regulations for Preventing Collisions at Sea (COLREG) and lack of communication between ships.

Ung in 2018 highlighted the crucial role of human error in ship safety management [8]. Complex assessments due to a lack of data were addressed by introducing new approaches. Six navigation experts provide professional assessments of the navigation of oil tanker vessels, using logical safety structures with FTAs and CREAM models based on Bayesian Fuzzy Network Analysis. With this method, the results strengthened the correlation between Common Performance Conditions (CPC) and Contextual Control Mode (COCOM), while sensitivity analysis identified fatigue and violation of Collision Rules (COLREG) as the trigger for ship sinking. This research has contributed to understanding the risks of grounding oil tanker vessels due to human error.

The research was then improved in Ung in 2019 to overcome the literature limitations in analyzing human error in the crash of oil tankers [9]. The method used involved a combination of Fault Tree Analysis (FTA) and the Fuzzy Cognitive Reliability Error Analysis Method (CREAM) model, which was then evaluated using Bayesian Reasoning to ensure the success of the analysis. Research results strengthen the relationship between Common Performance Conditions (CPC) and Contextual Control Mode (COCOM). FTA was also employed in the fire/explosion type of accident [10-12].

Table 2 shows the accidents that occurred in Indonesia in the last 20 years, along with their causes based on the report study.

Table 2. Maritime collision accidents in Indonesia obtained from the Republic of Indonesia KNKT investigation report.

No.	Accident	Reason
1	Uni Chart – Mandiri Nusantara	Faulty Procedure; Narrow and crowded; and Violation of Local Regulations
2	Tanto - Mitra Ocean	Faulty Procedure; Lack of Experience; Lack of Training; Narrow and crowded; and Disobeying Agreement
3	Soechi Chemical XIX - Dian no 1	Deficient Equipment; Faulty Procedure, Lack of Communication Equipment; Misuse of Communication Equipment; Narrow and crowded; and Violation of Local Regulations
4	Bosowa VI - Shinpo 18	Erroneous Decision; Lack of Experience; and Lack of Training
5	Indimatam V - Trisal Pratama	Erroneous Decision; Inattention; Lack of Experience; and Lack of Training
6	Gloria Sentosa – Jukung Irpansya	Communication Failures; Lack of Situational Awareness; Rute Precipitate; and Too Much Jobdesk
7	Marina Nusantara – Pulau Tiga	Bridge Management Failures; Lack of Experience; Lack of Training; Narrow and crowded; and Too long Working
8	Alken Pesat – Alpine	Absence of Tugboat; Lack of Experience; Lack of Training; Lack of VTS Coordination; and Too long Working
9	Lintas Bahari Utama – Lintas Bengkulu	Absence of Tugboat; Communication Failures; Mechanical Failures; Rute Obstacle
10	Journey – Fatima III – Lambelu	Absence of Tugboat; Erroneous Decision; Narrow and crowded; and Strong Current
11	Thorco Cloud – Stolt Commitment	Violation of COLREG Regulations
12	Victory Prima – Jaya II	Deficient Equipment and Mechanical Failures
13	Kartika Segara – JBB De Rong 19	Faulty Procedure; Lack of Equipment; Lack of Licence; Lack of Training; and Language Problems
14	Elisabeth – Bhaita Jaya Samudra	Lack of Equipment; Lack of Licence; and Narrow and crowded
15	Bunga Melati 79 – Tongkang Golden Way 3310	Lack of Communication Equipment and Misuse of Communication Equipment
16	Antea – Star Centurion	Faulty Procedure; Lack of Experience; and Lack of Training
17	Mellinda – Rezeki Penuh 1	Lack of Training and Violation of COLREG Regulations
18	Prince Soya – Cattleya Express	Erroneous Decision and Strong Current
19	Virgo – Windu Karya Dwitya	Erroneous Decision

20	Ocean River 01 – Tugboat	Mechanical Failures
21	Habco Pioneer – Barokah Jaya	Communication Failures and Erroneous Decision
22	Gerbang Samudra 2 – Trisila Bhakti II	Communication Failures and Erroneous Decision

3. Methodology

The analysis is done by employing the Fault Tree Analysis (FTA), which presents a structured, logical structure like a tree that facilitates the visualization of cause-effect relationships and enables the identification of causes and tracks of accidents clearly [4], [13]. The FTA helps in risk management by assessing the probability of accidents, enabling priority action for effective prevention. Another advantage is the FTA reproducibility, which allows repeated analysis to validate findings and make improvements based on additional information or changes in the system [14]. In general, the research methodology of the present study is shown in Figure 3.

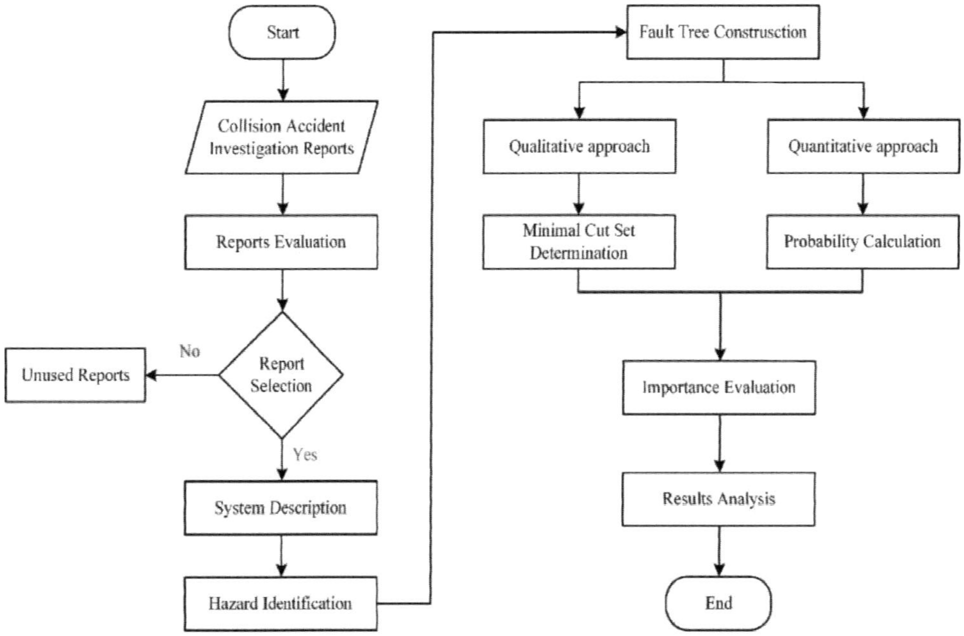

Figure 3. Research methodology.

Initially, the collision accident investigation reports are acquired from the government database of the National Transportation Safety Committee on Marine Accidental Reports. The reports are then examined to ensure that the minimum components required for FTA analysis are stated clearly. From the initial screening, 22 collision accident reports were assessed in the study from 2003 to 2023. The selected reports are then analyzed to obtain the general description of the accident, including the initial route, time and location of the accident, the accident scenario, and the accident circumstances. Based on the accident general description, the main events and the contributing factors are determined as a basis to propose the Fault Tree Diagram. The analysis is conducted qualitatively to obtain the minimum cut sets and quantitatively to get the probability of respected contributing factors.

The nomenclature of the FTA diagram is shown in Figure 4. There are two main components, which are the "event" and the "gate". The "event" is the particular scenario step of collision accident contributing to the "top event" or "upper event," denoted by a rectangular for the "top event" and the "intermediate event" and a circle for the "basic event". The "gate" points out the correlation between each contributing factor, including the "and" and "or" gates.

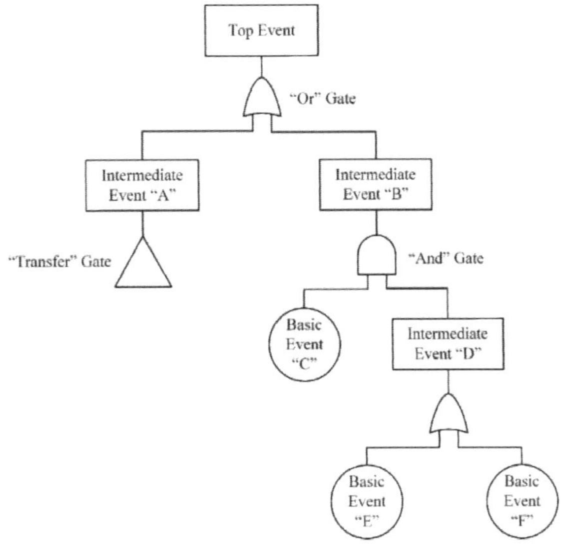

Figure 4. Descriptions of FTA diagram.

The occurrence probability of the top event, P_T, can be calculated using Equation 1 where P_i indicates the probability of the basic event i for the number of basic events n. The mathematical operator indicating the correlation between basic events are determined based on the "and" or "or" gates conditions [8].

$$P_T = \prod_{i=1}^{n} P_i \tag{1}$$

The probability of the basic event P_i is estimated based on the description of the initial case, as in Equation 2.

$$P_i = \sum_{l=1}^{m} \frac{N_l}{T_n \cdot T_t} \tag{2}$$

N_l is the rate contribution to the initial event l where the respective factor is involved. m indicates the number of primary events where the factor contributes to the initial event. T_n is the total number of accidents being investigated and T_t is the total duration of the accident data. After the probability of each factor is obtained, the analysis is then continued by investigating how significant the factor contributes to the top event. The significance factor is calculated based on the Fussel-Vesely Method [15].

For example, considering the FTA scheme of diagram 2, the cat sets from the tree are defined as $A + CxE + CxF$. Thus, the probability of the top even can be calculated as Equation 3.

$$P = P_A + P_C x P_E + P_C x P_F \tag{3}$$

IOP Conf. Series: Earth and Environmental Science **1423** (2024) 012003 doi:10.1088/1755-1315/1423/1/012003

4. Results and Discussion

4.1 Qualitative approach

The fault tree analysis of the initial event is conducted for all 22 initial events, which are listed in Table 2. Based on the failure analysis of initial event, the global FTA was constructed and formed to yield the combination of the basic events. The, MOCUS algorithm was applied to get the cut sets of the events.

Figure 5 shows the results of FTA analysis as a fault tree diagram. Based on the initial assessment of the Collison accident report, it was noticed that human-related factors dominated the failure type. Thus, for simplicity, the first sub-events of the contributing factors were divided into two first sub-events, which are human-related factors and other factors. Human-related failures consist of maneuvering, navigation, perception, and individual failures, while the other failures cover equipment and mechanical failures as well as external factors. For clarity, the diagram was partitioned using the 'transfer' gate into several parts, including Figure 5(a) for the main fault tree diagram, Figure 5(b), 5(c), and 5(d) respectively, for the maneuvering, navigation, and individual failure composing the human-related failure mode.

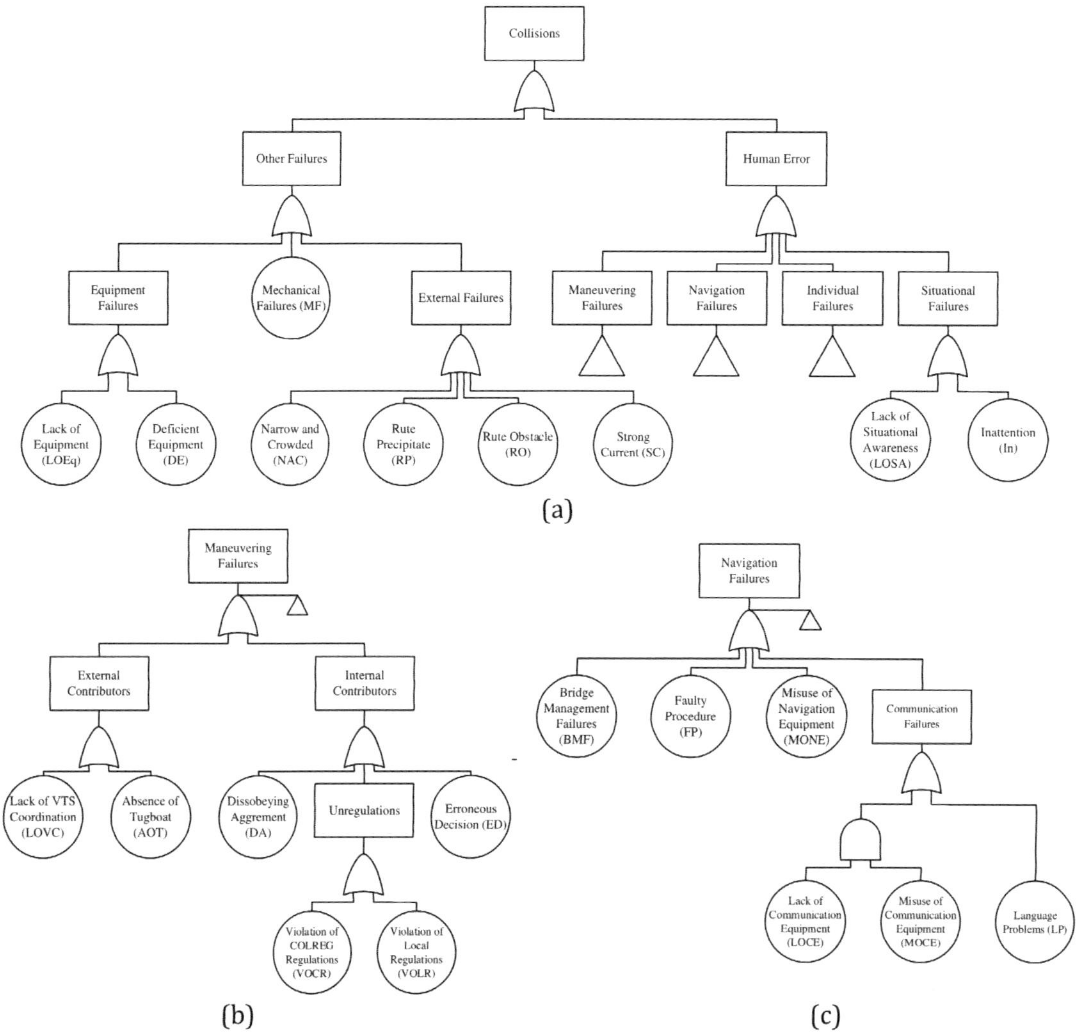

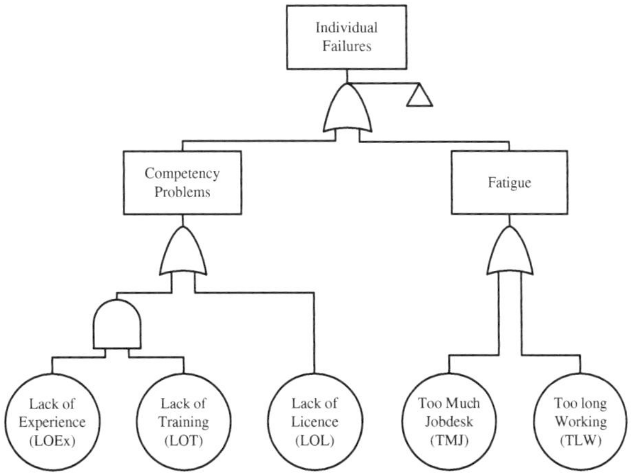

(d)

Figure 5. Main fault tree diagram; (a) Entire diagram, (b) Manouvering failures, (c) Navigation failures, and (d) Individual failures.

The maneuvering failures in Figure 5(b) are categorized based on the contributor source, which is the internal and external contributors. The internal contributors refer to the faults arising from the encountering ships' maneuvering system, including disobeying agreements, violation of regulations (COLREG and local regulations), and erroneous decisions. The external factors indicate that the faults do not come from the maneuvering systems of encountering ships but from external parties, such as the absence of tag boats and the lack of Vessel Traffic Services (VTS).

Figure 5(c) states the structure of the navigation failures consisting of bridge management failures, faulty procedures, communication failures, and misuse of navigation equipment. The navigation failures usually take place in one of the encountering ships, causing the ship to be out of the track and collide with the other ship. The consequences of this failure are crucial, especially on the crowded route, since the other ships cannot detect the missnavigated condition. In these failures, communication parameters play a significant role, emphasizing the importance of communication tools, not only communication equipment but also language.

Figure 5(d) breaks down the individual failures covering the competency problems and tiredness. The competency problems cover the lack of personal competency due to lack of experience, training, and license. The fatigue problems are mainly caused by the overwhelming job desk and working hours. The individual failures are mainly caused by the improper personal management.

4.2 Quantitative approach

The main aims of the quantitative analyses are to determine the probability of occurrence and the contribution of the basic events to the occurrence of the top event. Table 3 shows the list of the basic events and their abbreviation extracted from the fault tree of Figure 5. The list of initial events in Table 2 corresponds with the numbering of analyzed initial events in Table 3, indicating the inclusion of the respective factor to the initial events. Table 3 also shows the probability of the basic events which was calculated using equation 2 based on the total contribution of the basic and the duration of the obtained data.

From Table 3, it can be seen that the probability assessment, it can be seen that the basic events having the highest probability are related to the human factor. The basic events with the highest probability are the "Lack of Training" and "Lack of experience," which are considered individual failures, with probability of 0.0137 and 0.0137 respectively. The third is the "Erroneous Decision," with the probability of 0.00918 which is maneuvering failure rooted in personal improper decisions during the ship-to-ship encountering condition. The fourth highest probability is "Faulty Procedure" in the navigation failures with the value of 0.00796. The external factor of "Narrow and crowded" places the fifth highest probability with the value of 0.00783.

Table 3. List of the investigated initial collision events.

Basic Event	Code	No	Prob	Basic Event	Code	No	Prob
Absence of Tugboat	AOT	8, 9, 10	0.00459	Lack of Training	LOT	2, 4, 5, 7, 8, 13, 16, 17	0.0137
Bridge Management Failures	BMF	7	0.00171	Lack of VTS Coordination	LOVC	8	0.00153
Communication Failures	CF	6, 9, 21, 22	0.0054	Language Problems	LP	13	0.00182
Disobeying Agreement	DA	2	0.00171	Mechanical Failures	MF	9, 12, 20	0.00477
Deficient Equipment	DE	3, 12	0.00261	Misuse of Communication Equipment	MOCE	3, 15	0.00306
Erroneous Decision	ED	4, 5, 10, 18, 19, 21, 22	0.00918	Narrow and crowded	NAC	1, 2, 3, 7, 10, 14	0.00783
Faulty Procedure	FP	1, 2, 3, 13, 16	0.00796	Rute Obstacle	RO	9	0.00153
Inattention	In	5	0.00072	Rute Precipitate	RP	6	0.00108
Lack of Communication Equipment	LOCE	3, 15	0.00306	Strong Current	SC	10, 18	0.00261
Lack of Equipment	LOEq	13, 14	0.00261	Too long Working	TLW	7, 8	0.00324
Lack of Experience	LOEx	2, 4, 5, 7, 8, 13, 16, 17	0.0137	Too Much Jobdesk	TMJ	6	0.00171
Lack of Licence	LOL	13, 14	0.003	Violation of COLREG Regulations	VOCR	11, 17	0.00324

IOP Conf. Series: Earth and Environmental Science **1423** (2024) 012003 doi:10.1088/1755-1315/1423/1/012003

Lack of Situational Awareness	LOSA	6	0.00171	Violation of Local Regulations	VOLR	1, 3	0.00306

5. Conclusions

In this study, the main parameters of the collision accident in Indonesia were analyzed using Fault Tree Analysis including the assessment of the main factor contributing to the occurrence of the top events. The investigation was conducted qualitatively to get the minimal cut sets of the system and quantitively to obtain the percentage of top event failure probability due to the respective factors.

Considering the probability of the basic events in Table 3 and the fault tree diagram in Figure 5, the collision probability was estimated to be 0.06896. The results state that the collision accident is mainly contributed by human-related factors, including maneuverability failure, navigation failure, individual failures, and situational failures. The non-human related factors are mechanical failures, equipment failures, and external factors. This phenomenon is almost similar to the condition of collision from a global collision perspective, as analyzed in [15]. In Indonesia's case, the human factors contributing to the collision are mainly the lack of training, lack of experience, and erroneous decisions with a frequency of almost 40% of the total occurrence of the basic events. This condition shows the importance of developing mitigation that can be carried out including increasing training and education, implementing the latest safety technology, strict maintenance and inspection procedures, as well as strengthening safety policies and regulations.

References

[1] Christiansen M, Fagerholt K, Nygreen B and Ronen D 2007 Maritime Transportation 14 189–284
[2] Zhou X, Ruan X, Wang H and Zhou G 2023 Exploring spatial patterns and environmental risk factors for global maritime accidents: A 20-year analysis Ocean Eng. 286 115628
[3] Komite Nasional Keselamatan Transportasi 2017 Tubrukan Elisabet (IMO9041875) dengan Bhaita Jaya Samudra (IMO8713108) Di Sekitar Perairan Pulau Damar, Kepulauan Seribu, DKI Jakarta Republik Indonesia
[4] Kurniawan R 2022 Jurnal Teknologi 9 pp 26–36
[5] Cem Kuzu A, Akyuz E and Arslan O 2019 Application of Fuzzy Fault Tree Analysis (FFTA) to maritime industry: A risk analysing of ship mooring operation Ocean Eng. 179 pp 128–134
[6] Arslan Ö, Zorba Y and Svetak J 2018 Fault Tree Analysis of Tanker Accidents during Loading and Unloading Operations at the Tanker Terminals Journal of Eta Mar. Sci. 6 pp 3–16
[7] Uğurlu O, Yildirim U and Basar E 2015 Analysis of Grounding Accidents Caused by Human Error Journal of Marine Sci. and Tech. 23 pp 748-760
[8] Ung S-T 2018 Human error assessment of oil tanker grounding Safety Sci. 104 pp 16–28
[9] Ung S-T 2019 Evaluation of human error contribution to oil tanker collision using fault tree analysis and modified fuzzy Bayesian Network based CREAM Ocean Eng. 179 pp 159–172
[10] Zhao-mei X 2011 Research on FTA of Fire and Explosion in the Crude Oil Gatheringtransport Combination Station Procedia Eng. 11 pp 575–582
[11] Wang L, Wang J, Shi M, Fu S and Zhu M 2021 Critical risk factors in ship fire accidents Mar. Policy & Management 48 pp 895–913
[12] Lutfi Tunçel A, Bal Beşikçi E, Akyuz E and Arslan O 2023 Safety analysis of fire and explosion (F&E) accidents risk in bulk carrier ships under fuzzy fault tree approach Safety Sci. 158 p 10597
[13] Wu P, Meng X, Song L and Zuo W 2019 Transportation Research Record 2673 pp 403–416
[14] Wirandoro H B, Pratikto W A and Silvianita S 2022 Offshore and Coastal Eng. 6 p 30
[15] Ugurlu H and Cicek I 2022 Analysis and assessment of ship collision accidents using Fault Tree and Multiple Correspondence Analysis Ocean Eng. 245 p 110514

IOP Conf. Series: Earth and Environmental Science **1423** (2024) 012004 doi:10.1088/1755-1315/1423/1/012004

Numerical study of engine room air ventilation systems for landing ship tank class warships in supporting crew comfort and safety

Anton Nugroho[1], Wawan Aries Widodo[2*], Is Bunyamin Suryo[3]

[1,2,3] Department of Mechanical Engineering, Institut Teknologi Sepuluh Nopember (ITS), Surabaya (60111), Indonesia

*E-mail: wawanaries@me.its.ac.id

Abstract. The air ventilation system is one of the supporting systems in the machinery of a warship. High temperatures in the engine room can lead to an unpleasant atmosphere and, in particular, to heat stress for the operating personnel. This heat stress was caused by the design of the engine room ventilation system not meeting the requirements due to the supply and exhaust fan. Under the given conditions, the air temperature reached 50 °C, higher than the standard 45 °C. Therefore, it was necessary to conduct a thermal analysis and redesign the ventilation system using a 3D CFD (Computational Fluid Dynamic). The parameters included temperature distribution, flow distribution, and flow velocity in the engine room of the Landing Ship Tank class warship when the engine load reached 50%, 75%, and 100%. The results obtained from the simulation were in the form of quantitative data such as temperature gradients and iso contours of the air in the engine room inside the warship. The data comprised velocity contours, temperature, velocity vectors, streamlines, and particle trajectories in space. The results of the CFD approach showed that the quality of the ventilation system in the engine room influences extreme conditions in the engine room.

1. Introduction

The engine rooms of warships are critical areas that house the machinery necessary for ship propulsion and operation. These spaces are inherently susceptible to high temperatures and the accumulation of potentially harmful fumes and gases, primarily due to the combustion processes and mechanical operations of diesel engines, generators, and other equipment. Effective ventilation systems are critical to maintaining machine efficiency and ensuring that environmental conditions in these confined spaces remain within safe and manageable limits [1]. Proper air circulation helps dissipate heat and remove air pollutants, which is critical to equipment longevity and operational readiness. Thermal management in battleship class-Landing Ship Tank (LST) engine rooms presents significant challenges. The dense arrangement of engines generates significant heat, which can increase the ambient temperature to levels that, if not adequately controlled, can exceed safety limits [2]. High temperatures can affect the performance of critical components and systems, resulting in failures and possible operational disruptions [3]. For crew members working in these conditions, prolonged exposure to excessive heat can result in heat stress, reduced cognitive and physical performance, and an increased risk of heat-related illnesses, affecting crew efficiency and safety [4].

In addition to heat, machine rooms are environments in which exhaust gases and fumes predominate. These emissions can contain various harmful substances, including carbon monoxide, sulfur oxides, and unburned hydrocarbons, which pose serious health risks if inhaled [5]. These contaminants can accumulate without adequate ventilation at toxic levels and create a hazardous work environment. The health effects range from short-term effects, such as dizziness and respiratory irritation, to long-term consequences, such as cardiovascular and respiratory diseases. Ensuring a constant supply of fresh air through effective ventilation systems is of utmost importance to mitigate these risks.

The environmental conditions heavily influence the overall performance and safety of the crew aboard LST-class warships in the engine room. Optimal ventilation is a crucial factor that supports attention and physical performance by maintaining air quality and a comfortable temperature [6]. In addition, effective ventilation is a key component in ensuring compliance with international maritime safety standards, which require specific air quality and temperature conditions to ensure safety at sea. Therefore, a well-designed ventilation system is critical to operational efficiency and compliance and crew well-being and performance, ultimately impacting the ship's readiness and ability to complete its mission.

1.1 Literature Review

In the engine room, there are machines and devices such as the ship's propulsion engine or the so-called main engine, devices as an energy source for generating electricity in the form of the ship's electric generator, pumps, and various other devices that support the ship's operation. When the machines and equipment are operating, the condition of this room becomes hot. To overcome this heat, a ventilation system is required that supplies fresh air to the engine room and also circulates the hot air emitted by the engine and equipment to the outside of the engine room so that the temperature in the engine room rises to 45°C or more than 5°C do not exceed the outside air temperature [7]. Several ventilation standards must be adhered to in the ship's engine room. Some of these standards include: Based on ISO 8861:1998 regarding " Shipbuilding, Engine-Room Ventilation in Diesel, Engine Ships, Design Requirements, and Basis ff Calculations," the condition for the design of the engine room is that the outside air ambient temperature for the engine room can be approximately +35°C-RH 70% and pressure 101.3 kPa. The temperature rise from the intake air to the air passed from the engine compartment to the intake housing is assumed to be approximately +12.5 K (the increase in engine compartment air temperature, i.e., the difference between the inlet and exhaust temperatures measured at design). Conditions). Under normal conditions, when the machine and equipment in the machine room are turned on, the temperature in the machine room is 10-12°C higher than the ambient temperature outside, or about 45-47°C [8].

According to the Indonesian Classification Bureau) regarding "Rules for the Classification and Construction of Seagoing Steel Ships for Refrigerating Installation." The conditions required for machine rooms containing machines, equipment, and other auxiliary equipment and operated under working conditions in tropical areas are as follows [9]:

a. Maximum engine room temperature: 45°C or 5ºC more than the outside air temperature
b. Maximum sea water temperature: 32°C
c. Relative humidity: 50%
d. Barometric pressure: 76 cm Hg

Meanwhile, based on IACS (International Association of Classification Societies) 1978: rule M28, the reference conditions that apply to ship engine rooms are:

a. Total barometric pressure: 1,000 mbar
b. Air temperature: +45°C
c. Relative humidity: 60%
d. Seawater temperature: 32°C (Charge air coolant inlet).

The physical work environment includes everything about personnel that can influence them in their tasks. One factor that affects the working environment is the room temperature. Elevated temperature may impair the performance of personnel or harm their health. The following describes the conditions of humans when they are in an environment with a specific temperature [10]:

a. 37°C - Normal body temperature (36-36.5°C)
b. 38°C - sweaty, very uncomfortable, a little hungry
c. 39°C - Sweating, red wet skin, fast breathing and heartbeat, fatigue, stimulates epilepsy.
d. 40 °C - Fainting, dehydration, weakness, headache, vomiting, dizziness and sweating.
e. 41°C - emergency situation. Fainting, dizziness, confusion, headache, hallucinations, shortness of breath, drowsiness, blurred eyes, palpitations.
f. 42°C - pale, red, wet skin, coma, dark eyes, vomiting, severe disturbances, increased/decreased blood pressure, and rapid heartbeat.
g. 43°C - commonly, fatalities, brain damage, severe disorders and shocks, and respiratory function collapses.
h. 44°C - definitely dead, but some patients can survive above 46°C.

The success of a workplace depends on the comfort level of the crew measures. Therefore, a room must always be equipped with a ventilation system (HVAC). Measuring thermal comfort is difficult because it is very subjective [11]. This measurement depends on air temperature and humidity, jet temperature, flight speed, activity level, and type of clothing [12]. However, each person experiences this feeling very differently depending on their physiology and circumstances. The sensation of coolness is pleasant when the body is hot but unpleasant when the body is cold. At the same time, the skin's surface temperature in different parts of the body is also not uniform due to different blood circulation and fat thickness. The insulating properties of clothing also significantly influence the temperature level and skin temperature distribution [13]. Therefore, the perception of any part of the skin depends on the time, place, and clothing, as well as the environment's temperature. Six variables are believed to influence thermal comfort, including [14]:

a. Metabolic rate (the amount of energy released from the human body)
b. Clothing insulation (material used to retain or dissipate body heat.
c. Air temperature (Ambient temperature)
d. Radiant temperature (surrounding surface temperature)
e. Air velocity (the speed at which air moves and comes into contact with the skin)
f. Relative humidity (percentage of water vapor in the air-water mixture)

One of the dangers in the CCM (Continuous Casting Machine) sector is heat stress. Heat stress is the working climate stress that engine room crew are exposed to, especially on warships. Additional stress in the form of ambient heat can lead to physiological stress, e.g., B. cardiac output increases, and profuse sweating occurs, leading to dehydration [15]. Machines and work environment factors can influence an unpleasant working atmosphere and temperature

IOP Conf. Series: Earth and Environmental Science **1423** (2024) 012004 doi:10.1088/1755-1315/1423/1/012004

differences in the workplace [16]. The instrument used to measure heat stress in this research is the Heat Stress Apparatus Quest Temp 36°C [17]. Heat stress is generally caused by a combination of the human body's metabolism results when the human performs activities [8]. As long as the human body is active, it naturally adjusts the temperature it receives through heat loss. The effects of hot temperatures and associated excessive sweating may lead to the loss of sodium salts [18]. After several weeks of this illness, muscle cramps occur. If a person loses less than 1.5% fluid, there will still be no symptoms, but fatigue will appear early, with a parched mouth. A comfortable temperature for Indonesians at work is between 24 and 26 °C [19]. Exposure to hot temperatures can affect working dexterity, increase reaction time, slow down decision time, affect brain work accuracy, disrupt sensory-motor nerve coordination, and provoke emotions [20].

CFD is a calculation method that specializes in liquids. Starting from fluid flow, heat transfer, and chemical reactions that occur in fluids [21]. Calculations can be carried out using CFD based on fluid mechanics, energy, momentum, mass, and species conservation principles. In simple terms, the calculation process carried out by the CFD application is that with the calculation controls carried out, the calculation controls use the equations involved. These equations are generated by inputting all the parameters within the domain. For example, if a model to be analyzed includes temperature, this means that the model includes an energy equation or the conservation of that energy. The initial initialization of the equation is the boundary condition. Boundary conditions are conditions where computational controls are defined as initial definitions incorporated into adjacent computational controls via the equations involved [22].

2. Methodology

The research method used is a numerical simulation method using Computational Fluid Dynamic (CFD) software. This method was chosen to complete the distribution case of the engine room ventilation system using the CFD method and to improvise/modify the engine room ventilation system by employing analytical considerations and refinement/optimization. 3D model simulations are carried out in unsteady conditions with a turbulence model realizable k-e (RNG) wall function, and the peripheral equation used is the Semi-Implicit Method for Pressure Linked Equations (SIMPLE). The parameters included temperature distribution, flow distribution and flow velocity in the engine room of the Landing Ship Tank class warship when the engine load reached 50%, 75%, and 100%. The results obtained from the simulation were in the form of quantitative data in the form of temperature gradients and isocontours of the engine room air inside the warship. Qualitative data, on the other hand, consists of velocity contours, temperature, velocity vectors, streamlines and particle trajectories within the room.

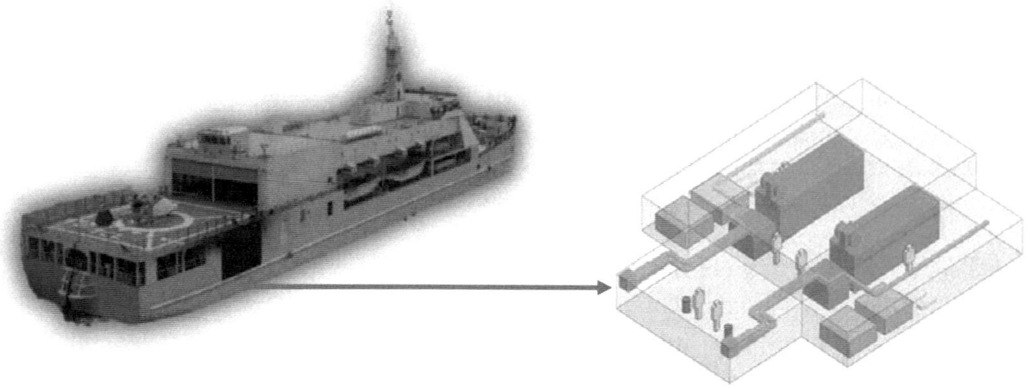

Figure 1. Geometric Model

Figure 1 shows a naval vessel alongside a schematic representation of its internal systems, probably its power generation or propulsion system. This illustration shows a three-dimensional view and explains the ventilation system in the engine room. It shows that this ship has two engines as its propulsion system. In addition, several auxiliary support systems support the main propulsion system, which provides the current heat flow in the engine room. The solution method uses the second-order gradient, pressure, momentum, turbulent kinetic energy, and energy. Setting up a numerical simulation in Computational Fluid Dynamics involves several steps. They are preprocessing, processing, and postprocessing. Each setup varies depending on the specific goals and complexity of the simulation. A detailed understanding of the physical processes and appropriate modeling decisions are crucial for accurate and meaningful results.

Table 1. Numerical Set-Up

Boundary Name	***Boundary Type***
Inlet	*Velocity-inlet*
Wall	*Stationary Wall*
Material Name	*Aluminium*
Outlet	*Pressure-outlet*

Table 1 outlines an overview of the boundary conditions and the material used in a flow simulation. It specifies an inlet with a boundary type of "Velocity Inlet", meaning that the fluid's velocity at the entry point is defined. The wall boundary is stationary, indicating a solid, non-moving surface in contact with the fluid. The material used in the simulation is aluminium, which means that properties such as density and thermal conductivity of aluminium are transferred to the relevant components. Finally, the outlet is defined with a pressure outlet limit type, which means that the fluid's pressure at the domain's exit point is specified.

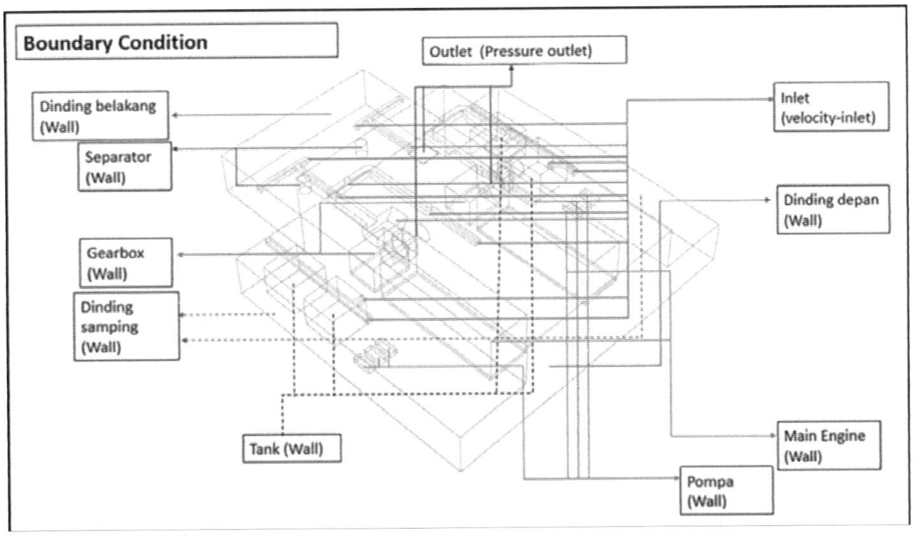

Figure 2. Boundary Condition

In Figure 2, inlet and outlet boundary conditions represent the boundaries or zones for fluid entering and exiting the volume control area, while the walls on the motor, pump, gearbox, and separator represent the boundary fields or zones for Heat Flux in the volume control area. This setup is critical to accurately define how the fluid interacts with the ship's internal structures and to ensure realistic simulation results.

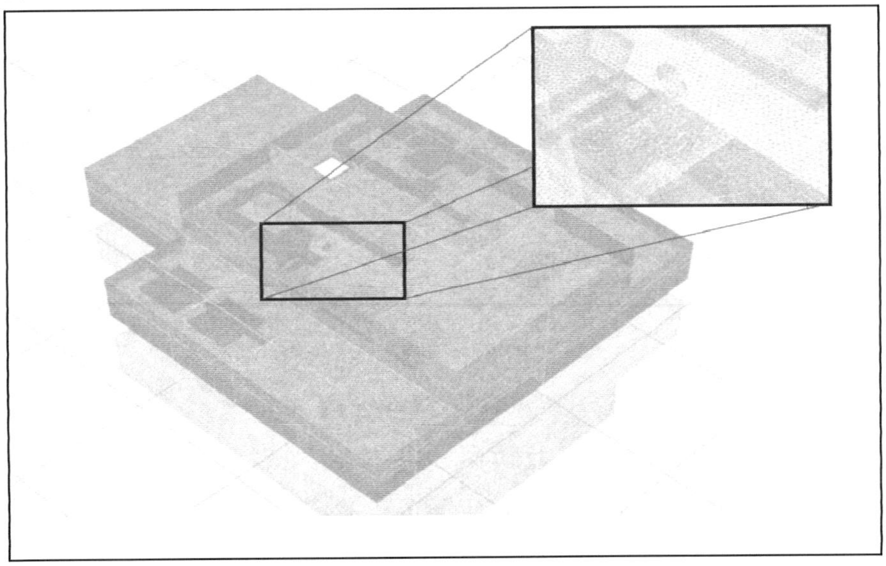

Figure 3. Meshing

Figure 3 shows a 3D model of a structure, likely part of the interior layout of a ship, with emphasis on meshing in CFD. The enlarged section highlights the mesh applied to part of the geometry. This network consists of numerous small, regularly shaped elements and shows how complex geometry is discretized into smaller, manageable volumes. The meshing modeling in the picture above shows that the meshing process performed as a simulation consists of more than 4,087,325 cells as a reference for continuously using fluid domain devices.

Table 2. Boundary Conditions

No	*Mesh Refinement*	*Total Element*	*Temperature*	*Relative error*
1	*Very Coarse*	113.163	35,28	-
2	*Coarse*	358.449	34,84	1,26%
3	*Normal*	2.671.841	34,66	0,52%
4	***Fine***	**4.087.325**	**34,6**	**0,17%**
5	*Very Fine*	5.489.944	34,55	0,14%

Table 2 shows the results of the meshing we used, and the number of cells significantly obtained impacted the research results. A grid dependency test was carried out to ensure that the meshing used in the research does not depend on density. This test is also intended to determine the correct number of meshes so that the results do not change significantly depending on the number of mesh elements. The table above shows that with a total of 4,087,325 elements. The temperature test value does not change significantly. Therefore, this research uses this mesh setting. Continuous airflow is modeled by averaging the Navier-Stokes equations over time. The general form of the equation contains the conservation of mass, momentum, and energy.

$$\frac{\partial(\rho\phi)}{\partial t}+\frac{\partial(\rho u\phi)}{\partial x}+\frac{\partial(\rho v\phi)}{\partial y}+\frac{\partial(\rho w\phi)}{\partial z}=\frac{\partial}{\partial x}\left(\Gamma_\phi\frac{\partial\phi}{\partial x}\right)+\frac{\partial}{\partial y}\left(\Gamma_\phi\frac{\partial\phi}{\partial y}\right)+\frac{\partial}{\partial z}\left(\Gamma_\phi\frac{\partial\phi}{\partial z}\right)+S_\phi+S_{p\phi} \quad (1)$$

3. Result and Discussion

The numerical study of the engine room air ventilation systems for Landing Ship Tank (LST) class warships revealed several critical airflow distribution and velocity findings. The simulations showed that the current ventilation design creates high and low airflow velocity regions within the engine room. Areas near the air inlets experienced high velocities exceeding 5 m/s, which can cause discomfort for the crew due to excessive air movement and noise. Conversely, regions farther from the inlets had significantly lower velocities, below 1 m/s, which may lead to insufficient ventilation and potential equipment overheating.

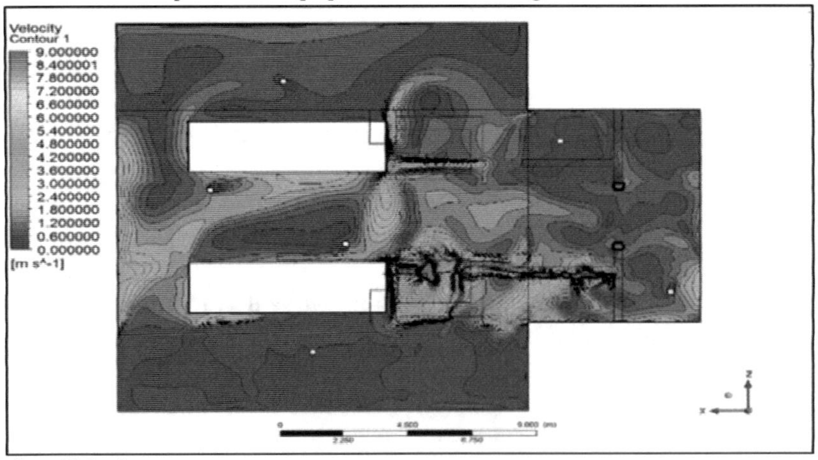

Figure 4. Airflow Velocity

Figure 4 shows the pressure distribution in the engine room. The results showed that the ventilation system generates a pressure gradient from the air inlets to the exhaust outlets. This gradient is essential for ensuring continuous air movement through the engine room. However, the study found that several sections of the engine room were exposed to negative pressure, which could potentially draw in contaminated air from adjacent compartments. Ensuring a positive pressure environment in the engine room is crucial for maintaining air quality and crew safety. The temperature distribution analysis indicated that the ventilation system effectively maintains the engine room temperature within acceptable limits under normal operating conditions. The average temperature within the engine room was around 35°C, with some hot spots reaching up to 45°C near the engines. If not adequately managed, these hot spots can impact crew comfort and equipment performance. The results suggest that improving the airflow uniformity could help reduce these extreme temperatures.

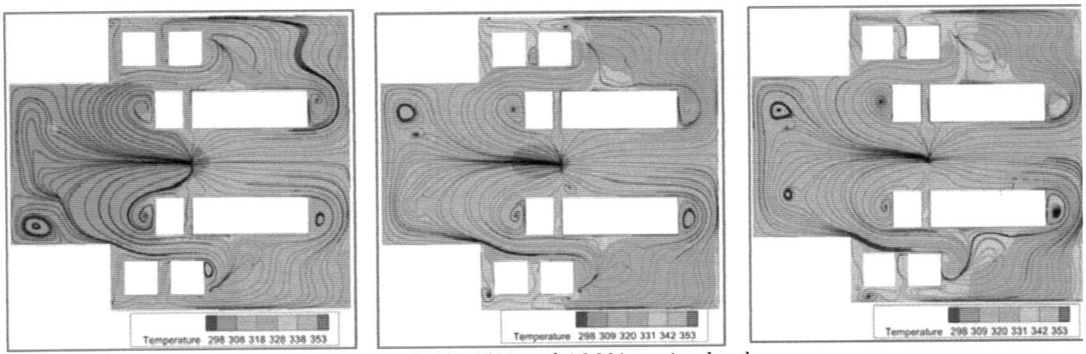

Figure 5. Temperature Distribution at 50%, 75% and 100% engine load

Figure 5 consists of three contour plots, each depicting the temperature distribution within a fluid flow simulation in a defined geometric domain. The color scale at the bottom of each plot ranges from blue (lower temperatures) to red (higher temperatures), indicating the temperature variation within the domain at three engine load variations. Each plot shows different temperature patterns and gradients, highlighting how the temperature distribution evolves under different conditions or configurations. The visual representation of temperature contours helps understand the thermal behavior and heat transfer characteristics within the given geometry. Figure 5 illustrates that when the engine load is 50%, the heat distribution within the engine room results in a temperature range of 30°C-50°C with an average temperature of 39.54°C. When the engine load is at 75%, the heat distribution in the engine room is such that the room temperature falls within the 30°C-60°C range. When the engine load is 100%, the heat distribution in the engine room results in a room temperature of 35°C-65°C. However, particular areas in the middle of the engine room still maintain the standard temperature, specifically staying below 45°C.

The study highlighted that the current ventilation system impacts crew comfort and safety in several ways. One of the primary concerns is the high airflow velocities near inlets. These high velocities can cause discomfort among the crew, potentially leading to health issues over time. This discomfort can manifest as physical irritations or respiratory problems, which can impair the crew's ability to perform their duties effectively. Another significant issue identified in the study is the insufficient air circulation in certain areas. Poor air circulation can lead to the buildup of contaminants and a lack of fresh air, which can adversely affect the crew's health and performance. Stagnant air can also contribute to the spread of airborne illnesses, further compromising the safety of the working environment. Additionally, some areas' temperature extremes and insufficient air circulation may lead to a less safe working environment. By optimizing the ventilation system, it is possible to create a more comfortable and safer environment for the crew, which is crucial for maintaining operational efficiency and crew well-being.

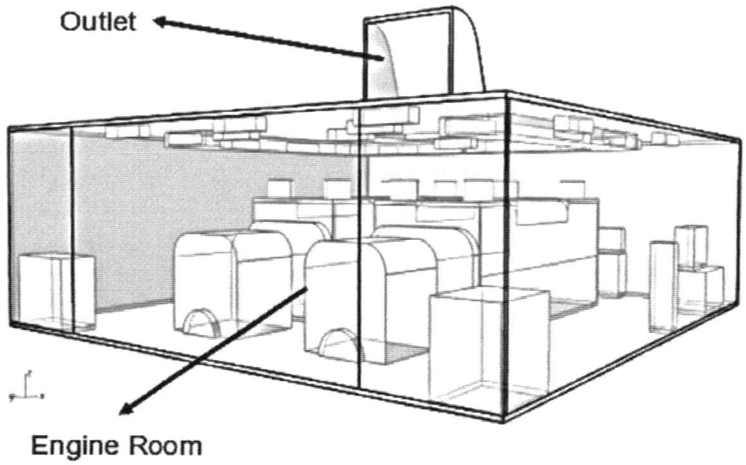

Figure 6. Ventilation Design

Figure 6 demonstrates that the evaluation of the ventilation design's effectiveness was based on its capacity to remove heat and maintain air quality. The simulations showed that the current design is moderately effective, but there is potential for improvement. More specifically, improving the placement and design of air inlets and outlets could enhance the overall ventilation efficiency, resulting in more consistent temperature and pressure distributions. The findings of this study align with standard requirements for marine ventilation systems, such as those outlined by the International Maritime Organization (IMO) and classification societies like ABS and Lloyd's Register. These standards emphasize the necessity for maintaining acceptable temperature, pressure, and airflow conditions within engine rooms to ensure safety and comfort. The discrepancies identified in the current system highlight the need for design modifications to meet these standards more effectively.

Based on the results, several design recommendations can be made to enhance the ventilation system of battleship-class LST:

a. Repositioning of Air Inlets and Outlets: Adjusting air inlets and outlet locations to ensure more uniform airflow distribution can help mitigate high-velocity areas and reduce temperature extremes.
b. Improvement of Airflow Paths: Introducing baffles or deflectors to guide airflow more effectively can enhance ventilation efficiency and prevent stagnant air zones.
c. Implementation of Variable Speed Fans: Using variable speed fans can help regulate airflow and maintain optimal ventilation under varying operational loads.

Regular Maintenance and Monitoring: Ensuring regular maintenance and monitoring of the ventilation system to prevent performance degradation over time and address any emerging issues promptly.

Recent studies employing CFD simulations have examined the correlation between air ventilation and fires in the engine room, presenting a comparative analysis. In 2023, Amir et al. [23] studied fluid flow and heat transfer in a ship's engine room. The simulation findings illuminate the substantial impact of radiation effects on wall temperatures. The air temperature adjacent to the ship's engine increases by up to 3K, while the temperature of the engine wall can rise by up to 14K. Therefore, optimizing the enhancement of the air ventilation system, maintaining the average temperature in the engine room, and distributing air within the engine room is crucial. Lan et al. [24] developed a design and simulation for ship ventilation to mitigate fires in the engine room. The findings of the simulation demonstrate that the rate of ventilation has an impact on the cooling rate of the engine room, as well as the configuration of the central ventilation system with top and bottom exhaust outlets. Various factors leading to fires, such as those originating from fuel oil in ship engines or from oil tanks during ship accidents, have been examined by previous researchers [25], [26]. Kang et al. [27] have developed a simulation of the ship's fire framework, specifically focusing on the ship's engine room. Extensive experimental trials conducted in the ship's engine room revealed that the primary cause of ship fires is the escalating temperature gradient within the engine room [28].

4. Conclusion

The numerical study of the engine room air ventilation system for battleship-class LST has provided valuable insights into its current performance and areas for improvement. By addressing the identified issues and making the recommended design changes can significantly improve crew comfort and safety. This study underscores the importance of continuously evaluating and optimizing ventilation systems to meet evolving needs. Conduct numerical analysis on the impacts of air condition changes in the engine room, such as temperature, pressure, and airspeed at any given time. Additionally, consider altering the placement of outlets in the engine room, particularly in areas not directly beside ventilation grilles and directly above the main engine, which generates heat. This setting will help decrease the heat stress experienced by the ship's crew in the engine room, ultimately enhancing safety and comfort.

Acknowledgments

Authors wishing to acknowledge assistance or encouragement from colleagues, special work by the Department of Mechanical Engineering, Faculty of Industrial Technology, and System Engineering, Institut Teknologi Sepuluh Nopember (ITS), Surabaya, especially to Fluid Mechanic and Machine Laboratory.

References

[1] A. K. Mishra, M. G. L. C. Loomans, and J. L. M. Hensen, "Thermal comfort of heterogeneous and dynamic indoor conditions — An overview," Build. Environ., vol. 109, 2016, doi: 10.1016/j.buildenv.2016.09.016.

[2] A. Franco and E. Schito, "Definition of optimal ventilation rates for balancing comfort and energy use in indoor spaces using CO2 concentration data," Buildings, vol. 10, no. 8, 2020, doi: 10.3390/BUILDINGS10080135.

[3] A. S. Rohman, U. Nurbaiti, and F. Fianti, "ANALISIS KENYAMANAN SUHU RUANG," EnviroScienteae, vol. 17, no. 1, 2021, doi: 10.20527/es.v17i1.11346.

[4] ANSI/ASHRAE, "ANSI/ASHRAE Standard 55-2017: Thermal Environmental Conditions for Human Occupancy," ASHRAE Inc., vol. 2017, 2017.

[5] İ. H. Yılmaz and A. Mwesigye, "Modeling, simulation and performance analysis of parabolic trough solar collectors: A comprehensive review," Applied Energy, vol. 225. 2018. doi: 10.1016/j.apenergy.2018.05.014.

[6] I. L. Organization, Ambient factors in the workplace, vol. 15, no. 2. 2010.

[7] I. B. Volume II, Rules for the classification and construction of seagoing steel ships. Volume I Rules For Hull Edition, 2009.

[8] ISO 8861:1998, Shipbuilding — Engine-room ventilation in diesel-engined ships — Design requirements and basis of calculations. International Standard under systematic review, 1998.

[9] M. V. Callahan, "Heat-Associated Illness," in Hunter's Tropical Medicine and Emerging Infectious Diseases, 2020. doi: 10.1016/b978-0-323-55512-8.00024-7.

[10] E. M. Shammout, "THE IMPACT OF WORK ENVIRONMENT ON EMPLOYEES PERFORMANCE," Int. J. Curr. Res. Sci. Eng. Technol., vol. 03, 2022.

[11] M. Amoadu, E. W. Ansah, J. O. Sarfo, and T. Hormenu, "Impact of climate change and heat stress on workers' health and productivity: A scoping review," Journal of Climate Change and Health, vol. 12. 2023. doi: 10.1016/j.joclim.2023.100249.

[12] N. Muhammad Ramadhani, "Optimasi Tata Letak Ventilator Untuk Menjaga Temperatur Ruang Kamar Mesin Kapal General Cargo 8000 DWT Menggunakan Pendekatan CFD," Institut Teknologi Sepuluh Nopember, 2016.

[13] M. B. P. Campa Portela, Rosa Mary de la, and Julio Louro Rodríguez, "Heat stress on board: risk and prevention," in International Conference on Maritime Transport-8th International Conference on Maritime Transport (Maritime Transport VIII), Barcelona: Universitat Politècnica de Catalunya. Departament de Ciència i Enginyeria Nàutiques, 2020.

[14] P. Aparicio-Ruiz, E. Barbadilla-Martín, J. Guadix, and J. Nevado, "Analysis of Variables Affecting Indoor Thermal Comfort in Mediterranean Climates Using Machine Learning," Buildings, vol. 13, no. 9, 2023, doi: 10.3390/buildings13092215.

[15] R. Viswanathan and J. Stringer, "Failure mechanisms of high temperature components in power plants," J. Eng. Mater. Technol. Trans. ASME, vol. 122, no. 3, 2000, doi: 10.1115/1.482794.

[16] S. Karthick, S. Kermanshachi, and A. Pamidimukkala, "Impact Analysis of Heat on Physical and Mental Health of Construction Workforce," in International Conference on Transportation and Development 2022: Application of Emerging Technologies - Selected Papers from the Proceedings of the International Conference on Transportation and Development 2022, 2022. doi: 10.1061/9780784484340.027.

[17] S. M. Sihaloho, F. Saftarina, and S. MA, "Analysis of Factors Causing Air Pollution on Ships," J. Soc. Res., vol. 2, no. 7, 2023, doi: 10.55324/josr.v2i7.1128.

[18] T. Lipinski, D. Ahmad, N. Serey, and H. Jouhara, "Review of ventilation strategies to reduce the risk of disease transmission in high occupancy buildings," Int. J. Thermofluids, vol. 7–8, 2020, doi: 10.1016/j.ijft.2020.100045.

[19] R. T. Oğulata, "The effect of thermal insulation of clothing on human thermal comfort," Fibres Text. East. Eur., vol. 15, no. 2, 2007.

[20] J. Wang, Z. Wang, R. de Dear, M. Luo, A. Ghahramani, and B. Lin, "The uncertainty of subjective thermal comfort measurement," Energy Build., vol. 181, 2018, doi: 10.1016/j.enbuild.2018.09.041.

[21] X. R. Tan et al., "Elevated brain temperature under severe heat exposure impairs cortical motor activity and executive function," J. Sport Heal. Sci., vol. 13, no. 2, 2024, doi: 10.1016/j.jshs.2023.09.001.

[22] Y. Kukus, W. Supit, and F. Lintong, "SUHU TUBUH: HOMEOSTASIS DAN EFEK TERHADAP KINERJA TUBUH MANUSIA," J. BIOMEDIK, vol. 1, no. 2, 2013, doi: 10.35790/jbm.1.2.2009.824.

[23] A. Salmani, S. F. Hosseini, and H. Khani, "Modelling heat transfer and fluid flow in a ship engine room using CFD analysis to evaluate and improve ventilation," Proc. Inst. Mech. Eng. Part M J. Eng. Marit. Environ., vol. 237, no. 2, 2023, doi: 10.1177/14750902221118485.

[24] Q. Lan, F. Han, Y. Liu, W. Li, and Z. Wang, "Effects of ventilation system design on flame behavior and smoke characteristics for mitigating marine engine room fire hazards," Ocean Eng., vol. 281, no. June, p. 114890, 2023, doi: 10.1016/j.oceaneng.2023.114890.

[25] C. Li, H. Zhang, Y. Zhang, and J. Kang, "Fire Risk Assessment of a Ship's Power System under the Conditions of an Engine Room Fire," J. Mar. Sci. Eng., vol. 10, no. 11, 2022, doi: 10.3390/jmse10111658.

[26] C. Li, J. Mao, Z. Kang, S. Zhao, and H. Ren, "Influence of Firefighting Intervention on Fire Spread Characteristics in Ship Engine Room," J. Mar. Sci. Eng., vol. 11, no. 4, 2023, doi: 10.3390/jmse11040877.

[27] H. J. Kang, J. Choi, D. Lee, and B. J. Park, "A framework for using computational fire simulations in the early phases of ship design," Ocean Eng., vol. 129, 2017, doi: 10.1016/j.oceaneng.2016.11.018.

[28] J. Wang et al., "An experimental and non-dimensional study on the vertical temperature distribution of a sealed ship engine room fire," Ocean Eng., vol. 165, 2018, doi: 10.1016/j.oceaneng.2018.07.018

IOP Conf. Series: Earth and Environmental Science **1423** (2024) 012005 doi:10.1088/1755-1315/1423/1/012005

Building resilience in LNG marine terminal operations: a comprehensive socio-technical risk analysis utilizing the resilience assessment grid (RAG) method

P I Mahatrisna[1], R O Saut Gurning[2], A A B Dinariyana[3], Rocky S J Makapuan[4]

[1]School of Interdisciplinary Management and Technology ITS, Jl. Cokroaminoto 12A Surabaya 60264
[2]Marine Engineering Department, Sepuluh Nopember Institute of Technology, Kampus ITS Sukolilo - Surabaya
[3]School of Interdisciplinary Management and Technology ITS, Jl. Cokroaminoto 12A Surabaya 60264
[4]Shipping and Maritime Department SKK Migas, Wisma Mulia, Jl, Jend. Gatot Subroto No. 42 Jakarta

*E-mail: pimahatrisna@skkmigas.go.id

Abstract. In the face of rising global energy demand, LNG terminals play an essential role in the energy supply chain. However, the complexity of terminal operations, combined with the risk of catastrophic events, necessitates a robust resilience framework. This paper presents a comprehensive socio-technical resilience analysis that uses the Resilience Assessment Grid (RAG) methodology to measure resilience in LNG terminal operations. We examine the complex interplay of human elements, organizational culture, and technological systems in the terminal environment. Our approach uses the RAG methodology to identify vulnerabilities, assess the potential impact of disruptions, and assess the effectiveness of current safety and risk management practices. The results of this study show that the implementation of the safety concept at the LNG Marine Terminal is quite optimal, as indicated by the compliance figures for each resilience aspect, which exceed three as a benchmark. The highest score in the entire system is found in the ability to respond aspect, namely 4.38, which shows that the organization, in implementing this concept, still focuses on preventing and controlling accidents. This condition differs from the ability to monitor aspect, which has the lowest value, 3.97. However, this item cannot be called an inhibiting factor because its value exceeds the expected standard. In response, there is still room for improvement from existing achievements in the future.

1. Introduction

Indonesia's oil marine terminal operations have experienced numerous accidents over the past ten years, leading to substantial financial losses and interruptions. The record from the Special Task Force for Upstream Oil and Gas Business Activities (SKK Migas) from 2013 to 2022 shows that there have been at least 7.7 accidents per year. The impact ranged from several days of operational disruption to requiring millions of dollars for the terminal to recover. On the other hand, there are only limited standards, and research has exclusively reviewed resilience in oil and gas marine terminal operations. Consequently, this condition has sparked worries about the decreasing operational efficiency of Indonesia's marine terminal and the possibility of extended closures, leading to extensive disruptions in the country's energy supply chain networks.

The inherent risks associated with maritime operations, such as inclement weather, mechanical failures, and human errors, necessitate a robust safety management framework. The International Safety Management (ISM) Code and the Safety of Life at Sea (SOLAS) Convention are two pivotal regulatory instruments that aim to enhance maritime safety and security. At the same time, the concept of resilience has emerged as a crucial factor in ensuring the long-term sustainability and adaptability of the maritime transportation system [1]. While the ISM Code has been instrumental in improving safety standards, recent research has revealed that compliance with the code has deteriorated over time, resulting in an increase in accidents, fatalities, and environmental pollution [2]. This observation underscores the need for a more resilient approach to maritime safety management, one that can adapt to changing circumstances and unexpected challenges. Despite the efforts to improve maritime safety through the ISM Code and SOLAS Convention, the dynamic nature of the maritime industry constantly presents new challenges and threats, requiring a shift towards a more resilient safety management approach. Resilience-based strategies not only focus on mitigating risks but also emphasize the ability to learn from incidents, adapt to changing conditions, and recover swiftly from disruptions.

The concept of resilience was first introduced in ecology by Holling [3], refers to the ability of a system to absorb disturbances, reorganize, and maintain its core functions and structures in the face of adversity [4]. In the context of maritime transportation, resilience encompasses the capacity of the system to withstand and recover from disruptive events, such as natural disasters, cyberattacks, and supply chain disruptions. Resilience-based approaches to maritime safety management emphasize the importance of adaptive capacity, organizational learning, and the integration of multiple stakeholders to enhance the overall resilience of the transportation system. Moreover, as a part of the maritime system, it is essential for oil marine terminals as it allows these vital infrastructure systems to endure and bounce back from disturbances. This study aims to measure the level of resilience of an LNG marine terminal by considering the socio-technical factors that influence it, to identify vulnerabilities, assess the potential impact of disruptions, and assess the effectiveness of current safety and risk management practices.

2. Methodology

2.1 The Resilience Assessment Grid (RAG)

The Resilience Assessment Grid (RAG), developed by Erik Hollnagel [5], is a tool within the resilience engineering space to measure an organization's potential for resilient performance in complex socio-technical systems. The RAG theory suggests that organizations cannot 'be' resilient but can have the potential for resilient performance, which is essential for systems to perform reliably under varying conditions [6], [7].

The Resilience Assessment Grid (RAG) offers several advantages for measuring terminal resilience by considering socio-technical aspects compared to other risk assessment methods. RAG adopts a proactive approach, focusing on a system's ability to adapt and endure under various conditions, unlike traditional risk assessments that are more reactive. It evaluates four fundamental potentials for resilient performance: the ability to respond, monitor, learn, and anticipate, providing a comprehensive view of the system's resilience. Additionally, RAG takes a holistic approach by considering the interactions between human and technical factors, allowing for a more integrative assessment. Using a Likert scale for performance evaluation offers flexibility and adaptability to different contexts and operational conditions. Furthermore, RAG emphasizes enhancing the system's ability to succeed under varying conditions, known as the Safety-II approach, rather than merely reducing adverse events. This comprehensive and proactive approach provides deeper insights into port resilience and aids in developing more effective strategies to enhance it.

The RAG theory and its associated methods have shown promise in various applications, contributing to developing resilience assessment and improvement strategies in various sectors. This method also provides a structured approach to understanding and managing complexity in everyday work [8], such as assessing the vulnerability of a power transmission network using complex network theory [9], and evaluating the resilience of interdependent infrastructure systems [10]. Despite its utility, the application of RAG in healthcare has encountered challenges, particularly in implementing suggested improvements and measuring the direct impact on quality enhancement [8].

Organizational resilience, as described by resilience engineering, is a paradigm that emphasizes the ability to cope with complexity and achieve success amidst conflicting goals. An analytic framework combining the resilience analysis grid and the analytic hierarchy process has been proposed to measure organizational resilience, providing a structured approach to identify system strengths and weaknesses and potential actions to increase adaptive capacity [11]. However, there is a need for more research to bridge the theory and practice divide on how organizations can pragmatically measure their current performance against four resilience potentials using the RAG [6].

In today's complex and interconnected world, socio-technical systems, which encompass social elements (people, organizations) and technical elements (machinery, software), are increasingly exposed to various shocks and stresses. These can range from natural disasters and technological failures to market conditions or regulatory environment changes. The ability of these systems to anticipate, withstand, recover from, and adapt to these challenges – in other words, their resilience – is of paramount importance. The RAG provides a structured and systematic approach to assess this resilience. By focusing on the four critical potentials of resilience – anticipating, monitoring, responding, and learning – the RAG allows organizations to identify strengths and weaknesses in their current resilience capabilities and to develop strategies for enhancing their resilience in the face of future challenges. Furthermore, the RAG is designed to provide a snapshot of operational safety system resilience concepts by measuring individual potentials and their combination [6]. It is not only a tool for assessment but also a guide for action, making it highly relevant and essential in managing the resilience performance of socio-technical systems.

2.2 The Four Basis of Resilient Performance

The RAG is built around four basic potentials that underpin resilient system performance [5]:

1. Anticipating refers to an organization's ability to envision and prepare for future conditions. This potential involves identifying challenges and changes impacting the organization's functioning and taking proactive steps to mitigate potential risks.
2. Monitoring continuously observes operations and the environment to detect changes or signs that may indicate emerging problems. It involves tracking what is happening inside and outside the organization to ensure that performance remains within safe boundaries.

3. Responding is about the capacity to act effectively when challenges or disruptions occur. It includes the ability to adapt and adjust actions to maintain or restore the desired level of performance. This potential is significant when the organization faces immediate and unexpected situations.
4. Learning is how an organization reflects on past experiences, including successes and failures, to improve future performance. It involves understanding what happened, why it happened, and how similar events can be prevented or better managed in the future.

These potentials are interdependent, meaning that improving one potential often requires or benefits from improvements in the others. For example, the potential to respond can benefit from the potential to monitor, and the potential to learn can improve the potential to monitor and respond [5].

2.3 Quantification of the Capacity for Resilient Performance

'Resilience' pertains to the actions of a system rather than its possessions. However, it encompasses multiple aspects rather than being defined by a single attribute or feature. Resilience does not exist; hence, there is no measure, quantity, or degree of resilience. The existing body of research on Resilience Engineering (RE) demonstrates various perspectives regarding the defining characteristics of resilient performance. Scholars who have contributed to this discourse are Hollnagel et al. [5]. Instead of contemplating the definition of resilient performance, we should focus on identifying the factors that promote resilient performance and those that hinder it if absent [5].

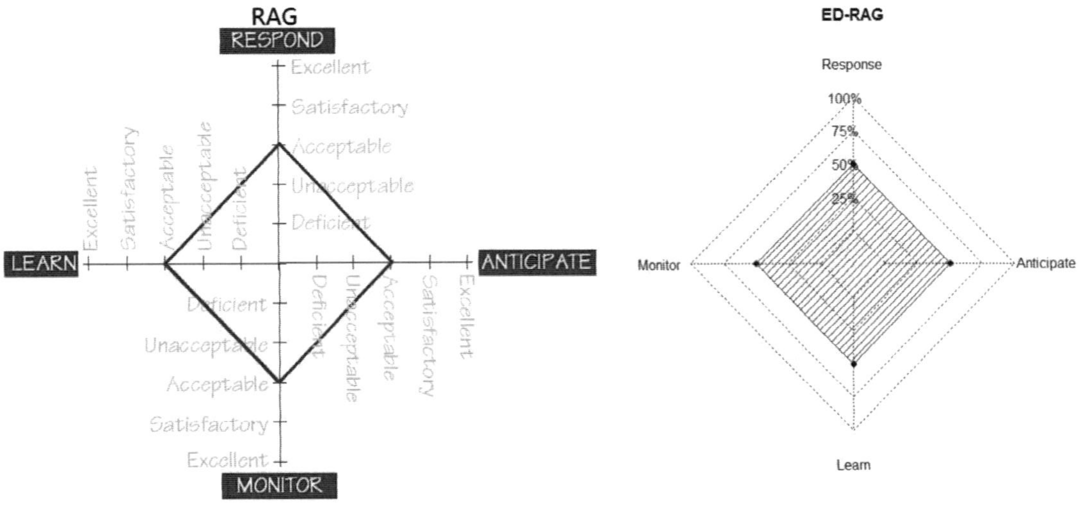

Figure 1. The RAG score for the four elements [5]

As seen in the picture above, we can see an example of a graph measuring the resilience potential of each aspect in implementing the RAG method, where each aspect is measured based on several detailed questions packaged in a questionnaire that each member of the organization must complete from a system that we are measuring.

For the RAG to be an effective tool, it is imperative that a rating may be assigned to each item. The rating can utilize the Likert-type scale provided below as an illustrative example:

IOP Conf. Series: Earth and Environmental Science **1423** (2024) 012005 doi:10.1088/1755-1315/1423/1/012005

Table 1. The RAG Response Scale [5]

The Response Scale		Description
Excellent	(5)	The system on the whole exceeds the criteria addressed by the specific item.
Satisfactory	(4)	The system fully meets all reasonable criteria addressed by the specific item.
Acceptable	(3)	The system meets the nominal criteria addressed by the specific item.
Unacceptable	(2)	The system does not meet the nominal criteria addressed by the specific item.
Deficient	(1)	There is insufficient capability to meet the criteria addressed by the item.
Missing	(0)	There is no capability whatsoever to address the specific item.

Furthermore, by using the responses to such particular questions as a (combined) proxy for resilience in this manner, a profile of the potential for each ability and, by extension, the potential for resilient performance as a whole can be constructed. The acronym for this proximate metric is RAG or Resilience Analysis Grid.

2.4 Overview of the LNG Terminal

The company being object in this study is a significant foreign investor in Indonesia, having been present in the country for 70 years. Their primary business activities in Indonesia focus on exploration and production of oil and gas, particularly the LNG facility. Additionally, they are involved in downstream and petrochemical industries[7].

The marine terminal starts in operation in 2009, just four years after receiving government approval. While specific safety records are not provided in the available sources, it's essential to note that safety is a top priority for any oil and gas facility. Company adhere to rigorous safety protocols, conduct regular safety audits, and invest in training programs to ensure the well-being of their workforce and the environment.

As an object in this study, the LNG Terminal is located on the south side of Bintuni Bay, West Papua Province, Indonesia. It has three jetties: the LNG Jetty, the Construction Jetty, and the Combo Dock. The LNG jetty is used for exporting LNG and is supported by a construction jetty that is used exclusively for barges and LCTs, and other jetty is primarily used for exporting condensate, general cargo, heavy lift, and offshore support vessels. The terminal also provided a combo dock with a tug mooring and floating pontoon, which is used for personnel transfer to and from the terminal. The Bintuni Bay and its waters have been declared a Marine Park by the Department of Environmental Protection.

The terminal receives natural gas from two unmanned platforms via two high-pressure sub-sea trunk lines approximately 17 km offshore; there are currently two trains supplying two LNG storage tanks, each of around 150,000 m^3 capacity, and one condensate storage tank of around 21,000 m^3 capacities. The terminal operates twenty-four hours every day per year. It is self-sufficient in all utility systems (such as instrument air, gaseous nitrogen for purging, fuel, power generation, venting, water, fire water, and potable water) that are required for the terminal's operation, maintenance, and safety.

2.5 Implementation of Occupational Health and Safety (OHS) Management System at LNG Marine Terminal

Given the inherent risks associated with handling and storing liquefied natural gas, the safe operation of an LNG marine terminal is of paramount importance. To mitigate these risks, the implementation of a robust Occupational Health and Safety (OHS) management system is crucial. However, the increasing complexity of such projects necessitates a more comprehensive management system to effectively implement the risk-based approach.

In addition to hazard identification and risk assessment, the OHS management system should also emphasize developing and implementing robust safety barriers. The system should ensure that critical safety systems such as emergency shutdown, fire and gas detection, and active and passive fire protection are designed, installed, and maintained to the highest standards. Periodic testing and inspection of these safety-critical systems is essential to verify their continued functionality and reliability. The establishment of a strong safety culture and the empowerment of the workforce to identify and mitigate hazards are not just elements, they are the backbone of an effective OHS management system. Comprehensive training programs, clear communication of safety policies, and a blame-free reporting environment are not just tools, they are the catalysts that can foster a proactive safety mind set among employees.

3. Discussion

In this study, we used a total of seventeen respondents who are a collection of management and each discipline leads as operators at the LNG marine terminal. Moreover, the respondents come from various positions such as Pilot Loading Master, Marine Specialist, Crisis & Continuity Management Advisor, Planner, field and Production Technician, HSE advisor and officer, Fire Officer, and Security team leader. They have an average educational background with five to twenty years of work experience in marine operations in the oil and gas industry. The respondents were selected as those who are responsible for the overall operation of the LNG marine terminal and who have the authority in the field to make decisions in the event of an operation disruption at the terminal.

In compiling the questionnaire, as explained by Hollnagel [5], it includes several socio-technical aspects that can be used as criteria for measuring the terminal resilience conditions of the ability to respond, the ability to anticipate, the ability to monitor, and the ability to learn. Each aspect is then described in a questionnaire matrix, as shown in the appendix file below.

3.1 Ability Analysis of Safety Resilience Elements

3.1.1 Response Ability

Figure 2 shows RAG study with the result of the average response ability value is 4.381, with the highest value in the Event list item, 4.688. This condition was demonstrated by most respondents who stated that within the organization, there was a list of potential events, both negative (risks) and positive (opportunities), covering several systems or activities. Meanwhile, the lowest value is item relevance 3.938, which relates to the duration of the review of risk register documents updated by a particular function involving related functions.

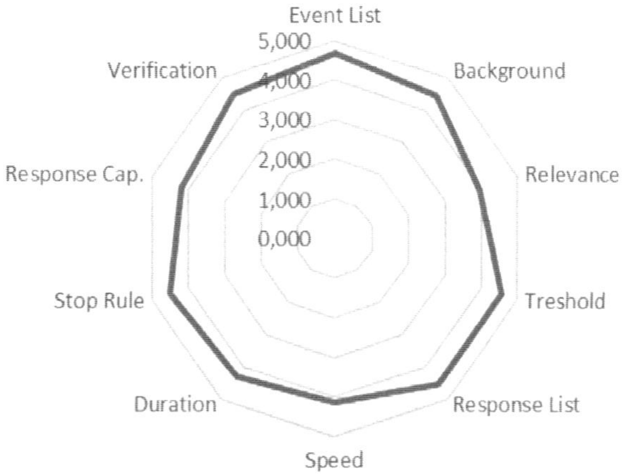

Figure 2. Ability of respond

3.1.2 Monitor Ability

Referring to the following figure 3 above, the average value of the ability to monitor is 3.969, with the highest value found in the validity aspect of 4.5, where the majority of respondents stated that in the organization, the resilience strategy is explicit, namely by carrying out calculations based on history (quantitative). The lowest value in this category is the measurement frequency aspect, which is 3.438. This condition is related to assumptions about the state of lifting operations that will occur in the future, where threats are mitigated, and operations exceed the maximum design capacity.

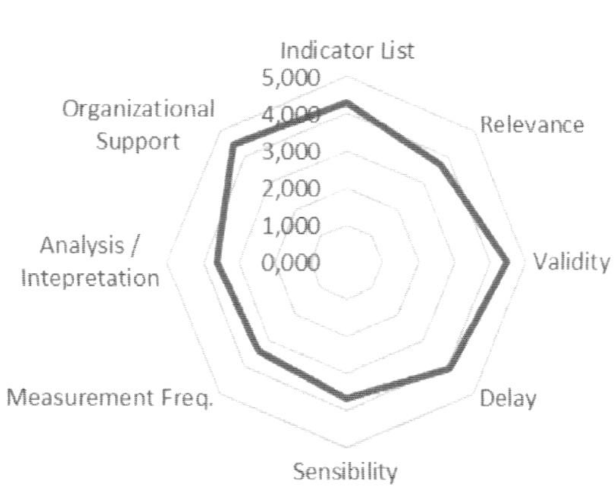

Figure 3. Ability of monitoring

3.1.3 Learning Ability

Referring to the figure 4 above, we can see that the average value for the ability to learn category of the terminal used as the object of study is 4.292, with the highest value found in the Learning base and Frequency/Learning Style aspect, namely 4.938. This item is related to the opinion of most respondents, who stated that companies learn from the successes and failures of internal and external parties and formalize them to improve operational performance. Meanwhile, the lowest value is for the Implementation aspect, namely 3.688, which is limited to implementing training for team members but has yet to reach the reorganization stage.

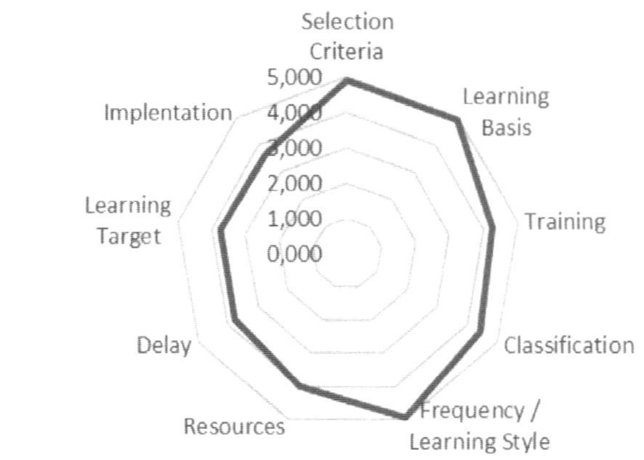

Figure 4. Ability of learning

3.1.4 Anticipation Ability

Referring to the figure 5 shown above, we can see that the average value for the ability to anticipate category is 4.340, with the highest value found in the Acceptability of Risk aspect, namely 4.938. These aspects relate to the basis of which risks are considered acceptable and unacceptable. Most respondents state that risks are acceptable based on quantitative research based on historical data using statistical/calculation methods with valid standards. The lowest value is for the Implementation item, namely 3.688, which relates to how often threats and opportunities in the future are assessed. Some respondents stated that the evaluation was carried out more than three times a year based on discussions with the head office.

Figure 5. Ability of anticipation

3.2 Level of Implementation of Safety Resilience

As shown in figure 6 above, overall, from the results of the resilience system measurements at the LNG terminal that have been carried out, it can be seen that the highest category is the ability to learn, with a value of 4.29, and the lowest value is the ability to monitor, namely 3.97. Of the four measured factors, only one factor, namely the ability to monitor, was slightly below 4, while the other three had values above. This condition shows that the level of terminal resistance is quite optimal, exceeds minimum compliance expectations, namely level 3.0.

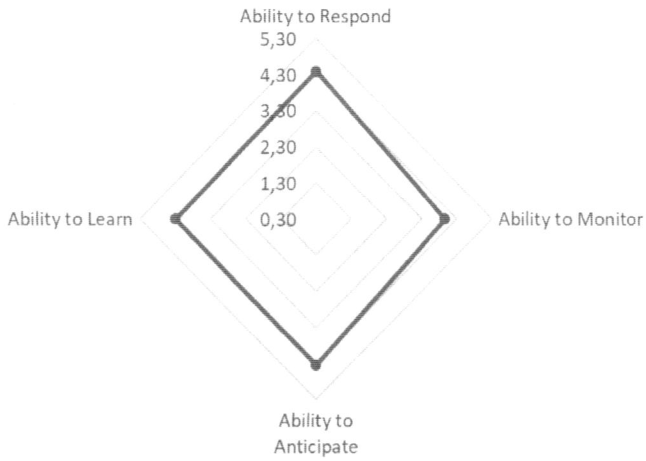

Figure 6. Four aspect of safety resilience implementation

Based on the analysis's results, companies are progressing towards a proactive level in the safety resilience maturity model. Company are in the initial stages of learning from negative experiences (Safety-I). Therefore, the learning process must encompass both success and failure stories comprehensively. Generally, the LNG terminal is primarily concerned with maintaining the operational process. The company's high response ability indicates that it views risk as the main factor contributing to accidents. Conversely, the low monitoring ability suggests that the monitoring element responsible for ensuring the smooth operation of the process needs to be fixed and requires proper improvement. Moreover, the assessment outcomes indicate that the company's low anticipating ability suggests a suboptimal implementation of an improvement program.

The highest score in the entire system is found in the ability to respond aspect, namely 4.38, which shows that the organization, in implementing this concept, still focuses on preventing and controlling accidents. Thus, the ability to respond factor can be considered a supporting factor for the resilience concept, which is worth maintaining. This condition differs from the ability to monitor, which has the lowest value, 3.97. However, this item cannot be called an inhibiting factor because its value exceeds the expected standard. In response, there is still room for improvement from existing achievements.

4. Conclusion

The results of this study show that the implementation of the safety concept at the LNG Marine Terminal is quite optimal, as indicated by the compliance figures for each resilience aspect, which exceed three as a benchmark. The highest score in the overall system is observed in the responsiveness category, precisely 4.38, indicating that the organization prioritizes accident prevention and control when implementing this idea. Therefore, the ability to respond aspect is a supporting factor for the concept of resilience, which is valuable to uphold. This condition is distinct from the monitoring ability, which has the lowest value of 3.97. Nevertheless, this item cannot be classified as inhibiting factor due to its value surpassing the anticipated benchmark. There is still potential for further enhancement based on the accomplishments made thus far.

Based on these findings, some recommendations might be suggested: The LNG terminal should not solely perceive risk as the sole cause of accidents but also as a potential opportunity to enhance future performance. Training and lessons from other terminals can facilitate the process of learning. Additionally, the company should regularly enhance the monitoring program to generate data that can inform the response and anticipation program.

Reference

[1] W. Qiao, X. Ma, Y. Liu, and W. Deng, "Resilience evaluation of maritime liquid cargo emergency response by integrating FRAM and a BN: A case study of a propylene leakage emergency scenario," *Ocean Eng.*, vol. 247, no. November 2021, p. 110584, 2022, doi: 10.1016/j.oceaneng.2022.110584.

[2] S. Baştuğ, E. Asyali, and T. Battal, "Beyond the ISM code: a conceptual proposal for an integrated system within the Seven C's approach," *Marit. Policy Manag.*, vol. 48, no. 3, pp. 354–377, 2021, doi: 10.1080/03088839.2020.1770884.

[3] C. S. Holling, "Resilience and stability of ecological systems," *Annu. Rev. Ecol. Syst.*, vol. 4, no. 1973, pp. 1–23, 1973, doi: 10.12987/9780300188479-023.

[4] J. Griffioen, M. van der Drift, and H. van den Broek, "Enhancing maritime crew resource management training by applying resilience engineering: A case study of the bachelor maritime officer training programme in rotterdam," *Educ. Sci.*, vol. 11, no. 8, 2021, doi: 10.3390/educsci11080378.

[5] E. Hollnagel, "RAG – Resilience Analysis Grid," *Resil. Eng. Pract. A Guideb.*, pp. 275–295, 2015.

[6] K. Klockner and P. Meredith, "Measuring resilience potentials: A pilot program using the resilience assessment grid," *Safety*, vol. 6, no. 4, p. 51, 2020, doi: 10.3390/safety6040051.

[7] S. Chuang, J. C. Ou, and H. P. Ma, "Measurement of resilience potentials in emergency departments: Applications of a tailored resilience assessment grid," *Saf. Sci.*, vol. 121, no. 291, pp. 385–393, 2020, doi: 10.1016/j.ssci.2019.09.012.

[8] M. Safi, B. R. Thude, F. Brandt, and R. Clay-Williams, "The application of resilience assessment grid in healthcare: A scoping review," *PLoS One*, vol. 17, no. 11 November, 2022, doi: 10.1371/journal.pone.0277289.

[9] D. Bose, C. K. Chanda, and A. Chakrabarti, "Vulnerability assessment of a power transmission network employing complex network theory in a resilience framework," *Microsyst. Technol.*, vol. 26, no. 8, pp. 2443–2451, 2020, doi: 10.1007/s00542-020-04785-x.

[10] J. Kong, S. P. Simonovic, and C. Zhang, "Resilience Assessment of Interdependent Infrastructure Systems: A Case Study Based on Different Response Strategies," *Sustain.*, vol. 11, no. 23, 2019, doi: 10.3390/su11236552.

[11] R. Patriarca, G. Di Gravio, F. Costantino, A. Falegnami, and F. Bilotta, "An Analytic Framework to Assess Organizational Resilience," *Saf. Health Work*, vol. 9, no. 3, pp. 265–276, 2018, doi: 10.1016/j.shaw.2017.10.005.

System Theory and Human Factors Hazard Identification Approach for Marine Survey Operation

Dwitya Harits Waskito[1*], Muhammad Faishal[2], Tris Handoyo[1], Indra Kurniawan[1], and Julianto Saut Hamonangan[3]

[1] Research Center for Transportation Technology, National Research and Innovation Agency, South Tangerang, Indonesia
[2] Marine Engineering Department, Institut Teknologi Sepuluh Nopember, Surabaya, Indonesia
[3]National Research and Innovation Agency, Jakarta, Indonesia

*E-mail: dwit001@brin.go.id

Abstract. Marine survey operation is one of the most frequent and essential activities in enhancing ocean knowledge. Some dangerous tasks and activities involve scientists, marine crews, survey equipment, and sensors, such as deploying equipment near the seabed, collecting the sediment sample, and towing the equipment with the ship's movement. Since it consists of several controllers and components, the comprehensive system theory must be applied to analyse the risk, and the effect of human error must be incorporated as the equipment's controller within the system. This study provides a holistic hazard identification of marine survey operation by using System Theory Process and Analysis (STPA) and integrating it with the Human Factors Classification System (HFACS) to define the unsafe control actions (UCA) and failure scenarios. According to the STPA, 194 UCAs could occur. The main causal factors of the UCAs are human and followed by technical errors. The STPA-HFACS analysis indicated that additional human and equipment actions would have a detrimental effect on the failures of the operation. This study will benefit stakeholders in marine survey operations with an alternative method for risk analysis.

1. Introduction

Marine survey operations are essential to improve understanding of the ocean and support various activities in the maritime industry. These surveys are essential for accurate seabed mapping, providing important data for navigation safety [1], environmental monitoring and exploration of marine resources to support the maritime industry, which is one of the main factors in the world economy, which includes oil, gas and mineral industry [2]–[5], and seafood industry [6] as the global demand for marine resources and offshore development increases, The need for accurate and reliable marine survey data is becoming increasingly important [7]. Marine survey activities require human action at sea supported by research vessels and relevant survey equipment. The data collected from marine surveys is essential for mapping the seabed, identifying hazards, and ensuring safe and efficient operations of maritime activities. Research on the marine environment is still a big challenge for humanity. Each survey campaign is a complex project using research vessels and relevant survey equipment. These operations involve systems and equipment susceptible to failure due to various factors [8].

Moreover, marine surveys are essential for understanding and managing the marine environment. They collect data on various aspects of the marine environment, such as water quality, marine life, and geological formations. This data is critical for making informed decisions about marine conservation, resource management, and environmental policy [9], [10].

Consequently, marine surveys involve working in challenging environments with many hazards. Equipment failures, human error, and environmental conditions can pose significant risks to personnel and survey success. For example, equipment failure can lead to delays or even abandonment of surveys, while human error can result in accidents or injuries [8]. Marine survey operations involve complex and potentially dangerous tasks, such as deploying equipment near the seafloor, collecting sediment samples, and pulling equipment along with ship movements. These tasks expose scientists, marine crews, and survey equipment to significant risks. Accidents can result from human error, equipment failure, or poor environmental conditions, resulting in injury, equipment loss, and environmental damage. In addition, the performance efficiency of the device depends on the changing sea and weather conditions, which affect the sensitivity of the device and the vessel [11]. The success of the offshore project depends on many factors [12] and the efficient implementation of the planned research campaign. Therefore, a quality defect in the equipment may cause the error itself, but it can also be caused by extreme research conditions and, ultimately, human error that can occur under such conditions.

Research on marine survey safety has been conducted, and several studies have focused on identifying the most common causes of equipment failure and developing risk assessment and mitigation methods. Studies have shown that human factors, such as fatigue and lack of training, significantly contribute to equipment failure[8], [10]. This study successfully identified the most common failure of measuring equipment working at sea, found the cause based on the results of surveys conducted, assessed and measured the risks, and proposed a response plan for each identified risk. Technical problems, such as equipment breakdown and environmental conditions, such as bad weather, can also contribute to failure [8], [9]. Other research on marine survey safety has also proliferated in recent years, focusing on several important aspects such as risk assessment, human factors, equipment conditions, and weather. An integrative risk approach that combines species-specific biological and contextual aspects to evaluate risks to marine mammals from specific industrial activities, such as seismic airgun surveys and pile guidance [13]. The risk assessment in this study integrates species-specific populations, life history, behavioural sensitivity, and spatio-temporal contextual aspects into the assessment of response probability, species vulnerability, and impact severity to provide resource managers with easy-to-understand and objective decision-making tools to assess the relative risk to marine species of seismic survey activities with systematic methods, Quantitative, transparent, repeatable, and simple, with a focus on the evaluation of biological risks to marine mammals from various operational scenarios.

A formal safety assessment framework for marine survey operations has also been conducted using risk matrix and error tree analysis to improve the operational safety of marine surveys [14]. This study proposes a unique application of formal safety assessment (FSA) to reduce accidents in marine seismic surveys and improve seismic vessel operations' safety in the context of a viable economic strategy. This study observed hazards in offshore seismic surveys to identify critical risk factors that cause marine seismic survey accidents. The risk factors that cause undesirable events are analysed both qualitatively and quantitatively. A risk matrix was introduced to screen out undesirable events that have been identified by using Fault Tree Analysis (FTA) to investigate and analyse the most critical risks in seismic survey operations, taking into account the lack of historical data. The results show that people falling into the sea (MOB) is a major risk factor in marine seismic survey operations. Other studies examine various studies related to maritime accidents caused by human factors, the importance of human factor risk management in maritime operations [15], human factor analysis and the application of corrective actions to marine accidents using the Human Factor Analysis and Classification System for Maritime Accidents (HFACS-MA) [16]. The study focused on the analysis of the human factor in marine accidents, given that human error accounts for 80-85% of all marine accidents. The study examined 135 marine accident reports recorded in the UK Marine Accident Investigation Branch (MAIB) database from 2010 to 2019, categorising the causative factors using the Human Factor Analysis and Classification System for Maritime Accidents (HFACS-MA).

Multiple linear regression is used to determine the relationship between the number of accidents and the most common causative factors of HFACS-MA. The study found that the causes of marine accidents mainly depend on two categories of human factors, and by influencing those categories of human factors, the number of marine accidents can be reduced, and shipping safety in general can be improved. Other articles discuss the importance of the reliability of survey equipment and suggest risk assessment and management protocols to reduce the impact of equipment failures and improve operational safety [8]. Other studies show that unpredictable weather conditions are a significant challenge in planning work schedules for offshore projects, and statistical weather data analysis is used to reduce weather-related risks during marine survey operations [12]. These studies show that safety in marine surveys can be improved with comprehensive risk management, attention to human factors, equipment reliability, and monitoring and mitigating weather risks.

Several methods have been used to identify safety in marine surveys. These include risk assessment, error tree analysis, and human reliability analysis. Risk assessment involves identifying potential hazards and evaluating the likelihood and impact of those hazards. Error tree analysis is a method used to identify possible causes of equipment failure. Human reliability analysis focuses on human factors that contribute to equipment failure[8], [17]. While current methods are not yet effective enough in identifying safety issues in marine surveys, there is still a need for more comprehensive risk analysis. This is because the marine environment constantly changes, and new dangers may arise. Additionally, the complexity of marine survey operations requires a more detailed understanding of the interconnected relationships between various systems and factors.

Systems theory is essential for understanding the complex relationships between various systems and factors in marine survey operations. It provides a framework for analysing the interconnected relationships between human, technical, and environmental factors. This approach can help identify potential hazards and develop more effective risk assessment and mitigation methods.

This study aims to develop a systems theory-based approach to identify hazards in marine survey operations. This study provides a holistic hazard identification of marine survey operation by using System Theory Process and Analysis (STPA) and integrating it with the Human Factors Classification System (HFACS) to define the unsafe control actions (UCA) and failure scenarios. The results from this study indicated that additional human and equipment actions would have a detrimental effect on the failures of the operation. This study will benefit stakeholders in marine survey operations with an alternative method for risk analysis. The study will also examine the current state of research on marine survey safety and identify gaps in current methods. The findings of this study will provide a comprehensive understanding of the hazards associated with marine survey operations and inform the development of more effective risk assessment and mitigation methods.

IOP Conf. Series: Earth and Environmental Science **1423** (2024) 012006 doi:10.1088/1755-1315/1423/1/012006

2. Methodology

2.1. General Overview

This study focused on comprehensively identifying hazards in marine cable route survey operations using STPA in conjunction with the Human Factors Analysis and Classification System HFACS. Based on studies conducted by [18], expanding STPA to consider human factors more thoroughly can significantly enhance the methodology's ability to analyse and improve safety in complex systems. The proposed method is structured into five steps, as shown in **Figure 1** and explained in the following paragraph.

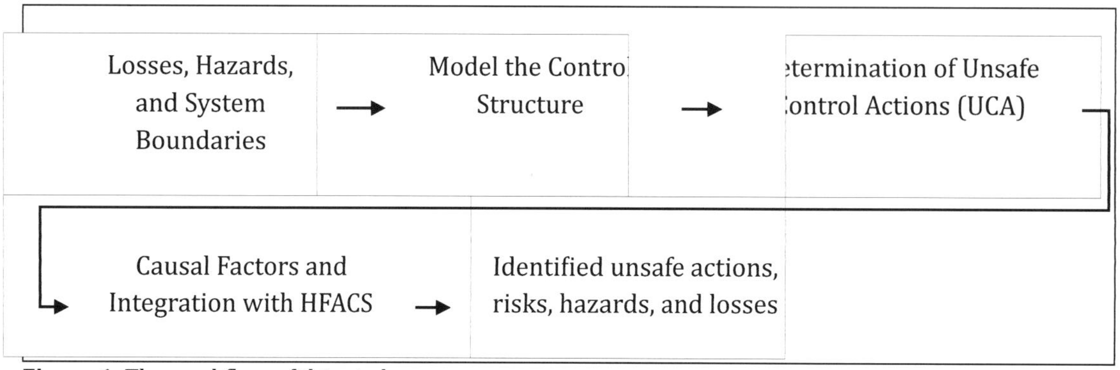

Figure 1. The workflow of this study

2.2. System Theoretic Process and Analysis (STPA)

STPA is a practical methodology for identifying and managing risks in marine survey activities. According to [19], by using STPA, we can identify that accidents can occur due to unsafe interactions between system components, even if none of the individual components fail. In this context, STPA is used to analyse control and operational processes to identify potential hazards during marine cable route surveys.

The process begins with defining the survey objectives and understanding the operational context, including the area and environmental conditions. Next, potential hazards and risks were identified, followed by defining the existing system controls and processes. The STPA analysis was then conducted to uncover control weaknesses that could lead to hazards. Below is a detailed explanation of the workflow represented in **Figure 1**:

Step 1 **Losses, Hazards, and System Boundaries**
The first step involved setting the research objectives and defining the system boundaries, which includes identifying potential losses (Ls) and levels of hazards (Hs). In this study, the system boundary was the marine survey operations, including the towfish deployment, using the research vessel operated by a research institution.

Step 2 **Control Structure**
An interconnected structure was developed at this stage, incorporating inputs and outputs from various system components, including controllers, control actions, and feedback loops. Each controller is assigned the responsibility of ensuring the effective execution of survey operations.

Step 3 **Determination of Unsafe Control Actions (UCA)**
This stage involves identifying control actions that could pose a danger. The goal was to investigate how these Unsafe Control Actions (UCAs) might lead to the identified hazards. In

this study, only three potential unsafe actions were considered: "not providing", "providing causes hazard", and "providing too early, too late, or out of order".

Step 4 **Causal Factors and Integration with HFACS**

This study outlined general causal factors, such as human, management, and technical. The integration helps identify underlying human errors and system deficiencies that may lead to incidents. By categorising these factors into tiers, such as organisational influences, unsafe supervision, pre-conditions for unsafe acts, and the unsafe acts themselves, HFACS provides a structured approach to uncovering the root causes of accidents.

Step 5 **Identified unsafe actions, risks, hazards, and losses.**

By focusing on control structures and their interactions, STPA examines how different components and their interrelations between humans and equipment in marine surveys can contribute to unsafe scenarios. Additionally, STPA evaluates risks by assessing the likelihood and impact of these hazardous states and identifying potential losses that may result from these risks.

2.3. Integration of HFACS into STPA

Combining HFACS's framework for analysing human errors with STPA's method for evaluating system-wide interactions and control structures enhances the ability to pinpoint the root causes of unsafe actions. As research conducted by [20] shows, this integration demonstrates an improvement in identifying and mitigating risks associated with human errors. This synergistic approach allows for a more thorough understanding of how human factors contribute to system failures, ultimately improving safety and reliability in various domains, including marine cable route surveys. In the research applications by [21], STPA has demonstrated effectiveness in evaluating risks not only at the component level but also at the operator level within the system[22]–[24].

3. Results and Discussion

3.1. Losses, Hazards, and System Boundaries

The first step of the STPA was to identify the system boundaries. In this study, the system boundary was the marine survey operations, including the towfish deployment, using the research vessel operated by a research institution. In this study, there were five losses, which are the consequences of the accidents, as stated in **Table 1**. L1 is considered necessary in terms of marine surveys since the price of the equipment is a high-cost investment. L2 and L4 are the repercussions of the L1 since the loss of equipment can lead to loss of mission and data. Marine survey operations involve numerous pieces of equipment that, in its deployment to the ocean, there will be a risk to the personnel working on deploying and un-deploy survey equipment. Therefore, human loss and injury are also incorporated, as stated in L3 and L5.

Table 2 shows the six possible hazards that can cause the loss and their relation to their respective losses. Collision, contact, and grounding of the ship (H.3) were found to be the hazards that possibly caused every loss. Since the system boundary emphasises marine survey operation that includes the towing operation of the towfish, there are three hazards related to towfish deployment (H4, H5, and H6).

Table 1. The losses as a consequence of hazards in marine survey operation

Loss ID	Loss Name
L.1	Loss of equipment
L.2	Loss of mission
L.3	Loss of life
L.4	Loss of data
L.5	Injury to people

Table 2. The hazards that can lead to losses in marine survey operation

Hazard ID	Loss Name
H.1	Oversight risk assessment (L2, L3,L5)
H.2	Navigation failure (L1, L3, L4)
H.3	Collision, Contact, Grounding of ship (L1,L2,L3,L4,L5)
H.4	Uncontrollable towing line (L1, L2, L4, L5)
H.5	Towfish contact with obstacles (L1, L2, L4)
H.6	Improper data output (L3, L4)

3.2. Control Structure

Figure 2 shows the control structure for the system defined in the previous step. The structure was based on several sources of literature, such as research papers [14], [17] and Standard Operational Procedures (SOPs) for marine survey operations. All stakeholders from the research institution management (Head of Laboratory), marine crew, onboard scientist, and survey equipment were involved. Fourteen controllers are involved: ten human-related controllers and four non-human controllers. There are 19 general control actions (CA) and 66 detailed actions, as a detailed structure, from the general CA denoted in Figure 1. The detailed CAs denote a command and action from the controller to a particular destination. The failure of each CA will be assessed in the form of UCA in the next step.

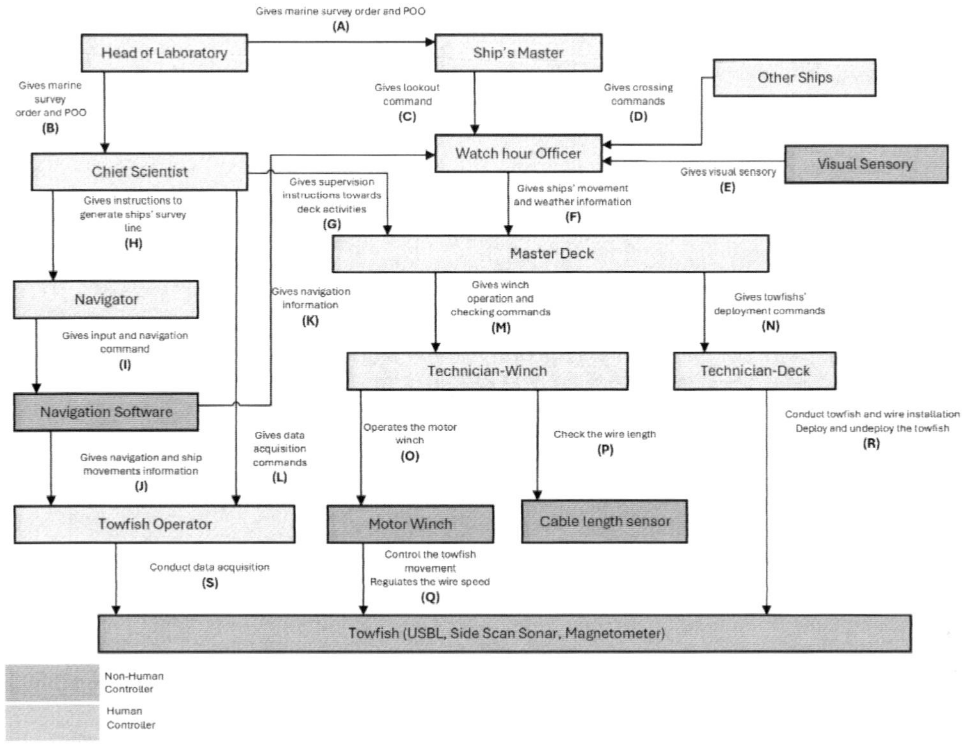

Figure 2. The control structure for towfish deployment in marine cable route survey

3.3. Determination of Unsafe Control Actions (UCA)

After deriving the CA with the possibility of unsafe action, 194 UCAs were identified. This study only applied three possible unsafe actions: "not providing", "providing causes hazard", and "providing too early, too late, or out of order". **Table 3** shows the example of the UCAs and associated hazards from each of the five phases in marine survey operations.

To delve more into the UCAs and associated hazards, **Figure 3** shows the distribution of hazards by the particular controller after identifying each UCA. Regarding the human controller, the Chief Scientist holds the highest number of hazards, with 87, followed by a Master Deck with 85 associated hazards, for the non-human controller. Navigation software has the highest number of possible hazards, with 116, mainly contributing to the H5 (Towfish contact with obstacles). Regarding the general system, H5 is the highest number at 135, followed by H6 at 125 occurrences after the identification of UCAs.

Table 3. Example of Unsafe Control Action within the marine survey operation for every phase

Controller	Control Action	Not providing	Providing cause hazard	Providing too early, late, or out of order
Head of Laboratory	Order the survey and give the plan of operations (POO) for the Research Vessel	A.1.1 Head of the laboratory does not give orders and POO to the master	A.1.2 Head of laboratory gives unclear POO to the master	A.1.3 Head of laboratory gives late order and POO; the ship already sailed
Other ships	Give crossing commands and information	D.1.1 Other ships do not give crossing information to the research vessel	D.1.2 Other ships give unclear and wrong crossing information to the research vessel	D.1.3 Other ships give too early or too late wrong crossing information to the research vessel

Controller	Control Action	Not providing	Providing cause hazard	Providing too early, late, or out of order
Chief Scientist	Give instructions to check the towfish connection	G.1.1 The Chief scientist does not instruct the master deck to check the towfish connection	G.1.2 The Chief scientist gives wrong and unclear instructions to the master deck to check the towfish connection	G.1.3 Chief Scientist gives too late instruction to the master deck to check the towfish connection
Chief Scientist	Give instructions for data acquisition	L.1.1 Chief scientist does not instruct data acquisition	L.1.2 Chief scientist gives wrong and unclear instructions for data acquisition	N/A
Master Deck	Give instructions to record wire length	M.4.1 Master deck does not give instructions to record wire length	M.4.2 Master deck gives wrong and unclear instructions to record wire length	M.4.3 Master deck gives too-early and too-late instructions to record wire length

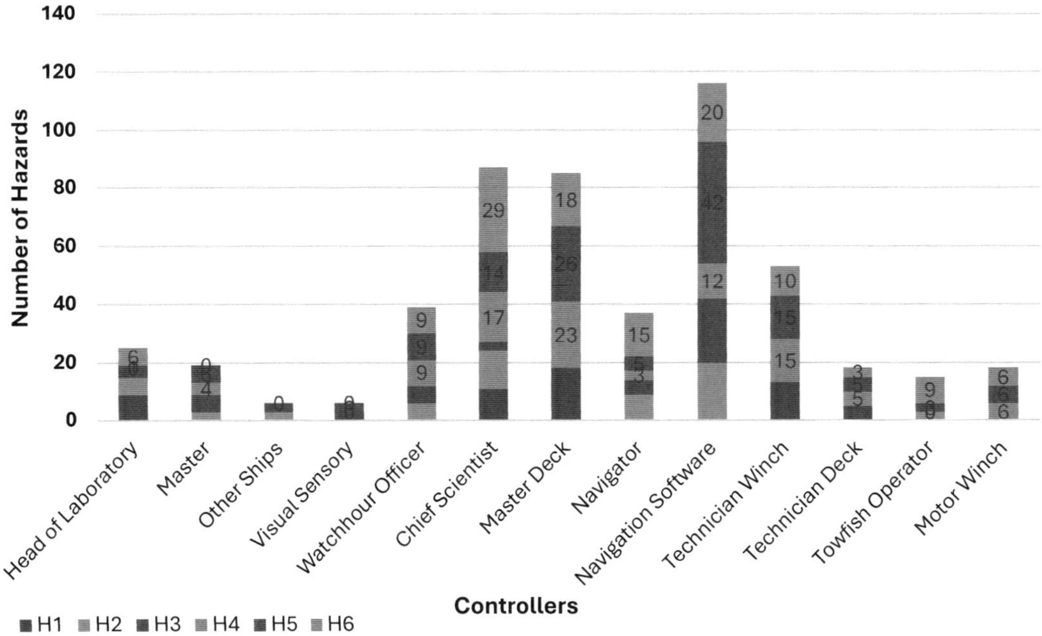

Figure 3. The distribution of hazards according to several controllers

3.4. Causal Factors and Integration with HFACS

It is imperative to identify the causal factors that lead the UCA to generate the loss scenario of the respective UCA. This study outlined general causal factors, such as human, management, and technical. Humans are factors that result from human conditions, capabilities in carrying out a task, and situational awareness's influence. Meanwhile, management influence is the effect of improper management activities between the shipping operator, laboratory, scientist, and marine crew. **Figure 4** shows the number of UCAs for every marine survey operation phase. Generally, human factors are the highest contributing factors to the UCAS, with 140 unsafe actions followed by technical errors, with 89. Regarding the survey phase, the Deploying and lifting is the activity with the most potential UCA regarding human error. Meanwhile, during the survey preparation phase, software errors and equipment influence the activities with the highest number of technical errors that can cause the UCA, such as inputting the coordinates, line survey, and maps.

Human errors should be categorised and detailed as the most causal factors to obtain more comprehensive root causes and detailed failure modes. HFACS is a method that can be used to classify human errors based on four main levels, namely Organizational Influence (OI), Unsafe Supervision (US), Pre-Conditions of Unsafe Acts (PCUA), and Unsafe Acts (UA). **Table 4** details UCAs involved in human errors categorised by particular HFACS levels. It is possible that one UCA is categorized into two different levels. For instance, UCA R.1.2, i.e., the towfish is not appropriately deployed, can be caused by two different HFACS levels. The first is the lack of skill, categorised in the unsafe acts level by the technician, and the second is the problem with the equipment and the lack of readiness of the technician, which is categorised in pre-conditions of unsafe acts level. The integration of HFACS into STPA is helpful in determining causal factors in detail because there is a complex relationship between human and machine interactions in marine survey operations. Therefore, the UCA can also be caused by the resonance between management, technology, and humans.

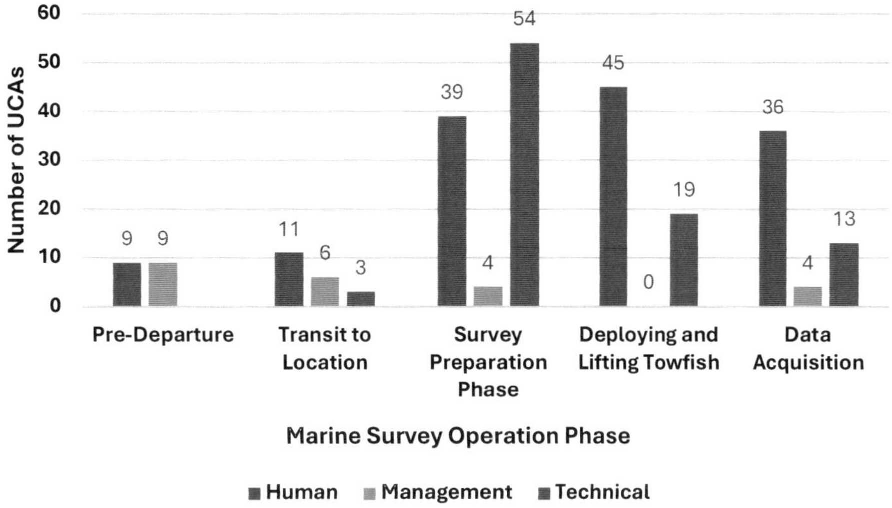

Figure 4. Distribution of number of UCAs according to marine survey operation phase

Table 4. Determination of causal factors by integrating HFACS

Unsafe Control Actions	Marine survey operation phase	(Human, technical, management)	Detailed causal factors (HFACS Level)
A.1.2 Head of laboratory gives unclear plan of operations to the master	Pre-Departure	Human, and Management	OI – Organizational Process UA - Violations
D.1.1 Other ships do not give crossing information to the research vessel	Transit to location	Human	PCUA – Adverse mental states UA- Perceptual error
G.1.1 Chief scientist does not give the instruction to the master deck to check the towfish connection	Survey preparation phase	Human	US – Inadequate supervision UA – Decision error
L.1.2 Chief scientist gives unclear instruction for data acquisition	Data acquisition	Human	PCUA – Technological environment UA- Skill-Based error

Unsafe Control Actions	Marine survey operation phase	(Human, technical, management)	Detailed causal factors (HFACS Level)
M.4.1 Master deck does not give instructions to record wire length	Deploying and lifting towfish	Human	US – Inadequate supervision UA- Decision Errors
R.1.2 The towfish is not appropriately deployed	Deploying and lifting towfish	Human, and Technical	PCUA – Technical Environment, Personal readiness UA – Skill-Based Error

After combining the UCA with the causal factors from HFACS, the loss scenario from UCA R.1.2. can be stated by:

"The towfish is not appropriately deployed into the water by the deck technician due to lack of skill (Unsafe Action – Skill Based Error), which can cause uncontrollable towing line, and the towfish could come into contact with obstacles (H4 and H5). If these hazards cannot be constrained, it will increase the chance of loss of equipment, mission, data, and injury to the people (L1,L2, L5, and L4)."

3.5. General discussions and implications

Generally, STPA has successfully identified numerous possible unsafe actions, risks, hazards, and losses in marine survey operations. The advantage of the STPA is suitable for the complex system interactions between humans and equipment in marine surveys, especially that involves the deployment of towfish equipment. With the assumption of system theory that sees the marine survey as a whole system, the detailed interconnected relation between components that construct the system has been comprehensively identified. Furthermore, three unsafe action categories can provide a detailed overview of failures for each control action, which will be beneficial in generating risk control measures for respective UCAs.

Compared to other studies of risk identifications in marine surveys, this study identified more risks than other studies using the conventional risk identification method, such as Fault Tree Analysis (FTA). STPA, in this analysis, managed to identify 194 UCAs compared to [17] with 24 basic events of FTA and 17 basic events of man overboard risk during marine survey operations [14]. These differences indicated that STPA focused on in-depth analysis of probable failures, while the FTA focused on immediate failures. Moreover, STPA covered several aspects such as human factors, technical, equipment, and organisational, while FTA only covered technical and component failures.

Human-related errors are the main causal factor in the marine survey operation after STPA was conducted. These findings are similar to the results from several research studies on STPA in the maritime industry [25], [26]. This result can be caused by the higher number of human controllers in the control structure. Another significant result is that technical factors such as software and equipment play an essential role during the preparation phase of marine survey operations.

This study also proposed a straightforward method to integrate human errors into the causal factors of the UCA. The integration of HFACS was executed in the last step of the STPA, the loss scenario, of the UCA to delve into more detail in the categorisation of human error, whether there is an influence from the organisation or it is due to the adverse mental condition of the human controller. Compared to related STPA studies [27], [28], this STPA-HFACS provides a more comprehensive analysis of the scenario and causal factors that can induce the UCA to occur, especially for human-related errors. This method can be used by stakeholders with basic risk analysis skills to increase the safety level in institutions and shipping operators.

With the number of UCA that have been generated, it is beneficial for the further analysis of the risk. However, with the amount of UCAs, it might be inefficient to assess many UCAs without priority and criticality analysis. Because some UCAs will have severe consequences and frequent occurrences, and others only have minor ramifications and are probably unlikely to happen. Therefore, the risk priority evaluation of the UCAs should be conducted to enhance the focus of risk management.

4. Conclusion

The safety consideration in marine survey operations should be enhanced by incorporating the system-theory aspect to identify risk. This study aims to identify and map all risks and hazards related to marine survey operations, especially those requiring towing the equipment under the sea level, such as side scan sonar and marine magnetometer. The STPA method was employed since marine survey operations involve numerous interactions between humans and equipment. From the STPA, 194 UCAs have been identified, and 55% are related to human factors. The chief scientist found the highest number of possible UCA for human controllers; meanwhile, navigation software is pivotal for non-humans with the highest number of UCA. Since human- error is still a general term, HFACS was utilised to provide details and classifications for each UCA. From the integration between STPA and HFACS, it can be concluded that in the UCA, there will be an influence from the organisation and supervision that can affect the operator's unsafe acts. This study provides a detailed approach that can benefit stakeholders in marine survey operations and enhance their safety level.

References

[1] I. Golub Medvešek, J. Šoda, I. Karin, and M. Maljković, "The State of the Hydrographic Survey and Assessment of the Potentially Risky Region for Navigation Safety," *J. Mar. Sci. Eng.*, vol. 11, no. 8, 2023, doi: 10.3390/jmse11081498.

[2] J. A. P. Chandler, "Developing offshore petroleum to meet socio-economic objectives: Lessons from Australia, Norway and the United Kingdom," *Energy Policy*, vol. 144, no. May, p. 111618, 2020, doi: 10.1016/j.enpol.2020.111618.

[3] C. Yin, "International law regulation of offshore oil and gas exploitation," *Environ. Impact Assess. Rev.*, vol. 88, no. January, p. 106551, 2021, doi: 10.1016/j.eiar.2021.106551.

[4] X. Wang, K. F. Yuen, Y. D. Wong, and K. X. Li, "How can the maritime industry meet Sustainable Development Goals? An analysis of sustainability reports from the social entrepreneurship perspective," *Transp. Res. Part D Transp. Environ.*, vol. 78, no. December 2019, p. 102173, 2020, doi: 10.1016/j.trd.2019.11.002.

[5] O. M. H. C, M. Shadman, M. M. Amiri, C. Silva, S. F. Estefen, and E. La Rovere, "Environmental impacts of offshore wind installation, operation and maintenance, and decommissioning activities: A case study of Brazil," *Renew. Sustain. Energy Rev.*, vol. 144, no. February, p. 110994, 2021, doi: 10.1016/j.rser.2021.110994.

[6] V. Venugopal and A. Sasidharan, "Seafood industry effluents: Environmental hazards, treatment and resource recovery," *J. Environ. Chem. Eng.*, vol. 9, no. 2, p. 104758, 2021, doi: 10.1016/j.jece.2020.104758.

[7] K. Yan, Y. Wang, W. Wang, C. Qiao, B. Chen, and L. Jia, "A System-Theory and Complex Network-Fused Approach to Analyze Vessel–Wind Turbine Allisions in Offshore Wind Farm Waters," *J. Mar. Sci. Eng.*, vol. 11, no. 7, 2023, doi: 10.3390/jmse11071306.

[8] M. Kubacka, L. Gajewski, M. Burchacz, M. Matczak, P. Janowski, and J. Piotrowicz, "Many faces of survey equipment failures during marine research at sea—Risk analysis," *PLoS One*, vol. 17, no. 8 August, pp. 1–23, 2022, doi: 10.1371/journal.pone.0272960.

[9] Norwegian Maritime Authority, "Results of the survey on maritime safety," 2023. https://www.sdir.no/en/news/news-from-the-nma/results-of-the-survey-on-maritime-safety/ (accessed Jun. 12, 2024).

[10] M. Sha, "A study of the human factors in maritime safety," 2020. [Online]. Available: https://commons.wmu.se/msem_dissertations

[11] T. Gintautas, J. D. Sørensen, and S. R. Vatne, "Towards a Risk-based Decision Support for Offshore Wind

Turbine Installation and Operation & Maintenance," *Energy Procedia*, vol. 94, no. January, pp. 207–217, 2016, doi: 10.1016/j.egypro.2016.09.225.

[12] M. Kubacka, M. Matczak, M. Kałas, L. Gajewski, and M. Burchacz, "Weather risk management in marine survey campaigns for the offshore investment projects in the polish exclusive economic zone," *Weather. Clim. Soc.*, vol. 13, no. 4, pp. 899–911, 2021, doi: 10.1175/WCAS-D-20-0168.1.

[13] B. L. Southall, D. Tollit, J. Amaral, C. W. Clark, and W. T. Ellison, "Managing human activity and marine mammals: A biologically based, relativistic risk assessment framework," *Front. Mar. Sci.*, vol. 10, no. February, pp. 1–15, 2023, doi: 10.3389/fmars.2023.1090132.

[14] G. Asuelimen, E. Blanco-Davis, J. Wang, Z. Yang, and D. B. Matellini, "Formal Safety Assessment of a Marine Seismic Survey Vessel Operation, Incorporating Risk Matrix and Fault Tree Analysis," *J. Mar. Sci. Appl.*, vol. 19, no. 2, pp. 155–172, 2020, doi: 10.1007/s11804-020-00136-4.

[15] F. X. Martínez de Osés and N. P. Ventikos, "A critical assessment of human element regarding maritime safety," 2003.

[16] N. Hasanspahić, S. Vujičić, V. Frančić, and L. Čampara, "The role of the human factor in marine accidents," *J. Mar. Sci. Eng.*, vol. 9, no. 3, pp. 1–16, 2021, doi: 10.3390/jmse9030261.

[17] A. Muhtadi, D. H. Waskito, and D. F. Prasetyo, "Improving Safety of Marine Cable Survey Operation Through Safety Assessment using the Formal Safety Assessment Method (Case Study RV Baruna Jaya)," in *IOP Conference Series: Earth and Environmental Science*, IOP Publishing, 2023. doi: 10.1088/1755-1315/1166/1/012031.

[18] M. Lower, J. Magott, and J. Skorupski, "A System-Theoretic Accident Model and Process with Human Factors Analysis and Classification System taxonomy," *Saf. Sci.*, vol. 110, no. November 2016, pp. 393–410, 2018, doi: 10.1016/j.ssci.2018.04.015.

[19] X. Yang *et al.*, "Integrating Software FMEA and STPA to Develop a Bayesian Network-Based Software Risk Model for Autonomous Ships," *J. Mar. Sci. Eng.*, vol. 12, no. 1, 2024, doi: 10.3390/jmse12010004.

[20] M. Elizabeth, F. B. S. Human, M. E. France, and Y. M. Marzouk, "Engineering for Humans: A New Extension to STPA Signature of Author," 2017.

[21] B. Rachel Cabosky, J. P. Thomas, T. Supervisor, and J. Rubin, "Application of Hierarchy to STPA: A Human Factors Study on Vehicle Automation," no. 2018, 2020.

[22] Y. Zhang, C. Dong, W. Guo, J. Dai, and Z. Zhao, "Systems theoretic accident model and process (STAMP): A literature review," *Saf. Sci.*, vol. 152, no. April 2021, p. 105596, 2022, doi: 10.1016/j.ssci.2021.105596.

[23] C. Li, T. Tang, M. M. Chatzimichailidou, G. T. Jun, and P. Waterson, "A hybrid human and organisational analysis method for railway accidents based on STAMP-HFACS and human information processing," *Appl. Ergon.*, vol. 79, no. May 2017, pp. 122–142, 2019, doi: 10.1016/j.apergo.2018.12.011.

[24] C. Dong, Y. Zhang, Z. Wang, J. Liu, and J. Zhang, "The hybrid systems method integrating STAMP and HFACS for the causal analysis of the road traffic accident," *Ergonomics*, vol. 0, no. 0, pp. 1–24, 2023, doi: 10.1080/00140139.2023.2270783.

[25] S. Basnet, A. BahooToroody, M. Chaal, J. Lahtinen, V. Bolbot, and O. A. Valdez Banda, "Risk analysis methodology using STPA-based Bayesian network- applied to remote pilotage operation," *Ocean Eng.*, vol. 270, no. December 2022, p. 113569, 2023, doi: 10.1016/j.oceaneng.2022.113569.

[26] S. Il Ahn, R. E. Kurt, and O. Turan, "The hybrid method combined STPA and SLIM to assess the reliability of the human interaction system to the emergency shutdown system of LNG ship-to-ship bunkering," *Ocean Eng.*, vol. 265, no. September, p. 112643, 2022, doi: 10.1016/j.oceaneng.2022.112643.

[27] K. Wróbel, J. Montewka, and P. Kujala, "Towards the development of a system-theoretic model for safety assessment of autonomous merchant vessels," *Reliab. Eng. Syst. Saf.*, vol. 178, no. May 2017, pp. 209–224, 2018, doi: 10.1016/j.ress.2018.05.019.

[28] M. Chaal, O. A. Valdez Banda, J. A. Glomsrud, S. Basnet, S. Hirdaris, and P. Kujala, "A framework to model the STPA hierarchical control structure of an autonomous ship," *Saf. Sci.*, vol. 132, no. August, p. 104939, 2020, doi: 10.1016/j.ssci.2020.104939.

IOP Conf. Series: Earth and Environmental Science **1423** (2024) 012007 doi:10.1088/1755-1315/1423/1/012007

Identification of Ship Domain Shape to Degree of Ship Collision Risk on Head-on Encounter Situation

Yuda Apri Hermawan[1*], Fernanda Wahyu Pratama[2*], Setyo Nugroho[3*], Sjarief Widjaja[4*], Totok Yulianto[5*], and Dedi Budi Purwanto[6*]

[1, 2, 4, 5, 6] Department of Naval Architecture, Faculty of Marine Technology, Institut Teknologi Sepuluh Nopember, Surabaya, Indonesia
[3] Department of Marine Transportation Engineering, Faculty of Marine Technology, Institut Teknologi Sepuluh Nopember, Surabaya, Indonesia

[1]E-mail: yuda.apri@its.ac.id, [2]E-mail: 6018221001@mhs.its.ac.id, [3]E-mail: snugroho@seatrans.its.ac.id, [4]E-mail: sjarief@na.its.ac.id, [5]E-mail: totoky@na.its.ac.id, [6]E-mail: debudip00@na.its.ac.id,

Abstract. Ship collisions are one of the marine accidents that can occur at any time, causing significant material and cargo losses. Lack of awareness of collision impacts and human factors are primary issues for ship collisions. Consequently, ship captains and technology have devised an imaginary area known as the ship domain. This ship domain is an imagined region used to define safe conditions around the ship. With the presence of a ship domain, collision risk can be reduced. This paper investigates the influence of ship domains on collision risk assessment during head-on encounters. To simulate realistic ship movement conditions, the manoeuvring mathematical group (MMG) model is used to predict ship movements. Additionally, fuzzy logic calculations and the closest point of approach (CPA) are employed to help identify the effectiveness of ship domains. Numerical simulations are conducted using three different ship domain shapes: egg, circle, and ellipse. The results indicate that the shape of the ship domain does not significantly influence collision risk assessment in head-on encounters. Instead, the division of the ship domain plays a more crucial role in evaluating ship collision risk.

Keyword: Ship collision, Ship Domain, Collision Risk.

1. Introduction

In maritime transportation, navigation safety is a top priority [1]. Ensuring safe navigation requires continuous analysis and assessment to better understand navigational situations [2]. One approach to enhancing the safety of navigation is through ship collision risk analysis. This analysis is crucial in evaluating the effectiveness and reliability of risk management in navigation safety [3]. As ship collisions become more frequent due to increasing traffic volume, they are considered a significant hazard in maritime transportation [4]. One approach to improving ship collision risk analysis is through the use of the ship safety domain (ship domain) concept. [5].

The concept of the ship domain is essential and practical in maritime traffic engineering, playing a key role in maritime traffic design and collision risk assessment [6]. A ship domain is typically defined as a safety buffer around a ship (whether it be the own ship (OS) or a target ship (TS)) that other vessels are advised not to enter, and this concept has evolved through contributions from numerous researchers over past year [7]. Despite its over 50-year history, the ship domain concept continues to be essential in maritime traffic engineering [8]. Therefore, identifying and refining the ship domain for improving collision risk assessment is important.

This paper aims to identify the most suitable ship domain shape for assessing the degree of ship collision risk in head-on encounters. Various ship domain shapes exist, but this paper focuses on analyzing three specific shapes using numerical simulations. The methodology includes predicting ship movements, calculating collision risk, determining the CPA, and defining the ship domain. The identification of the appropriate ship domain shape will be calculated using these methodologies and applying Convention on the International Regulations for Preventing Collisions at Sea (COLREGs) regulations for head-on situations. Finally, the results will estimate whether the ship domain significantly influences collision risk assessment.

2. Research methodology

2.1 Maneuvering mathematical group

In this study, to address the problem of predicting ship movements, the MMG (Maneuvering Mathematical Group) model is utilized to describe the hydrodynamic forces acting on the ship. The depiction of ship movement in the MMG consists of the hull, propeller, rudder, and other external components. This subsection will briefly explain how the MMG calculations are performed.

A ship has 6 degrees of freedom in motion; however, in this calculation, only 3 degrees of freedom are used: surge, sway, and yaw, which are applied on the horizontal plane. To describe the ship's movement, we can use a coordinate system as shown in Figure 1. The $o - x_o y_o$ coordinate system represents the horizontal plane (global), and the $G - xy$ coordinate system is a local coordinate system where G is the midpoint of the ship. In the local coordinate system, the x-axis points towards the fore of the ship, and the y-axis points towards the starboard side. The heading angle, denoted as ψ, is the angle between x_o and x. The rudder angle is represented by δ The velocities in the x and y directions are u and v respectively, and they represent the components of the ship's speed U in the coordinate system. The drift angle β is the angle between the ship's velocity vector and the longitudinal axis of the ship. N represents the yaw moment of the ship about the vertical axis through point G. X and Y are the forces acting on the ship along the x and y axes, respectively. The equations of motion for a ship can be expressed as shown in Equation 1 below:

$$\left.\begin{aligned} (m + m_x)\dot{u} - (m + m_y)vr &= X, \\ (m + m_y)\dot{v} - (m + m_x)ur &= Y, \\ (I_{zz} + i_{zz})r' &= N. \end{aligned}\right\} \quad (1)$$

Where X, Y, N are the components of forces in x_0 and y_0 axes and moment around the z-axis respectively. m, m_x, m_y are the mass and x- and y-axis components of the added mass of the ship. I_{zz}, i_{zz} are the moment and added moment of inertia of a ship. u, v are x- and y-axis speed components. And r is ship yaw rate. The right side of Equation 1 can be expressed as shown in Equation 2 below, derived from the MMG mathematical model:

$$\left.\begin{aligned} X &= X_H + X_P + X_R \quad , \\ Y &= Y_H + Y_R \quad , \\ N &= N_H + N_R \quad . \end{aligned}\right\} \tag{2}$$

Where the subscripts 'H', 'P', and 'R' represent the hull, propeller, and rudder, respectively. More detailed explanations regarding the calculation of ship motion can be found in the paper by Yasukawa H and Yoshimura Y [9].

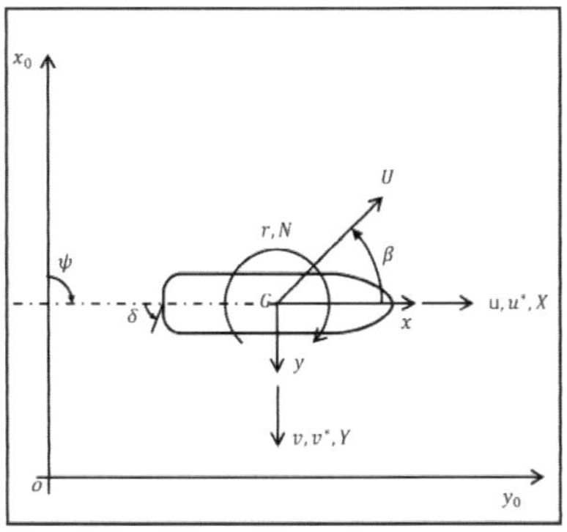

Figure 1. Ship coordinate system.

2.2 Calculation of collision risk

In this study, a parameter is used to determine the collision risk of a ship. This parameter is the Collision Risk (CR), which assigns a value to assess the risk of collision, with CR = 0.0 indicating safety and CR = 100 indicating danger. This parameter is used to evaluate the CR between ships. CR is calculated based on the positions and velocities of other ships around the own ship.

In this study, the CR is calculated using fuzzy inference because fuzzy functions can be defined by humans using linguistic variables. Therefore, the problem can be solved with limited parameters and a short computation time. This method is considered very suitable for calculating CR, especially when computation time is limited. An example of how fuzzy inference works can be seen in Equation 3:

$$\mu_A : X \rightarrow [0,1] \tag{3}$$

In Equation 3, there is a fuzzy set A with a membership function μ_A. Within the function $\mu_A(x)$, there is a value x that belongs to the set X, indicating that the element x is part of the fuzzy set. The membership value takes a real number between 0.0 and 1.0, and the closer the value of $\mu_A(x)$ is to 1.0, the higher the probability that the element x belongs to the fuzzy set A.

2.3 Closest point approach

The Closest Point of Approach (CPA) is the point at which two or more moving objects reach their closest distance based on their trajectories and current velocities, assuming no changes in trajectory or speed. CPA refers to the nearest point that the two moving objects will reach if they remain on their paths.

Generally, the formula for the CPA can be divided into two parts: Time to Closest Point of Approach (TCPA) and Distance to Closest Point of Approach (DCPA). These can be expressed in Equation 4 below:

$$\left.\begin{aligned} TCPA &= -(r.v)/v^2 \,, \\ DCPA &= r + TCPA.v. \end{aligned}\right\} \tag{4}$$

Where r is the distance between two objects and v is the speed between two objects.

TCPA and DCPA are used as parameters for calculating collision risk. With the development of these parameters, an approach using TCPA and DCPA can be created. The explanation of the TCPA and DCPA approach can be seen in Figure 2. According to Figure 2, DCPA is the closest distance of the own ship to the relative speed of the two ships. TCPA is the time it takes for the ship to reach the DCPA meeting position. It is assumed that both ships do not change their speed

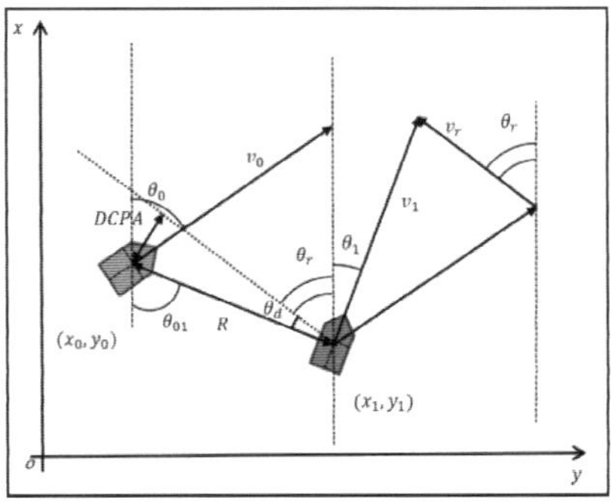

Figure 2. Determination of TCPA and DCPA in an encounter situation.

and course. The calculation of TCPA and DCPA can be expressed in Equation 5.

$$\left.\begin{aligned} V_{xr} &= V_1 \cos\theta_1 - V_0 \cos\theta_0 \,, \\ V_{yr} &= V_1 \sin\theta_1 - V_0 \sin\theta_0 \,, \\ R &= \sqrt{(y_1 - y_0)^2 + (x_1 - x_0)^2} \,, \\ \theta_{01} &= \tan^{-1}\frac{y_1 - y_0}{x_1 - x_0} \,, \\ \theta_r &= \tan^{-1}\frac{V_{yr}}{V_{xr}} \,, \\ V_r &= V_0 \cos(\theta_0 - \theta_r) - V_1 \cos(\theta_0 - \theta_r) \,, \\ DCPA &= R \sin(\theta_{01} - \theta_r) \,, \\ TCPA &= \frac{R\cos(\theta_{01} - \theta_r)}{V_r} \,. \end{aligned}\right\} \tag{5}$$

Where V_0 is OS speed and V_1 is TS speed. R indicates the distance between OS and TS, and θ_{01} is bearing angle of R. θ_0 and θ_1 are angles between x -axis and the course of OS and TS, respectively. V_{xr} and V_{yr} are components of relative speed V_r between two ships in the horizontal plane and θ_r is the bearing angle if the two components.

2.4 Ship domain

The concept of blocking area was initially used in the field of engineering marine traffic as a parameter to measure the amount of ship traffic around the own ship. This can be used to present the evaluation index for preventing ship collisions, with the value of the blocking area estimated by the ship operator. In this study, the blocking area, also referred to as the ship's domain, is used as one of the important parameters in calculating collision risk.

IOP Conf. Series: Earth and Environmental Science **1423** (2024) 012007 doi:10.1088/1755-1315/1423/1/012007

Research on ship domain has been conducted for a long time. The shape of the ship domain currently used is based on the development of research by Fujii [10], Goodwin [11], and Davis [12]. There are three common shapes often used: ellipse, circle, and three sections, as shown in Figure 3. In this study, these three shapes will be used with further developments based on previous research.

The egg ship domain, based on the research conducted by Daichi [13]. In this model, the domain is divided into three parts: fore, side, and aft. It is called the ship domain three section

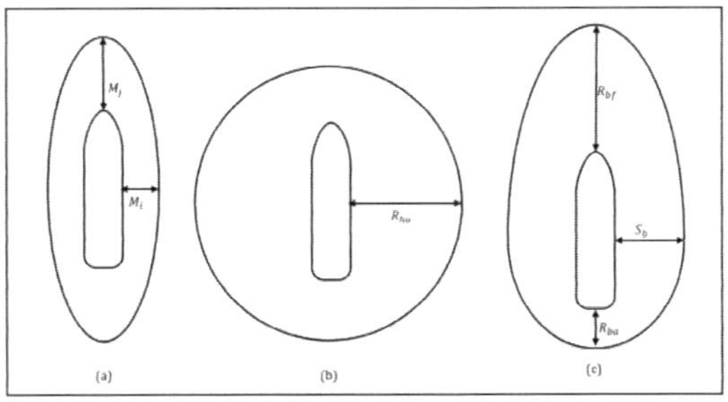

Figure 3. (a) ellipse ship domain, (b) circle ship domain, and (C) egg ship domain.

because each part does not have the same dimensions. The calculation of the egg ship domain can be described in Equation 6. Here, R_{bf} and R_{ba} are parameters used for the fore and aft regions, respectively. S_b is the common transverse radius for both the fore and aft regions. L is the length of the ship, B is the ship's breadth, and U is the ship's speed. T_{90} and D_T are the time to 90 degrees heading and the tactical diameter of the ship, respectively.

$$\left.\begin{aligned} R_{bf} &= L + T_{90} * U/2 \ , \\ R_{ba} &= L + T_{90} * U/4 \ , \\ S_b &= B + D_T \ . \end{aligned}\right\} \qquad (6)$$

The circular ship domain used in this study is based on the research conducted by Jinxian [14]. The shape of the circular ship domain is the same as shown in Figure 3, with the difference being the size of the circle. The radius length used can be seen in Equation 7, with each encounter have a different radius.

$$R_{Ho} = (5.625 - 0.612S_o - 0.602S_{B-G} - 0.561T_D - 0.495V_s - 0.245V_m)l_{ij} \qquad (7)$$

Where the subscripts 'Ho' represent head-on. S_o and S_{B-G} indicate the type of ship, T_D denotes the time condition, T_D and V_m represent the relative speed of the two ships, and l_{ij} is the average length of ships i and j.

The elliptical ship domain used in this study is based on the research conducted by Hee-Jin Lee [15]. This shape is derived from the experiences of mariners, making the ship domain more accurate. The shape of the domain is the same as in Figure 3. The dimensions of the ellipse to be used have a M_j = 5.9 * L and a M_i = 4.1 * L , where M_j is the major axes of the ellipse and M_i is the minor axes of the ellipse. The L refers to the length of the ship. With the parameters defined, the next step is to perform simulations according to the specified conditions.

3. Simulation

In section 3, the simulation conditions and the illustration of the simulations from this study will be explained.

3.1 Simulation conditions

For the collision risk simulation, it is essential to consider the ships used in the study as the OS and the TS. The sr108 cargo ship is chosen as both the OS and the TS. The principal dimensions of the ship can be seen in Table 1.

Table 1. Ship principal dimensions

	Container ship
Length (m)	3.0
Breadth (m)	0.435
Draft (m)	0.163
Block coef. (C_b)	0.5717

In addition to introducing the ships used, it is also necessary to outline the simulation conditions. In this study, simulations of encounters between the two ships will be conducted under head-on with the variations in ship domain explained in sub-section 2.4. Table 2 provides the initial positions, heading angles, and speeds of the OS and the TS.

Table 2. Initial condition for numerical simulations

	Own ship	Target ship
Position (x_0/L, y_0/L)	(-20, 0)	(20, 0)
Course (deg.)	0	180

3.2 Calculation of collision risk using fuzzy inference

Collision risk is evaluated using fuzzy inference in this paper. There are three variables involved: DCPA, TCPA, and collision risk. DCPA and TCPA are used as the antecedent parts of the fuzzy inference, while collision risk is the consequent part. Figure 4 shows the fuzzy set for the antecedent part, and Figure 5 shows the fuzzy set for the consequent part.

The fuzzy rules for the fuzzy inference system are shown in Table 3. The parameters used in the fuzzy inference are defined as follows: N (Negative value), ZO (Near zero), PS (Positive small value), PM (Positive medium value), and PB (Positive big value). The collision risk (CR) levels are: CR1 (Safe encountering situation, CR = 0.0), CR2 (Relatively safe encountering situation, CR = 25.0), CR3 (Relatively dangerous encountering situation, CR = 50.0), CR4 (Dangerous encountering situation, CR = 75.0), and CR5 (Situation in which ships may collide, CR = 100.0).

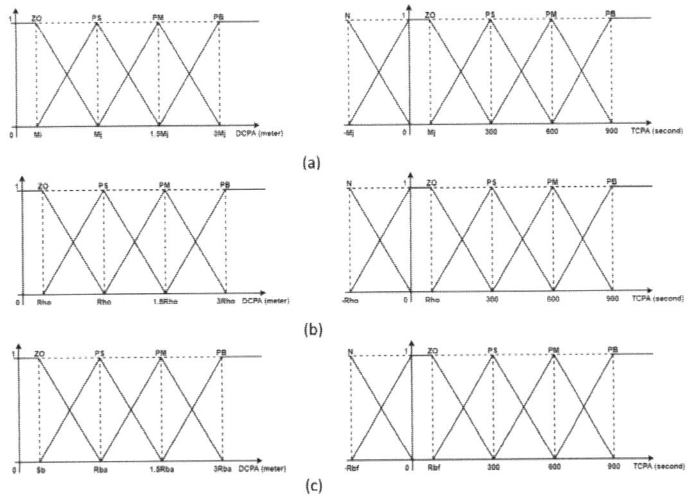

Figure 4. Membership functions of antecedent part (a) ellipse ship domain, (b) circle ship domain, and (c) egg ship domain

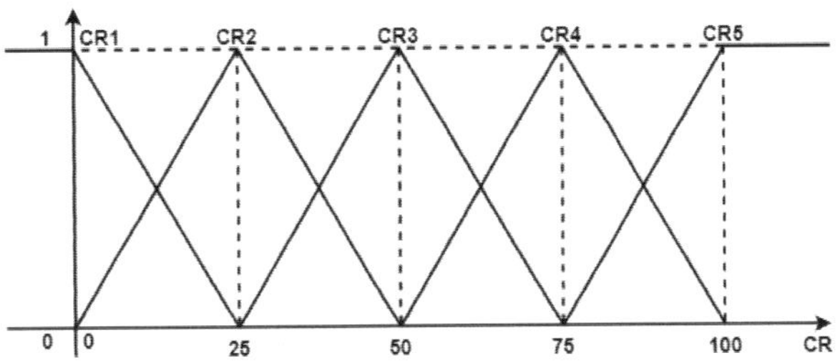

Figure 5. Membership functions of consequent part (CR)

Table 3. Fuzzy rule for obtaining the collision risk

Head - on		TCPA				
		N	ZO	PS	PM	PB
DCPA	ZO	CR1	CR5	CR3	CR1	CR1
	PS	CR1	CR3	CR2	CR1	CR1
	PM	CR1	CR1	CR1	CR1	CR1
	PB	CR1	CR1	CR1	CR1	CR1

These rules help determine the collision risk based on the values of DCPA and TCPA, ensuring a more accurate assessment of potential maritime collisions.

3.3 Head-on encounter

In this sub-section, an illustration of the simulation conditions for a head-on encounter between the two ships will be shown. Figure 6 depicts the simulation illustration. The OS will move with a heading angle of 0 degrees, while the TS will move with a heading angle of 180 degrees. The two ships are set on a collision course in this simulation to evaluate the collision risk values.

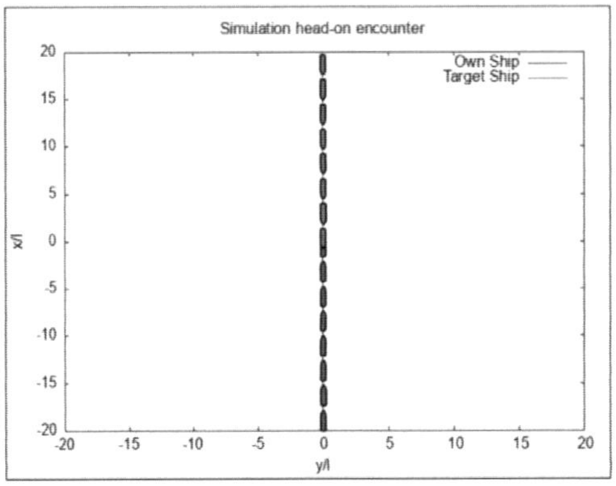

Figure 6. Simulation head-on encounter.

4. Simulation results and discussion

In section 4, the results of the simulations will be described. The simulation results include collision risk, ship positions, and a comparison of the results from each ship domain. This will allow for an evaluation of which ship domain is more effective in assessing collision risk.

Table 4 shows the dimensions of each domain obtained from the simulation calculations. For the elliptical domain, the calculated values are larger than those of the other ship domains. The elliptical domain shows larger dimensions compared to the circular and egg domains, indicating a potentially different collision risk assessment.

Table 4. Ship domain size

Ship domain	Major (m)	Minor (m)	Radius (m)
Egg	16.50 [(a)]	9.75[(b)]	4.98
Circle	-	-	11.90
Ellipse	17.70	12.30	-

[a]fore, [b]after.

4.1 Simulation results head-on encounter

Figure 7 shows the time history of CR, with the y-axis representing the CR values and the x-axis representing the time as the ships move. According to the number of ship domains, there are three lines representing the results from the simulations.

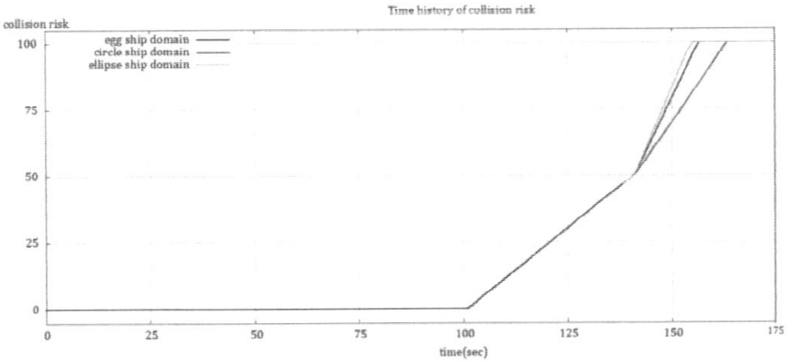

Figure 7. Time history of collision risk head-on encounter.

Collision risk values remain unchanged for each ship domain until around 100 seconds. As the ships approach each other, these values increase. At 140 seconds, differences between the domains become evident, as shown in Figure 8. The elliptical domain shows a faster increase in collision risk compared to the circular domain, which increases more slowly due to its smaller dimensions. Figures 9-11 illustrate the ship positions where collision risk differences occur and when collisions begin at a head-on encounter angle of 30 degrees.

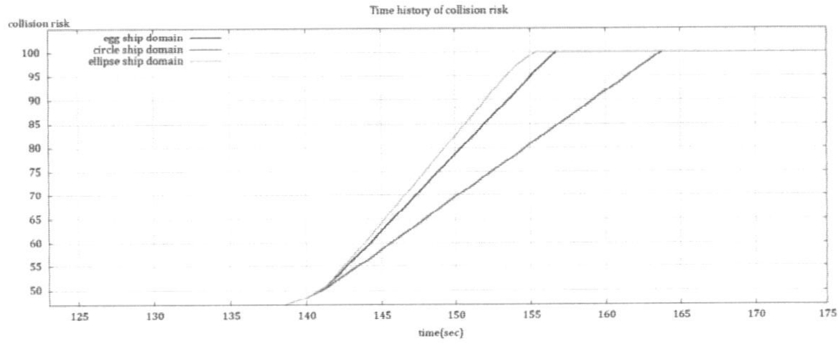

Figure 8. Time history of collision risk head-on encounter at 125 s – 175 s.

To better understand the influence of ship domains on collision risk assessment, we must consider both the increase in risk values and the positions of the vessels. Figure 9 shows a scenario where CR values rise simultaneously due to the calculation parameters. Larger TCPA values yield the same result, with significant changes occurring as TCPA values decrease. In scenarios with elliptical and egg ship domains, both vessels have already entered each other's domains. In contrast, for the circular domain, only parts overlap. This event occurred at 102.2 seconds, with the own ship at (0, -8.84) and the target ship at (-8.84, 0), resulting in a collision risk value of 1.54.

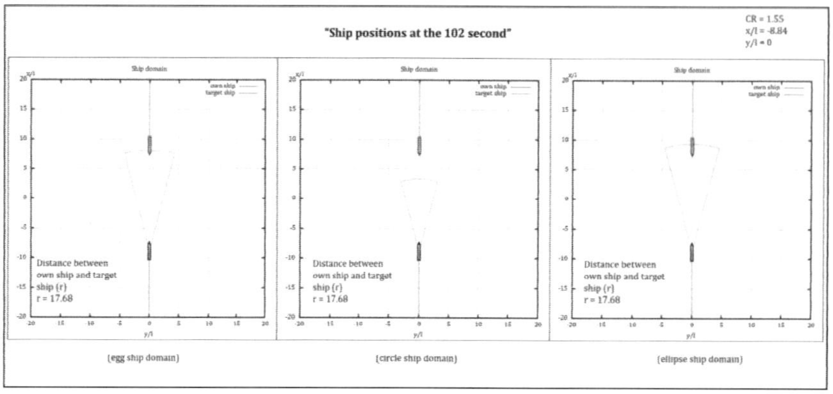

Figure 9. Ship position at the 102 sec.

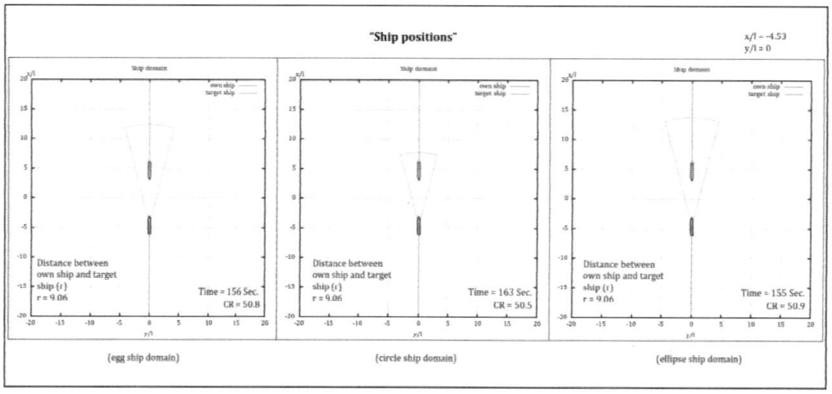

Figure 10. Ship position when value collision risk different.

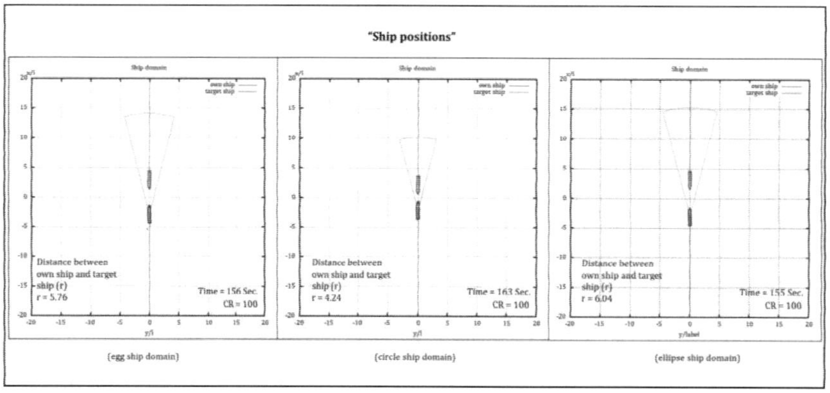

Figure 11. Ship position when value collision risk 100.

Figure 10 illustrates the ship positions at the moment when the CR value starts to rise for the first time, which varies across each domain. The elliptical domain experiences a quicker increase in risk due to its longer dimensions, leading to smaller margins for TCPA parameters. This occurs at 141.4 seconds, with the own ship at (0, -4.53) and the target ship at (-4.53, 0). At

IOP Conf. Series: Earth and Environmental Science **1423** (2024) 012007 doi:10.1088/1755-1315/1423/1/012007

this point, the collision risk values for the egg, circular, and elliptical domains are 50.8, 50.5, and 50.9, respectively. Although these differences are initially small, they become more noticeable over time, as shown in Figure 8.

In Figure 11 (egg ship domain), a collision appears inevitable as there's insufficient time for avoidance; the target ship has penetrated deep into its own ship's domain. In Figure 11 (circular ship domain), the ships are closer, increasing the likelihood of collision compared to the egg domain. Figure 11 (ellipse ship domain) shows the target ship still relatively distant, yet with a collision risk value already at 100.

Comparing these figures results at a collision risk value of 100, we find that the egg and circular domains offer more accurate assessments, given the proximity of the ships. However, the ellipse domain allows decision-making even at a collision risk value of 100. Each domain has its strengths and weaknesses. This simulation suggests a threshold parameter for collision risk calculation.

4.2 Discussion

The simulation results, shown in Figure 7, indicate that using the same parameters for every domain shape is ineffective, as it leads to the collision risk values rising simultaneously. This issue is illustrated in Figure 4, where the fuzzy sets for DCPA and TCPA have the same approach for all domain shapes. Each shape requires a different size-based approach. In a head-on encounter, only the front or major axes of the domain are involved. Hence, identifying major axis length divisions is crucial for accurate risk assessment in the fuzzy set of TCPA. In Namkyun Im paper [7], four zones are identified: safe, cautious, dangerous, and very dangerous. By establishing these thresholds, zones can be defined according to the divisions shown in Figure 12.

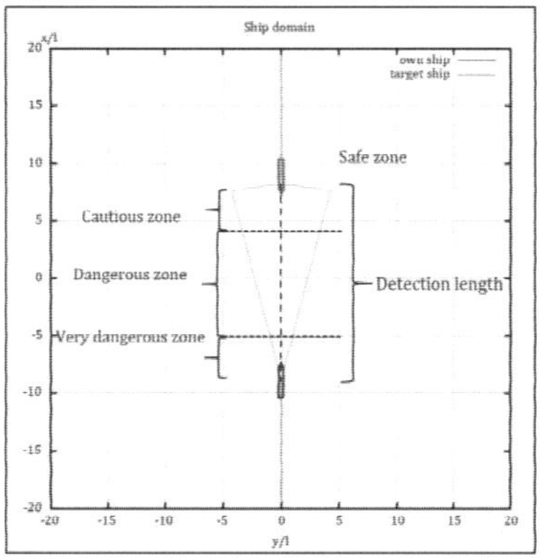

Figure 12. Ship domain zone.

Figure 7 shows increasing collision risk values, but only the ellipse and egg domains make contact with the target ship in Figure 9, while the circular domain does not. These results help establish threshold parameters for collision risk assessment, as illustrated in Figure 13 and Table 5. The principle used is that as TCPA and DCPA values decrease, CR for the target ship increases. When the target ship penetrates the dangerous or very dangerous zones, the TCPA and DCPA values are low. Using this basic principle, an approach can be developed to produce a collision risk assessment.

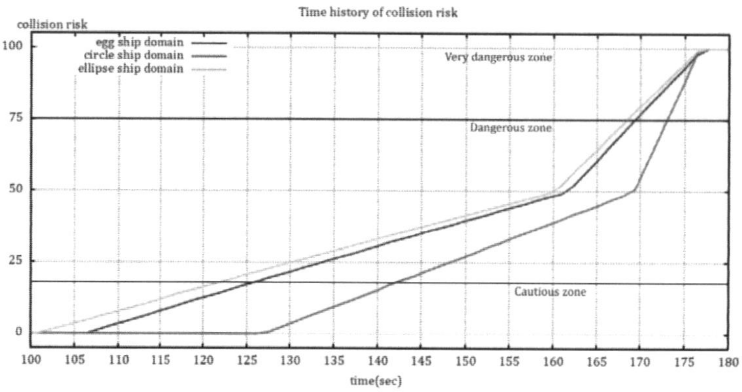

Figure 13. Ship domain zone.

Table 5. Ship domain thresholds zone.

Ship domain	Cautious zone	Dangerous zone	Very dangerous zone
Egg	$R_{bf} - 0.65R_{bf}$	$0.65R_{bf} - 0.18R_{bf}$	$0.18R_{bf} - 0$
Circle	$R_{ho} - 0.65R_{ho}$	$0.65R_{ho} - 0.20R_{ho}$	$0.20R_{ho} - 0$
Ellipse	$M_j - 0.64M_j$	$0.64M_j - 0.19M_j$	$M_j - 0$

Figure 14 shows the threshold zones for each ship domain shape. The range for these thresholds can be found in Table 6. The visualization demonstrates that the entry detection varies for each ship domain shape, leading to differences in collision risk values, as seen in Figure 13. This study indicates that in a head-on encounter, the longer the domain area in front of the ship, the quicker the detection. However, in terms of efficiency, the egg-shaped domain is more optimal because it is neither too long nor too short. Therefore, it is better to measure the appropriate size for the domain to be used in collision risk assessment.

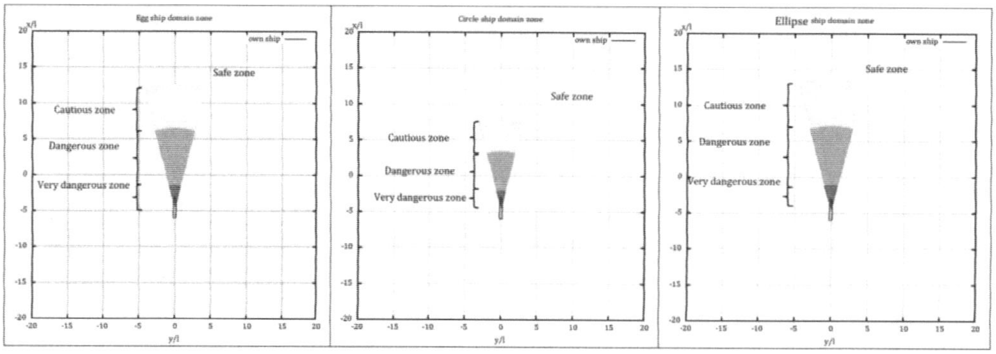

Figure 14. Ship domain zone for every shape.

Table 6. Collision risk value.

Ship domain	Cautious zone (sec.)	Dangerous zone (sec.)	Very dangerous zone (sec.)
Egg	106 – 133	133 – 169	169 – 182
Circle	127 – 147	147 – 172	172 – 182
Ellipse	100 – 130	130 – 168	168 – 182

5. Conclusion

Based on the numerical simulations conducted and the results presented in this paper, ship domain is a crucial parameter for accurately calculating collision risk. The shape of ship domain has a minor influence on collision risk assessment in head-on encounter; the primary factor is the length of major axes or the radius ship domain. The difference in the length of the major axes or radius is significant because, according to COLREGs regulations, a head-on encounter is defined by a 30-degree angle from the own ship's heading. This study found that the most influential factor in identifying collision risk is the division of each zone, where each zone provides a different assessment of the collision risk. Additionally, it was found that for safer detection of collision risk, the ellipse ship domain is advantageous due to its larger size in head-on encounters. For efficiency, the egg ship domain is recommended as it provides a more accurate assessment. The circular ship domain may be suitable for smaller and more agile vessels due to its balanced size on all sides.

Acknowledgement

This research is supported by Institut Teknologi Sepuluh Nopember (ITS) through ITS research fund with contract numbers: 1752/PKS/ITS/2023

References

[1] R. Fiskin, "An advanced decision-making model for determining ship domain size with a combination of MCDM and fuzzy logic," *Ocean Engineering*, vol. 283, Sep. 2023, doi: 10.1016/j.oceaneng.2023.114976.

[2] Z. Pietrzykowski and M. Wielgosz, "Effective ship domain – Impact of ship size and speed," *Ocean Engineering*, vol. 219, Jan. 2021, doi: 10.1016/j.oceaneng.2020.108423.

[3] K. Liu, Z. Yuan, X. Xin, J. Zhang, and W. Wang, "Conflict detection method based on dynamic ship domain model for visualization of collision risk Hot-Spots," *Ocean Engineering*, vol. 242, Dec. 2021, doi: 10.1016/j.oceaneng.2021.110143.

[4] J. Montewka, S. Ehlers, F. Goerlandt, T. Hinz, K. Tabri, and P. Kujala, "A framework for risk assessment for maritime transportation systems - A case study for open sea collisions involving RoPax vessels," *Reliab Eng Syst Saf*, vol. 124, pp. 142–157, 2014, doi: 10.1016/j.ress.2013.11.014.

[5] R. Szlapczynski and J. Szlapczynska, "A ship domain-based model of collision risk for near-miss detection and Collision Alert Systems," *Reliab Eng Syst Saf*, vol. 214, Oct. 2021, doi: 10.1016/j.ress.2021.107766.

[6] Z. Jingsong, W. Zhaolin, and W. Fengchen, "Comments on Ship Domains."

[7] N. Im and T. N. Luong, "Potential risk ship domain as a danger criterion for real-time ship collision risk evaluation," *Ocean Engineering*, vol. 194, Dec. 2019, doi: 10.1016/j.oceaneng.2019.106610.

[8] R. Szlapczynski and J. Szlapczynska, "Review of ship safety domains: Models and applications," *Ocean Engineering*, vol. 145, pp. 277–289, Nov. 2017, doi: 10.1016/j.oceaneng.2017.09.020.

[9] H. Yasukawa and Y. Yoshimura, "Introduction of MMG standard method for ship maneuvering predictions," *Journal of Marine Science and Technology (Japan)*, vol. 20, no. 1, pp. 37–52, Mar. 2015, doi: 10.1007/s00773-014-0293-y.

[10] Y. Fujii and K. Tanaka, "NO. 4 STUDIES IN MARINE TRAFFIC ENGINEERING £43 Traffic Capacity," 1971.

[11] E. M. Goodwin, "A Statistical Study of Ship Domains," 1975.

[12] P. V Davis, M. J. Dove, and C. T. Stockel, "A Computer Simulation of Marine Traffic Using Domains and Arenas," 1980.

[13] D. Ota, T. Masuyama, Y. Furukawa, and H. Ibaragi, "Development of Automatic Collision Avoidance System for Ships Using Reinforcement Learning," 2016.

[14] J. Weng, J. Du, K. Shi, and S. Liao, "Effects of ship domain shapes on ship collision risk estimates considering collision frequency and severity," *Ocean Engineering*, vol. 283, Sep. 2023, doi: 10.1016/j.oceaneng.2023.115070.

[15] H. J. Lee, Y. Furukawa, and D. J. Park, "Seafarers' awareness-based domain modelling in restricted areas," *Journal of Navigation*, vol. 74, no. 5, pp. 1172–1188, Sep. 2021, doi: 10.1017/S0373463321000394.

IOP Conf. Series: Earth and Environmental Science **1423** (2024) 012008 doi:10.1088/1755-1315/1423/1/012008

Probabilistic Modeling of Ship Collision in The Lombok Strait, Indonesia: A Bayesian Network Approach

Iis Dewi Ratih [1] Ketut Buda Artana [2], Heri Kuswanto [3], Emmy Pratiwi [4] and Athaya Raihan Zaki [5]

[1,2,4] Department of Marine Engineering, Institut Teknologi Sepuluh Nopember , Surabaya, Indonesia
[3] Department of Statistics, Institut Teknologi Sepuluh Nopember , Surabaya, Indonesia
[5] Depatment of Business Statistics, Institut Teknologi Sepuluh Nopember , Surabaya, Indonesia

* Email: kb.artana@gmail.com

Abstract. Ship collisions are a type of maritime accident that can result in significant losses due to their direct impact on the structural integrity of the vessels involved. Collisions can occur between two or more ships crashing into each other. Such accidents can cause severe damage, threatening the safety of passengers and crew on all vessels involved, and generally result in greater losses than single-ship accidents. According to KNKT (National Transportation Safety Committee of the Republic of Indonesia) investigations, ship collisions occurred in the Lombok Strait from 2007 to 2019. The Lombok Strait is part of ALKI (Indonesian Archipelagic Sea Lanes) II, which connects Australia and East Asia. Therefore, assessing the frequency of collisions in these waters is necessary to determine the prevalence of such incidents. If the rate is high, efforts can be made to minimize risks and ensure safety in the shipping lanes, particularly in ALKI. A Bayesian Network can provide information about the likelihood of ship collisions, facilitating decision-making to prevent or mitigate the risks posed by such incidents. The model's formation involves calculating the formed structure's prior, conditional, and joint probability of the formed structure. Thus, conclusions were drawn based on the model's predictions of ship collision chances in Indonesia, yielding an accuracy rate of 96.97%, a specificity of 90.00%, and a sensitivity of 100%. Based on Bayesian Network analysis and modeling, the causal probability values for the Lombok Strait are as follows: Pc Head-on, Pc Overtaking, and Pc Crossing are 2.85 x 10^-4, 1.03 x 10^-5, and 6.24 x 10^-5, respectively. Based on the Bayesian Network modeling results, the frequency of ship collisions in the Lombok Strait for each type of collision are 0.000026 collisions/year, 0.0000031 collisions/year, and 0.0000015 collisions/year, respectively.

1. Introduction

The number of ship visits from domestic and international origins at ports in Indonesia from 2013 to 2021 demonstrates that maritime transportation has been widely used for people's movement and the distribution of logistics and societal needs. However, marine transportation in Indonesia cannot be separated from a long history of ship accidents. According to investigation reports from the National Transportation Safety Committee of the Republic of Indonesia (KNKT), from 2007 to 2021, there were 156 reports of ship accident cases. Hasigian et al. [1] explained that the factors that determine the magnitude of losses experienced in ship accidents include the type or category of the accident, the cause of the accident, operating conditions, and the characteristics of the ship. Collisions, capsize, and sinking accidents are estimated to cause more severe ship damage than groundings. According to KNKT investigation reports, ship accidents caused by sinking, fire, and collisions are the most frequent in Indonesian waters. Approximately 24.56% of KNKT's investigation reports are collision cases, making it the type of accident with the highest percentage of investigations compared to other types of accidents. According to KNKT investigations, two ship collisions occurred in the Lombok Strait from 2007 to 2019. The Lombok Strait is part of ALKI II, which connects Australia and East Asia. Therefore, assessing the frequency of collisions in these waters is necessary to understand the prevalence of such incidents. If the rate is high, efforts should be made to minimize risks.

Dzikron and Yulianto [2] state that ship collisions are a type of ship accident that causes significant losses because they directly impact the ship's structural integrity. According to statistical data from the International Oil Pollution Compensation Fund (IOPCF) in 2006, collisions and sinkings accounted for 50% of environmental damage due to oil spills. Collisions can involve two or more ships crashing into each other. Collisions resulting in severe damage can threaten the safety of passengers and crew on board each vessel, leading to greater losses than single-ship accidents.

Research conducted by Rahman et al. [3] explains that the occurrence of ship collisions can be influenced by three main factors: natural factors (force majeure), human error, and other factors. The factors influencing ship collisions have indicators that can be used to determine the probability of collisions. Information obtained from factors suspected to influence the likelihood of ship collisions can be used to build a Bayesian network model. Jiang and Lu [4] explained that Bayesian Networks allow for complex risk analysis and prediction with many parameters. It can provide information about the probability of ship collisions, enabling decision-making to prevent or reduce the risks posed by ship collisions. In modeling the probability of ship collisions, a series of factors influencing ship accidents are arranged as network nodes, which will be analyzed for their relationships based on the chronology in KNKT investigation reports. Each node has a probability based on historical ship accident data, following the concept of conditional probability.

Research conducted by Ung [5] explains that the proposed Bayesian Network model has the strength to predict risks by considering specific ship characteristics and environmental conditions based on Posterior Probability analysis. According to research by Nivolianitou et al. [6], compared to other modeling techniques such as fault tree analysis and fuzzy modeling, Bayesian Networks have been accredited for several factors, such as their ability to combine data with expert knowledge, handle missing data, avoid over-fitting, and present causal relationships while also providing an easily understandable graphical representation. Kaptan et al. [7] also explained that, unlike regression and similar methods, Bayesian Network methods do not depend on a single output variable and can be inferred for all variables in the network. These features make it an effective tool for decision-making and analysis.

In conclusion, this study makes a significant contribution by employing a Bayesian Network approach to estimate the causation probability values for ship collisions in the Lombok Strait. This approach addresses the necessity to tailor these probabilities according to each maritime location's specific characteristics, as Montekwa et al. argued [8]. Adapting these values is crucial for a deeper understanding of the dynamics of ship accidents in diverse maritime regions.

2. Literature Review

2.1 Bayesian Network

Usadha [9] explains that a Bayesian Network is a straightforward Probabilistic Graphical Model (PGM) based on probability and graph theories. Probability theory directly relates to the data, while graph theory is associated with the desired representation. The Bayesian Network method is an effective approach in machine learning that relies on training data, using Conditional Probability as its foundation. Figure 1 illustrates an example of a Bayesian Network structure comprising several nodes and edges.

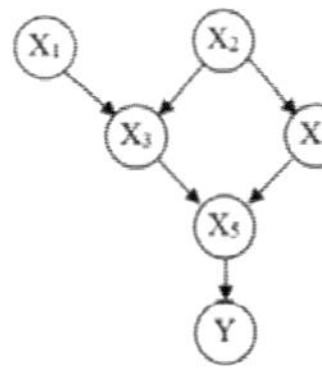

Figure 1. Example of Bayesian Network Structure.

A complete Bayesian Network model contains both qualitative and quantitative components. According to Wu et al. [10] the qualitative aspect of a Bayesian Network is a directed acyclic graph (DAG), in which nodes and edges represent system variables and their causal relationships. As illustrated in Figure 1, the qualitative components indicate that X_1 and X_2 are root nodes, meaning they do not have parent nodes. Subsequently, X_4 is a child node of the parent node X_2, and X_3 is a child node of the parent nodes X_1 and X_2. There are intermediate nodes, which can function as both parent and child nodes, such as X_3, X_4, and X_5.

The edge from X_1 to X_3 encodes the assumption that there is a direct causal dependency of X_1 on X_2; thus, X_1 is referred to as the parent node of X_3. No cycles occur within the graph (e.g., if there is an arc from X_1 to X_3 and from X_3 to X_5, then there is no arc from X_5 to X_1). This structure is used to avoid circular reasoning. According to Fenton and Neil [11], each node X_p has an associated probability, known as the node probability of X_p. The likelihood of an event X_p (prior probability) can be calculated using a simple probability function, as shown in Equation 1.

$$P\left(X_{p(A)}\right) = \frac{n_{p(A)}}{N} \tag{1}$$

Equation 1 shows the probability of the occurrence of category A for the p-th ship collision cause factor, where np(A) represents the number of occurrences of category A for the p-th ship collision cause factor, while N indicates the total number of occurrences for all categories of the p-th ship collision cause factor. The quantitative component of the Bayesian Network model is presented with a set of Conditional Probabilities for each child node, given the parent nodes in the network using Bayes' theorem. Conditional Probability refers to the calculation of the likelihood of event $X_p(A)$ given the occurrence of category B in the parent node, denoted as $P(X_p(A)|(parents(X_p(A))B)$. This theorem is utilized to determine the probability that a data set belongs to a particular class based on inference from existing data, as illustrated by the formula in Equation 2.

$$P\left(X_{p(A)}\middle|\left(parents\ X_p\right)_{(B)}\right) = \frac{n\left(X_{p(A)}|\left(parents\ X_p\right)_{(B)}\right)}{n\left(\left(parents\ X_p\right)_{(B)}\right)} \tag{2}$$

Suppose there are p random variables X_1, X_2, X_3 ...., X_p, a directed acyclic graph with ppp numbered nodes, and suppose node p−1 of the graph is associated with the variable X_{p-1}. In That case, this is called Joint Probability, the probability of the simultaneous occurrence of all events. The Joint Probability in a Bayesian Network is represented in Equations 3 and 4.

$$P(X_1, X_2, X_3, \dots, X_p) = \prod_{p=1}^{31} P(X_p | Parents(X_p)) \quad (3)$$

$$P(X_1, X_2, X_3, \dots, X_p) = P(X_1 | X_2, X_3, \dots, X_p) P(X_2 | X_3, \dots, X_p) P(X_{p-1} | X_p) P(X_p) \quad (4)$$

If additional information indicates that event Xp has occurred, there may be a change in the Prior estimate regarding the probability of the occurrence of parents (Xp). This is known as Posterior Probability, which can be calculated using Equation 5.

$$P\left((parents\ X_p)_{(B)} | X_{p(A)}\right) = \frac{P\left(X_{p(A)} \middle| (parents\ X_p)_{(B)} * (parents\ X_p)_{(B)}\right)}{P\left(X_{p_{(A)}}\right)} \quad (5)$$

2.2 Holdout Validation

Hermawan and Yoannita [12] explained that the holdout validation method randomly splits the data into two groups: training data and testing data. The Bayesian Network algorithm uses the training data to construct the classification tree, while the testing data is used to evaluate the created classification model. Estimation and prediction are then performed to determine the reliability of the classification model based on the training data.

The holdout validation method is selected to assess the Bayesian Network model. This method is the most commonly used approach for evaluating machine learning models. However, as Maleki et al. [13] argue, when the dataset used for training and evaluating the machine learning model is small, performance metrics derived from the validation and testing sets are sensitive to their composition, often resulting in unreliable performance measures.

2.3 Performance Evaluation Model

Imron [14] said the Performance Evaluation Model (PEM) is a phase designed to assess the system's performance. PEM is performed using the training data to evaluate the developed model. It is usually represented in a table called a confusion matrix, which specifically demonstrates the model's performance. Also explained by Saputro and Sari [15], each row of the matrix represents the actual class, while each column represents the predicted class of the data.

Accuracy is the ratio of correct predictions to the total number of target data points, representing the overall frequency with which the model correctly classifies instances. Sensitivity, also known as recall, is the ratio of correct positive predictions to the actual positive data points. Sensitivity indicates how often the model predicts a positive when the actual class is positive. Specificity measures the reliability of the model in correctly detecting negative-labeled data. Specificity describes the ratio of correctly predicting negative data points to the total number of negative data points. The formulas for accuracy, sensitivity, and specificity are as follows: Accuracy = (TP + TN) / n , Sensitivity = TP / (TP + FN) and Specificity = TN / (TN + FP)

2.4 Sensitivity Analysis

According to Anggela and Subekti [16], sensitivity analysis investigates how the outcomes of changes in parameter values can result in alterations in results or decisions. When parameter value changes occur, it can be interpreted that the decision is sensitive to parameter changes. In determining the level of sensitivity of a decision to the parameters influencing it, sensitivity analysis provides an overview of the extent to which a decision remains consistent despite changes in influencing parameters. One method of sensitivity analysis that can be used is the tornado diagram.

The tornado diagram in sensitivity analysis is a graphical representation used to visualize the impact of variations in input parameters on the output of a model or system. This diagram helps identify the parameters that have the most significant effect on the outcome, whether profit, loss or the performance of a process.

Byrne et al. [17] explained that the bars in a tornado diagram are stacked in descending order of width, indicating that variations at the top have the greatest effect on the population estimate, while variations in the input at the bottom have a relatively minor impact on the population estimate. The color of the bars in a tornado diagram indicates the direction of change in probability due to parameter variation: green can indicate a positive increase, while red can signify a negative decrease, as argued by Anik and Habib [18].

2.5 Frequency Analysis

According to research by Čorić et al. [19] estimating ship collision frequency can be done through statistical approaches, expert opinions, and models. However, statistical methods have weaknesses, mainly because past maritime traffic conditions may not reflect current or future conditions due to traffic volume changes or new safety regulations. Additionally, insufficient or non-standardized historical data in many navigation areas makes maritime accident statistics unreliable for valid risk assessments. New maritime technologies also add to the inaccuracy of using historical data. Therefore, quantitative models have become the more dominant and accepted method for estimating ship collision frequency because they address these weaknesses and provide measurable quantitative results for risk assessment and mitigation.

The IALA Waterway Risk Assessment Program (IWRAP MK II) is often used to study ship collision frequency. IWRAP is a comprehensive program that analyzes various aspects of ship traffic movements, including hydrographic conditions, the use of navigation channels, and collision risks. This program helps map ship movements to illustrate traffic density and calculate the number of ships at risk of collision. The calculation of collision frequency involves geometric probability values obtained through the IWRAP software and analyzed causal probabilities. The mathematical model used is

$$F = N_G \times P_c \tag{6}$$

where F represents the collision frequency, NG represents the number of ships at risk, and Pg is the geometric probability from the spatial distribution of ship traffic.

3. Research Methodology

3.1 Data Source

Data sources for engineering a ship collision prediction dashboard using the Bayesian Network include secondary data obtained from investigation reports by the National Transportation Safety Committee (KNKT) accessible at https://knkt.go.id/, and verdict reports from the Maritime Court session outcomes accessible at https://mahpel.dephub.go.id/. These sources provide data on 44 cases of ship accidents involving collisions from 2007 to 2021, involving 94 ships.

3.2 Research Variables

The research variables, operational definitions, and standard categorization of variables are sourced from investigation reports by the National Transportation Safety Committee (KNKT), verdict reports from the Maritime Court, COLREGS 1972 (International Regulations for Preventing Collisions at Sea), and Ministerial Decisions on transportation, as indicated in Table 1.

Table 1. Research Variables

Variable	Information	Variable	Information	Variable	Information
Y	Collision	X_{11}	Dual task	X_{22}	Maneuverability
X_1	Good seamanship	X_{12}	Crew Fatigue	X_{23}	Technical failure
X_2	Crew competence	X_{13}	Situational awareness	X_{24}	Ship type
X_3	Crew Leadership	X_{14}	Visual observation	X_{25}	Sealing Area
X_4	Crew communication	X_{15}	Daylight	X_{26}	Wind velocity
X_5	Understanding ship characteristics	X_{16}	Master presence on bridge	X_{27}	Wave height
X_6	Understanding environment	X_{17}	Understanding of navigation and communication sign	X_{28}	Ship length
X_7	Inexperience	X_{18}	Proper utilization of Navigation and communication	X_{29}	Ship breadth
X_8	Capacity in Decision making	X_{19}	Standarize Navigation and communication equipment	X_{30}	Ship draft
X_9	Crew health	X_{20}	Crew Responsiveness	X_{31}	Ship speed
X_{10}	Number of crew	X_{21}	Presence of Pilot		

3.3 *Analysis Steps*

The steps undertaken to determine the ship collision probability model using a probabilistic network framework are as follows: First, divide the ship collision data in Indonesia into two parts: training data and testing data using the holdout validation method. Next, a Bayesian network structure will be constructed using a directed acyclic graph with parent, root, and child nodes. Then, parameter estimation is performed by calculating initial, conditional, joint, and posterior probabilities for each variable node suspected to influence the ship collision probability. After obtaining the parameter estimates, validate the formed model with the testing data based on the confusion matrix. Conduct sensitivity analysis to determine whether the model is sensitive to changes in input variables. Finally, interpret the analysis results.

4. Result and Discussion

4.1 Modeling Ship Collision Cases Using Bayesian Network

Modeling ship collision incidents in Indonesia through Bayesian Network methodology encompasses several essential phases: constructing the Bayesian Network framework, determining Bayesian Network parameters, assessing model performance, and conducting sensitivity analysis.

The construction of the Bayesian Network structure for predicting ship collision incidents involves a thorough analysis of chronological data to identify influential factors and their interdependencies. This model development leverages factors from KNKT investigation reports, Maritime Court rulings, related studies, and expert assessments. The node representing collisions is the final element in the Bayesian Network, used to predict collision probabilities. Investigations and rulings indicate that critical factors influencing collision likelihood include seamanship

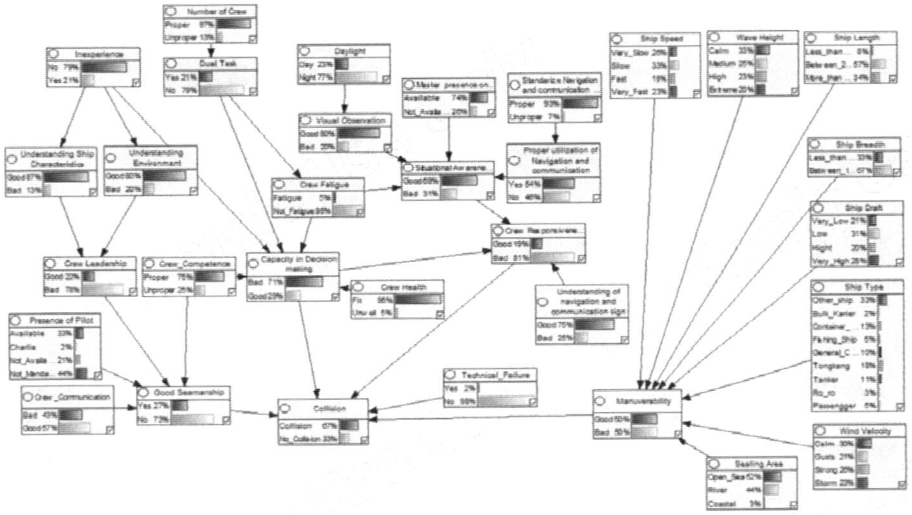

Figure 2. Ship Collison Opportunities Model Using Bayesian Network

proficiency, decision-making capacity, maneuverability, technical reliability, and crew responsiveness.

Before estimating the parameters of the Bayesian Network, the data is partitioned into training and testing sets using the Holdout Validation method, with a 70:30 ratio resulting in 61 training data points and 33 testing points. The parameter estimation process begins by calculating Prior probabilities for each root node, followed by Conditional Probabilities, Joint Probabilities, and Posterior Probabilities. After estimating these parameters, a Bayesian Network model is constructed to predict the probability of ship collisions. According to the model illustrated in Figure 3, the probability of a ship collision occurring is 67%, while the probability of a collision not occurring is 33%.

The performance evaluation of the Bayesian Network model uses training data to assess the constructed model by comparing testing data with model predictions. The comparison reveals one mismatch: the model predicted a collision for TK. Nova III & KT. Robi, which did not occur. The model's prediction results are summarized in a confusion matrix (Table 3), showing accurate collision predictions for 23 ships and accurate non-collision predictions for nine ships, with one incorrect collision prediction. The model's evaluation, based on accuracy, sensitivity, and

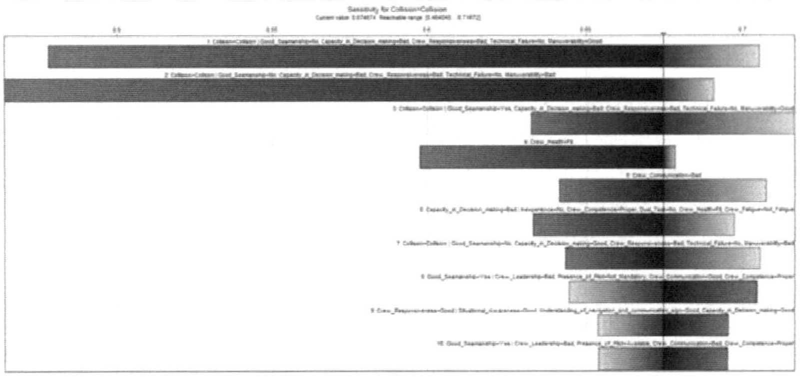

Figure 3. Tornado Diagram on Sensitivity Analysis

specificity values calculated from equations 6 to 8, indicates an accuracy of 96.97%, a sensitivity value showing no misclassified collision cases, and a specificity of 90%, reflecting precision in identifying non-collision ships accurately.

Sensitivity analysis is conducted to understand how parameter value changes affect decision probability. This analysis will use a tornado diagram to illustrate the top 10 parameter variations that most significantly affect the decrease or increase in the probability of ship collisions. The collision status node, categorized as a collision, will be set as the target node with a ±100% change in value for each scenario. The changes in probability values for each scenario result in the tornado diagram visualized in Figure 3.

Figure 3 displays the top 10 parameter variations that most significantly affect the probability of ship collisions. The upper part of the tornado diagram shows the parameter variations that most significantly impact the probability of ship collisions, indicated by variations in poor seamanship, poor decision-making, poor preventive actions, absence of engine failure, and good maneuverability. Increasing the probability variation to 100% will cause the probability of ship collisions to increase from 67.467% to 70.569%.

4.2 Causation Probability and Frequency Analysis

Causation probability (Pc) and frequency analysis constitute foundational tools for examining the relationships between cause and effect, alongside their associated probabilities in diverse contexts. Pc denotes the likelihood of an event occurring due to specific actions or conditions, while frequency analysis elucidates the patterns of occurrences within extensive datasets. Integrating these methodologies enables the exploration of intricate cause-and-effect dynamics and facilitates the estimation of outcome probabilities based on empirical evidence.

This research applies these principles to analyze ship collisions in the Lombok Strait, a crucial waterway in Indonesia. The study utilized Automatic Identification System (AIS) data on ship movements in 2022. The dataset underwent rigorous processing to generate a density map illustrating maritime traffic distribution within the Lombok Strait. This spatial representation was pivotal for assessing the frequency of different ship collisions. The analysis involved mapping waterway density, delineating maritime traffic patterns, and employing IWRAP MKII software to compute collision frequencies across various ship categories.

Table 2. Frequency Values of Ship Collisions in the Lombok Strait

Collision Type	Pc Default IWRAP	Pc Analysis	Frequency (Pc default IWRAP)	Frequency Analysis
Head-On	5 x 10^{-5}	2,85 x 10^{-4}	0,0000045	0,000026
Overtaking	1,1 x 10^{-4}	1,03x 10^{-5}	0,000033	0,0000031
Crossing	1,3 x 10-4	6,24 x 10^{-5}	0,0000032	0,0000015

From the mapping and density distribution analysis of the shipping lanes in the Lombok Strait, the collision frequency for each type of ship collision was obtained. Table 3 indicates that among the three types of ship collisions in the Lombok Strait, the most frequent collision type is Head-on, with a frequency of 0.000026.

5. Conclusion

The research conclusions are as follows based on the results described in the analysis and discussion.

1. Based on Bayesian network analysis and modeling, the causal probability values for the Lombok Strait are obtained as follows: Pc Head-on, Pc Overtaking, and Pc Crossing are 2,85 x 10^{-4}, 1,03x 10^{-5}, dan 6,24 x 10^{-5} with a model accuracy level of 96.97%, specificity of 90.00% and sensitivity of 100%.
2. The frequency of ship collisions in the Lombok Strait for each type of collision (Head-on, Overtaking, and Crossing) using the default values of Pc from IWRAP, which is 0.0000045 collisions/year, 0.000033 collisions/year, and 0.0000032 collisions/year, respectively. Meanwhile, based on the results of Bayesian Network modeling, the frequency values are 0.000026 collisions/year, 0.0000031 collisions/year, and 0.0000015 collisions/year, respectively.

Based on the analysis, discussion, and conclusions obtained, recommendations are provided to decision-makers for ships at sea and for future research: For decision-makers, to mitigate the

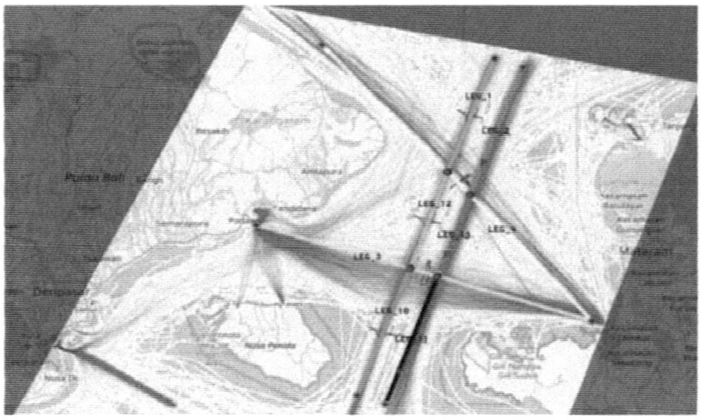

Figure 4. Traffic Density Mapping in Sunda Strait Waters Using IWRAP MKII

risk of ship collisions, it is essential to pay attention to good seamanship practices, decision-making processes, preventive actions, machinery failure, and maneuvering capabilities.

Reference

[1] S. Hasugian, A. A. I. Sri Wahyuni, M. Rahmawati, and A. Arleiny, 'Pemetaan Karakteristik Kecelakaan Kapal di Perairan Indonesia Berdasarkan Investigasi KNKT', *warlit*, vol. 29, no. 2, pp. 229–240, Jul. 2018, doi: 10.25104/warlit.v29i2.521.
[2] N. A. Dzikron and T. Yulianto, 'Analisis Tegangan Haluan Kapal Akibat Tubrukan', *JURNAL TEKNIK ITS*, 2019.
[3] H. Rahman, A. Satria, B. H. Iskandar, and D. A. Soeboer, 'PENENTUAN FAKTOR DOMINAN PENYEBAB KECELAKAAN KAPAL DI KESYAHBANDARAN UTAMA TANJUNG PRIOK', *CORE*, vol. 1, no. 3, pp. 277–284, Jul. 2018, doi: 10.29244/core.1.3.277-284.
[4] M. Jiang and J. Lu, 'The analysis of maritime piracy occurred in Southeast Asia by using Bayesian network', *Transportation Research Part E: Logistics and Transportation Review*, vol. 139, p. 101965, Jul. 2020, doi: 10.1016/j.tre.2020.101965.
[5] S. T. Ung, 'Navigation Risk estimation using a modified Bayesian Network modeling-a case study in Taiwan', *Reliability Engineering & System Safety*, vol. 213, p. 107777, Sep. 2021, doi: 10.1016/j.ress.2021.107777.
[6] Z. S. Nivolianitou, I. A. Koromila, and T. Giannakopoulos, 'Bayesian network to predict environmental risk of a possible ship accident'.
[7] M. Kaptan, Ö. Uğurlu, and J. Wang, 'The effect of nonconformities encountered in the use of technology on the occurrence of collision, contact and grounding accidents', *Reliability Engineering & System Safety*, vol. 215, p. 107886, Nov. 2021, doi: 10.1016/j.ress.2021.107886.
[8] J. Montekwa, F. Goerlandt, and P. Kujala, 'New Definition of a Collision Zone for a Geometrical Model for Ship-Ship Collision Probability Estimation, TransNav, the International Journal on Marine Navigation and Safety of Sea Transportation', vol. Vo.5, No.4, pp.497–504, 2011.
[9] N. R. Usadha, 'Implementasi Bayesian Network Untuk Perhitungan Probabilitas Pada Penilaian Risiko Pipa Bawah Laut oleh Faktor Kapal', *LIMITS*, vol. 14, no. 1, p. 61, May 2017, doi: 10.12962/limits.v14i1.2250.
[10] Z. Wu, Y. Shen, H. Wang, and M. Wu, 'Assessing urban flood disaster risk using Bayesian network model and GIS applications', *Geomatics, Natural Hazards and Risk*, vol. 10, no. 1, pp. 2163–2184, Jan. 2019, doi: 10.1080/19475705.2019.1685010.
[11] N. Fenton and M. Neil, 'Risk Assessment and Decision Analysis with Bayesian Networks (2nd ed.)', *Chapman and Hall/CRC*, 2018, doi: https://doi.org/10.1201/b21982.
[12] H. Hermawan and Y. Yoannita, 'Komparasi Metode Evaluasi Pada Credit Scoring Data Mining', *JTKSI*, 2018.
[13] F. Maleki, N. Muthukrishnan, K. Ovens, C. Reinhold, and R. Forghani, 'Machine Learning Algorithm Validation: From Essentials to Advanced Applications and Implications for Regulatory Certification and Deployment.', *Neuroimaging Clin N Am*, vol. 30, no. 4, pp. 433–445, Nov. 2020, doi: 10.1016/j.nic.2020.08.004.
[14] A. Imron, 'Analisis Sentimen Terhadap Tempat Wisata Di Kabupaten Rembang Menggunakan Metode Naive Bayes Classifier', 2019.
[15] I. W. Saputro and B. W. Sari, 'Uji Performa Algoritma Naïve Bayes untuk Prediksi Masa Studi Mahasiswa', *citec*, vol. 6, no. 1, p. 1, Apr. 2020, doi: 10.24076/citec.2019v6i1.178.
[16] I. C. Anggela and R. Subekti, 'ANALISIS SENSITIVITAS MODEL BLACK-LITTERMAN PADA PORTOFOLIO REKSA DANA SENSITIVITY ANALYSIS OF BLACK-LITTERMAN MODEL ON MUTUAL FUND PORTFOLIO'.
[17] A. W. Byrne, A. Parnell, J. O'Keeffe, and J. M. Madden, 'The challenge of estimating wildlife populations at scale: the case of the European badger (Meles meles) in Ireland', *Eur J Wildl Res*, vol. 67, no. 5, p. 85, Oct. 2021, doi: 10.1007/s10344-021-01528-9.
[18] M. A. H. Anik and M. A. Habib, 'COVID-19 and Teleworking: Lessons, Current Issues and Future Directions for Transport and Land-Use Planning', *Transportation Research Record: Journal of the Transportation Research Board*, p. 036119812311663, May 2023, doi: 10.1177/03611981231166384.
[19] M. Čorić, S. Mandžuka, A. Gudelj, and Z. Lušić, 'Quantitative Ship Collision Frequency Estimation Models: A Review', *JMSE*, vol. 9, no. 5, p. 533, May 2021, doi: 10.3390/jmse9050533.

IOP Conf. Series: Earth and Environmental Science **1423** (2024) 012009 doi:10.1088/1755-1315/1423/1/012009

Formal Safety Assessment (FSA) for ship collision in Musi River

TF Nugroho[1], KB Artana[2*], AAB Dinariyana[3], S Gurning[4], MAA Mas'ud[5]

Department of Marine Engineering, Institut Teknologi Sepuluh Nopember, Surabaya, Indonesia

*E-mail: taufikfajar@its.ac.id

Abstract. Accident data from the Palembang Port Authority Office Class II shows that the level of traffic accidents in the waters of the Musi River in Palembang is relatively high. Hazard identification was conducted on the Musi River, resulting in 25 ship accidents caused by collisions and 4 ship accidents due to grounding. Subsequently, risk assessment was performed by analysing the collision frequency and its consequences on human casualties. The collision frequency was calculated using IWRAP, yielding results of 8.18 x 10-2, 9.74 x 10-3, and 8.18 x 10-4 for the head-on, overtaking, and crossing scenarios, respectively. An analysis of the consequences in terms of human casualties estimated an average of 11 victims. Six options were proposed as risk control options (RCOs): public awareness and regulation implementation, Automatic Identification System (AIS) installation, crew education and training, standardization of safety equipment, utilization of pilotage services, and channel widening. Each RCO will be evaluated based on its cost and benefits. The final stage of the FSA involves recommendations aimed at selecting the best RCOs, considering the cost-benefit analysis (CBA). The optimal choice to minimize ship collisions and human casualties in the waters of the Musi River is to conduct crew education and training. This option is the most cost-effective due to its lowest Gross Cost value. The benefits gained amount to USD 890,273.02 with a 68% reduction in risk value.

1. Introduction

Navigation channels play a crucial role as access points for entering or exiting a port. They are vital elements of a port, enabling smooth operational processes and facilitating the functions and activities of the port (Arfian, 2017). According to the Republic of Indonesia Law No. 21 of 1992 concerning shipping, a navigation channel refers to a natural or man-made part of the water that is considered safe for navigation based on its depth, width, and other navigational obstacles. On the other hand, a narrow channel is defined as one that is smaller than the required size or width but still considered safe for navigation.

Palembang has a close relationship with the navigation channel of the Musi River. This river is the second longest river on the island of Sumatra. In the city of Palembang itself, the Musi River divides the city into two parts: Seberang Ilir in the north and Seberang Ulu in the south. The Musi River, serving as a backbone of water transportation in Palembang, holds significant potential and value for its maritime traffic.

The importance of river transportation in Palembang is evident from the activities of passenger and cargo ships sailing along the Musi River, facilitating the mobility of people and goods. The role of the Musi River as a transportation route cannot be underestimated. The Palembang City government continues to develop and expand the river transportation system, transforming it from merely an alternative mode of transportation to a simultaneous mode that grows alongside road transportation. Some navigation routes have naturally emerged due to the need for connectivity between different areas (Arianto, 2014).

In line with the successful development in South Sumatra province, particularly in the economic sector, new activity centres emerge, and existing ones expand, leading to increased mobility of people and goods. The Musi River, as the primary water transportation infrastructure in South Sumatra province, plays a crucial role in the region's development, particularly in the distribution of goods through river routes. Its significance is even greater now, as the Musi River has become a vital transportation route, especially for maritime shipping spanning approximately 111.12 km from the Port of Palembang to the river mouth (outer threshold) (Sugeng, 2010).

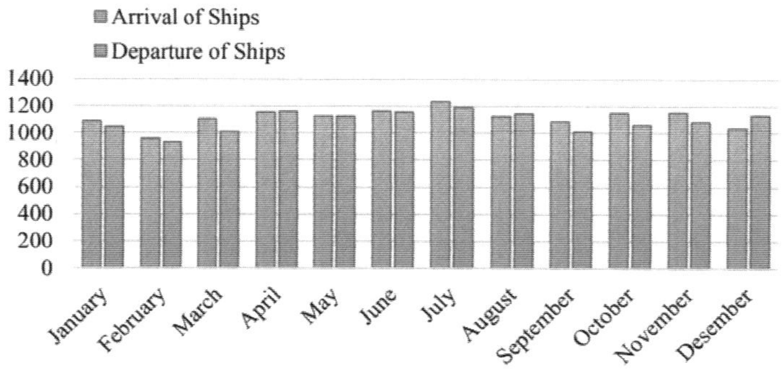

Figure 1. Ship Traffic Data in Musi River 2022 (KSOP Palembang).

The above image shows that the Musi River has a high traffic density. Based on data from the Class II Port Authority Office in Palembang, in 2022 there were 26,445 vessel movements consisting of incoming and outgoing ships. Besides passenger, cargo, and tanker ships, there are also barges carrying coal that pass through the Musi River. The combination of traffic density and the limited width and depth of the Musi River exposes vessels to the possibility of accidents, leading to emergency situations on board. The traffic density, coupled with the limited width and depth of the Musi River, exposes ships to a constant possibility of accidents that can lead to emergencies onboard. Emergency situations on ships are generally classified into 6 types, namely collision, fire or explosion, grounding, leakage, pollution, and man overboard (Gutanto, 2019) This traffic condition allows the potential for collisions on ships passing through and crossing the Lombok Strait which is very likely caused by navigation errors, human errors, weather conditions, and other technical problems (Dinariyana et al., 2020).

Ship collisions can have detrimental impacts on the environment, society, and economy (Dong & Frangopol, 2015). The level of traffic accidents in the waters of the Musi River in Palembang is relatively high.

According to merdeka.com, in 2018, there were seven accident cases resulting in 22 fatalities and eight injuries in the Musi River. The number of fatalities in 2018 showed an increase compared to the previous year, which recorded six deaths and 25 severe injuries. In 2021 and 2022, several accidents also occurred, and the number of accidents increased in 2022 compared to 2021.

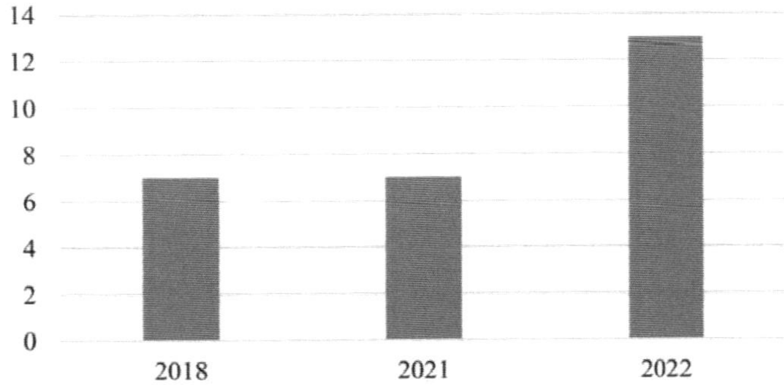

Figure 2. Number of Ship Accident in Musi River (KSOP Palembang).

2. Method

Formal Safety Assessment (FSA) is a structured and systematic methodology aimed at enhancing safety in the maritime world, including the protection of life, health, environment, and property, using risk analysis and cost-benefit assessment (IMO, 2018).

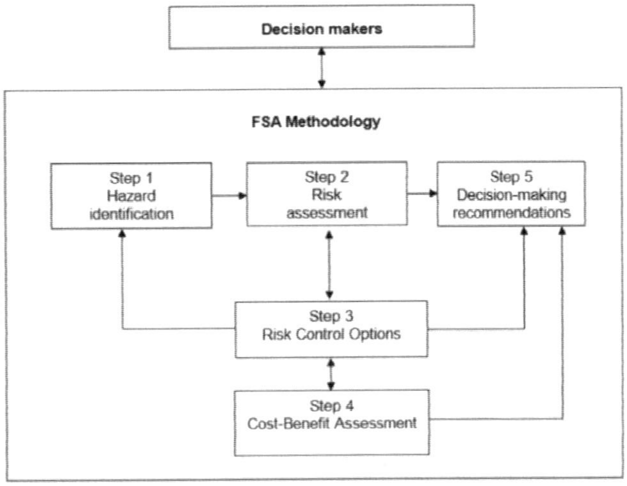

Figure 3. FSA Methodology (IMO, 2018)

The methodology which consists of 5 main steps, in which the steps are carried out in sequence and already consist of the evaluation of risk, mitigation of risk, and the economical assessment of risk mitigations. The outcome of an FSA study is recommendations which obtained from the study and could be implemented for decision makers in the practice. The flowchart of FSA as shown in Figure 3 would assess the implementation of routeing as Risk Control Option (RCO) are applied on to the same condition as before, where each reduced frequency of collision can be compared to one another. The primary parameter in comparing all RCOs is only based on the frequency, in which the higher the initial frequency could be reduced indicates the higher the effectivity of the RCO. But as FSA study also should give output in the form of recommendation, the selection of RCOs is based on the result of Cost Benefit Assessment (CBA) as to reflects which of the RCO is more applicable for the practice seen from economic perspective.

IOP Conf. Series: Earth and Environmental Science **1423** (2024) 012009 doi:10.1088/1755-1315/1423/1/012009

3. Result and Discussion

3.1 Hazard Identfication (FSA Step 1)

The Musi River spans about 111.12 km from its outer threshold to Boom Baru Port in Palembang and has a limited width due to its river nature. Despite this, it experiences high traffic density, with at least 13,385 incoming vessel movements. These vessels include passenger ships, container ships, tankers, tugboats, and barges.

Table 1. Ship Traffic Data in Musi River 2022 (KSOP Palembang)

Month	Arrival of Ships	Departure of Ships
January	1091	1049
February	956	933
March	1105	1010
April	1154	1160
May	1124	1123
June	1163	1156
July	1236	1193
August	1123	1145
September	1089	1012
October	1150	1062
November	1153	1086
December	1041	1131

The Musi River is approximately 111.12 km long, running from its outer threshold to Boom Baru Port in Palembang, with limited width due to its river nature. Despite this, it sees high traffic density, with at least 13,385 incoming vessel movements, including passenger ships, container ships, tankers, tugboats, and barges. Refer to Table 3 for detailed data on incoming and outgoing movements. Fig. 4 compiles data on each accident, while the Class II Sea Port Authority in Palembang provides reports for each accident, enabling identification of their causes. Fig. 5 displays the data on the causes of these accidents.

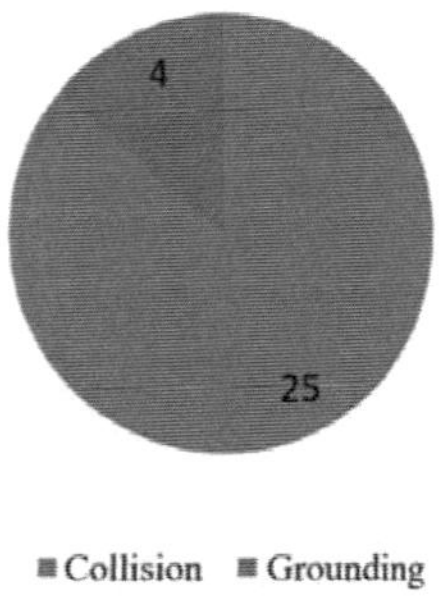

Figure 4. Graph of Ship Accident Type in Musi River.

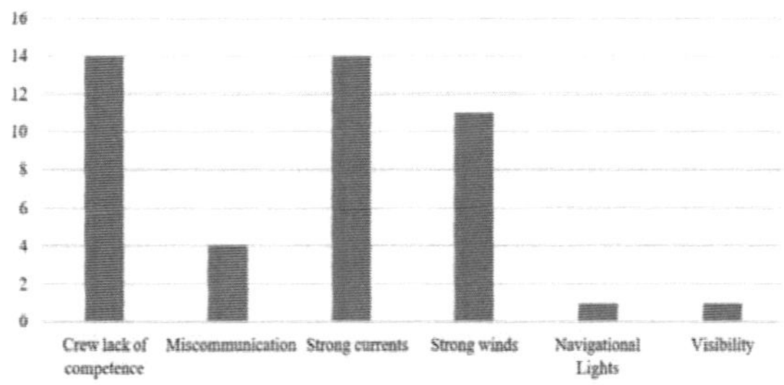

Figure 5. Graph of Ship Accident Type in Musi River.

Based on Fig. 5, which depicts the causes of accidents, it can be observed that the most significant factors contributing to accidents in the Musi River are strong currents and a lack of crew competence. Both factors are equally responsible for accidents in terms of their frequency.

IOP Conf. Series: Earth and Environmental Science **1423** (2024) 012009 doi:10.1088/1755-1315/1423/1/012009

3.2 Risk Assessment (FSA Step 2)

Risk Analysis is a systematic approach to assess system safety levels and recommend Risk Control Options (RCOs) using both frequency and consequence estimation (Khaled et al., 2018). IWRAP is a tool used for frequency estimation, combining calculation concepts into a single application to calculate collision frequencies accurately.

The next step after inputting AIS data from the Musi River is to design legs. The frequency research area spans from Benteng Kuta Besak Pier to Sungai Lais Pier, with leg designs shown in Figure. 6 and Figure. 7. Figure 6 shows leg designs based on Musi River ship traffic. The design includes several locations specifically intended for crossing conditions. This is due to the presence of anchor mooring areas located across from the destination port and intersecting the navigational channel of the Musi River, making it potentially hazardous for crossing.

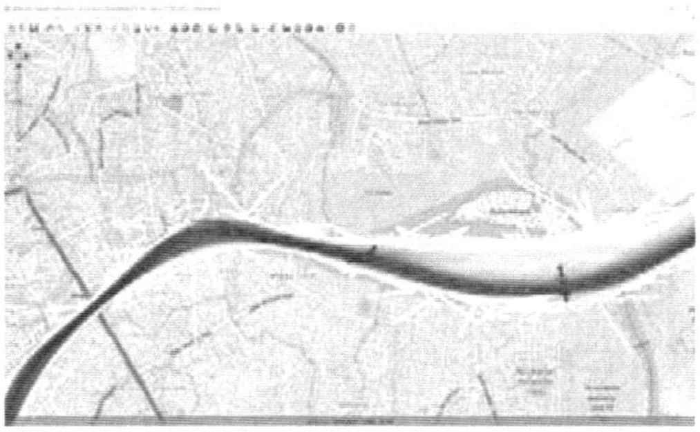

Figure 6. Leg Design

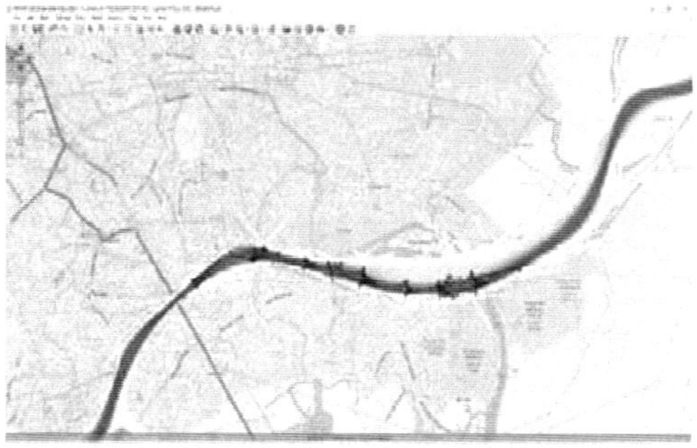

Figure 7. Leg Design and Distribution

The calculation process is assisted by the IWRAP application. The frequency values for collisions can be seen in Table 4.

Table 2. Collision Frequency

Collision Type	Frequency	Category
Head-on	8.18×10^{-2}	High Likelihood
Overtaking	9.74×10^{-3}	Moderate Likelihood
Crossing	8.18×10^{-4}	Low Likelihood

Each collision type is categorized based on its frequency value using the standards provided by the Biro Klasifikasi Indonesia (BKI) in the Guidelines for Domestic Ship Risk Assessment Part 8, Vol. A (2023). The estimation of fatalities can be obtained as shown in Table 3.

Table 3. Potential Number of Fatalities

Hull Damage	Survivability	Estimated Number of Fatality
Minor	Stay afloat	0
	Sink slowly	1
	Sink rapidly	30
Critical	Stay afloat	0
	Sink slowly	1
	Sink rapidly	30

From the total potential number of fatalities in each category of hull damage and survivability, the average potential number of fatalities is determined to be 11 people. The frequency values for each collision type and the estimated number of fatalities are inputted into a risk matrix that follows the standards provided by the BKI in the Risk Assessment Guidelines for Domestic Ships Part 8, Vol. A.

The frequency value for head-on collision type is 0.0818, rounded to 0.1, which, according to the risk assessment guidelines for domestic vessels by the Indonesian Classification Bureau, falls into the category of high probability. As for the estimation of casualties, based on risk acceptance criteria, it falls into the catastrophic category. The frequency value for overtaking collision type is 0.009743, rounded to 0.01, which, according to the risk assessment guidelines for domestic vessels by the Indonesian Classification Bureau, falls into the category of moderate probability. As for the estimation of casualties, based on risk acceptance criteria, it falls into the catastrophic category.

Table 4. Head-On Risk Matrix

Risk Index					
FI	Frequency	1	2	3	4
		Minor	Significant	Severe	Catastrophic
7	Very Frequent	8	9	10	11
6	Frequent	7	8	9	10
5	High Likelihood	6	7	8	HEAD ON
4	Moderate Likelihood	5	6	7	8
3	Low Likelihood	4	5	6	7
2	Remote	3	4	5	6
1	Extremely Remote	2	3	4	5

3.3 Risk Control Option (FSA Step 3)

Based on the risk assessment conducted in the previous stage, it was determined that head- on collisions and overtaking fall into the red category, which is unacceptable. Therefore, the proposed risk control option is as follows:

1. Implementation and socialization of regulations and law enforcement.
2. Utilizing the Automatic Identification System (AIS) integrated with the Vessel Traffic System (VTS) in narrow waters.
3. Conducting education and training for the crew. Education and training for the crew ensure that the ship's crew receives adequate education and training regarding ship navigation in narrow waters. It is expected that this risk control measure will help improve situational awareness, the ability to operate navigation equipment, and make informed decisions to avoid collision risks.
4. Implementing strict maritime safety regulations and standards by adopting and adhering to the regulations and standards established by maritime authorities and international organizations. Ensuring that the vessel has sufficient safety equipment, follows prescribed procedures, and undergoes regular inspections and examinations.
5. Utilizing experienced ship pilots to assist vessels navigating through narrow waters. Ship pilots possess in-depth knowledge of water conditions and safe routes.
6. River widening will provide greater manoeuvring space for vessels, thus reducing the risk of collisions between passing ships. Widening will be carried out only in the two legs with the highest frequency values.

3.4 Cost-Benefit Assessment (FSA Step 4)

In this stage, a cost analysis needs to be conducted, where each risk control option has its own implementation costs, that can be found in Table 5.

Table 5. Cost of Each Risk Control Option

No	Risk Control Option	Cost
1	Implementing and socializing regulations.	USD 4,733.13
2	Automatic Identification System (AIS)	USD 2,996.50
3	Crew education and training	USD 6,263.34
4	Safety equipment standardization	USD 37,868.47
5	Utilizing pilotage services	USD 51,829.73
6	Widening channel legs 1 and 3	USD14,897,318.98

There are several references used to determine the needs and prices for each detail. The references for the requirements include the standardization of safety equipment based on PM 25 of 2015 regarding the Safety Standards for River, Lake, and Crossing Transportation. The cost for each detail of safety equipment standardization is based on the prices listed on online marketplaces. The cost of crew education and training is based on the training tariff set by the Centre for Human Resources Development in Maritime Transportation, Regional I (West), according to Government Regulation No. 11 of 2015 regarding Non-Tax State Revenue (PNBP) at the Ministry of Transportation. The cost of using pilotage services is based on the applicable tariff on the Musi River, as agreed upon by the Indonesian National Shipowners Association (INSA), Palembang Branch. The cost of channel widening includes land clearance and dredging expenses, based on the Ministry of Transportation Regulation No. PM. 83 of 2011 regarding the Standard Cost of the Ministry of Transportation for the Fiscal Year 2012.

In addition to the cost analysis, it is important to determine the benefits obtained from each risk control option. Therefore, it is necessary to first understand the economic benefits received by the vessel if accidents resulting from the risk control options are avoided. The economic benefits received by the vessel in this study are based on the example of KM. Bahari Express 3B (170 passengers), and the calculations include annual revenue and compensation costs for loss of life. The revenue for each class per trip can be seen in Table 6.

Table 6. Total Revenue for each class

Class Types	Price	Revenue (per trip)
VIP (34 people)	USD 20.50	USD 697.10
Executive (51 people)	USD 17.17	USD 875.90
Business (85 people)	USD 15.11	USD 1284.43

Table 7. Estimated Revenue

Estimated Total Revenue	Amount
One trip	USD 2,857.43
One week (6 trips)	USD 17,144.56
One year (52 weeks)	USD 891,517.26

Compensation Cost Calculation Estimated Loss of Life = 11 people

Compensation per person = USD 3,328.39

Estimated Compensation = USD 36,612.34

Total Economic Benefit = Estimated Revenue (one year) + Estimated Compensation = USD 928,129.60

By knowing the value of economic benefits received by the vessel, we can also determine the benefits received by the vessel from each proposed risk control option. The benefit itself is the economic gain received by the vessel minus the cost of the risk control option.

Table 8. Estimated Revenue

Risk Control Option	ImplementationCost	Economic Benefit	Benefit
Implementing and socializingregulations.	USD 4,733.13	USD 928,129.60	USD 923,397.96
Automatic Identification System (AIS)	USD 2,996.50	USD 928,129.60	USD 925,094.11
Crew education and training	USD 6,263.34	USD 928,129.60	USD 921,868.23
Safety equipment standardization	USD 37,868.47	USD 928,129.60	USD 890,273.02
Utilizing pilotage services	USD 51,829.73	USD 928,129.60	USD 876,316.14
Widening channel legs 1 and 3	USD 14,897,318.98	USD 928,129.60	-USD 13,964,513.61

Each risk control option has a risk reduction value. The risk reduction value in this calculation is obtained from the implementation of each risk control option based on historical data obtained from the Class II Harbor Master's Office in Palembang. As for the channel widening, the frequency value is obtained from the new frequency data obtained from the IWRAP application. Risk reduction value can be seen in Table 9

Table 9. Estimated Revenue

Risk Control Option	Reduction in Collision Risk and Loss of Life
Implementing and socializing regulations.	43%
Automatic Identification System (AIS)	4%
Crew education and training	68%
Safety equipment standardization	21%
Utilizing pilotage services	7%
Widening channel legs 1 and 3	48%

In determining the cost-effectiveness, the Gross Cost of Averting Fatality (GCAF) is considered. GCAF is used to assess the cost-effectiveness of a risk control option in reducing the likelihood of collisions and preventable loss of life. Additionally, there is the Net Cost of Averting Fatality (NCAF), which considers the economic benefits of a risk control option in reducing the likelihood of collisions and preventable loss of life. The GCAF and NCAF values for each risk control option can be found in Table 10.

Table 10. Estimated Revenue

Risk Control Option	Gross CAF	Net CAF
Implementing and socializing regulations.	USD 11,040.51	-USD 2,154,595.23
Automatic Identification System (AIS)	USD 83,875.55	-USD 25,903,753.25
Crew education and training	USD 9,227.29	-USD 1,358,542.64

Risk Control Option	Gross CAF	Net CAF
Safety equipment standardization	USD 176,664.06	-USD 4,154,607.40
Utilizing pilotage services	USD 725,388.53	-USD 12,268,425.87
Widening channel legs 1 and 3	USD 31,026,339.94	USD 29,092,736.61

3.5 Cost-Benefit Assessment (FSA Step 4)

According to the Formal Safety Assessment (FSA) guidelines based on the International Maritime Organization (IMO), in principle, GCAF (Gross Cost of Averting Fatality) and NCAF (Net Cost of Averting Fatality) can be used as considerations. However, it is recommended to first consider GCAF before NCAF.

Overall, the recommended risk control option is crew education and training, as it has the lowest GCAF value. The reduction in collision risk and loss of life will certainly have an impact on the new risk matrix.

Table 11. Head-On Risk Matrix After RCO

Risk Index					
FI	Frequency	1	2	3	4
		Minor	Significant	Severe	Catastrophic
7	Very Frequent	8	9	10	11
6	Frequent	7	8	9	10
5	High Likelihood	6	HEAD ON	8	9
4	Moderate Likelihood	5	6	7	8
3	Low Likelihood	4	5	6	7
2	Remote	3	4	5	6
1	Extremely Remote	2	3	4	5

Table 11 is a new risk matrix for the head-on collision scenario. As seen in Table 11, the values for the head-on collision scenario fall within the yellow region, indicating that it is categorized as "As Low as Reasonably Practicable" (ALARP).

The new risk matrix for the overtaking collision scenario falls within the yellow region, indicating that it is categorized as "As Low as Reasonably Practicable" (ALARP), also the new risk matrix for the crossing collision scenario fall within the yellow region, indicating that it is categorized as "As Low as Reasonably Practicable" (ALARP).

4. Conclusion

The conclusions from the research findings are as follows:

1. Hazard identification was conducted using historical data obtained from the Class II Port Authority Office in Palembang. The results revealed several hazards associated with ship traffic density in the waters of the Musi River. The available data recorded 29 ship accidents, of which 25 were ship collisions and 4 were ship groundings. The accidents on the Musi River were predominantly caused by lack of competence among the ship's crew and weather factors such as strong currents.
2. Based on the risk matrix from Biro Klasifikasi Indonesia (BKI), the head-on scenario fell into the high-risk category, the overtaking scenario fell into the moderate-risk category, and the crossing scenario fell into the low-risk category. In terms of consequences, according to the risk

matrix from Biro Klasifikasi Indonesia (BKI), was classified as a catastrophic. When combining frequency and consequences, the risk level for ship collisions on the Musi River was considered unacceptable for the head-on and overtaking scenarios, while the crossing scenario fell into the As Low as Reasonably Practicable (ALARP) risk level.

3. The best three proposed risk control options are as follows:
   a. Implementation and socialization of regulations and law enforcement.
   b. Use of an integrated Automatic Identification System (AIS) with the Vessel Traffic System (VTS) in narrow waters.
   c. Education and training of crew members to ensure they receive adequate education and training regarding ship navigation in narrow waters. This is expected to improve situational awareness, navigation equipment operation skills, and decision-making abilities to avoid collision risks.
   d. Utilization of experienced ship pilots to assist vessels navigating through narrow waters. Ship pilots have in-depth knowledge of water conditions and safe routes.
4. Each risk control option has its associated implementation cost. Each risk control option also has its own risk reduction value. Socialization and regulation implementation option has a risk reduction value of 43%, AIS option has a risk reduction value of 4%, crew education and training option has a risk reduction value of 68%, standardization of safety equipment has a risk reduction value of 21%.
5. Based on the cost-benefit analysis conducted, the best recommendation is crew education and training. This option provides the most cost-effective solution as it has the lowest Gross Cost of Averting Fatality (GCAF) value. The benefit value of implementing crew education and training is USD 921,868.23 with a risk reduction value of 68%.

References

[1] Arfian, Z. (2017). Penilaian Risiko Tubrukan Kapal Akibat Instalasi Anjungan Lepas Pantai di Dekat Alur Pelayaran Barat Surabaya

[2] Arianto, S. B. (2014). EVALUASI KEBUTUHAN RAMBU DI SUNGAI MUSI KOTA PALEMBANG. 16.

[3] Biro Klasifikasi Indonesia. (2023). Petunjuk Penilaian Risiko Kapal Domestik.

[4] Dinariyana, A. A. B., Avesina, M. A., Sambodho, K., & Fahrirozan, F. (2020). Collision Risk Analysis and Oil Spill Dispersion Modelling in Lombok Strait. IOP Conference Series: Earth and Environmental Science, 557(1), 012002. https://doi.org/10.1088/17551315/557/1/012002

[5] Dong, Y., & Frangopol, D. M. (2015). Probabilistic ship collision risk and sustainability assessment considering risk attitudes. Structural Safety, 53, 75–84. https://doi.org/10.1016/j.strusafe.2014.10.004

[6] Gutanto, M. B. (2019). Analisa Penyebab Kandas pada Kapal MT. Success Altair XLII di Outer Buoy Sungai Musi. Politeknik Ilmu Pelayaran Semarang.

[7] International Maritime Organization. (2018). Revised Guidelines for Formal Safety Assessment (FSA) for Use in the IMO Rule-Making Process.

[8] Kementerian Perhubungan. (2015). Tarif Diklat Keterampilan Pusat Pengembangan SDM Pehubungan Laut Wilayah I (Barat) dalam PP No.11 Tahun 2015 tentang Penerimaan Negara Bukan Pajak (PNBP) Pada Kementerian Perhubungan.

[9] Khaled, M. E., Kawamura, Y., Abdullah, M. S. B., Banik, A., Faruk, A. K., & Khan, M. R. (2018). ASSESSMENT OF COLLISION & GROUNDING RISK AT.

[10] Menteri Perhubungan Republik Indonesia. (2015). Peraturan Menteri Perhubungan Republik Indonesia Nomor: PM 25 Tahun 2015 tentang Standar Keselamatan Transportasi Sungai, Danau, dan Penyeberangan.

[11] Sugeng, S. (2010). DAMPAK PELAYARAN KAPAL LAUT DI ALUR SUNGAI MUSI.

IOP Conf. Series: Earth and Environmental Science **1423** (2024) 012010 doi:10.1088/1755-1315/1423/1/012010

Maintenance Priority: A Literature Review of Equipment Criticality Analysis in the Oil and Gas Industries

D Priyanta[1], M B Zaman[1] and Semin[1]

[1] Department of Marine Engineering, Institut Teknologi Sepuluh Nopember (ITS), Surabaya 60111, Indonesia

*E-mail: priyanta@its.ac.id

Abstract. Criticality analysis is a fundamental tool in the oil and gas industry to ensure the efficiency and safety of equipment. This paper presents a comprehensive literature review on current methodologies for identifying safety critical equipment (SCE) and critical equipment (CE). The review begins by discussing the initial step of identifying SCE, followed by the identification of CE. Various methodologies are examined, ranging from simple, quick screening techniques to complex, quantitative analyses. Simple methodologies are often used for initial assessments, while complex methods are applied to equipment posing high risks that require detailed, quantitative calculations. The review highlights the strengths and limitations of each approach, providing a comparative analysis of their effectiveness in different scenarios. Finally, the paper suggests a structured methodology for developing maintenance tasks tailored specifically for SCE and CE, ensuring optimal maintenance strategies that enhance safety and operational efficiency. This review aims to guide practitioners in selecting appropriate methodologies for criticality analysis and maintenance planning in the oil and gas sector.

1. Introduction

The oil and gas (O&G) industries commonly process crude and natural gas into customer-ready products. This process can take place in the central processing plant, involving many equipment ranging from static, rotating, and automatic control systems. This complex production operation shall be maintained by managing the integrity of all equipment. Therefore, implementing asset integrity management (AIM) is strongly recommended. AIM will identify and address potential risks, develop steps to manage and reduce the risks, monitor the condition of the assets, and ensure the safety and efficiency of the assets [1].

Equipment maintenance is one of the major ways to get the machine to operate properly. Equipment screening serves as a basis for prioritizing tasks according to their criticality or importance level and helping to manage risks within an organization. For example, equipment can be classified into safety-critical (SCE), critical (CE), or non-critical (NCE) categories, after which maintenance tasks may be assigned accordingly.

Maintenance tasks are administered at the level of failure modes, with certain equipment exhibiting a prevalent failure mode. An appropriate methodology is determined through a systematic process, exemplified by using a flowchart to discern the optimal method for formulating maintenance tasks predicated upon these failure modes [2]. This paper reviews various methodologies for identifying SCE and CE and developing maintenance tasks. It aims to provide in-depth knowledge of current and best practices in categorizing tools in the oil and gas industry and formulating maintenance strategies to ensure safe and reliable critical equipment.

2. Methodology

The literature review was centered on the safety-critical equipment (SCE), critical equipment (CE), and maintenance tasks in the oil and gas industries. A rigorous approach was adopted to examine relevant materials and reduce bias. The searched databases include Scopus, Google Scholar, and ResearchGate to find influential studies, recent advancements, and emerging trends. The first query framework had broad terms such as SCE, CE, and related maintenance tasks. The focus then narrowed to SCE-specific CE identification and maintenance task development methodologies. This approach emphasized the studies, frameworks, and techniques developed to measure equipment criticality and enhance maintenance strategies. It sought to present a variety of literature on equipment criticality and maintenance practices across different sectors.

The following are a few keywords that are employed in literature searches.

- SCE AND Methodology
- SCE AND Multi-Criteria Decision Making
- CE AND Methodology
- Criticality AND Failure Modes Effects and Criticality Analysis

Papers were chosen from conferences and journals and were not restricted to a specific year. The abstract, title, and conclusion were examined to ascertain a paper's relevance. The selected papers will be grouped into two papers discussing SCE and CE. Papers in the SCE and CE group will be reviewed and regrouped again into subcategories based on the methodology used to determine SCE or CE. Each methodology in each grouped paper will be reviewed and compared for its applicability level for practical purposes without leaving the academic sense. The attributes to determine the practical applicability are the resolution of data requirements, complexity, assessment tools, analysis resolution, duration, and necessary level of expertise in implementing each reviewed methodology. The result is a comparison matrix between methods that shows the advantages and disadvantages of each methodology so that users can determine which method should be used according to their needs and available resources.

3. Results

A comprehensive review of 45 published papers discussing the methodology for Safety-Critical Equipment (SCE) and Critical Equipment (CE) have been conducted. There are SCE methodologies such as logic tree analysis, risk assessment, Hazard and Operability Study (HAZOP), Layers of Protection Analysis (LOPA), and Fuzzy-Based LOPA, which would ensure equipment getting damaged when they fail would lead to significant hazards, thereby making industrial processes safe and reliable. The CE methodologies consist of risk priority number (RPN), consequence-based approaches, risk-based methods, fuzzy-consequence-based techniques, fuzzy-based methods, and Bayesian network analysis.

Each review publication has been examined in terms of key aspects such as methodology, level of analysis, type of analysis, difficulty level, time and resources required, documentation involved, and methodology applicability. Figure 1 displays the Sankey diagram of the analysis flows before determining appropriate maintenance tasks for the equipment.

3.1. Safety Critical Equipment (SCE)

Safety-critical equipment (SCE) and safety-critical elements (SCEs) have different roles in high-risk industries. This understanding will help take care of all safety actions and property that should address every possible risk by recognizing these differences. SCEs have a wider set of systems or units capable of catastrophic disasters than SCE, whose items are physical devices or machines that prevent accidents or bring down their severity whenever they occur. SCEs range from procedural components to barriers against potential mortals while there are SCE which is only used on valves, alarms and temperature regulators. Knowing these two terms apart makes it possible to aim at safety measures and perform maintenance correctly so that all forms of danger are dealt with broadly.

Various methodologies are used to determine SCE, listed in Table 1. The methodologies are reviewed and compared to determine their advantages and disadvantages.

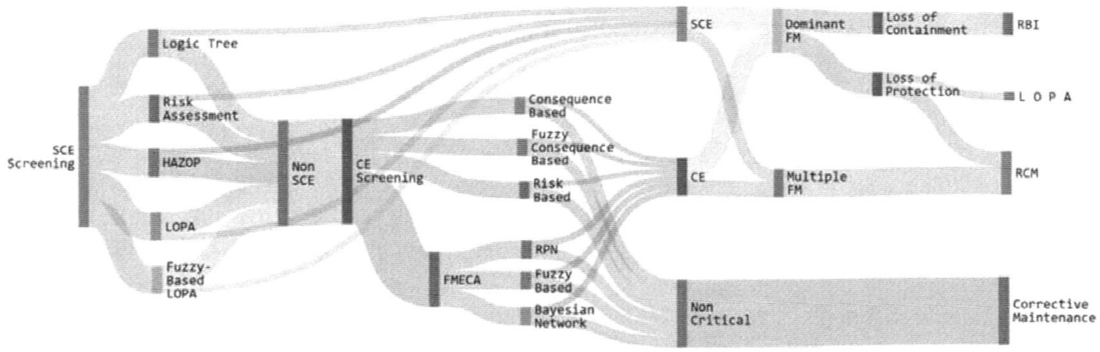

Figure 1. The analysis flows to determine maintenance tasks.

Table 1. Comparison of various methodologies to identify SCE

	Logic Tree	Risk Assessment	HAZOP	LOPA	Fuzzy-Based LOPA
Methodology	A series of YES / NO questions in a logic tree applied to the equipment in question.	Identifies and evaluates potential risks in a process or system. It involves assessing the likelihood and impact of various hazards to determine the overall risk level.	A systematic hazard analysis technique involves a team of experts analyzing each operational stage of a facility to identify potential hazards and their consequences.	The technique identifies various layers of protection and assesses their effectiveness in mitigating risk, thereby enhancing individuals' overall safety and security.	An extension of LOPA that uses fuzzy logic to handle uncertainty and imprecision in risk assessment.
Level of Analysis	At equipment level	At the process or system level	At the process or system level	At the process or system level	At the process or system level
Type of analysis	Qualitative	Qualitative and Quantitative	Primarily qualitative	Semi-Quantitative	Utilizes fuzzy logic for quantitative risk assessment.
Level of difficulty	Low	Medium to High	High	High	High
Time and resources	It can vary and depend on the complexity of the process	It can vary and depend on the complexity of the process	Typically, it is time-consuming and resource intensive.	Generally, it is less time-consuming than HAZOP but requires detailed data.	It may require more time and resources due to the complexity of fuzzy logic applications.
Documentation	Simple documentation	Required detail documentation	Generates extensive documentation of identified hazards and recommended actions.	Produces detailed reports on the effectiveness of protection layers and recommended risk reduction measures.	Requires documentation of fuzzy sets and rules used in the analysis.
Applicability	Commonly found in Oil and Gas Industries	Broadly applicable across various industries.	It is primarily used in industries with complex processes.	It is commonly used in industries with high-hazard processes.	Suitable for complex systems where traditional LOPA may not fully capture the nuances of risk.
Publication reviewed	[1] [3] [4] [5]	[6] [7] [8]	[9] [10] [11] [12] [13] [14]	[15] [16] [17] [18] [19]	[20] [21]

3.1.1. Logic Tree Analysis. A logic tree is a systematic tool for evaluating the equipment failure consequences of major accidents. It consists of binary questions that represent various aspects of the potential consequences of failure. The flowchart guides users through a step-by-step analysis that identifies Safety-Critical Equipment (SCE) or non-SCE components. This method aids in identifying critical components that need strong safety measures.

The SCE logic tree is a pragmatic method for determining Safety Critical Equipment (SCE), presented in reference [1][3]. This easy-to-understand binary-question system effectively assesses the role of machinery in reducing serious incidents. Because of its practicality and efficiency, it has gained popularity in the oil and gas industry, where safety remains an ongoing issue. References [4][5] contain discussions and elaborations on SCE logic tree methodology.

3.1.2. Risk Assessment. Risk assessment is critical for identifying and assessing Safety Critical Elements (SCEs), which can significantly impact a system's safety and integrity. This involves risk identification and analysis, which helps develop major hazard scenarios and pinpoint major hazards. High-risk major hazards are typically comprised of SCEs, and within these SCEs, specific Safety Critical Equipment (SCE) can be pinpointed [6]. In oil and gas facilities, SCEs are integral to barrier systems that protect against major hazardous events like fires, explosions, and collisions. A systematic approach to implementing SCEs involves risk assessment, scenario development, and systematic SCE management to enhance safety and prevent major accidents [7]. A thorough risk assessment is conducted to understand the risks associated with various activities within the facility, and major hazard scenarios are developed to determine which systems or equipment qualify as SCEs based on their criticality in preventing or mitigating major accidents [8].

3.1.3. Hazard and Operability (HAZOP) Study. HAZOP is a systematic approach used to identify risks in processes or systems, involving multi-disciplinary teams evaluating flowcharts, piping drawings, and sequence diagrams. This helps determine Safety-Critical Elements (SCEs) by estimating potential aftermath of technical malfunctions and hazards. SCEs are crucial as failures can lead to major accidents. Once identified, SCE can also be identified.

HAZOP's application in offshore platforms assesses the risk level of equipment components and identifies critical elements requiring intensive maintenance [9]. In a Crude Oil Production Unit (COPU), HAZOP systematically analyzes the system to identify 71 relevant risks and evaluate their environmental, health, and economic impacts [10]. An enhanced HAZOP analysis using Multilevel Flow Modeling (MFM) provides a comprehensive approach to process safety and risk mitigation [11].

FMECA and HAZOP Integrated Analysis (FHIA) methodologies are combined in LNG regasification plant storage systems for safety assessment, merging failure mode estimation with systematic hazard identification, facilitating the identification of Safety-Critical Elements (SCEs) and associated risks [12]. System complexity increases the complexity of HAZOP studies, making the process challenging. An example from a Sour Crude Oil Processing Industry (SCOPI) and a Measurement and Regulatory Station (MRS) in Greece suggests an integrated approach combining HAZOP with Decision-Matrix Risk Assessment (DMRA) and safety-color mapping to identify, rank, and address SCEs [13]. HAZOP analysis is a systematic approach to software failure analysis, utilizing the viewpoint concept from IEEE-1471. This approach simplifies data handling and helps analysts systematically identify SCEs and their risk factors [14]. This systematic approach enables complete evaluation and risk prioritization, enhancing efficiency and accuracy in data handling.

3.1.4. Layer of Protection Analysis (LOPA). The Layer of Protection Analysis (LOPA) is a risk management tool that helps to uncover security layers and assigns Safety Integrity Levels (SILs) for every layer in the process sector. The SIL measures how often safety instrumented functions (SIFs) fail and carry a meaning of risk reduction attained through SIFs. Thus, LOPA can be applied in identifying safety instrumented systems (SIS) along with safety-critical elements (SCEs) [15]. SIS comprises input components, logic solvers, and the final components, which are crucial in detecting and controlling harmful events in the oil and gas industries [16]. Additionally, LOPA finds out how effective SCEs are so that the correct identification and management of dangers can occur. Also, LOPA when combined with expert systems and fuzzy logic enhances safety assurance. Furthermore, LOPA studies have looked at fire explosion prevention in tank storage facilities where several protection levels have been established to reduce threats [17]. Moreover, an IOW bank covers equipment integrity when it combines LOPA, which would describe safe operating limits for pressure devices [18]. In addition, the combination between LOPA and Bayesian approach appeared prominently in risk evaluation conducted on crucial subsea gas compression systems (SGCS) that rendered a complete platform for assessing and controlling threats related to SGCS parts [19].

3.1.5. Fuzzy-Based LOPA. The study explores the integration of fuzzy logic and LOPA in addressing the failure of crude oil shipping pumps, reducing the risk reduction factor (RRF) for potential hazards [20]. This method provides a more accurate and precise risk assessment, enabling better-informed decision-making and potentially more cost-effective safety measures. The study also explores the application of fuzzy-based LOPA methodology in the natural gas industry, analyzing ammonia hazards [21]. The study proposes a quantitative risk assessment model based on fuzzy logic, assigning membership functions and ranges for likelihood, severity of consequences, and total risk levels. The approach is demonstrated through a case study on ammonia hazards, showcasing its potential application in resolving risk-ranking ties. Overall, the study highlights the potential of fuzzy logic in enhancing risk assessment and decision-making in various industries.

3.2. Critical Equipment (CE)

Critical equipment (CE) can be determined by conducting Equipment Criticality Analysis (ECA). ECA can be defined as a systematic process for evaluating and prioritizing equipment based on its potential impact on the organization. ECA assesses factors such as safety, reliability, quality, frequency, and cost to determine the criticality level of each piece of equipment, which is crucial for effective maintenance planning and resource allocation.

Numerous methodologies are employed in ECA, each with its strengths and weaknesses. These methodologies can be broadly categorized into three main approaches: consequence-based, reliability-based, and fuzzy-based. The following sections will review various methodologies for determining ECA. Table 2 summarizes the comparison of various methodologies used to identify CE.

3.2.1. Risk Priority Number (RPN). Failure Mode Effects and Criticality Analysis (FMECA) is a systematic methodology for criticality analysis pioneered by MIL-STD-1629A [22]. It aids in identifying potential failure modes, their consequences, and the strategic allocation of resources for mitigation and prevention. FMECA is particularly useful for analyzing equipment with multiple failure modes, prioritizing issues based on their potential impact. The Risk Priority Number (RPN) is a key tool in FMECA, combining severity (S), occurrence (O), and detection ratings (D) to quantify the risk level of each failure mode. A higher RPN indicates a more critical failure mode, enabling effective identification and prioritization of critical equipment.

The practical application of Risk Priority Number (RPN) and FMEA is crucial in assessing potential failures in pipeline ball valves [23] and reciprocating gas compressors [24]. RPN is calculated to prioritize risks associated with different components based on severity, occurrence probability, and risk priority. This approach is particularly useful in reducing failures and enhancing oil and natural gas industry performance. The risk priority number (RPN) is calculated by multiplying occurrence, detection, and severity, which is defined qualitatively without any underlying model. This method is useful for evaluating failure risk and identifying appropriate risk reduction actions, even when only qualitative data is available [25]. The modified method accommodates expert evaluations even when only qualitative data is available.

Table 2. Comparison of various methodologies to identify CE

	Risk Priority Number (RPN)	Consequence-Based	Risk-Based	Fuzzy Consequence Based	Fuzzy Based	Bayesian Network
Methodology	Assigns a numerical value to each potential failure mode based on its severity, occurrence, and detection ease.	Evaluates the potential consequences of equipment failure, focusing on safety, environmental impact, and production loss.	Assesses the risk of equipment failure by considering the probability and impact of potential failures.	Uses fuzzy logic to evaluate the consequences of equipment failure, providing a more nuanced assessment.	Utilizes fuzzy logic to rank equipment based on their criticality, often used in conjunction with other methods.	Models the relationships between different failure modes and their consequences using Bayesian networks.
Level of Analysis	Equipment level	Equipment level	Equipment level	Equipment level	Equipment level	Failure Mode Level
Type of Analysis	Semi-quantitative	Quantitative	Semi-quantitative	Semi-quantitative	Semi-quantitative	Quantitative
Level of Difficulty	Medium	High	Medium	High	High	High
Time and Resources	Moderate	High	Moderate	High	High	High
Documentation	Simple	Detailed	Simple	Detailed	Detailed	Complex
Applicability	Widely used in industries	Oil and gas, chemical, and other high-risk industries	General applicability across various industries	Industries requiring complex risk assessments	Industries requiring nuanced criticality assessments	Advanced risk assessments in complex systems
Publication reviewed	[22] [23] [24] [25] [26] [27] [28]	[29] [30] [31]	[32] [33] [5]	[34] [35]	[36] [37] [38] [39] [40]	[41] [42] [43] [44] [45]

The RPN enhancement method, proposed by [26], combines FMECA with the Dempster–Shafer (D-S) evidence theory to assess the risk and reliability of oil pump units. This approach addresses the limitations of traditional risk assessment methods, involving expert evaluations, multi-value representation of RPN, and optimized prioritization. However, the RPN approach has flaws, such as gaps in acceptable values, duplicate values, and sensitivity to minor changes. Alternative RPN formulations, such as those aligned with IEC 60812, are explored [27]. The proposed modified RPN calculation divides severity and occurrence into subfactors, providing a more realistic evaluation of risk and enhancing oil pump safety and reliability [28].

3.2.2. Consequence-Based Criticality Analysis. The NORSOK Z-008 standard is a widely recognized industry guideline that provides comprehensive instructions for risk-based maintenance and consequence classification [29]. This standard applies to all plant systems and items, including offshore topside systems, subsea production systems, and oil and gas terminals. The standard emphasizes the importance of assessing the criticality of equipment and evaluating the potential consequences of failure. This approach helps identify equipment that requires more stringent monitoring and maintenance to prevent catastrophic failures, which could lead to safety hazards, environmental damage, operational downtime, and financial losses. By prioritizing high-risk equipment, maintenance strategies can be optimized, ensuring that the most critical components receive the necessary attention and resources to mitigate the risk of failure. This risk-based approach to maintenance enhances operational reliability and safety, contributing to more efficient allocation of maintenance resources and reducing the likelihood of unplanned outages and associated costs.

The study focuses on prioritizing maintenance efforts for complex components in a gas central processing plant using risk-based methods can be found in [30]. Criticality rankings are obtained by assessing probability and consequence ratings, categorizing components as high, medium, or low criticality. The research vessel's machinery system is analyzed using modified NORSOK Z-008, revealing that 21% of equipment falls into the high-risk category and requires planned maintenance, while 79% is in the low-risk category and requires unplanned maintenance. This approach helps prioritize maintenance efforts in a complex system with limited resources [31].

3.2.3. Risk-Based Criticality Analysis. Risk-based equipment criticality analysis uses risk as a metric to determine the criticality of equipment, with higher risk indicating higher criticality. This risk is calculated by combining the probability of failure (PoF) and the consequences of failure (CoF). Risk-based modeling focuses on developing methods to model equipment failure's probability and consequence. Models can be qualitative, semi-quantitative, or quantitative, each offering varying levels of precision and complexity in evaluating equipment criticality. Quantitative risk analysis is preferred in the risk prioritization process due to its clarity and facilitation of decision-making. Qualitative risk analysis uses narrative criteria to evaluate probability and consequence ratings, while semi-quantitative risk analysis quantifies factors influencing PoF and CoF by scoring. Determination of PoF, CoF, and risk level ratings is based on an assessment

The Reference [32] introduced ECA protocols for offshore carbon steel static mechanical equipment, utilizing a 5 x 6 risk matrix based on probability and consequence ratings to determine the criticality level of each equipment item. Risk-scoring protocols are also essential in determining the criticality of electromechanical equipment in the Catalyst Preparation Unit [33]. Two scoring approaches are compared to rank the criticality of electromechanical equipment. A case study evaluates current maintenance strategies and suggests a revised framework prioritizing critical equipment based on failure impact. Using secondary data like OREDA, probability estimation determines the probability of failure rating and minimizes downtime [5]. This approach optimizes resource allocation and minimizes downtime.

3.2.4. Fuzzy Consequence-Based Criticality Analysis. Fuzzy logic, a mathematical framework that addresses uncertainty and imprecision, can enhance the NORSOK Z-008 framework by providing a more realistic representation of system importance. Traditional approaches may not fully capture uncertainties and expert reasoning, but incorporating fuzzy sets and linguistic variables allows for more nuanced and flexible risk evaluation. This criticality matrix assists decision-makers in prioritizing maintenance tasks effectively, overcoming the drawbacks of traditional approaches [34].

Unexpected failures and production losses are significant challenges in manufacturing systems. Risk-based maintenance (RBM) planning is crucial for mitigating these issues. Prioritizing maintenance tasks based on potential failures informs maintenance strategy, intervals, and spare parts inventory decisions. A fuzzy inference system (FIS) is introduced to minimize suboptimal classifications [35].

3.2.5. Fuzzy-Based Criticality Analysis. Conventional FMECA calculates an RPN for each failure mode by multiplying the ranks of three risk factors: Occurrence, Severity, and Non-Detection. Classical FMECA has limitations, such as homogeneity and challenges in predicting occurrence, severity, and detection. Fuzzy-FMECA uses linguistic terms and an if-then rule-based system to identify and prioritize failure modes, providing a more nuanced evaluation of potential failures and their criticality. Fuzzy rules encapsulate expert knowledge and reasoning.

The study to explore the application of fuzzy-based FMECA in assessing the criticality of LNG storage facilities may refer to [36][37]. The fuzzy logic addresses uncertainties and vagueness in risk assessments, enhancing the precision of risk evaluations. The other study focuses on the unloading facility of an LNG terminal, highlighting the importance of identifying safety-critical elements (SCEs) and proposing mitigation strategies [38]. Fuzzy-based FMECA has also been used to evaluate risks associated with offshore platforms, combining FMECA with fuzzy set theory and gray relational analysis (GRA). The study also employs the Fuzzy Borda Method (FBM) to aggregate expert evaluations, demonstrating its effectiveness in identifying and ranking risks more accurately than conventional methods [39].

The study discusses fuzzy-based FMECA in risk analysis of FPSO oil and gas processing systems [40]. The method, enhanced with fuzzy evidence reasoning and fuzzy TOPSIS, addresses conventional FMECA's weaknesses like equal factor weighting and subjective biases by using fuzzy logic to represent occurrence, severity, and detectability ratings, providing a robust tool for decision-making in complex marine environments.

3.2.6. Bayesian Network Criticality Analysis.

Bayesian networks (BN) provide a robust visual representation of dependencies among variables, enhancing FMECA by identifying critical failure modes, causes, and effects. FMECA, an enhanced method, integrates fuzzy logic and Bayesian networks to address traditional limitations like handling uncertain data and subjective evaluations. The fuzzy best-worst method (FBWM) determines the weights of severity, occurrence, and detection risk parameters, while the fuzzy Bayesian network (FBN) calculates the occurrence probabilities of failure modes, incorporating expert judgments to handle uncertainties and dependencies between failure events [41].

A Bayesian network (BN) approach is proposed to assess the risk of third-party damage (TPD) in oil and gas pipelines [42]. This approach considers pre- and post-failure risk factors, integrating risk indicators like pipeline location, material, and environmental conditions. This helps comprehensively identify the critical segment of the pipeline. A hybrid methodology integrating FMECA, BN, and dynamic principal component analysis (DPCA) is also proposed to enhance fault detection in industrial processes [43]. FMECA identifies and prioritizes potential failure modes based on their severity, occurrence, and detectability, while BN models probabilistic relationships between process variables and failure modes. This method effectively differentiates between non-critical and critical faults and, indirectly, the critical and non-critical equipment that accompany non-critical and critical faults.

A novel safety assessment model for oil and gas pipeline failures is proposed, incorporating fuzzy logic into Bayesian belief networks to deal with uncertainties [44]. The FBBN (fuzzy Bayesian belief network) model is useful in safety analysis due to its flexible structure, which allows it to fit various accident scenarios. The proposed method can identify the most significant causes of pipeline failures through sensitivity analysis, such as construction defects, overload, mechanical damage, bad installation, and worker quality.

Other applications of Bayesian networks in determining equipment criticality include comparing constant bottom-hole pressure drilling to conventional overbalanced drilling [45]. Bow-tie models are transformed into Bayesian networks to overcome the complexities of modeling dependencies and updating beliefs based on operational data.

3.3. Developing Maintenance Tasks

The oil and gas industry can utilize generic maintenance concepts (GMC) described in [29] to create maintenance procedures for their equipment. GMC refers to a collection of cost-effective maintenance actions, techniques, and specifics that sustain equipment in similar conditions. If a GMC is unavailable, it is suggested that equipment be treated individually as a separate generic class. It involves a detailed analysis to identify and document the maintenance actions, strategies, and details specific to that equipment. The detailed analysis will classify the equipment as SCE, CE, or non-critical. The method is outlined in section 3.1. and 3.2. can be adopted to classify the equipment.

Maintenance tasks of classified equipment can be developed by considering its failure modes. For practical purposes, reference [2] suggested that risk-based inspection (RBI) analysis could be adopted to develop an inspection plan for equipment with loss of containment as the dominant failure mode. Reliability-centered maintenance (RCM) could be adopted for equipment with multiple failure modes to develop maintenance tasks. RCM used FMEA to document the failure analysis and maintenance logic tree to determine the equipment's maintenance task based on its failure modes. The RBI and RCM extension analysis, incorporating sophisticated techniques such as fuzzy logic and Bayesian networks, could be implemented for SCE and CE.

4. Discussion

ECA is a crucial step in Risk-Based Maintenance (RBM), prioritizing maintenance efforts based on equipment risk and criticality. ECA assigns criticality levels using a three-tier classification system: C1 for high criticality, C2 for medium criticality, and C3 for low criticality. This helps identify equipment that requires more stringent maintenance and monitoring due to its potential impact on system performance and safety.

SCE is categorized as highly critical equipment, ensuring system safety and preventing catastrophic failures. Non-SCE equipment is classified into Critical Equipment (CE) or Non-Critical Equipment (NCE). CEs, classified as C1 or C2, have significant importance but may not be as critical as SCE, while NCEs, classified as C3, require less intensive maintenance efforts.

This paper provides a comprehensive overview of methodologies for identifying safety-critical equipment (SCE) and critical equipment (CE) in the oil and gas (O&G) industry. The logic tree approach is widely used in the O&G industry due to its simplicity and direct method for identifying SCE. This method is effective for equipment with loss of containment as the dominant failure mode, allowing for efficient safety management and resource allocation. Other methods include risk assessment, Hazard and Operability Study (HAZOP), and Layers of Protection Analysis (LOPA). Advanced techniques like Fuzzy-Based LOPA incorporate fuzzy logic for more nuanced risk evaluations. These methods typically do not directly point out SCE, as the analysis occurs at a level lower than the equipment, such as failure modes level. This approach ensures thorough evaluation of all potential failure modes, leading to a more accurate and reliable identification of critical safety equipment.

FMECA is a traditional method for assessing critical equipment (CE). It systematically examines potential failure modes, causes, and effects, evaluating their criticality using a risk-priority number (RPN). This method prioritizes maintenance and safety efforts based on criticality. The consequence-based criticality analysis identifies the equipment that poses the greatest risk to safety and operations, particularly in environments where the primary concern is mitigating the impact of failures. Risk-based criticality analysis integrates the probability and consequences of failure, providing a balanced view of equipment risks and enabling informed decision-making. FMECA can be expanded by incorporating fuzzy logic or Bayesian networks for more quantitative and detailed analysis. These techniques handle uncertainties and imprecisions in data, enhancing the precision and robustness of FMECA and enabling a deeper understanding of equipment failure risks.

5. Conclusions

RCA is crucial in RBM in prioritizing maintenance efforts based on equipment risk and criticality. ECA identifies equipment requiring stringent maintenance and monitoring due to its potential impact on system performance and safety. SCE is critical for ensuring system safety and preventing catastrophic failures.

Various methodologies are used in the O&G industry to identify SCE and critical equipment. The logic tree approach is a popular method for identifying SCE, especially for equipment with loss of containment as the dominant failure mode. Alternative methods like Risk assessment, HAZOP, and LOPA provide nuanced risk evaluations. Advanced techniques like Fuzzy-Based LOPA enhance the precision of criticality assessments.

FMECA is another method for prioritizing maintenance and safety efforts based on criticality. Enhancements with fuzzy logic or Bayesian networks improve the accuracy of criticality assessments. In conclusion, rigorous criticality assessments are essential for ensuring the safety and reliability of critical equipment in the O&G industry. Combining traditional and advanced methodologies can optimize system performance.

6. Acknowledgments

The authors express their gratitude to the Department of Marine Engineering, Institut Teknologi Sepuluh Nopember (ITS), Surabaya, Indonesia, for supporting this research.

7. References

[1] American Institute of Chemical Engineers. Center for Chemical Process Safety., Guidelines for Asset Integrity Management. American Institute of Chemical Engineers, 2016.

[2] D. Priyanta, M. B. Zaman, and Semin, "The development of a risk-based maintenance flowchart to select the correct methodology to develop maintenance strategies of oil and gas equipment," IOP Conf Ser Mater Sci Eng, vol. 1052, no. 1, p. 012042, Jan. 2021, doi: 10.1088/1757-899x/1052/1/012042.

[3] OCIMF, "Safety Critical Equipment and Spare Parts Guidance Safety Critical Equipment and Spare Parts Guidance," London, 2018. Accessed: Jun. 07, 2024. [Online]. Available: https://www.ocimf.org/publications/information-papers/safety-critical-equipment-and-spare-parts-guidance

[4] M. P. Broadribb and E. Freiburger, "Do you feel lucky? or do you want to identify and manage safety critical equipment?," Process Safety Progress, vol. 37, no. 3, pp. 340–346, Sep. 2018, doi: 10.1002/PRS.11963.

[5] D. Priyanta, M. B. Zaman, and Semin, "The application of OREDA (offshore reliability data) in developing valve criticality analysis criteria," in IOP Conference Series: Earth and Environmental Science, Institute of Physics, 2024. doi: 10.1088/1755-1315/1298/1/012015.

[6] R. Ghasemi Pirbalouti, B. Behnam, and M. Karimi Dehkordi, "A risk-based approach to identify safety-critical equipment in process industries," Results in Engineering, vol. 20, p. 101448, Dec. 2023, doi: 10.1016/J.RINENG.2023.101448.

[7] F. W. Deddy, Z. Djunaidi, and M. Tejamaya, "Safety Critical Element Methodology in an Oil and Gas Company," Indian J Public Health Res Dev, vol. 10, no. 9, p. 1956, 2019, doi: 10.5958/0976-5506.2019.02743.8.

[8] T. Ai, "The identification and analysis of safety critical elements in safety management of oil and gas fields," in 2016 International Field Exploration and Development Conference (IFEDC), IEEE, Aug. 2016.

[9] N. Cahyono, M. B. Zaman, N. Siswantoro, and D. Priyanta, "Hazard and Operability (HAZOP) Study at Separation System in the Offshore Platform," International Journal on Engineering Applications, vol. 10, no. 1, 2022, doi: 10.15866/irea.v10i1.20951.

[10] A. de J. Penelas and J. C. M. Pires, "Hazop analysis in terms of safety operations processes for oil production units: A case study," Applied Sciences (Switzerland), vol. 11, no. 21, Nov. 2021, doi: 10.3390/app112110210.

[11] J. Wu et al., "Hazard Identification of the Offshore Three-Phase Separation Process Based on Multilevel Flow Modeling and HAZOP," in Recent Trends in Applied Artificial Intelligence. IEA/AIE 2013. Lecture Notes in Computer Science, Berlin: Springer, 2013, pp. 421–430.

[12] M. Giardina and M. Morale, "Safety study of an LNG regasification plant using an FMECA and HAZOP integrated methodology," J Loss Prev Process Ind, vol. 35, pp. 35–45, May 2015, doi: 10.1016/J.JLP.2015.03.013.

[13] P. K. Marhavilas, M. Filippidis, G. K. Koulinas, and D. E. Koulouriotis, "Safety Considerations by Synergy of HAZOP/DMRA with Safety Color Maps—Applications on: A Crude-Oil Processing Industry/a Gas Transportation System," Processes, vol. 9, no. 8, Jul. 2021.

[14] Z. Hu, "Use Viewpoints to Control Effort of HAZOP Analysis for Safety Critical Systems," IFAC Proceedings Volumes, vol. 42, no. 8, pp. 1521–1526, Jan. 2009, doi: 10.3182/20090630-4-ES-2003.00248.

[15] M. A. Lundteigen and M. Rausand, "Common cause failures in safety instrumented systems on oil and gas installations: Implementing defense measures through function testing," J Loss Prev Process Ind, vol. 20, no. 3, pp. 218–229, May 2007, doi: 10.1016/J.JLP.2007.03.007.

[16] A. S. Markowski, "A review of layer of protection analysis techniques for oil and gas industry," International Journal of Oil, Gas and Coal Technology, vol. 5, no. 1, Dec. 2011.

[17] K. Sotoodeh, "A case study demonstrating the use of Layers of Protection Analysis (LOPA) in order to prevent fire and explosion in storage tanks due to overfilling," Safety in Extreme Environments, vol. 6, pp. 161–172, 2024.

[18] P. T. Wilson, "Managing the mechanical integrity of pressure equipment using a layers of protection framework and incorporating integrity operating windows," J Loss Prev Process Ind, vol. 79, p. 104821, Oct. 2022, doi: 10.1016/J.JLP.2022.104821.

[19] A. O. Ifelebuegu, E. O. Awotu-Ukiri, S. C. Theophilus, A. O. Arewa, and E. Bassey, "The application of Bayesian – Layer of Protection Analysis method for risk assessment of critical subsea gas compression systems," Process Safety and Environmental Protection, vol. 113, pp. 305–318, Jan. 2018, doi: 10.1016/J.PSEP.2017.10.019.

[20] F. M. Eltahan, M. Toderas, M. S. Mansour, E. S. Z. El-Ashtoukhy, M. A. Abdou, and F. Shokry, "Applying a semi-quantitative risk assessment on petroleum production unit," Sci Rep, vol. 14, no. 1, Dec. 2024, doi: 10.1038/s41598-024-57600-2.

[21] M. Khalil, M. A. Abdou, M. S. Mansour, H. A. Farag, and M. E. Ossman, "A cascaded fuzzy-LOPA risk assessment model applied in natural gas industry," J Loss Prev Process Ind, vol. 25, no. 6, pp. 877–882, Nov. 2012, doi: 10.1016/J.JLP.2012.04.010.

[22] United States Department of Defense, MIL-STD-1629A MILITARY STANDARD PROCEDURES FOR PERFORMING A FAILURE MODE, EFFECTS AND CRITICALITY ANALYSIS. 1977.

[23] K. Sotoodeh, "Failure Mode and Effect Analysis (FMEA) of Pipeline Ball Valves in the Offshore Industry," Journal of Failure Analysis and Prevention, vol. 20, no. 4, pp. 1175–1183, Aug. 2020, doi: 10.1007/s11668-020-00924-8.

[24] K. Ratendra and N. Virender, "Failure Modes and Effect Analysis of a Gas Compressor in a Gas Processing Complex: A Case Study Based Approach," in Advances in Fluid and Thermal Engineering (FLAME 2022). Lecture Notes in Mechanical Engineering., B. S. Sikarwar, S. K. Sharma, A. Jain, and K. M. Singh, Eds., Singapore: Springer, Jul. 2023, pp. 167–179.

[25] K. O. Kim and M. J. Zuo, "General model for the risk priority number in failure mode and effects analysis," Reliab Eng Syst Saf, vol. 169, pp. 321–329, Jan. 2018, doi: 10.1016/J.RESS.2017.09.010.

[26] X. Zhang, R. Wei, Z. Wu, L. Dong, and H. Liu, "Risk Assessment and Reliability Analysis of Oil Pump Unit Based on D-S Evidence Theory," Energies (Basel), vol. 16, no. 13, Jul. 2023, doi: 10.3390/en16134887.

[27] L. Ciani, G. Guidi, and G. Patrizi, "A Critical Comparison of Alternative Risk Priority Numbers in Failure Modes, Effects, and Criticality Analysis," IEEE Access, vol. 7, pp. 92398–92409, 2019, doi: 10.1109/ACCESS.2019.2928120.

[28] J. Fabis-Domagala, M. Domagala, and H. Momeni, "A concept of risk prioritization in fmea analysis for fluid power systems," Energies (Basel), vol. 14, no. 20, Oct. 2021, doi: 10.3390/en14206482.

[29] NORSOK, "NORSOK Z-008:2017 Risk based maintenance and consequence classification," 2017. [Online]. Available: www.standard.no/petroleum

[30] D. Priyanta, N. Siswantoro, M. B. Zaman, and F. Prasetyo, "The Implementation of Norsok Z-008 for Equipment Criticality Analysis of Gas Central Processing Plant," International Journal of Marine Engineering Innovation and Research, vol. 4, no. 1, pp. 2548–1479, 2019.

[31] F. Maulana Noor, D. Fajar Prasetyo, and R. Dwi Sakti Wijaya, "Criticality Analysis for Research Vessel Machinery System Maintenance Strategy Study Case: RV. Baruna Jaya H) Check for," Kapal:Journal of Marine Science and Technology), vol. 20, no. 1, pp. 44–59, 2023, doi: 10.14710/kapal.

[32] D. Priyanta, "The Development of Equipment Criticality Analysis (ECA) Protocols of Offshore Carbon Steel Static Mechanical Equipment," Asian Journal of Applied Sciences, vol. 4, no. 6, pp. 2321–0893, Dec. 2016, [Online]. Available: www.ajouronline.com

[33] P. Maphosa, P. Nziu, and L. Masu, "Criticality Analysis of Electromechanical Equipment Maintenance. A Case Study in Sasol Synfuels Catalyst Preparation Unit," International Journal of Engineering Trends and Technology, vol. 72, no. 1, pp. 93–100, Jan. 2024, doi: 10.14445/22315381/IJETT-V72I1P110.

[34] R. M. C. Ratnayake, "Plant systems and equipment maintenance: Use of fuzzy logic for criticality assessment in NORSOK standard Z-008," in IEEE International Conference on Industrial Engineering and Engineering Management, IEEE Computer Society, Nov. 2014, pp. 1468–1472. doi: 10.1109/IEEM.2013.6962654.

[35] R. M. C. Ratnayake and K. Antosz, "Development of a Risk Matrix and Extending the Risk-based Maintenance Analysis with Fuzzy Logic," Procedia Eng, vol. 182, pp. 602–610, 2017, doi: 10.1016/j.proeng.2017.03.163.

[36] M. J. Kalathil, V. R. Renjith, and N. R. Augustine, "Failure mode effect and criticality analysis using dempster shafer theory and its comparison with fuzzy failure mode effect and criticality analysis: A case study applied to LNG storage facility," Process Safety and Environmental Protection, vol. 138, pp. 337-348, Jun. 2020, doi: 10.1016/J.PSEP.2020.03.042.

[37] V. R. Renjith, M. Jose kalathil, P. H. Kumar, and D. Madhavan, "Fuzzy FMECA (failure mode effect and criticality analysis) of LNG storage facility," J Loss Prev Process Ind, pp. 537–547, Nov. 2018, doi: 10.1016/j.jlp.2018.01.002.

[38] J. J. George, V. R. Renjith, P. George, and A. S. George, "Application of fuzzy failure mode effect and criticality analysis on unloading facility of LNG terminal," J Loss Prev Process Ind, vol. 61, pp. 104–113, Sep. 2019, doi: 10.1016/J.JLP.2019.06.009.

[39] S. Ni, Y. Tang, G. Wang, L. Yang, B. Lei, and Z. Zhang, "Risk identification and quantitative assessment method of offshore platform equipment," Energy Reports, vol. 8, pp. 7219–7229, Nov. 2022, doi: 10.1016/j.egyr.2022.05.159.

[40] L. Wang, L. Sun, J. Kang, Y. Wang, and H. Wang, "Risk identification of FPSO oil and gas processing system based on an improved FMEA approach," Applied Sciences (Switzerland), vol. 11, no. 2, pp. 1–19, Jan. 2021, doi: 10.3390/app11020567.

[41] M. Yucesan, M. Gul, and E. Celik, "A holistic FMEA approach by fuzzy-based Bayesian network and best–worst method," Complex and Intelligent Systems, vol. 7, no. 3, pp. 1547–1564, Jun. 2021, doi: 10.1007/s40747-021-00279-z.

[42] X. Guo, L. Zhang, W. Liang, and S. Haugen, "Risk identification of third-party damage on oil and gas pipelines through the Bayesian network," J Loss Prev Process Ind, vol. 54, pp. 163–178, Jul. 2018, doi: 10.1016/J.JLP.2018.03.012.

[43] B. L. Tarcsay, Á. Bárkányi, S. Németh, T. Chován, L. Lovas, and A. Egedy, "Risk-Based Fault Detection Using Bayesian Networks Based on Failure Mode and Effect Analysis," Sensors, vol. 24, no. 11, p. 3511, May 2024, doi: 10.3390/s24113511.

[44] G. Kabir, R. Sadiq, and S. Tesfamariam, "A fuzzy Bayesian belief network for safety assessment of oil and gas pipelines," Structure and Infrastructure Engineering, vol. 12, no. 8, pp. 874–889, Aug. 2016, doi: 10.1080/15732479.2015.1053093.

[45] M. Abimbola, F. Khan, N. Khakzad, and S. Butt, "Safety and risk analysis of managed pressure drilling operation using Bayesian network," Saf Sci, vol. 76, pp. 133–144, Jul. 2015, doi: 10.1016/J.SSCI.2015.01.010.

IOP Conf. Series: Earth and Environmental Science **1423** (2024) 012011 doi:10.1088/1755-1315/1423/1/012011

The Effect of Port Growth on the Safety Index in Madura Strait

DW Hardhita[1]*, IPS Asmara[1] and Y Widiarti[1]

[1] Safety and Risk Engineering Department, Shipbuilding Institute of Polytechnic Surabaya Jl. Teknik Kimia, Kampus ITS Sukolilo Surabaya, Indonesia

*E-mail: widyahardhita23@student.ppns.ac.id

Abstract. This research examines ship safety level in port development through environmental stress method, which involves assessing the safety level in current conditions and simulating the addition of new ports to enhance shipping efficiency. This method evaluates safety levels based on each ship's COG degree and identifies obstacle characteristics. Analysis results are visualized using the perception safety index matrix. These findings determine the permissible speed tolerance for ships navigating the *Madura* Strait under current and planned conditions. As per sample, simulation will be using MV *PIS Mahakam* for this research. The simulation indicates that safety index in the existing condition or scenario X is at a Marginal level, with the highest value being 749. Furthermore, in planned scenario with addition of 2 new ports and combination of 2 ports which called scenario A, B, and C, safety index values are 749 thus all considered marginal. The maximum speed tolerance under planned conditions ranges from 1.5 – 1.6 knots for a 114m ship.

1. Introduction

This study deals with the modelling of ship manouvering on a new jetty by looking at the environmental stress model & environmental level condition aspects. The development of mining in eastern Indonesia has resulted in an increase in the volume of ships passing through the Surabaya West Access Channel (SWAC). As is well known, SWAC is the gateway to the supply chain of western Indonesia to eastern Indonesia or vice versa. The large logistical flows of goods and services passing through the SWAC line have caused problems in the accumulation of the arrival line to the Port of *Tanjung Perak* and its surroundings. In addition to the government's program of building a new smelter in *JIIPE Gresik* so that the intensity of ships passing through the Surabaya West Access Channel (SWAC) increases. [1]. Therefore, a new jetty was added that connects directly from the JIIPE smelter to the new *PKG Gresik* Jetty to detract from the volume of ships that are heading and going to carry copper mines at the Port of *Tanjung Perak*. With the increasing volume of ships, taking advantage of good ship maneuvering can reduce the risk of a ship crashing while backing in the port, that the difficulty of the ship's manoeuvre can increase exponentially in ports and maritime areas with shipping flows of much higher densities [2]. The vulnerability of a ship to a collision at the port is assessed using various types of indicators, which, if properly regulated and monitored, can reduce the risk of collision or collision, and tend to ensure safe operation of the ship [3]. This study will examine the safety of ship maneuvering when entering the old and new docks under construction, by assessing ship speed and safety indicators through calculating the Environmental Stress value land and ship obstacles within ±0 - ±90^0 eyes from the ship's stern tube will be considered then mapped into the marine judgment's risk matrix. During the comprehensive study of the total summation, the safety of ships will be assessed at 3 different docks: the old jetty / jetty X, jetty A, and jetty B. This will result in 4 traffic scenarios being analyzed - the current scenario (jetty X), scenario A (new jetty A/construction phase), scenario B (new jetty B/construction phase), and scenario C (new jetty A & B/construction phase), illustration of each jetty can be seen in the Figure 1.

Once the research area is identified, the main issue is determining the safe speed for ships entering jetty X, A, and B. Once the maximum allowed speed is determined, evaluate the safety of the ship entering the old / existing jetty and jetty A and B during construction to determine if the safety level remains safe, high risk, or intermediate at specific speeds. The fact that jety A and B are newly built means there is ongoing construction in the vicinity. This leads to numerous obstacles and areas of high risk.

2. Methodology

2.1. Environmental Stress Model

Environmental stress model is one of the risk assessments that considers the narrow environmental topography and dense traffic in the port/restricted area. This method evaluates the maneuvering processes of ships caused by topographic constraints [4]. The calculation of voltage values on the Environmental Stress method includes [5]:

a. Evaluation of the difficulty of handling vessels arising from the restriction of the area of water available for manoeuvre. A quantitative index indicating the level of stress experienced by seafarers due to topographic constraints (ES values for land - ESL values) calculated since TTC (Time for Collision) with any obstacle.
b. Evaluate the difficulty of handling vessels arising from the restriction of freedom to perform a maneuver to avoid collisions. A quantitative index indicating the level of pressure experienced by seafarers because of traffic congestion which calculated since TTC's with ships.

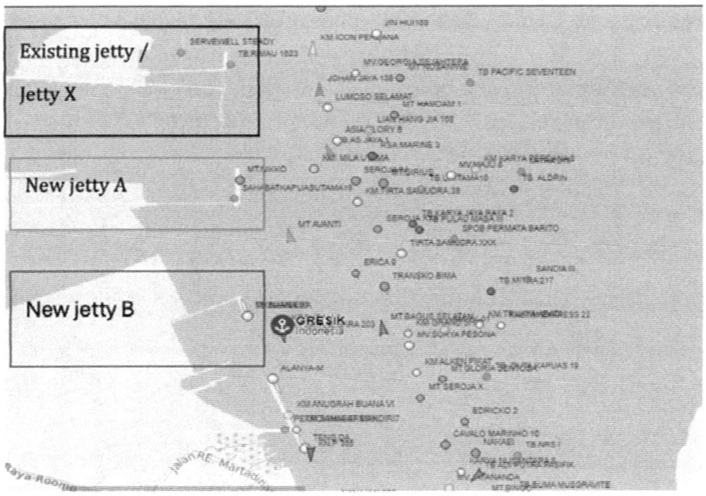

Figure 1. Location of old jetty (jetty X), jetty A, and jetty B

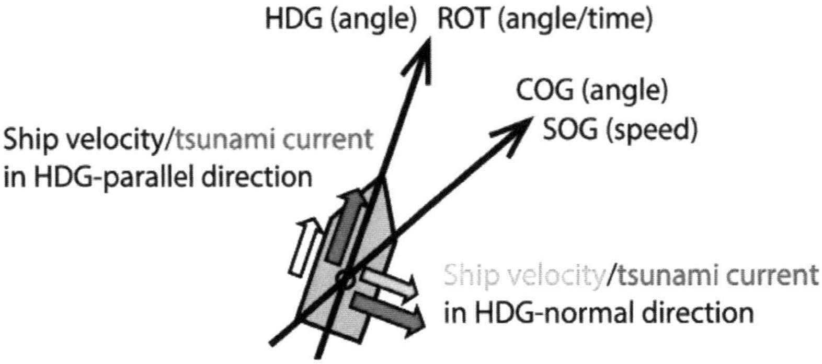

Figure 2 Illustration of Course Over Ground (COG) direction (Inazu, D., et al. 2020)

c. Aggregate assessment of ship handling difficulties caused by topography and traffic environment, where the voltage value (ES value for ESA aggregate values) is obtained by overlapping ESL and ESS values in the same route.

The Environmental Stress Model method simulates the condition of a ship with degrees between -90 and +90 along the ship's tracks. The environmental stress model values are stated from 0 to 1000 determined by the risk of collision with other vessels. The ES value is expressed as the sum of the Subjective Judgment (SJ) for the movement and traffic of ships, the subjective judgment itself is divided into 2 :

$$SJL = \alpha(R/V) + \beta \quad (1)$$

Known as :
SJL= Subjective Judgement Land (If the nearest obstacle is land)
α = 0,00092 log 10 GT + 0,0099 , for GT ≤ 10000 GT
α = 0,00671 exp (-7x10-6) GT, for GT ≥ 10000 GT
β = -3,82 GT is called as Gross Tonnage of vessel
R = distance topographic barrier (meter)
V = ship speed

$$SJS = \alpha\,(R/Lm) + \beta \quad (2)$$

Known as :
SJS= Subjective Judgement Ship (If the nearest obstacle is vessel)
α = 0,0019 Lm , for the ship of the starboard : β ln (Lm) – 2,07
(Lm) – 2,07 , for the ship of the starboard : β ln (Lm) – 2,35
(Lm) – 2,35 , head condition : β ln (Lm) – 2,07
(Lm) – 2,07 , overtaking condition : β ln (Lm) – 0,85
(Lm) – 0,85 Lm = Average length between the subject of the ship and the targeted ship
R = intership distance

Course Overground (COG) is simulated for changes between -90 and +90 of the ship's stern point. Environmental Stress (ES) summarizes the aggregate subjective assessment by adding equation 1 and equation 2 to the following equation 3.:

$$ES\,(t) = \sum SJL\,j + \sum_j SJS\,j \quad (3)$$

Known as :
where j is the ship's barrier for every 1° COG (Course Overground)

Course Over Ground (COG) notation as an additional angle variable to calculate the distance between the ship and obstacles at each 1^0 degree intervals on either side of the ship (angles $\pm 0^0$ – $\pm 90^0$) [6].

2.2. *Environmental Level and Condition*

In addition to using the concept of an environmental stress model, it is also necessary to interpret the results calculations in the risk matrix scale. The Environmental Stress rating is classified based on hazardous situations that cause certain SJ values in the range of 0 to ±90° from the horizontal direction of the ship at the stern point. The environmental stress risk matrix model is set and classified between 0 and 1000 as shown in Table 1 [7]. There are four categories for Calculation of Environmental Stress (ES) Model and Environmental Level & Condition.

This study will discuss the calculation of the permitted speed levels when entering scenario X, A, B, and C. The course of this research is to determine the closest obstacle to the ship whether it is a particular land / ship topography, then mapping the distance of the ship to the obstacle each 1° COG which is marked as COG 0° to +90° and COG 0° to -90°. In addition, the ship data is needed to complete the topographic obstacle calculation, in this case the ship *PIS Mahakam* data can be seen in Table 2.

Table 1 Stress ranking and acceptance criteria

Criteria level	Marine Judgement	Total Environmental Barriers (ΣSJ)	Ranking Stress	Acceptance Criteria
0	Very Secure	0-100	Negligible	Acceptable
1	Secure	100-200	Negligible	Acceptable
2	Quite Safe	200-400	Negligible	Acceptable
3	Intermediete	400-500	Negligible	Acceptable
4	Quite unsafe	500-750	Marginal	Acceptable
5	Hazardous	750-900	Critical	Unacceptable
6	Extremely hazardous	900-1000	Catasthropic	Unacceptable

Table 2 PIS Mahakam Ship Data

Vessel name	PIS Mahakam
Gross Tonnage (Ton)	5667
Length of ship (meter)	114
Vessel speed when entering the port X (knot)	1.6

IOP Conf. Series: Earth and Environmental Science **1423** (2024) 012011 doi:10.1088/1755-1315/1423/1/012011

In addition to the ship data, it is necessary to know the position and topography of ports X and ports A, B, C. Ports X are old existing ports while ports B, and C are new ports in the process of construction. Figure 3 is an illustration of the *PIS Mahakam* ship entering the port of X at a speed of 1.6 knots. (SJL). After knowing the position and obstacles of the ship then calculate the safety value of the vessel PIS Mahakam by ES Model method. Since the obstacle PIS Mahakam is land, then use the formula subjective judgement land (SJL) as in the equation 1. Since the GT of the *PIS Mahakam* is 5667 Ton, then using the value α = 0,00092 log 10 GT + 0,0099. In more detail, the calculation of the ES level of *PIS Mahakam* can be seen in Table 3. In calculating the results of Table 3, supporting calculations and notation explanations are required as follows:

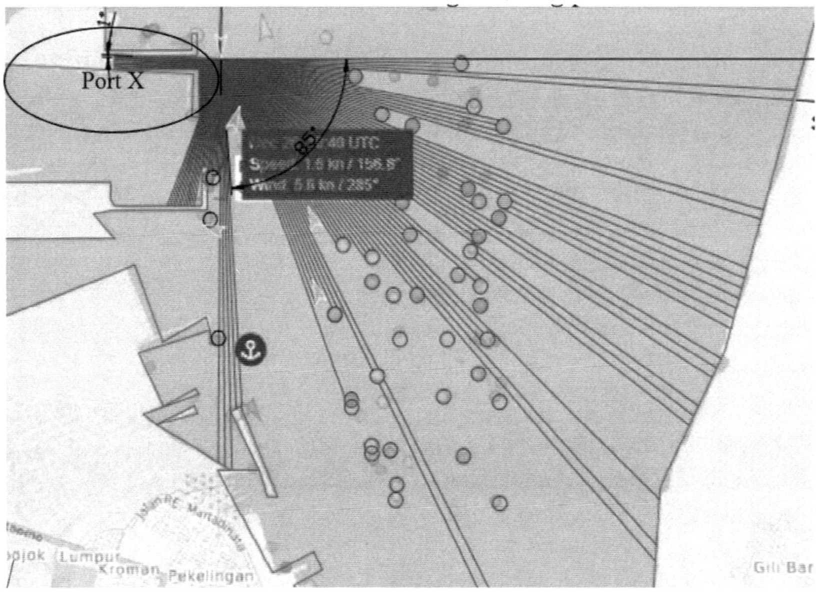

Figure 3 Position of PIS Mahakam when entering Port X

Table 3 Calculation of SJL and ES values *PIS Mahakam* enters existing port X

COG Degrees 1	R1	COG Degrees 2	R2	R/V1	R/V2	SJL1	SJL2	Criteria Level 1	Criteria Level 2
-1	312	1	839	195	524	-2.56	-0.44	5	3
-2	241	2	832	151	520	-2.85	-0.47	5	3
-3	387	3	840	242	525	-2.26	-0.43	5	3
-4	385	4	2048	241	1280	-2.27	4.43	5	0
-5	386	5	465	241	291	-2.26	-1.94	5	4
-6	386	6	459	241	287	-2.26	-1.97	5	4
...etc	...etc	...etc	...etc	...etc	...etc	...etc	...etc	...etc	...etc
-90	395	90	576	247	360	-2.23	-1.50	5	4
					Sum of Obstacles all degrees			489	266
			Values of Environmental Obstacles (ΣSJ) Scenario X					755	

Notation table 3:

a. Ship speed = 1.6 knots
b. α = 0,00092 log 10 GT + 0,0099, for GT ≤ 10000 GT
c. α = 0,00092 log 10 [5667 ton] + 0,0099 = 0.00644692

d. R1 = Distance of the ship to the topographic barrier in meters with a radius of -0° to -90° from stern point of the vessel (meter) (portside)
e. R2 = Distance of the ship to the topographic barrier in meters with a radius of +0° to +90° from stern point of the vessel (meter) (starboard)
f. R/V1 = Distance to the topographic barrier (meter) compared to the speed of the ship (Knot) with a radius of -0° to -90° from stern point of the vessel (portside)
g. R/V2 = Distance to the topographic barrier (meter) compared to the speed of the ship (Knot) with a radius of +0° to +90° from stern point of the vessel (starboard)
h. SJL1 = Subjective Judgement Land values with a radius of -0° to +90° from stern point of the ship (portside)
i. SJL2 = Subjective Judgement Land values with a radius of +0° to +90° from stern point of the ship(starboard)
j. Criteria Level 1 = Marine judgement values with radius -0° to -90° from stern point of the vessel (portside)
k. Criteria Level 2 = Marine judgement values with radius +0° to +90° from stern point of the vessel (starboard)

From table 3, the total environmental barrier value (SJ) of the *PIS Mahakam* at a speed of 1.6 knots is 755. If interpreted in table 2 marine judgement Environmental Level and Condition obtained critical / hazardous risk level. Thus, the author calculates how much speed is allowed to lower the critical level to marginal and parallel adding a new gate A or called scenario A. The results of the calculation of scenario A can be seen in table 4 and the description of the notation as same as previous calculation in table 3:

a. Ship Speed = 1.537 Knot
b. α = 0,00092 log 10 GT + 0,0099, for GT ≤ 10000 GT
c. α = 0,00092 log 10 [5667 ton] + 0,0099 = 0.00644692

Table 4 Calculation of SJL and ES values *PIS Mahakam* scenario A

COG Degrees 1	R_1	COG Degrees 2	R_2	R/V_1	R/V_2	SJL_1	SJL_2	Criteria Level 1	Criteria Level 2
-1	312	1	839	203	546	-2.51	-0.30	5	3
-2	241	2	832	157	542	-2.81	-0.33	5	3
-3	387	3	840	252	547	-2.20	-0.29	5	3
-4	385	4	2048	250	1332	-2.21	4.77	5	0
-5	386	5	465	251	303	-2.20	-1.87	5	4
-6	386	6	459	251	299	-2.20	-1.90	5	4
...etc	*...etc*	*...etc*	*...etc*	*...etc*	*...etc*	*...etc*	*...etc*	*...etc*	*...etc*
-90	395	90	576	257	375	-2.16	-1.40	5	4
						Sum of Obstacles all degrees		486	263
			Values of Environmental Obstacles (ΣSJ) Scenario A					749	

From table 4 is obtained the total environmental barrier value (SJ) of the ship *PIS Mahakam* at a speed of 1.537 knots that is 749 when entering the port X with the conditions of construction of the port A or called scenario A. If interpreted in table 2 marine judgement Environmental Level and Condition obtains a marginal risk level / quite dangerous. An illustration of the *PIS Mahakam* with scenario A table 4 can be seen in figure 4.

IOP Conf. Series: Earth and Environmental Science **1423** (2024) 012011 doi:10.1088/1755-1315/1423/1/012011

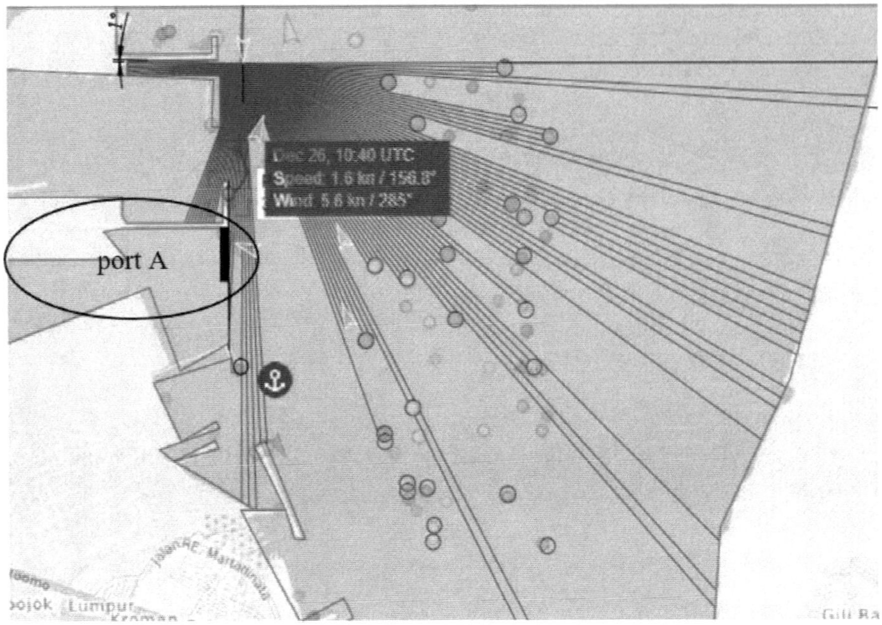

Figure 4 PIS Mahakam with scenario A

Then proceed to scenario B, which is the position of the *PIS Mahakam* entering port X by considering the construction of port B. An illustration of scenario B can be seen in figure 5, while the calculation of the scenario b can be found in table 5 with the description of the notation as follows :

a. Ship Speed = 1.6 Knot
b. α = 0,00092 log 10 GT + 0,0099, for GT ≤ 10000 GT
c. α = 0,00092 log 10 [5667 ton] + 0,0099 = 0.00644692

Table 5 Calculation of SJL and ES values *PIS Mahakam* scenario B

COG Degrees $_1$	**R_1**	**COG Degrees** $_2$	**R_2**	**R/V_1**	**R/V_2**	**SJL_1**	**SJL_2**	**Criteria Level** $_1$	**Criteria Level** $_2$
-1	312	1	839	195	524	-2.56	-0.44	5	3
-2	241	2	832	151	520	-2.85	-0.47	5	3
-3	387	3	840	242	525	-2.26	-0.43	5	3
-4	385	4	2048	241	1280	-2.27	4.43	5	0
-5	386	5	465	241	291	-2.26	-1.94	5	4
-6	386	6	459	241	287	-2.26	-1.97	5	4
...etc	*...etc*	*...etc*	*...etc*	*...etc*	*...etc*	*...etc*	*...etc*	*...etc*	*...etc*
-90	770	90	576	481	360	-0.72	-1.50	3	4
Sum of Obstacles all degrees								483	266
Values of Environmental Obstacles (ΣSJ) Scenario B								749	

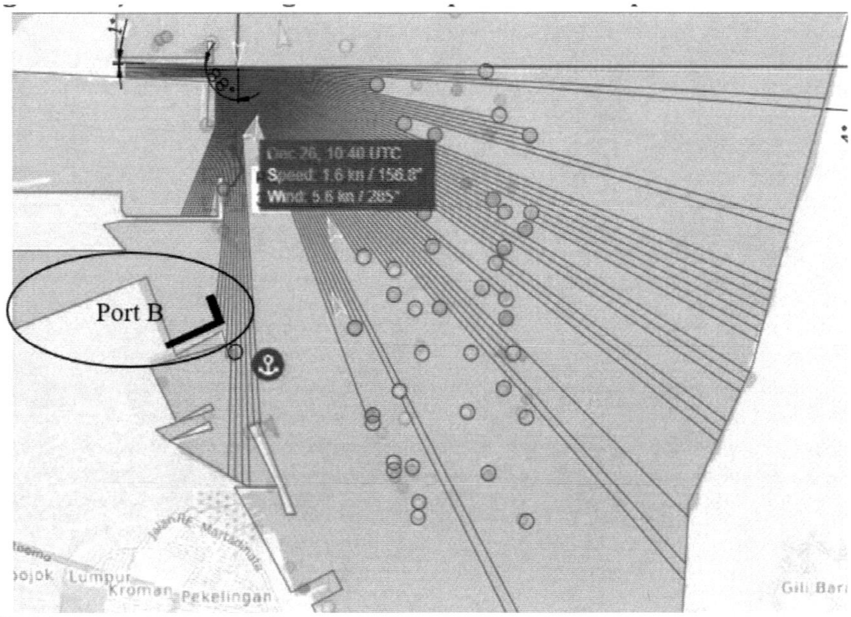

Figure 5 PIS Mahakam with scenario B

Knowing the total environmental barrier value (SJ) of scenario B is 749. Referring to table 2 of the marine judgment, SJ 749 is marginal with a permissible speed of 1.6 knots. Then calculate the ES Model value and the marine judgement scenario C, where the position of the *PIS Mahakam* enters the port X by considering the construction of the port A & B. An illustration of the scenario C can be seen in Figure 6 whereas the results of the calculation of scenarios C are seen in Table 6 with the description of the notation as follows:

a. Ship speed = 1.6 Knot
b. α = 0,00092 log 10 GT + 0,0099, for GT ≤ 10000 GT
c. α = 0,00092 log 10 [5667 ton] + 0,0099 = 0.00644692

Table 6 Calculation of SJL and ES values *PIS Mahakam* scenario C

COG Degrees 1	R1	COG Degrees 2	R2	R/V1	R/V2	SJL1	SJL2	Criteria Level 1	Criteria Level 2
-1	312	1	839	195	524	-2.56	-0.44	5	3
-2	241	2	832	151	520	-2.85	-0.47	5	3
-3	387	3	840	242	525	-2.26	-0.43	5	3
-4	385	4	2048	241	1280	-2.27	4.43	5	0
-5	386	5	465	241	291	-2.26	-1.94	5	4
-6	386	6	459	241	287	-2.26	-1.97	5	4
...etc	...etc	...etc	...etc	...etc	...etc	...etc	...etc	...etc	...etc
-90	770	90	576	481	360	-0.72	-1.50	3	4
					Sum of Obstacles all degrees			483	266
		Values of Environmental Obstacles (ΣSJ) Scenario C						749	

Interpretation of the calculation of scenario C is the total environmental barrier value (SJ) of scenarios C is 749. Referring to table 2 marine judgement, SG 749 is a marginal with the permitted speed of the ship is 1.6 knots. Based on table 2, marine Judgement Environmental Level and Condition obtained marginal risk level.

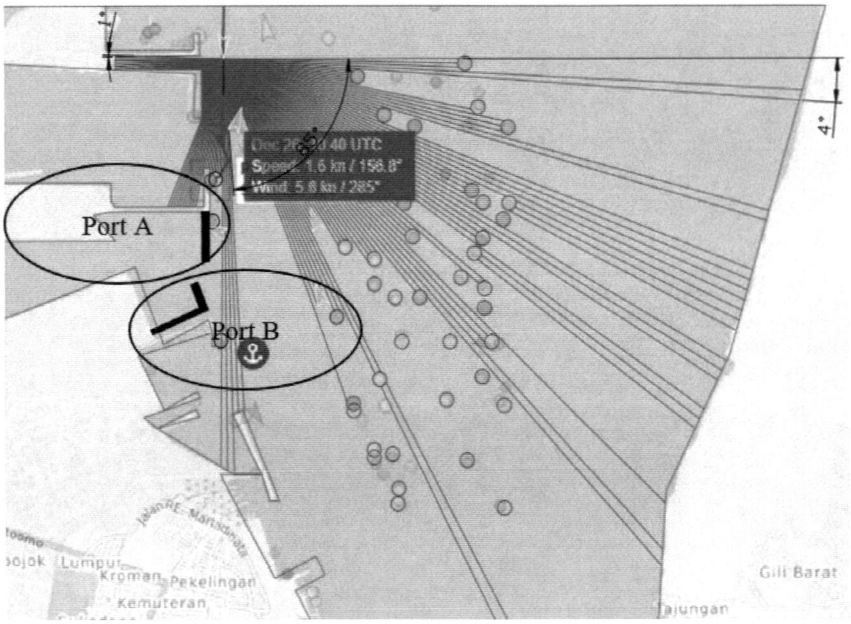

Figure 6 PIS Mahakam with scenario C

3. Conclusion

From the results of the calculations of the four models in the previous chapter, obtained the result of the tolerable speed for the scenario of adding ports and differences of location as well as the combination of 2 ports obtains the speed value of 1.5 – 1.6 knots. A ship with a length of 114 meters and a speed of 1.5 – 1.6 knots is relatively safe even though it is in the marginal category. It means that marginal limits can still be tolerated when ships approach ports. The development of ports cannot be ruled out as an alternative to pressuring dwelling time, so efforts are needed in controlling the risk of maritime traffic. Although the four scenarios are at a marginal level, it is still allowed to squeeze the fuel budget and drive fast ships to rely on ports so as not to slow down shipping traffic. The Environmental Stress (ES) Model and Environments Level & Condition methods can provide an efficient overview of current conditions related to maritime traffic risks, considering the topography of the surrounding environment. With an average ES value estimated at 749 means a marginal risk situation can reach a critical level can occur at any time if the restriction of the area of manoeuvre or something else supports. Surveillance and experienced crew are required to undermine the level of risk when ships are maneuvering on Port X or headed to port A & B, which is sometimes technically overdue. Otherwise, the marginal level can reach unimpedable critical numbers [8].

4. References

[1] A Fadillah, A. Pusaka, S. Manullang, et al, "Strategy for Reducing Pollutant Emissions from Ship Activities at the Port of Tanjung Perak, Surabaya", WSEAS Transactions on Environment and Development, Vol.11, page 155-162, 2015.

[2] Hu, Y., Zhang, A., Tian, W., Zhang, J., & Hou, Z., "Multi-ship collision avoidance decision-making based on collision risk index", Journal of Marine Science and Engineering, Vol. 8(9), page 640, 2020.

[3] Montewka, J., Manderbacka, T., Ruponen, P., Tompuri, M., Gil, M., & Hirdaris, S., "Accident susceptibility index for a passenger ship-a framework and case study", Reliability Engineering & System Safety, Vol. 218, page 108145, 2022.

[4] Inoue, Kinzo, Park, Young Soo, Usui, Hideo, Sera, Wataru and Masuda, Kenji.,"Safety management of vessel traffic in ports and waterways", 30th PIANC-AIPCN Congress Sydney N.S.W. Institution of Engineers, page 805-816, 2002.

[5] Park Y. S., Jong J. Y. and Inoue K., "A Study on Assessment of Vessel Traffic Safety Management by Marine Traffic Flow Simulation", Journal of Korea Society of Simulation, Vol. 11, No. 4, page 43-55, 2002.

[6] Inazu, D., Ikeya, T., Iseki, T., & Waseda, T., "Extracting clearer tsunami currents from shipborne Automatic Identification System data using ship yaw and equation of ship response", Earth, Planets and Space, Vol. 72, page 1-13, 2020.

[7] Park, Y. S., Lee, M. K., Kim, J. K., Lee, Y. S., & Park, M. J., "Consideration of the Possibility of Excursion Ship Passage in Busan North Port using Marine Traffic Assessment Index", Journal of the Korean Society of Marine Environment & Safety, Vol. 25(3), page 298-305, 2019.

[8] Yurtören, C., Duru, O., & Satir, T., "The environmental effects of projected container terminal to the safely manoeuvring", TransNav, International Journal on Marine Navigation and Safety of Sea Transportation, Vol. 2(1), page 93-97, 2008.

Determining The Causation Probability of Ship Collision in the Barito River Using Fault Tree analysis (FTA) & Bayesian Network Modelling

Diyah Purwitasari[1,2*], Ketut Buda Artana [2], Dhimas Widhi Handani [2] and Zulfan Ardiansyah K.E[2]

[1] Technique Department, Politeknik Pelayaran Surabaya, Surabaya, Indonesia
[2] Department of Marine Engineering, Institut Teknologi Sepuluh Nopember, Surabaya, Indonesia

*E-mail: purwitasari.diyah@gmail.com

Abstract. In Indonesia, five of the twenty-five vital ports are situated along rivers, with the Port of Banjarmasin having the highest visit value. Over the previous five years from 2017 to 2021, 20 collision-type accidents have occurred on the Barito River shipping channel. With the large number of ship collision accidents and fatalities occurring on the Barito River, it is necessary to make initial efforts to determine the probability of the main causal factors in order to carry out a risk assessment. This paper aims to determine the causation probability of ship collisions on the Barito River so that it can be used in risk assessments to formulate recommendations for mitigating the risk of ship collisions. The causes for ship collisions are identified using fault tree analysis (FTA) and modelled using bayesian networks, including the technical factors, environment factors, and human factors. Based on FTA result and BN sensitivity analysis, the main factor causing ship collision accidents with the largest probability value is environmental factors.

1. Introduction

Rivers and lakes serve as a network for both the transportation of people and goods, allowing the neighbouring cities to grow quickly and have a sizable fleet of vehicles. The surrounding population's economic survival depends on this system of rivers and lakes, with most economic activities still occurring along rivers and lakes. With the growth of river and lake transportation infrastructure, efficiency has become increasingly crucial, especially when considering transportation safety as a supporting component. Five of Indonesia's twenty-five vital ports, including Banjarmasin Port, are situated near rivers, the Barito River is one of them.

The number of ships visiting the five strategic ports in the river area has increased overall, particularly the Banjarmasin port, which receives the most visits, according to data from the Central Statistics Agency's recording of ship visits from within the nation and from abroad. The traffic reached 33,498 ships in 2020, an 11.94% increase from the year before. But at 2023, the frequency of visits by ships at ports in Central Kalimantan region decreased by 20,72 percent, from 782 visits (January 2023) to 620 visits (February 2023). The number of sea transportation passengers fell by 11,47 percent, from 31.194 people (January 2023) to 27.617 (February 2023). Goods traffic using sea transportation fell 18,36 percent, from 1,61 million tonnes

(January 2023) to 1,31 million tonnes (February 2023) [1], as shown in Figure 1. Even though the number of ship visits in 2023 decrease, this cannot deny the fact that the shipping traffic route via the Barito River from the estuary to Trisakti Port, Banjarmasin is prone to accidents. This occurs because of the narrow shipping lanes and high silting due to mud sedimentation [2].

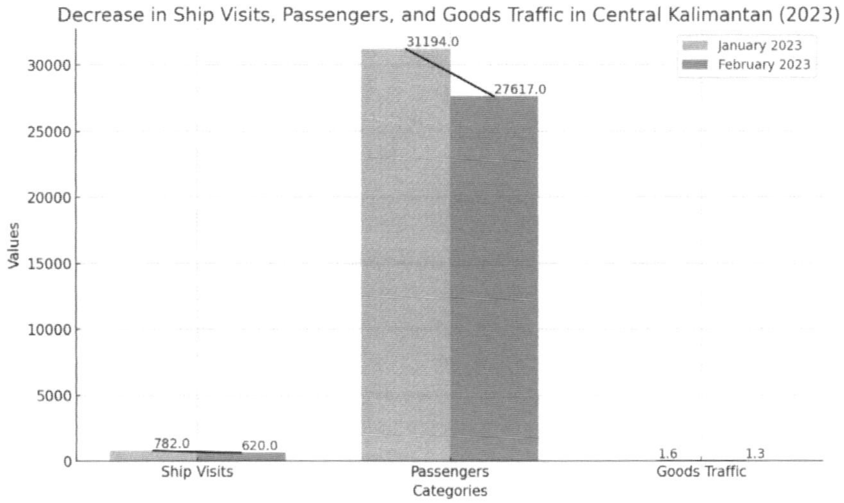

Figure 1. Bar Chart of Ship Visits, Passengers, and Good Traffic Data in Central Kalimantan from January until Februari 2023.

Since the reign of the Sultanate of Banjar, the Barito River has served as a major conduit for trade and transportation. This river has a length of about 900 km that can be travelled by ships and has its ends in Banjarmasin, South Kalimantan Province [3]. The Barito River has a relatively long distance and is home to numerous river piers, crossing ports, and special terminals run by various associated businesses. The ship's crew must work effectively with other passing ships and the marine pilot from the closest port because the channel at the Barito river's threshold is 100 meters wide and the guaranteed depth is up to a maximum of 5 LWS [4].

Figure 2. The Barito River's Marine Traffic in September 8th, 2023 [11]

IOP Conf. Series: Earth and Environmental Science **1423** (2024) 012012 doi:10.1088/1755-1315/1423/1/012012

Due to the high quantity of shipping, ship accidents are quite likely to occur, which will affect how smoothly the 15 kilometres along the Barito River threshold are travelled, as previous number of accidents shown in Table 1. In the event of an accident that has the potential to restrict shipping channels with a constrained area and river depth, the supply chain of the cities in question will be considerably impacted. A risk assessment is required as a future preventive strategy to reduce risk, as in earlier research [5], [6], [7], [8], [9], [10]. It is possible to take preventive measures before a ship accident happen that may cause losses that many linked parties will have to endure, including both material and physical losses. Figure 2 shows the Barito River's inland waterways from Jelapat Pier in the north to the connection with Tamban Muara Village in the south and the busy of ship traffic at Barito River on September 8th, 2023 [11].

A risk is an area where something or someone could experience potential danger, damage, or unfavorable health impacts. The process of locating and documenting any risks that could exist at work is known as hazard identification [12], [13]. Table 1 below shows the number of ship accidents that took place in the Barito River shipping channel between 2010 and 2019. Table 1 shows that collisions are the most frequent type of ship accident on the Barito River, with the highest number of fatalities and consequences compared to other types of accidents.

Table 1. Ship Accidents in the Barito River Shipping Channel 2010-2019 [14]

No.	Year	Ship Accident Type	Fatalities	
			Died/ Disappeared	Injured
1.	2011	Collision	18	76
2.	2013	Collision	-	-
3.	2017	Sinking	-	-
4.	2018	Fatality On Board	5	-
		Total Amount	23	76

Based on National Transportation Safety Committee ship accident data from 2017 to 2022 [15], there are three factors that causing ship accidents, are technical factors (50.63%), human factors (48.10%), and environmental variables (weather, currents, shipping channel conditions, etc.) (1.27%). Figure 3 shows, based on National Transportation Safety Committee ship accident data from 2017 to 2022, that the main cause of ship collisions is human factors.

Fig. 3. Ship accidents causation factors on 2017–2022 [15]

In this context, determining causation probability is crucial for several reasons. Firstly, quantifying the probability of causation helps to assess the impact of technical, human, and environmental factors on ship accidents. This understanding allows for targeted interventions and safety protocol improvements, such as addressing human error through training programs or enhancing technical systems. Secondly, analyzing causation probability aids in identifying patterns and trends over time. By establishing causal relationships from historical data, authorities can implement preventive measures and policies to reduce overall ship accident rates. For example, if human factors consistently emerge as primary causes, policies could focus on stricter regulations, enhanced monitoring, or technology adoption to minimize errors. Thirdly, causation probability analysis supports evidence-based decision-making by enabling stakeholders—including policymakers, regulators, and industry experts—to allocate resources effectively. By prioritizing areas with higher causation probabilities, such as human factors in ship collisions, resources can be efficiently used to enhance maritime safety. Analytical tools like Fault Tree Analysis (FTA) and Bayesian Network Modelling (BNM) play a crucial role in this process. FTA breaks down causal pathways and identifies critical factors, while BNM integrates uncertainties to provide a comprehensive understanding of risk factors. Together, these tools create a robust framework for addressing the primary causes of ship accidents identified in data from the National Transportation Safety Committee, aiming ultimately to reduce their frequency and severity.

2. Methods

Starting at Jelapat Pier in the north and ending at the river crossing at Tamban Muara Village in the south, the Barito River was the object of this analysis. The river's breadth and depth vary along the length of the channel at different points. Thus, the Barito River must be divided into four sections according to the predominant features seen in each section, which may include bends, straights, or widths that are noticeably reduced or increased. In addition to the primary attributes, every section of a river needs to have an average water draft of five meters in order for ships, river ships, and crossings to pass over it. Given the aforementioned parameters, the width and length of each river segment will be studied, as outlined in Table 2.

Table 2. Object of study [16]

Segment	Estimated Length (m)	Average Width (m)	Initial Coordinate	Final Coordinate
1	4170	365	-3.257091, 114.554097	-3.296366, 114.565542
2	2190	191	-3.296366, 114.565542	-3.316264, 114.560943
3	2360	331	-3.316264, 114.560943	-3.335080, 114.548707
4	3460	384	-3.335080, 114.548707	-3.354502, 114.528940

The traffic statistics for ships, river vessels, and crossing traversing the Barito River between January and April 2019 were gathered from KSOP Class I Banjarmasin and te Banjarmasin City Transportation Service. These data were utilized as a foundation for computing the probability of collision occurrences. The selection of this data was based on the presence of both the lowest and highest number of incidents during the specified period in 2019. Additionally, ship traffic information included details such as vessel dimensions (overall length, depth, and gross tonnage), the origin and destination of each voyage which providing insight into locations where vessels were moored along the Barito River. Additional information such as speed and vessel width obtainable from the Indonesian Classification Bureau (BKI).

IOP Conf. Series: Earth and Environmental Science **1423** (2024) 012012 doi:10.1088/1755-1315/1423/1/012012

2.1 Estimation of Causation Probability

The Barito Kuala District Transportation Service's data was used to identify the accident in this article. The data provides information on sixteen incidents that happened on the Barito River between January 2011 until December 2021 including the cause of accidents, from 449398 ship that operate in Barito River n the same timeline. The data indicates that in 2011, there were 2 accidents caused by low river levels, miscommunication, navigation equipment issues, strong winds, limites visibility, strong river currents, and drifting. In 2012, there was one accident due to strong river cureents. In 2013, there were three accidents caused by high waves, drifting, strong winds, strong currents, limited visibility, captain miscommunication, and navigation equipment issues. In 2017, there was one accident caused by drifting. No accidents were recorded in 2018 until 2019. In 2020, there were four accidents caused by strong river currents in all cases, drifting, navigation light failure, high river levels, engine stoppage, and captain negligence. In 2021, there were five accidents, two of which were caused by strong river currents, four by navigation equipment issues, limited visibility, two by miscommunication, low river levels, and error on steering gear. From the data on the causes obtained from each accident, the frequency can be determined. The causal probability value in table 3 is obtained by dividing the sum of the causes of accidents by the total traffic of ships crossing the Barito River during the same period based on data from the Harbormaster's Office and the Banjarmasin Port Authority.

This equation can be used as a method to calculate the frequency [17] [18]:

$$F_{col} = Na \times Pc \qquad (1)$$

Which, F_{col} is Possible amount of collision (Frequency); *Na* is Collision Number Candidates (Geometric collision probability, which describes the probability of ships in encounter that has potential of collision in the assessed region, or the frequency of collision candidate within specific period of time), in this matter, Na can be considered as an estimate probability of collision events based on various related variables, which in this case is assumed to be the number of vessel passing through Barito River between January 2011 until December 2021, amounting to 449398 ship. And *Pc* is Probability of failing to avoid a collision (causation probability, which describes the probability of collision due to failures from various aspects, e.g., human reliabilities, human and organisational factors, mechanical failure, etc.).

Table 3. Calculation of Causation Probability

Cause of Accident	Frequency (*F*)	Causation Probability (*Pc*)
Strong Currents	8	1.780E-05
Strong Wind	2	4.450E-06
River Tides	3	6.676E-06
High Waves (Surface Currents)	2	4.450E-06
Limited Visibility	3	6.676E-06
Machine Failure (Machine Stop Working)	1	2.225E-06
Navigation equipment issues (navigation lights, etc.)	8	1.780E-05
Error Steering System (Automatic control error/steering gear problem)	1	2.225E-06
Drifting (Lack of Qualifications, Poor Watchkeeping)	4	8.901E-06
Master's Negligence	1	2.225E-06
Misscommunication	5	1.113E-05

The causal probability of failure refers to the likelihood of a ship and its crew successfully executing maneuvers to avoid potential accidents in critical situations [19]. Accidents can be attributed to three main categories of causes, environmental factors, human factors, and other factors [20]. In this paper, data obtained from the Barito Kuala District Transportation Service indicates that there are three categories of accident causes:

a. Technical factors pertain to the mechanical, electronic, and structural components of ships, as well as the design and maintenance practices that impact their operation. These factors encompass issues such as equipment failures (e.g., engine malfunction, navigation system errors), structural integrity (e.g., hull damage), propulsion systems (e.g., steering failure), and deficiencies in onboard systems (e.g., radar, communication equipment). Technical factors directly influence the reliability and performance of ship operations, potentially leading to collisions if not properly addressed.
b. Environmental factors encompass natural conditions and external forces present in the maritime environment that contribute to ship collisions. These factors include weather conditions (e.g., fog, storms, heavy rain), visibility limitations (e.g., reduced visibility due to darkness or mist), sea state (e.g., waves, currents), ice conditions (in colder regions), and navigational hazards (e.g., shallow waters, reefs, submerged objects). Environmental factors pose challenges to safe navigation and can increase the risk of collisions by affecting ship manoeuvrability, situational awareness, and response times of crew members.
c. Human factors refer to aspects of human behaviours, performance, and characteristics that contribute to accidents or incidents. In the context of ship collisions, human factors may include errors in judgment, decision-making, communication, coordination among crew members, fatigue, distraction, or inadequate training. These factors influence how individuals operate and interact within the maritime environment, affecting the safety and efficiency of navigation.

Table 4. Calculation of Causation Probability Variables

Cause of Accident	Frequency (*F*)	Causation Probability (*Pc*)	Percentage
Technical Factors	10	2.2252E-05	26.316%
Environmental Factors	18	4.0054E-05	47.368%
Human Factors	10	2.2252E-05	26.316%
Collision	38	8.4558E-05	100.00%

2.2 FTA (Fault Tree Analysis) to determined Collision Causation Probability

The effects of a ship collision on the Barito River include fatalities, injuries, damage to the ship's hull and cargo, damage to bridges, and marine pollution due to oil or fuel spill. As a result, the company will have to budget time and money to fix the damage and make up for the losses that occur.

Fault Tree Analysis (FTA) is a systematic, deductive method used to analyze the causes of system failures. Originating from reliability and safety engineering, FTA helps identify potential causes of failures and their interrelationships. The method focuses on breaking down complex processes into simpler, identifiable, and manageable components, facilitating risk management and mitigation strategies. This method is implemented utilizing a top-down methodology, which starts with the assumption of failure from a top event (top event), then specifically identifies the causes of a top event down to a fundamental failure (root cause). The failure circumstances that logic gates express can be a single condition or a combination of different criteria. Fault Tree Analysis (FTA) is built using logic gates, specifically AND gates and OR gates [21].

The causal likelihood of a collision on the Barito River Shipping Line can be established when all collision causes have been aggregated, as shown in Table 3 and Table 4 for the ship collision. The existence of one of the three groups can lead to collisions, particularly at a number of sites with thin grooves; hence, the logic gate "or" was chosen. Based on the data of the ship collision above, using Isograph Reliability Workbench as FTA software Figure 4 is obtained.

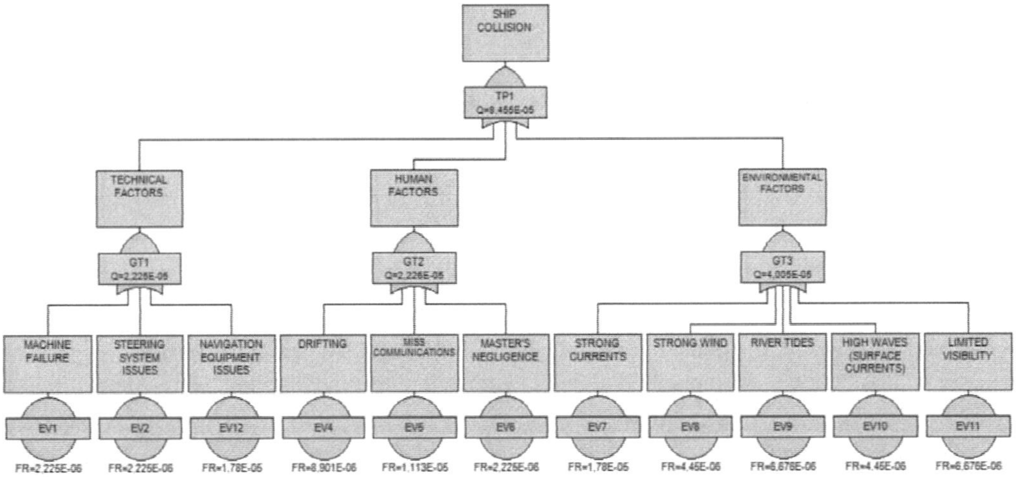

Figure. 4. Fault Tree of Ship Collision Causation on the Barito River between 2011 and 2021

The overall probability of a ship collision (TP1) is primarily driven by environmental factors, which have the highest combined probability (4.005E-05). Human and technical factors contribute nearly equally to the collision probability, with slight differences in their calculated probabilities (2.225E-05 and 2.226E-05). Within technical factors, Navigation Equipment Issues (EV12) have the highest individual failure rate (1.78E-05). Within human factors, Miscommunications (EV5) have the highest failure rate (1.13E-05). Within environmental factors, Strong Currents (EV7) are the most significant contributor (1.78E-05). The estimate of the collision probability in each collision scenario will be based on the causative probability of the Barito River Shipping Lane, which is 8.455E-05, either from equation calculations or fault tree analysis results. The environmental factors have the highest probability value as the main cause of ship collision accidents, which is 4.005E-05.

2.3 Collision Causation Modelling using Bayesian Network

Bayesian networks (BN) are commonly employed in quantitative risk assessment because of their intuitive visual layout and ability to quantitatively depict the connections among influencing factors. Bayesian Networks (BN) can be utilized as models for evaluating the safety of complex systems, offering a flexible and robust graphical structure using a probabilistic approach. In a BN, relationships between variables are depicted with directed acyclic arrows, with each variable represented as a node. The BN comprises a topological structure (including network nodes and directed links) and the Conditional Probabilities Table (CPT) for each node. This structure shows variables as nodes and their probabilistic causal dependencies as directed arrows. In these relationships, the child node points to the node it affects, while the parent node is the one from which the arrows originate. The quantitative aspect of the BN involves probability tables for the system's variables, including probabilities, conditional probabilities, and posterior probabilities derived from them. If a probability pertains to a root node, it is termed marginal (unconditional) probability. In theory, CPTs and marginal probabilities of root nodes can be formulated using statistical data, expert judgment, or a combination of both. The inference principle of Bayesian networks is based on Bayes' probability theory, fundamentally concerning probability calculation [22].

In this study, historical data on ship collision accident is used and the causes of collision which have been catagorized through fault tree analysis are used to determine node factors.

There are three root nodes based on environmental factors, human factors, and technical factors. The following in Table 5 is a description node factors used as models:

Table 5. Node factors of Ship Collision Causation

Node Factors	Description
Environmental	*Natural conditions and external forces present in the maritime environment that contribute to ship collisions, such as weather condition*
	YES: Strong Currents
	YES: Strong Wind
	YES: River Tides
	YES: High Waves (Surface Currents)
	YES: Limited Visibility
Technical	*Equipment failures, such as mechanical, electronic, and structural components of ships, as well as the design and maintenance practices that impact ship operation*
	YES: Machine Failure (Machine Stop Working)
	YES: Navigation equipment issues (navigation lights, etc.)
	YES: Error Steering System (Automatic control error/steering gear problem)
Human	*Refer to aspects of human behaviours, performance, and characteristics that contribute to accidents or incidents, such as errors in judgment, decision-making, communication, coordination among crew members, fatigue, distraction, or inadequate training.*
	YES: Drifting (Lack of Qualifications, Poor Watchkeeping)
	YES: Master's Negligence
	YES: Miscommunication

To calculate the probability of collision, this simple formula is used:

$$P(A) = \frac{n}{N} \qquad (2)$$

Which, P(A) is the probability that event A occurs, n is the number of times event A occurs, and N is the number of times all event A occurs [23].

Nodes influenced by other factors such as prior nodes, so that child nodes possess dependent characteristics. To calculate the probability of event A occurring given that event B has already occurred, the following formula is used:

$$P\,(A|B) = \frac{P(A,B)}{P(B)} \qquad (3)$$

Which, $P\,(A|B)$ is the probability of event A occurring given that event B has already occurred, $P(A,B)$is the probability of events A and B occurring, and $P(B)$ is the probability events B occurring.

In this part, Bayesian networks model the causes of collision accidents based on Table 5. The definitions, conditions, and relationships are created by considering the collision investigation reports. A GeNle program is employed to use the data for modelling the Bayes Network in Figure 5. In Figure 5, it can be seen which aspects contribute the most from each root node. Based on ship collision accident data from the Barito River between 2011 and 2021, a Bayesian network model identified that 57% occurrence probability of these accidents were due to three primary factors. Environmental factors contributed the most, with strong river currents accounting for 44%. Human factors followed, with miscommunication as the leading cause at 50%. Technical factors also played a significant role, with navigation equipment issues contributing 80%. The results can vary depending on how the causes of collision accidents are modelled in the Bayesian network. In this paper, the Bayesian network model is built based on limited data, where there are 16 ship collision data points with their causal factors released in

IOP Conf. Series: Earth and Environmental Science **1423** (2024) 012012 doi:10.1088/1755-1315/1423/1/012012

the accident report. The determination of conditional probability uses a 50-50 probability if the condition is not present in the data.

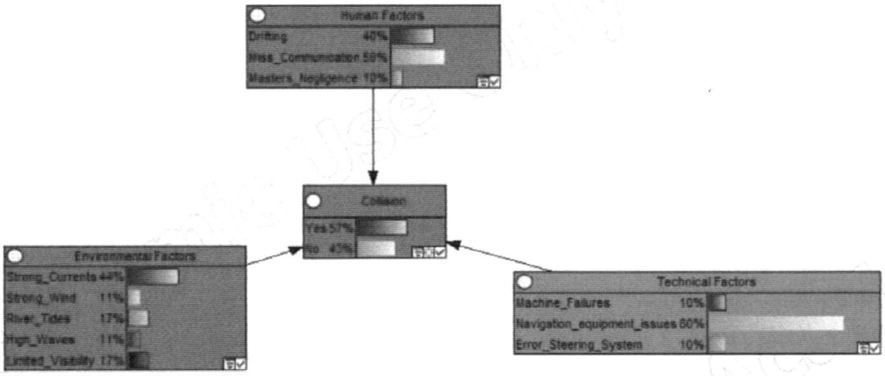

Figure 5. BN Structure of Collision Causation Factor on the Barito River between 2011 and 2021

Sensitivity analysis plays a vital role in probabilistic risk analysis by illustrating the impact of each risk factor on the likelihood of failure in a Bayesian structure. This analysis identifies which variables in the model have the most significant impact on the target node and ranks them according to their importance. In a Bayesian network, sensitivity analysis shows how changes in root and intermediate nodes affect higher-level child nodes or the leaf node. To further investigate the risk of collision, variables from different levels of the network structure are selected, and the values of their prior probabilities are adjusted. This allows for the observation of how changes in the model variables influence the overall probabilities [22]. Figure 6 below presents the results of the sensitivity analysis for the Bayesian network.

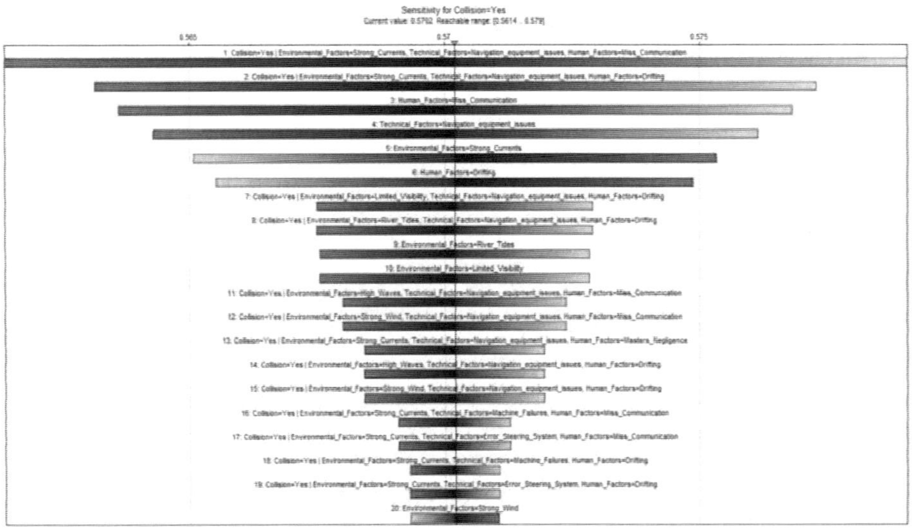

Figure 6. Tornado Sensitivity Analysis Result

The current value (0.5702) is the baseline or starting probability that a ship collision will occur, based on the initial set of conditions and inputs in the Bayesian Network model. This means there is a 57.02% chance of a ship collision happening under the specified conditions. Reachable range (0.5614 - 0.579) indicates the possible variation in the collision probability due to changes in the input variables. It shows that the probability can vary between 56.14% and

57.90%, depending on the specific values of the factors analyzed. Strong currents, navigation equipment issues, and miscommunication are highly influential in causing collisions within 0.565 - 0.575 range. Drifting also significantly impacts collision likelihood when combined with strong currents and equipment issues within the same range. Miscommunication alone is a crucial factor but less so than when combined with environmental and technical issues at moderate range (0.570 - 0.572). Equipment issues are critical, showing a significant range of moderate impact at 0.570 - 0.573. Strong currents are a major factor in the likelihood of collisions also in the same moderate range. River tides and Limited visibility have a lower impact compared to strong currents and equipment issues (Range: 0.570 - 0.572). Within the range 0.570 - 0.571 Limited visibility alone is a less significant (lower impact) factor compared to other factors like strong currents and equipment issues. High waves combined with equipment issues and drifting show a minor impact, within range 0.570 - 0.572. Strong winds also have a minor impact when compared with other more influential factors.

3. Result and Discussion

Using the Failure Fault Tree Analysis (FTA) method, the causal probability (Pc) value for the Barito River Shipping Line based on ship accident data was found to be 8.4558E-05. Utilizing methods such FTA has provided a quantitative understanding of ship collision risks. The "or" gate in the FTA diagram implies that the occurrence of any one of these causes alone can lead to a ship collision. However, the data shows otherwise. There are indeed two collision cases caused by a single factor, but in fourteen other collision cases between 2011-2021 in Barito River, they are caused by combinations of factors with several variables involved. Therefore, it would be better to modelling collision causation probability using Bayesian Networks, which are considered a method where the interactions between causes can be comprehensively observed. The BN sensitivity analysis indicates that the most significant factors contributing to the likelihood of ship collisions are strong currents, navigation equipment issues, and human factors like miscommunication and drifting. These factors have the highest impact on the collision probability and should be the primary focus for risk mitigation strategies. Lesser but still notable impacts are observed with limited visibility and river tides. Minor impacts are seen from high waves and strong winds. By addressing the top factors identified in this analysis, maritime safety professionals can develop more effective strategies to reduce the risk of ship collisions. The modelling of ship collision probabilities can affect the obtained probability values and produce different tornado sensitivity analysis results as well. Expert opinions are needed to validate the accident model to ensure that the results obtained are more reliable. Future work of this study is intended to include fuzzification in the Bayesian networks for better result.

References

[1] B. K. B. Selatan, "barselkab.bps.go.id," 2023. [Online]. Available: https://barselkab.bps.go.id/pressrelease/2023/04/03/1213/februari-2023--jumlah-penumpang-angkutan-udara-dan-angkutan-laut-mengalami-kenaikan-dibanding-januari-2023--masing-masing-sebesar-17-61-persen-dan-11-47-persen.html.

[2] A. Portal, "hubla.dephub.go.id," 2023. [Online]. Available: https://hubla.dephub.go.id/home/post/read/12656/perjanjian-konsesi-pengelolaan-alur-pelayaran-ambang-sungai-barito-ditandatangani.

[3] D. P. P. K. B. Utara, "Peta Alur Sungai Barito di Wilayah Kabupaten Barito Utara.," Dinas Perhubungan Pemerintah Kabupaten Barito Utara, Barito Utara, 2020.

[4] P. A. B. Nusapersada, "Barito River's Threshold Data," PT. Ambang Barito Nusapersada, Barito Utara, 2021.

[5] J. Montewka, P. Krata, F. Goerlandt, A. Mazaheri and P. Kujala, "Marine traffic risk modelling – an innovative

approach and a case study.," *Proceedings of the Institution of Mechanical Engineers, Part O: Journal of Risk and Reliability, „* vol. 225(3), pp. 307-322, 2011.

[6] S. Li, Q. Meng and X. Qu, "An Overview of Maritime Waterway Quantitative Risk Assessment Models," *Risk Analysis,* vol. 32(3), p. 496–512, 2011.

[7] M. E. Khaled, Y. Kawamura, E. Khaled and S. Member., "Application of Bayesian Belief Network to Estimate Causation Probability of Collision at Chittagong Port by Analyzing Accident Database of Bangladesh," in *Conference paper on ISOPE 2015,* Hawai, 2015.

[8] B. Wu, H. Tian, X. Yan and C. Guedes Soares, "A probabilistic consequence estimation model for collision accidents in the downstream of Yangtze River using Bayesian Networks," *Proceedings of the Institution of Mechanical Engineers, Part O: Journal of Risk and Reliability,* vol. 00(0), pp. 1-15, 2019.

[9] L. Sun, H. Zhang, W. Liu and F. Chen, "Research on risk assessment and control of inland navigation safety," *International Journal of System Assurance Engineering and Management,* vol. 9(3), p. 729–738, 2017.

[10] P. H. Purba, A. A. B. Dinariyana, D. W. Handani and A. F. Rachman, "Application of Formal Safety Assessment for Ship Collision Risk Analysis in Surabaya West Access Channel," in *Maritime Safety International Conference, IOP Conference Series: Earth and Environmental Science,* 2020.

[11] "earth.google.com," 2023. [Online]. Available: https://earth.google.com.

[12] Sitemate, "The Definition of hazard identification – and how you can better identify hazard," 2023. [Online]. Available: https://sitemate.com/resources/articles/safety/hazard-identification-definition. [Accessed 8 September 2023].

[13] I. A. K. K. H. P. E. &. N. M. Ratih, "A Bayesian Network for Classifying and Predicting Ship Collision," *Lecture Notes on Data Engineering and Communications Technologies. Springer Science and Business Media Deuttschland GmbH,* vol. 191, pp. 403-414, 2024.

[14] N. T. S. Committee, "Ship Accident Data in the Barito River Shipping Channel 2010-2019," 2021.

[15] N. T. S. Committee, "Ship Accident Data in Indonesia 2017-2022," 2023.

[16] Z. K. Ardiansyach, *Penilaian Risiko Tubrukan Kapal di Alur Pelayaran Sunagi Barito,* Surabaya: Institut Teknologi Sepuluh Nopember Surabaya, 2022.

[17] P. T. Pedersen, "Collision and grounding mechanics," in *Proceedings of WEMT 95,* 1995.

[18] P. Chen, Y. Huang, J. Mou and P. H. A. J. M. Van Gelder, "Probabilistic risk analysis for ship-ship collision: State-of-the-art," *Safety Science,* vol. 117, p. 108–122, 2019.

[19] P. Friis-Hansen, "IWRAP MK II Working Document Basic Modelling Principles For Prediction Of Collision And Grounding Frequencies," 2007. [Online]. Available: www.iseso.org. [Accessed 2023].

[20] S. A. I. B. S. D. Rahman. H, "Determination of the Dominant Factor Cause of Ship Accidents in Tanjung Priok," *Albacore,* vol. I, pp. 277-284, 2017.

[21] S. Kabir, "An Overview of Fault Tree Analysis and Its Application in Model Based Dependability Analysis," *Expert Systems With Applications,* vol. 77, pp. 114-135, 2017.

[22] E. A. O. T. O. A. Muhammet Aydin, "Validation of risk analysis for ship collision in narrow waters by using fuzzy bayesian networks approach," *Ocean Engineering,* vol. 231, no. 108973, pp. 1-14, 2021.

[23] H. Y. E. S. H. Heslina Fifani Maharani, "Analisis Frekuensi Tubrukan Kapal Menggunakan Causation Probability Berdasarkan Metode Bayensian Network : Studi Kasus APBS," *Jurnal Teknik Perkapalan,* vol. Vol. XX, no. No. X, pp. 1 - 11, 2020.

[24] Y. Fujii and R. Shiobara, "The Analysis of Traffic Accidents," *Journal of Navigation,* vol. 24(04), p. 534–543, 1971.

[25] D. Stamatis, Failure Mode and Effect Analysis: FMEA from Theory to Execution, Second Edition., Wisconsin: ASQ Quality Press Milwaukee, 2003.

[26] M. COCCIA, "The Fishbone diagram to identify, systematize and analyze the sources of general purpose technologies," *Social and Administrative Sciences,* vol. 4, no. 4, pp. 291-303, 2017.

IOP Conf. Series: Earth and Environmental Science **1423** (2024) 012013 doi:10.1088/1755-1315/1423/1/012013

Assessing post-fire material degradation: A ship structural analysis case study

S Komariyah[1], R Lesmana[1], M R F Hariadi[1] and S Anggara[1]

[1] Biro Klasifikasi Indonesia, Jl. Yos Sudarso 38-40, Jakarta, Indonesia, 14320

siti.komariyah@bki.co.id

Abstract. A bulk carrier ship caught fire while loading fertilizer at the dock, originating from generator of auxiliary engine No. 2 portside. It took approximately 12 hours to successfully extinguish the fire. As the consequence of this accident, the ship suffered significant damage to its structure, machinery, and electrical systems in various areas, including the engine room, superstructure and deck house. A comprehensive assessment was conducted to evaluate the structural integrity of the ship, particularly concerning the post-fire material strength degradation, as well as the impacts of using seawater to extinguish the fire. Destructive tests were carried out on eight material samples taken from parts of the structure heavily affected by fire-induced deformation and from areas that were visually affected by the fire accident. The test results show that there is a decrease in strength of the materials, particularly in areas experiencing slow cooling due to the annealing phenomenon. Areas of the structure exposed to water spray during the firefighting process did not exhibit material embrittlement, as no martensite phase was detected in the microstructure. Additionally, the hardness value remained significantly lower than that associated with the martensite phase.

1. Introduction

A fire on a ship is a serious accident that can not only cause loss of life and environmental pollution but also significant damage to the ship's structure due to exposure to high temperatures. In the context of seaworthiness related to the strength of the ship's structure, understanding the effect of high temperatures on the properties of materials is a crucial aspect in assessing the ship's structure after a fire. Knowing how high temperatures can degrade the strength of material and its impact on the structural integrity of the ship can help in determining mitigation measures related to the seaworthiness of the ship's structure after a fire to enhance maritime safety.

Studies related to material degradation due to fire have been conducted by several researchers through both laboratory testing and numerical analysis. Paik et al. [1][2] conducted full-scale tests on steel plate structures to provide valid data for analyzing structural failure due to fire and applied this model to analyze the top site structure of a VLCC [3]. Zhang et al. [4] analyzed the FPSO structure exposed to pool fire, where the temperature of the steel plate would increase during the fire. The mechanical properties of steel plate structures will decrease drastically when exposed to temperatures above 600°C [5]. A study by Melchers et al. [6] stated that exposure to seawater on steel can activate the corrosion process, which results in reduced structural strength [7]. In addition, there are several other studies related to material degradation after a fire [8][9][10].

By integrating knowledge from various related studies, we can develop a more holistic understanding of material degradation after a fire and apply appropriate measures to ensure the safety of ships after the accident. This research involves destructive testing on material samples taken from parts of the ship's structure that have experienced significant deformation due to fire and from areas visually affected by the accident. The primary objective of this research is to identify and understand the changes in material properties after exposure to fire and the impact of using seawater as a firefighting medium. By comprehending the effects of fire on material properties, we can make a better decision about whether the material needs to be re-plating, which is a crucial step in maintaining the structural integrity and overall safety of the ship.

2. In-site observation

A case study analysis of the ship's structure was conducted on a bulk carrier that experienced a fire while loading fertilizer at the dock. The fire originated from an auxiliary engine in the engine room and spread upwards through openings such as engine openings, stairs, doors, and other apertures. The fire did not spread to the cargo area, so the affected area of the fire accident was the stern of the ship, from the engine room to the wheel house, see figure 1(a). A quick action was taken to extinguish the fire and evacuate the ship, and after approximately 12 hours, the fire was declared extinguished by the authority.

(a) (b) (c)

Figure 1. Condition of the ship after fire, (a) stern of the ship (b) main engine (c) wheel house deck.

Visual observations of the ship's structure, machinery and electrical equipment, navigation and radio communication equipment were conducted. Additionally, measurements of the gap in deformed plates caused by high temperatures due to fire were taken to assess the severity of the damage. The focus was particularly on the bulkhead, wall and deck structures in the engine room, accommodation room and wheel house.

It found that all machinery and electrical equipment, piping, partitions, radio communication, navigation equipment were burned except those made of steel, see figures 1(b) and 1(c). The partitions on the accommodation deck were severely damaged, leaving only the steel walls, and similarly with the piping system in the engine room, see figure 2. The coating that protects the material, in a chipped and damaged condition. According to the manufacturer's coating product

IOP Conf. Series: Earth and Environmental Science **1423** (2024) 012013 doi:10.1088/1755-1315/1423/1/012013

catalogue [11][12], the temperature that can cause the coating to peel or become damaged is above 120°C.

Figure 2. Condition of piping system after fire.

The measurement of the gap revealed that the structures connected to the source of the fire had significant deformation. The average deformation of the plate structure is approximately 1-2 cm, although at some points on the engine room bulkhead, deformations of approximately 3-4 cm were found. The largest deformation, measuring 5 cm, was found on the steel wall of the accommodation room on the poop deck. These deformation values are still smaller than the maximum allowable deformation [13].

2.1. Impact of seawater to extinguish fire

The extinguishing of the fire on the ship was carried out using seawater. Seawater can cause rust or corrosion on the surface of affected plates because seawater as an electrolyte solution will oxidize iron metal. These can be seen visually at several points on the plates exposed to seawater where corrosion has occurred, as evidenced by the formation of dark yellow rust, see figure 3.

(a) (b)

Figure 3. The condition of plate due to the use of seawater to extinguish the fire, (a) bridge deck (b) deck plate

3. Mechanical properties

In order to determine the changes in material properties due to fire, destructive tests were conducted, including tensile tests, hardness tests and impact tests. Additionally, chemical composition tests and microstructure tests were carried out to complete the analysis. The test results were then compared with the class requirements [14] to assess the impact of the fire on the material properties. Material samples for testing were taken from areas with significant deformation and from areas visually affected by the fire accident. A total of eight material samples were taken from the locations shown in figure 4. Figure 5 shows pieces of plates taken from some structures affected by the fire.

IOP Conf. Series: Earth and Environmental Science **1423** (2024) 012013 doi:10.1088/1755-1315/1423/1/012013

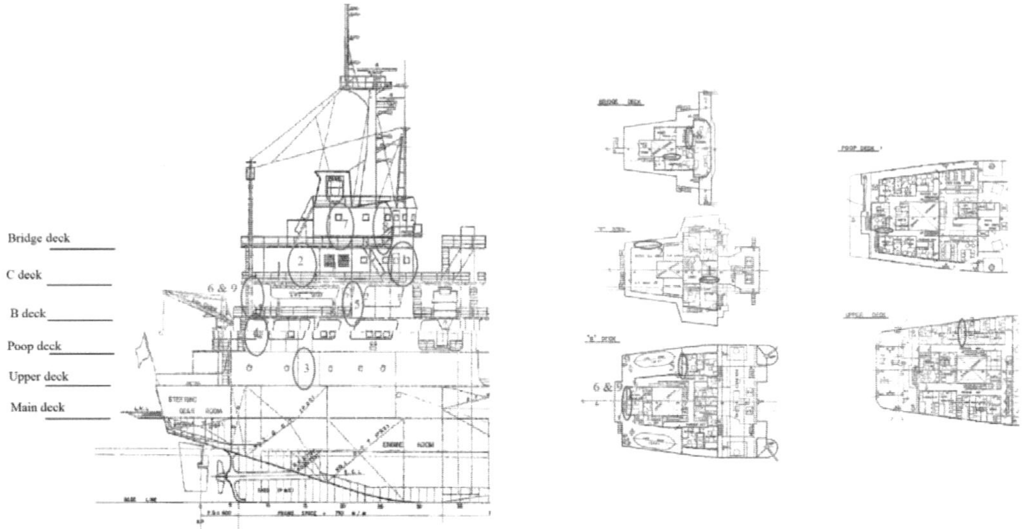

Figure 4. Location of material sampling

(a) (b) (c) (d)

(e) (f) (g) (h)

Figure 5. Pieces of plate taken from affected area

(a) Sample no 1, taken from C deck, captain room, longitudinal bulkhead, Fr 28 – 29
(b) Sample no 2, taken from C deck, Emergency Generator Room, longitudinal bulkhead, Fr 15 – 16
(c) Sample no 3, taken from Upper deck, Filter wash, transverse bulkhead, Fr 15P
(d) Sample no 4, taken from Poop Deck, Officer lounge, longitudinal bulkhead, Fr 8-9P
(e) Sample no 5, taken from B Deck, second officer room, transverse bulkhead, Fr 21P
(f) Sample no 6, taken from B Deck, second officer spare room, transverse bulkhead, Fr 8
(g) Sample no 7, taken from Bridge Deck, Radio officer room, longitudinal bulkhead, Fr 19-20
(h) Sample no 8, taken from Bridge Deck, wheel house, TBHD, Fr 24P

3.1. Yield strength and tensile strength

Table 1. Results of tensile test

Sample No.	Direction	Yield strength [N/mm²]	Min. Req. 235 [N/mm²]	Tensile strength [N/mm²]	Requirement 400 – 520 [N/mm²]
1	Transverse (A)	300.86	Pass	421.78	Pass
	Longitudinal (B)	281.59	Pass	407.07	Pass
2	Transverse (A)	368.14	Pass	480.26	Pass
	Longitudinal (B)	352.58	Pass	460.47	Pass
3	Transverse (A)	288.15	Pass	401.80	Pass
	Longitudinal (B)	285.86	Pass	405.39	Pass
4	Transverse (A)	334.69	Pass	437.11	Pass
	Longitudinal (B)	279.10	Pass	405.09	Pass
5	Transverse (A)	303.14	Pass	434.45	Pass
	Longitudinal (B)	306.62	Pass	425.59	Pass
6	Transverse (A)	314.54	Pass	450.32	Pass
	Longitudinal (B)	301.27	Pass	456.05	Pass
7	Transverse (A)	255.62	Pass	374.31	Fail
	Longitudinal (B)	256.63	Pass	371.04	Fail
8	Transverse (A)	361.75	Pass	464.63	Pass
	Longitudinal (B)	367.30	Pass	459.41	Pass

Table 1 shows the results of the tensile tests conducted on the material samples in the transverse (A) and longitudinal (B) directions. Out of the eight material samples tested, there was one sample that did not meet the requirements. Sample no. 7 had a tensile strength value that was below the required threshold. This sample was taken from the longitudinal bulkhead plate of the radio officer room, from frame number 19 to 20, which are located on the bridge deck. This plate experienced significant deformation which caused the tensile strength value of sample no. 7 to become small.

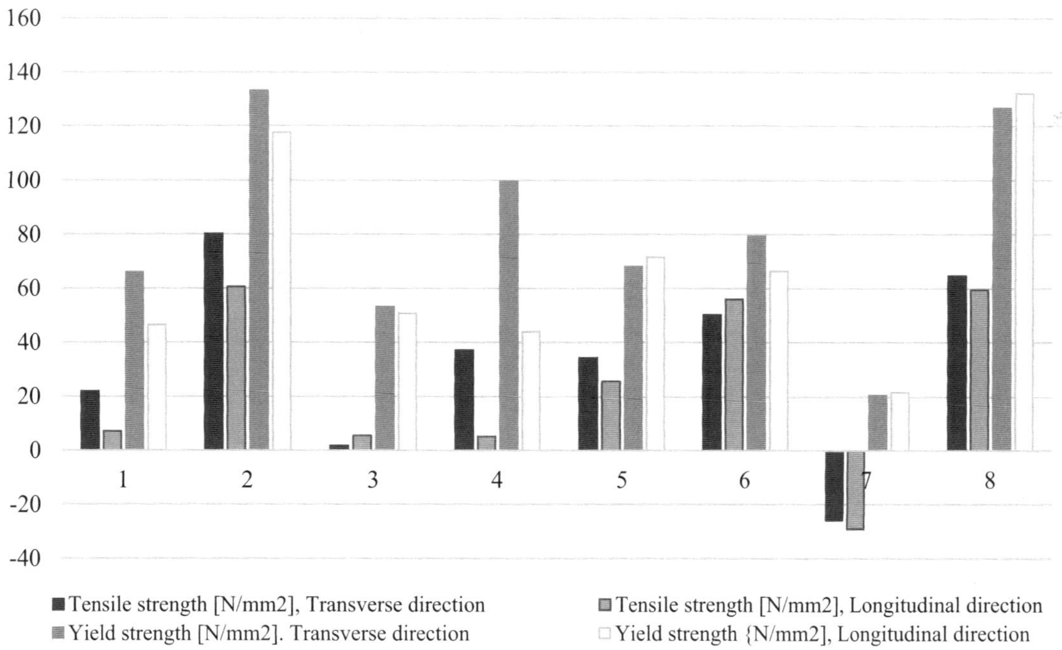

Figure 6. The differences between the actual sample and minimum requirement

When the difference between the actual tensile strength values of the samples and the minimum required values [14] is displayed in the bar charts, see figure 6, it shows that in the longitudinal direction (B), some samples have a small difference, namely samples no. 1, 3, and 4, while the other samples, except for sample no. 7 which has a negative difference, have a relatively large difference. For the yield strength values, it can be seen that all samples still have sufficiently high yield strength values, with samples no. 2 and 8 having the highest differences compared to the other samples.

It is observed that samples no. 1, 3, 4 in the longitudinal direction (B), and 7 have low strength, suggesting that the plates from which these samples were taken experienced a decrease in strength due to the fire, although for some samples, this reduction in strength still meets the minimum required.

The decrease in strength observed in some samples after the fire is suspected to be due to the softening of the material as a result of the annealing process that occurred during the fire and extinguishing process. Additionally, the reduction in strength could also be caused by the deformation experienced by the material during the fire, resulting in internal stress that can degrade the material's strength. The annealing process is suspected to have occurred in several parts of the structures, especially those inside the ship that were exposed to high temperatures from the fire but were not directly reached by water during extinguishing, leading to very slow cooling of the material in these structures.

Samples no. 2 were taken from the longitudinal bulkhead plate of the emergency generator room at Fr 15-16 located on the C-deck, sample no. 6 taken from the transverse bulkhead plate of the second officer spare room Fr 8 located on the B-deck, and sample no. 8 taken from the transverse bulkhead of the wheelhouse Fr 24P located on the bridge deck, still have high strength after the fire. This is likely because they were positioned on the outer sides of their respective structures, see figure 4, which were reachable by water during extinguishing, so the plates in these structures did not undergo slow cooling that could lead to softening or annealing, and deformation.

3.2. Hardness

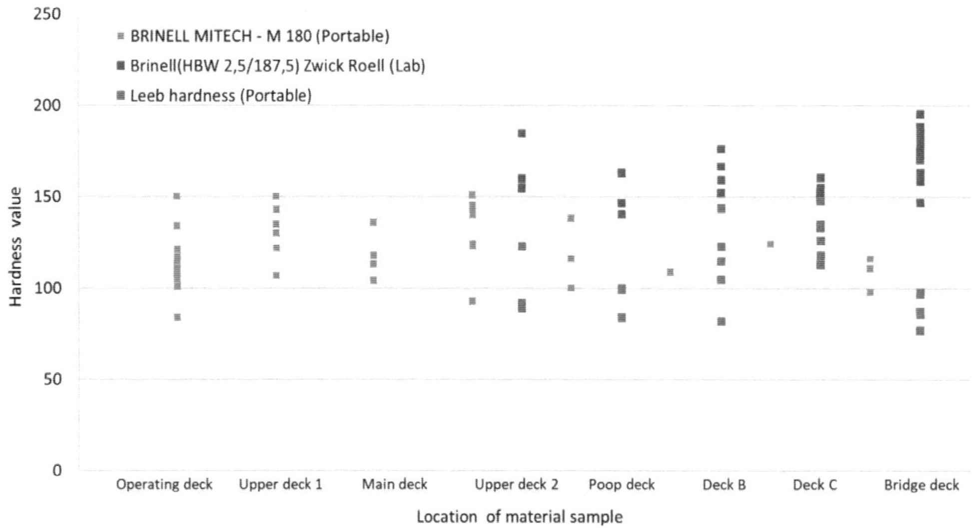

Figure 7. Results of hardness test

Hardness tests were conducted on the eight material samples using two pieces of laboratory equipment, i.e. Brinnel Zwick Roell and Brinel Mitech-M180. These results were compared with direct measurements on board using the Leeb hardness. It was found that the hardness values obtained from these three different testing equipment varied, especially with the portable equipment showing a significant range of differences compared to the fixed equipment, see figure 7.

The hardness values for each location ranged between 85 and 190 HB. These values are still far below the hardness of martensite, which is above 570 HB [15]. Additionally, hardness tests were also conducted on the daily oil tank plate located in the engine room to compare the hardness value of the outside of the tank, which was directly exposed to the fire, against the inside of the tank, which contains fuel, as shown in figure 8. It shows that materials directly exposed to the fire have higher hardness values compared to materials that were not exposed.

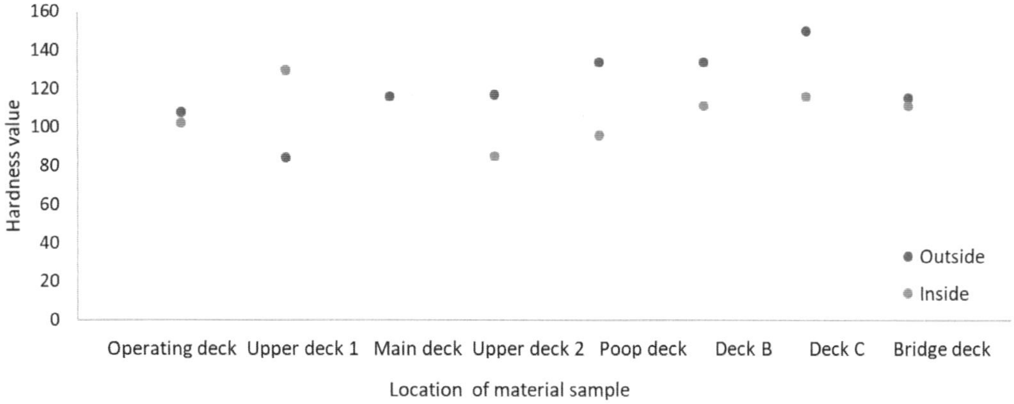

Figure 8. Results of hardness test for daily fuel tank

3.3. Toughness

Toughness was assessed through impact tests, a method used to evaluate the ability of metal to absorb energy by measuring the energy required to break a material sample through rapid impact or swinging. For the transverse direction, sample number 7 had toughness values below the minimum requirement of 20 J, and for the longitudinal direction, sample number 5 failed to meet the minimum requirement of 27 J, see Table 2.

Table 2. Results of impact test

Sample No.	Transverse direction [Joule]	Min. Req. 20 [Joule]	Longitudinal direction [Joule]	Min. Req. 27 [Joule]
1	27.308	Pass	54.414	Pass
2	22.877	Pass	65.679	Pass
3	30.438	Pass	66.655	Pass
4	28.424	Pass	64.408	Pass
5	47.399	Pass	21.374	Fail
6	55.136	Pass	69.454	Pass
7	19.174	Fail	49.253	Pass
8	20.488	Pass	52.901	Pass

Overall, the test results indicate that the samples tend to have low impact or toughness values due to the decrease in strength resulting from the fire. Additionally, this could be attributed to low manganese content, as confirmed by the chemical composition tests, represented by samples no 3 and 8, although these values still meet the requirements, see Table 3.

Table 3. Results of chemical composition test

Sample No.	C	Si	Mn	P	S	Cr	Mo	Ni	Al	Fe
3	0.163	0.104	0.630	0.008	0.026	0.063	0.005	0.053	0.006	98.900
8	0.173	0.119	0.511	0.006	0.015	0.075	0.004	0.066	0.007	00.000

4. Microstructure

With fire simulation, Zhang [4] predicted that the ship's construction near the fire source would be exposed to temperatures exceeding 1000°C, surpassing the critical temperature (723°C) for microstructural changes in the material. The microstructure test results are shown in Table 4.

Table 4. Results of microstructure test

Sample No.	Ferrite [%]	Pearlite [%]	Macro	
			200 x	500x
1	67.53	32.47		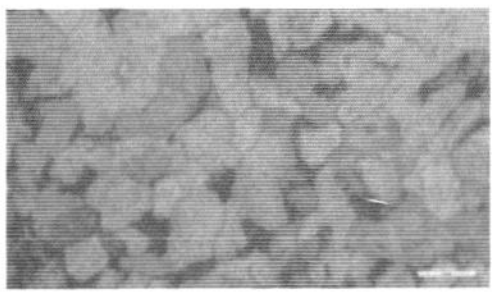
2	67.86	32.14	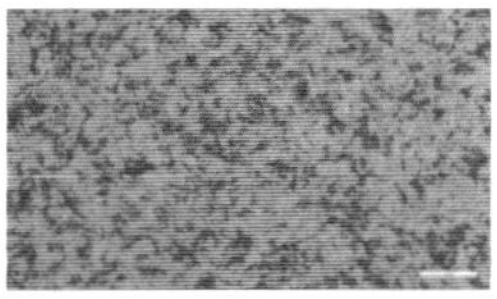	
3	61.92	38.08	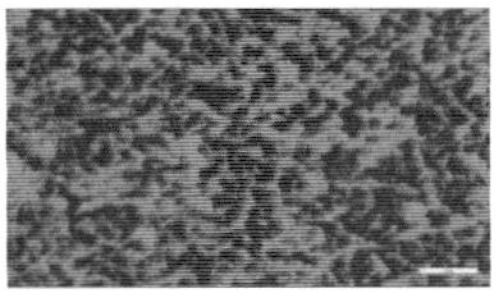	
4	61.40	38.60		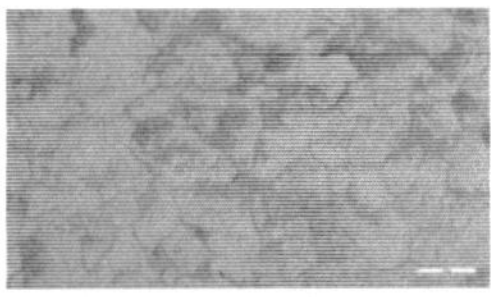
5	56.81	43.19	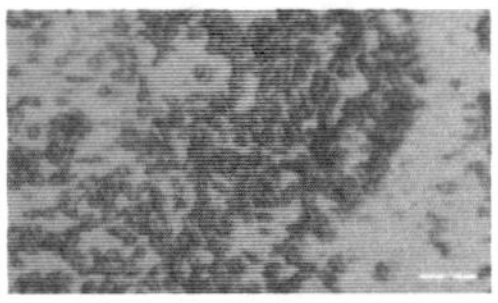	
6	66.38	33.64	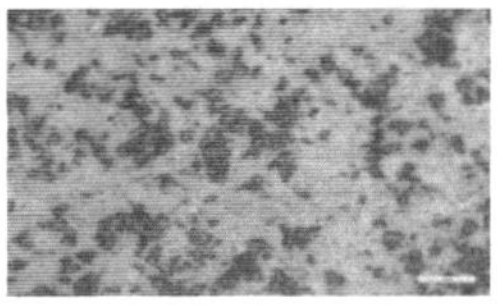	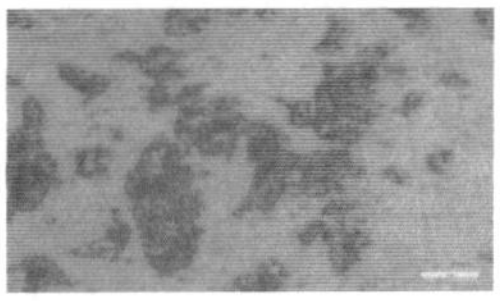

7	81.54	18.46	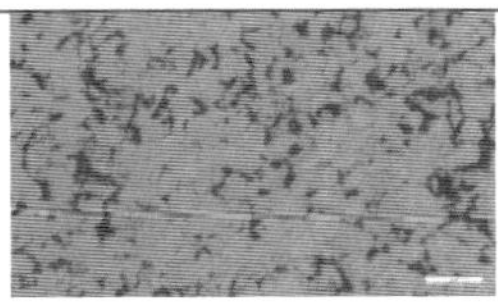	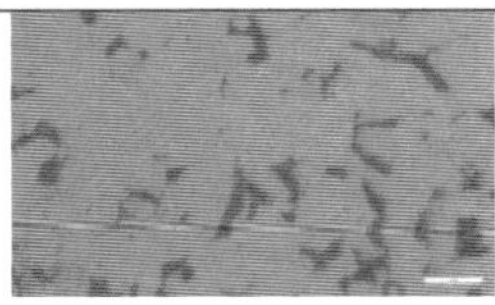
8	58.54	41.46	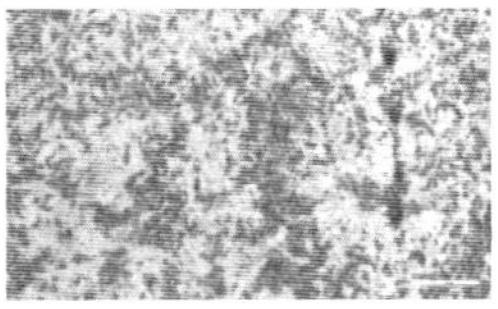	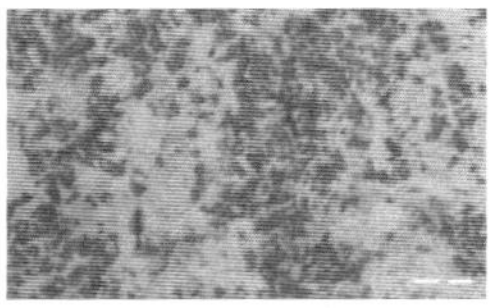

The phases formed in all samples are similar, consisting of the ferrite phase, which is characterized by a bright phase, and the pearlite phase, which is characterized by a dark phase. The percentage composition of each phase is almost the same between samples, except for one result from sample no. 7, taken from the longitudinal bulkhead of the radio officer room Fr 19-20 located on the bridge deck, which is the engine casing wall, shows a significant difference where the number of ferrite phases has increased. This is likely caused by prolonged heating of the plate during the fire, leading to phase transformation.

5. Conclusion

Based on the research findings, some important conclusions regarding the impact of fire on the strength of materials were drawn. Fire causes a degradation in material strength, especially in areas of the structure that experience slow cooling. This is due to the annealing process that occurs in the material during cooling. Additionally, the degradation in strength also results from the deformation experienced by the material due to the high temperatures from the fire.

In the parts of the structure that were exposed to water during the extinguishing process, there were no indications of quenching in the material. This is because the martensite phase was not detected in the microstructure of the tested material. Furthermore, the hardness values of the material in areas exposed to water were significantly lower than the hardness typically associated with the martensite phase.

6. References

[1] Paik J K, Ryu M G, He K, Lee D H, Lee S Y, Park D K and Thomas G 2021 Full-scale fire testing to collapse of steel stiffened plate structures under lateral patch loading (part 1) – without passive fire protection *Ships and Offshore Structures* 16(3) 227–242

[2] Paik J K, Ryu M G, He K, Lee D H, Lee S Y, Park D K and Thomas G 2021 Full-scale fire testing to collapse of steel stiffened plate structures under lateral patch loading (part 2) – with passive fire protection *Ships and Offshore Structures* 16(3) 243–254

[3] He K, Kim H J, Thomas G and Paik J K 2022 Analysis of fire-induced progressive collapse for topside structures of a VLCC-class ship-shaped offshore installation *Ships and Offshore Structures*

[4] Zhang D, Zhao B and Zhu K 2022 Structural Response Analysis of FPSO under Pool Fire *Buildings* 12 1665

[5] Wang W and Kodur V 2020 *Material Properties of Steel in Fire Condition* Elsevier

[6] Melchers R E and Jeffrey R 2005 Early corrosion of mild steel in seawater *Corrosion Science* 47 1678–1693

[7] Melchers R E 2005 The effect of corrosion on the structural reliability of steel offshore structures *Corrosion Science* 47 2391–2410

[8] Osman M H, Sarbini N N, Ibrahim I S, Ma C K, Ismail M and Mohd M F 2017 A case study on the structural assessment of fire damaged building IOP *Conf. Series: Materials Science and Engineering* 271 012100

[9] Winful D A, Cashell K A, Afshan S, Barnes A M and Pargeter R J 2017 Material properties of high strength steel under fire conditions *EUROSTEEL* Copenhagen, Denmark

[10] Tang Z, Wei T, Ma Y and Chen L 2019 Residual strength of steel structure after fire events considering materials damages *Arabian Journal for Science and Engineering*

[11] https://pdf.nauticexpo.com/pdf/hempel/hempadur-85671/21594-123518.html

[12] https://pdf.nauticexpo.com /pdf/jotun/jotamastic/31197-26882.html

[13] IACS 2017 *Rec. 47 Shipbuilding and repair quality standard* Rev.8

[14] BKI 2024 *Rules for material*
[15] Bramfitt B L 2002 Carbon and alloy steels *Handbook of Materials Selection* p 39

Acknowledgments
We would like to extend our deepest gratitude to the field surveyors, namely Ilham Syahrir, Adi Yudho Wijayanto, Miftah Faridy and Edwin Olaf for their unwavering support and dedication. Their efforts have been invaluable in ensuring the smooth progress and success of this research. We truly appreciate their contributions and support throughout this endeavour.

IOP Conf. Series: Earth and Environmental Science **1423** (2024) 012014 doi:10.1088/1755-1315/1423/1/012014

Data and transaction security as priority in operationalizing blockchain technology in ferry transportation: a literature review

Sri Sarjana

Marketing, Innovation and Technology Department, Politeknik Transportasi Darat Indonesia – STTD, Bekasi, 17520, Indonesia

Email: srisarjana@gmail.com

Abstract. Blockchain technology has a potential benefit of becoming valuable asset in ferry transportation industry to increase safety, transparency, efficiency and meet increasing market demands in meeting safety and sustainability standards. This study aims to find the important role of blockchain technology in the development of ferry transportation. Two main approaches in describing analytical techniques include bibliometrics to find derivatives of blockchain technology concept, and analytical hierarchy process to obtain alternative priorities in the development of ferry transportation. Journal publications within last ten years obtained through searches on Google Scholar are used as the main data in developing scientific literature analysis. The results found that several alternatives play an important role in the development of blockchain technology applied to ferry transportation include data and transaction security, supply chain tracking, identity and access management, operational efficiency, regulatory compliance, interoperability with external systems, transparency and accountability. Three alternatives in developing blockchain technology are main priority in this study and potential to have high novelty value, including data and transaction security, operational efficiency and supply chain tracking. The results of study recommend that the operationalization of ferry transportation really requires the development of blockchain technology to encourage increased safety and sustainability of ferry transportation.

1. Introduction

New challenges for ports that require transformation in information and communication systems to increase global competitiveness through technology optimization [1]. The application of blockchain technology can change consumer behaviour and increase business value in various sectors [2], to strengthen levels of authenticity, security and risk management [3]. With increasing dependence on digital systems and data, blockchain technology is being implemented as a solution to address the need to increase security, transparency and efficiency in the maritime industry [4]. A blockchain technology in ferry transportation plays a role in bringing a number of benefits to the industry and related stakeholders. The concept of blockchain technology in ferry transportation involves the application of blockchain principles to increase efficiency, transparency and security in the operation of the ferry transportation system to help optimize the operation of ferry transportation, increasing the efficiency, transparency and security of the system as a whole. The hope is that by integrating blockchain technology in ferry transportation operations to gain competitive advantage, increase efficiency, security and trust, as well as open up opportunities for innovation and long-term business growth. The application of blockchain technology in the shipping industry can provide transparency, security and reduced costs in financial transactions [5]. Blockchain uses a chain data structure aimed at verifying and recovering data, applies a distributed node consensus algorithm aimed at updating data, uses cryptography in a secure data transmission process and carries out smart contracts using automatic script code as an effort to manipulate data [6].

Even though blockchain technology produces many great benefits for the operationalization of ferry transportation and has been widely used in various countries including China, Ghana, Italy, Russia, there are several problems that need to be overcome, including the importance of using a diffusion mechanism that utilizes blockchain technology which is applied in the supply chain financial system. and the ability to verify the stability of the resulting system model [7]. Apart from that, there are problems in cross-shipping, delays in delivery, data hacking, transparency, intermediaries, and robbery which require the integration of blockchain technology in the implementation of traditional supply chains [8]. Other problems in integrating blockchain technology are due to institutional and regulatory incompatibility, incompatibility of technical integration and alignment, and inadequate stakeholder involvement [9]. Further issues that have not been explored in the use of blockchain technology are aimed at identifying obstacles, providing local solutions, increasing consumer confidence, having resilience to crises, understanding environmental implications and leading to fair market access [10]. Issues that are important to be followed up immediately and can be studied in more depth in the application of blockchain technology developed in the form of intelligent sustainable energy business models, the importance of studying sustainable energy system conversion, strengthening blockchain technology that prioritizes integrated energy services and building a sustainable energy transition model [11]. There are problems in storing information on a large scale and problems with storing data that cannot be changed or deleted, so various technological developments are needed, including blockchain technology [12].

Many solutions for utilizing blockchain technology have been proposed, including the development of a decentralized identity management system based on blockchain technology that is able to provide a high level of security and transparency for all parties involved in the transportation ecosystem, including the use of ferry transportation [13]. Blockchain technology is a solution for developing decentralized access control schemes in order to provide data access with transparency and traceability to ensure the privacy of policies and attributes as well as minimizing parameter sizes and increasing algorithm efficiency [14]. Furthermore, the use of Turing complete consortium blockchain is used as a solution in facilitating international trade to reduce the risk of transaction costs related to logistics chain management and contract arbitration by complementing trade financing which requires a stable banking system and an ever-evolving financial market [15]. IoT and blockchain are applied as a means of integration in monitoring, tracking and optimizing real-time traffic flows aimed at reducing congestion, increasing safety aimed at achieving overall transportation system efficiency [16]. Blockchain technology is able to improve business operational processes and the quality of supply chain management [17], to increase the efficiency of accounting management in safer and more transparent financial management [18]. Blockchain technology can take into account quality and quantity reductions in the product transportation process [19], be applied in digital waste management and the circular economy [20]. Other uses implemented through a data-driven chatbot framework based on blockchain technology can be applied to design and manage maintenance activities through the use of building information modelling which is useful for government authorities and asset managers for the preservation of historic buildings [21]. Blockchain can be applied in digital currency systems, the Internet of Things, smart networks, supply and financial systems [22], the importance of using blockchain technology in various applications that can support agriculture, education, government and voting systems [23]. Blockchain technology develops a distributed storage system that produces effective solutions in solving various problems by sharing information [7].

Study of the role of blockchain technology in the development of maritime transport could provide valuable insights into how innovative technologies can improve efficiency, safety and transparency in the maritime transport industry. Carrying out blockchain analysis can understand its potential use in speeding up logistics processes, ensuring data integrity, and improving risk management and compliance. Blockchain technology requires the diffusion of innovation to achieve competitive advantage [24]. The prospects for developing blockchain technology in the future can be carried out through standardization, platform development, regulatory system development, cost reduction and talent development [25]. It is necessary to make various efforts to explore and design new blockchain technology to face future challenges [26].

2. Method

This research uses a combined method to find innovations originating from blockchain technology in ferry transportation. This mixed method requires a thorough understanding of the problem under study, including data collection and analysis [27]. The analysis technique used involves a qualitative approach by developing Vosviewer as a bibliometric analysis tool to identify various new indicators or dimensions from the analysis results. A quantitative approach is carried out using the analytical hierarchy process (AHP) to determine priorities in decision making. The qualitative approach includes the development of Vosviewer to generate new indicators or dimensions from analysis results as well as a visualization tool that analyzes and displays the relationships between various elements in a data set, enabling the identification of new patterns or themes. Vosviewer is able to present data visualizations from journal databases in the form of knowledge mapping to facilitate understanding in an interesting and elegant manner [28], as well as functioning as an analytical tool to identify various new aspects related to concepts, theories or knowledge [29], with the ability to create network and density visualizations [30]. On the other hand, quantitative approaches use AHP to determine the priority of findings in decision making, solving complex problems by decomposing decisions into multi-level hierarchies, evaluating criteria and alternatives, and calculating the relative weight of each criterion to set priorities. AHP is applied as a complex approach in decision making by identifying relevant and alternative assessment criteria, then comparing them based on relative weights [31].

The data source for this research was obtained through literature searches from a collection of journals regarding the concept of blockchain technology in ferry transportation. A search was conducted on Google Scholar using the keywords "blockchain technology in ferry transportation" which includes publications from all over the world, with data collection carried out in May 2024. From the period 2014 to 2023, there were 668 related journal publications, with an average of 67 manuscripts published every year for the last ten years that is relevant to the topic under study. Literature searches in journal collections rely on trusted and verified information sources, such as Google Scholar, which is very useful for finding scientific journals from various scientific disciplines and regions throughout the world. By using the keywords "blockchain technology in ferry transportation," search results can be focused specifically on topics of interest. This allows the research to collect relevant texts from a variety of different perspectives and studies. With access to this data source, in-depth analysis can be carried out on the latest trends, methods and findings in blockchain technology in ferry transportation, and provide valuable insights for the development of solutions in the context of technology innovation in ferry transportation. This allows the collection of relevant texts from various points of view and studies, in-depth analysis of current trends and findings, as well as the development of innovative solutions in the context of ferry transport technology. This approach not only enriches research but also makes important contributions to the development of industry and academic literature.

AHP method [32], Analytic Network Process (ANP) [33], and Technique for Order Preference by Similarity to Ideal Solution (TOPSIS) [34] are three multi-criteria for decision making techniques that are often used, each of which has a different approach and characteristics. AHP approach develops a hierarchical structure, assumes independence between criteria and alternatives, utilizes pairwise comparisons and is able to measure the consistency of assessments [35]. The ANP method uses a network structure with linkages and feedback, uses pairwise comparisons and is able to measure the consistency of assessments. Meanwhile, the TOPSIS method uses a vector-based approach with positive and negative ideal solutions, utilizes measurements of closeness to the ideal solution, does not measure consistency of assessment and focuses on Euclidean distance. The approach used in this study has high accuracy by applying the AHP method as part of the analysis technique used.

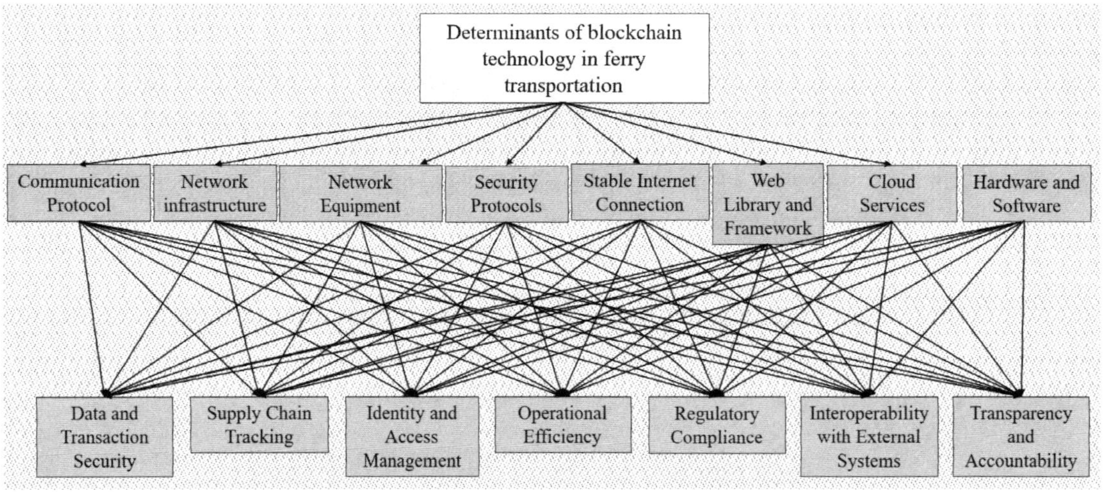

Figure 1. AHP Hierarchy on blockchain technology in ferry transportation

AHP implementation is used to analyze criteria and alternatives to recommend the resulting decisions is shown in Figure 1. The AHP method can be applied in decision making to solve complex problems using criteria and alternatives [36]. In this research, the criteria used to measure and determine the application of blockchain technology in ferry transportation include communication protocol, network infrastructure, network equipment, security protocols, stable internet connection, web libraries and frameworks, cloud services, hardware and software. Alternatives analyzed with AHP include data and transaction security, supply chain tracking, identity and access management, operational efficiency, regulatory compliance, interoperability with external systems, transparency and accountability. AHP assigns relative weights to each criterion for the purpose of finding the best application of blockchain technology in ferry transportation, then assesses each alternative based on these criteria and multiplies the criteria values by their weights to obtain an aggregate value. The goal is to understand which alternative is best suited to achieve the desired goal. AHP helps in overcoming the complexity of decision making by providing a structured and measurable framework, so that decisions taken are more informative in the context of blockchain technology in ferry transportation that considers various factors and alternatives.

AHP's important role in decision making lies in its ability to provide a systematic, objective and consistent method for evaluating and selecting the best alternative among the various existing options. AHP is able to help in defining and structuring problems hierarchically, enabling decision makers to see and understand all elements and relationships between elements in the problem being studied. In addition, AHP assessment criteria can be quantified using qualitative or quantitative criteria as an objective basis for assessment and for identifying priorities between criteria and alternatives through pairwise comparisons to provide appropriate weighting based on the importance of each criterion. AHP uses a questionnaire in the form of a pairwise comparison of criteria to produce a priority vector for each criterion which is expressed by a weight where the value is obtained from multiplying the weight with a weighting factor originating from the priority vector. AHP uses a questionnaire in the form of a pairwise comparison of criteria to produce a priority vector for each criterion which is expressed by a weight where the value is obtained from multiplying the weight with a weighting factor originating from the priority vector [37].

3. Result and Discussion

The concept of blockchain technology in ferry transportation is explained in detail through a network and density visualization presentation using Vosviewer. Study data was obtained by narrowing the keyword "blockchain technology in ferry transportation" to find journal articles through a search on Google Scholar. The data collected comes from publications in the last ten years, namely the 2014-2023 period. Visually illustrating the concept of blockchain technology in ferry transportation provides a clearer and more intuitive understanding of how the system works. Network and density visualizations show interactions between various elements in the system. By referring to publications published in the last ten years, this research has access to the latest information regarding the development of blockchain technology in ferry transportation. Network visualization displays the relationships between various entities in the system, while density visualization shows the distribution of blockchain technology development and helps in understanding the concept. The combination of visualization and data from this research provides a comprehensive picture of the importance of blockchain technology concepts in increasing the efficiency and effectiveness of ferry transportation. Combining these visualizations with data from research on blockchain technology's impact on ferry transportation can paint a holistic picture of how leveraging blockchain concepts can enhance efficiency and effectiveness in this domain. For instance, visualizing the connections between ferry operators, passengers, ticketing systems, logistics partners, and regulatory bodies can highlight potential areas for optimization or streamlining through blockchain solutions like smart contracts or transparent ledger systems. Additionally, density visualizations showing the adoption and development of blockchain solutions within the ferry transportation industry can indicate the level of readiness or interest in embracing these technologies.

Table 1. Characteristics of keyword publications on blockchain technology in ferry transportation

Publication	2014	2015	2016	2017	2018	2019	2020	2021	2022	2023

Papers	16	17	19	21	25	33	82	112	153	190
Citations	1407	606	1262	658	936	1465	3180	2438	1297	832
Cites/year	140.70	67.33	157.75	94.00	156.00	293.00	795.00	812.67	648.50	832.00
Cites/paper	87.94	35.65	66.42	31.33	37.44	44.39	38.78	21.77	8.48	4.38
Cites/author	870.92	246.00	392.97	194.70	412.20	822.92	941.82	874.76	441.51	313.59
Papers/author	8.00	7.50	8.27	10.04	14.15	15.89	39.83	54.99	74.75	80.84
Authors/paper	2.88	3.12	3.00	3.24	2.52	2.94	2.82	3.00	2.91	3.22
h-index	12	9	13	9	10	13	22	22	21	14
g-index	16	17	19	21	25	33	56	47	30	21
hA-index	4	5	5	4	6	7	12	14	14	14

Keyword analysis of blockchain technology in ferry transportation involves examining a database to identify derivative indicators and current trend patterns. According to the data presented in Table 1, this research topic has received 14,081 citations, averaging 1,408.1 citations per year. In the past decade, the annual number of citations has been 3,996.95, with an average of 399.695 citations per year. The increasing number of citations per year indicates a growing interest among researchers and academics in blockchain technology for ferry transportation. Particularly in recent years, there has been a significant rise in average annual citations, reflecting heightened interest in this subject. The data helps identify trends and patterns in research and development related to blockchain technology in ferry transportation, highlighting the growing focus on subtopics such as artificial intelligence usage and the implementation of smart traffic policies and regulations. Understanding these trends and indicators allows for more informed decision-making in guiding research and development in this field. This analysis can assist in planning more effective innovations, policies, and investments to enhance the overall ferry transportation system, with an emphasis on high-quality safety standards. Apart from that, prioritizing safety standards is the most important thing in ferry transportation. By emphasizing high quality safety measures the overall safety of ferry operations can be significantly improved. This not only protects passengers and crew but also increases confidence in the ferry transport system, encourages greater use and supports economic growth. Leveraging insights from trend and indicator analysis allows stakeholders to plan and allocate resources strategically, ultimately leading to the advancement of a ferry transportation system with a strong focus on safety and efficiency.

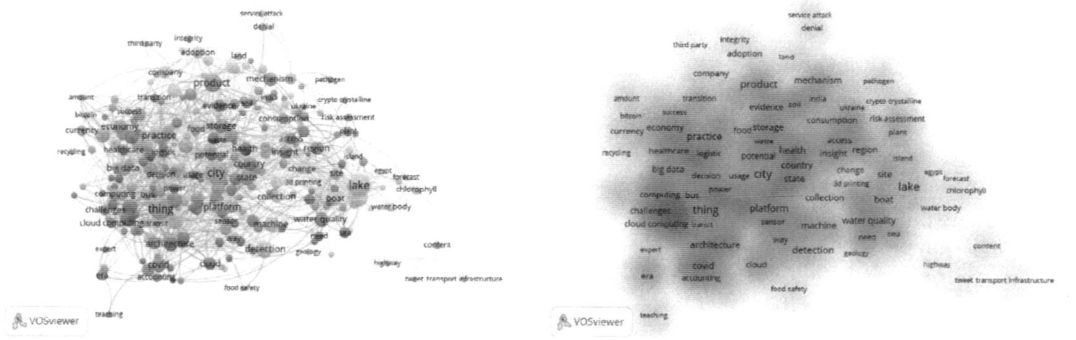

Figure 2. Network and density visualization on blockchain technology in ferry transportation

A visualization of the network and density of blockchain technology concepts in ferry transportation is presented in Figure 2 as a knowledge map of the study. This mapping helps in understanding the relationship between various related topics in the concept of blockchain technology in ferry transportation through network and density visualization. The identification of nodes and their names reflect various subtopics shown through variations in colour, node size, and connecting links between nodes. Similar topics are identified with the same colour, while the size of the node indicates how widely published the topic is. The existence of links between nodes shows the connection between the topics studied.

The nodes representing the studied topic are visualized with attributes such as colour and size to describe their nature and level of importance. Each colour indicates a different category or topic group, where topics that are closely related or have similar concepts are assigned the same colour to help identify related topic groups. The size of the node reflects the number of publications or research that has been conducted on the topic; the larger the node size, the more publications there are on the topic, indicating the breadth and importance of the topic in the existing literature. Links between nodes describe the relationship between topics, and the more links connecting two nodes, the closer the relationship between the two topics. This helps understand how different concepts are related and how information or research in one topic can influence other topics. There are several nodes that are still small in size and have the potential to develop blockchain technology in ferry transportation, including data and transaction security, supply chain tracking, identity and access management, operational efficiency, regulatory compliance, interoperability with external systems, transparency and accountability. By using network and density visualizations like these, researchers can more easily identify patterns, trends, and knowledge gaps in intelligent traffic management concepts, as well as direct future research to fill existing knowledge gaps.

Table 2. Top ten citation rankings on blockchain technology in ferry transportation

TC	APY	Title	Source Journal	Publisher	Authors (Year)
990	247.50	Industry 4.0 technologies assessment: A sustainability perspective	International journal of production economics	Elsevier	Bai et al (2020) [38]
588	58.80	Melanins: skin pigments and much more - types, structural models, biological functions, and formation routes	New Journal of Science	Hindawi	Solano (2014) [39]
465	116.25	IoT, big data, and artificial intelligence in agriculture and food industry	IEEE Internet of Things Journal	IEEE	Misra et al (2020) [40]
427	85.40	Artificial intelligence: a survey on evolution, models, applications and future trends	Journal of Management Analytics	Taylor & Francis	Lu (2019) [41]
413	51.63	Market-based emissions regulation and industry dynamics	Journal of Political Economic	The University of Chicago Press	Fowlie et al (2016) [42]
378	54.00	Design of secure user authenticated key management protocol for generic IoT networks	IEEE Internet of Things Journal	IEEE	Wazid et al (2017) [43]

294	36.75	A survey on chip to system reverse engineering	ACM Journal on Emerging Technologies in Computing Systems	ACM Digital Library	Quadir et al (2016) [44]
288	28.80	Review of cyber-physical system in healthcare	International Journal of Distributed Sensor Networks	Sage Publication	Haque et al (2014) [45]
287	95.67	Blockchain-based food supply chain traceability: a case study in the dairy sector	International Journal of Production Research	Taylor & Francis	Casino et al (2021) [46]
263	29.22	Monogenetic volcanism: personal views and discussion	International Journal of Earth Sciences	Springer	Németh & Kereszturi (2015) [47]

TC= Total Citation; APY= Average Per Year

Ranking data referring to the number of publication citations in the last ten years is presented in Table 2. The ten papers of the highest quality were selected based on the highest number of citations, with most having high annual averages. These papers generally come from favourite publishers such as Elsevier, IEEE, Taylor & Francis, Hindawi, ACM Digital Library, and Springer. Therefore, it is important to utilize high-quality publishers such as those mentioned above when publishing journal articles. Citation ranking is an important indicator in assessing the quality of scientific work, because frequently cited papers tend to be considered to have made significant and relevant contributions to the scientific community. A high average annual citation value indicates that the paper has had a sustainable impact over a long period of time, indicating its relevance and benefits for readers and researchers in related fields. The fact that most of the best papers come from well-known publishers emphasizes the importance of choosing a quality publisher in publishing journal articles. Leading publishers usually have a strict review and selection process and a wide reach in the scientific community, so they can increase the visibility and citations of published papers. Thus, utilizing quality publishers can help researchers increase their chances of publication success and gain wider recognition in the scientific community.

AHP is multi criteria decision making method used to determine priorities and make complex decisions. In the context of blockchain technology on ferry transportation, AHP can help assess and compare various alternatives based on the various criteria used. The important role of AHP analysis using performance sensitivity analysis measurements is used to understand how sensitive the final results are to changes in weight on different criteria. Blockchain technology in ferry transportation plays a role in evaluating the linkage criteria where blockchain has transparency, security and efficiency. The application of performance sensitivity analysis can see how changing weights on criteria such as cost, reliability, or security thus influence the final preference for various blockchain technologies. In addition, the performance sensitivity analysis measurement is used for better decision making which shows that the decisions made are very dependent on performance to help decision makers to focus more on the most influential criteria. This measurement can also be applied to risk area identification to find out how small changes to certain criteria affect the final results to help identify areas of risk or uncertainty that need attention.

Head-to-head sensitivity analysis measurements are used to directly compare two alternatives to understand how sensitive preferences are to changes in criteria weights. In the context of blockchain technology in ferry transportation it is necessary to choose the best solution where head-to-head sensitivity analysis allows to see directly how changes in assessment criteria can change preferences between two alternatives. This analysis can also help in achieving consensus among stakeholders by clearly showing how preferences change under different conditions and is useful in an environment where multiple parties are involved in decision making. In implementing new technology, strong justification is required for the selection of a particular solution that is able to provide quantitative evidence to support the decisions taken. Using AHP with performance sensitivity and head-to-head sensitivity analysis in evaluating blockchain technology for ferry transportation helps in clarifying and justifying decisions, considering various criteria and preferences, and identifying risk areas. This is important to ensure that the technology adopted is in line with existing needs and priorities and has a high chance of success in its implementation.

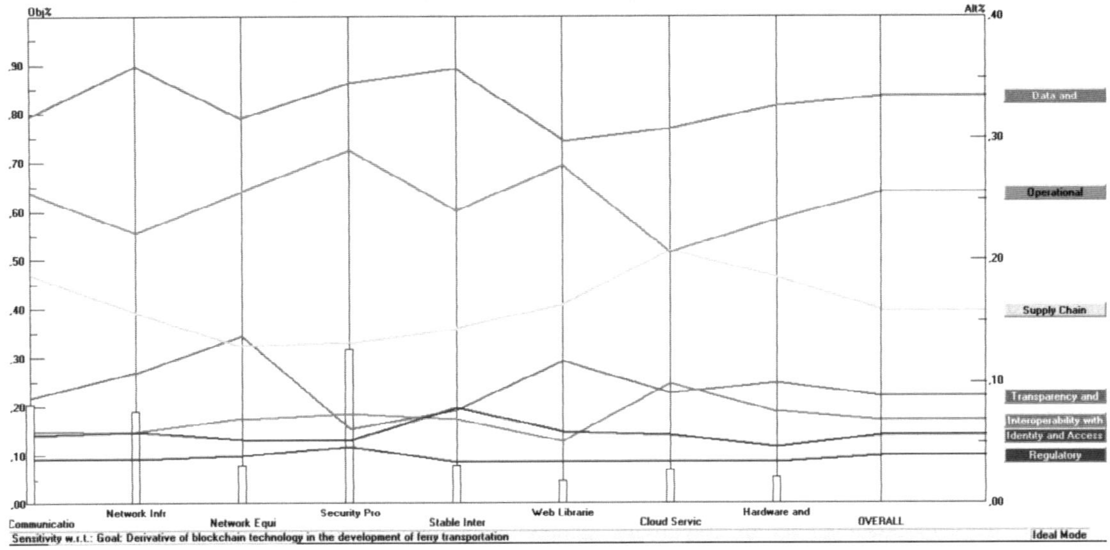

Figure 3. Performance sensitivity on blockchain technology in ferry transportation

Indicators for measuring AHP are identified by describing performance sensitivity in order to assess the weight of criteria in determining the selected alternative. The follow-up to this identification is to apply AHP measurements to ensure that the criteria weights are accurate and relevant, so that they can be used to select the most appropriate alternative based on in-depth analysis. AHP measurements can improve the quality of indicators by describing performance sensitivity to assess the weight of criteria in determining alternative choices. Weight measurements and comparison of criteria aim to calculate alternative rankings, so that the most sensitive factors can be identified based on priority values at the blockchain technology in ferry transportation concept. The measurement direction is to find out sensitive factors to assess priorities in deriving the concept of blockchain technology in ferry transportation. The three superior priorities in measuring performance sensitivity in this study include data and transaction security, operational efficiency and supply chain tracking. The reporting of performance sensitivity measurements is shown in Figure 3 which states that data and transaction security received the highest priority in the results of measuring a blockchain technology concept in ferry transportation. Alternative measurements are based on eight criteria used, of which the three criteria that have dominant categories include security protocols, communication protocols, and network infrastructure. The application of blockchain technology in ferry transportation is characterized by the most sensitive factors in assessing priorities being data and transaction security, operational efficiency, and supply chain tracking. Of these three priorities, data and transaction security receive the highest attention in measurement results. In addition, alternative measurements based on eight criteria identify that security protocols, communication protocols, and network infrastructure are the dominant categories that need attention. This emphasizes the importance of security and infrastructure in the implementation of blockchain technology in the ferry transportation sector.

Pairwise comparison assessment in AHP is a technique for determining priorities between various criteria and alternatives. In the context of assessing blockchain technology on ferry transportation using pairwise comparisons for various alternatives such as data and transaction security, supply chain tracking, identity and access management, operational efficiency, regulatory compliance, interoperability with external systems, transparency and accountability has several advantages and benefits. The importance of pairwise comparison assessment in AHP in increasing clarity and accuracy in decision making is to focus attention on two alternative elements. By comparing two aspects at a time, decision makers can be clearer in determining preferences and weights and reduce the possibility of bias and error. In the context of blockchain technology on ferry transportation, which uses pairwise comparisons, it is possible to decompose complexity by breaking down the problem into small, manageable pairs, the assessment process becomes easier and more structured. AHP uses a consistency ratio to ensure that the assessments given by decision makers are consistent. Each alternative has a different level of importance based on different contexts or objectives, so pairwise comparisons can determine which alternatives should be prioritized based on the weight and value given to each criterion and enable more effective and efficient resource allocation. Multiple perspectives can be integrated into a single framework and help in making more comprehensive and inclusive decisions. By using pairwise comparison assessments in AHP, decision makers can ensure that decisions regarding blockchain technology on ferry transportation are taken in the most rational, objective and transparent way.

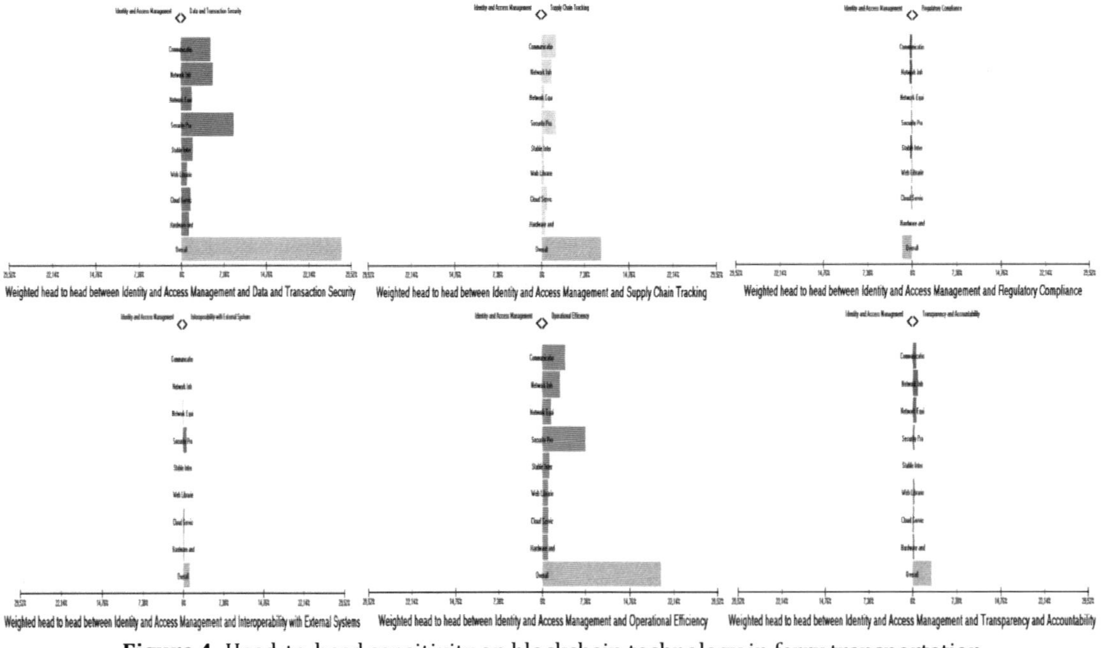

Figure 4. Head-to-head sensitivity on blockchain technology in ferry transportation

Implementation of head-to-head sensitivity through comparison of tested alternatives with those used. Measuring head-to-head sensitivity in blockchain technology in ferry transportation

IOP Conf. Series: Earth and Environmental Science **1423** (2024) 012014 doi:10.1088/1755-1315/1423/1/012014

focuses on ranking alternatives through comparison of tested alternatives. The results of the head-to-head sensitivity test determine the priority scale in decision making. Measurement results by comparing alternative assessments include data and transaction security, supply chain tracking, identity and access management, operational efficiency, regulatory compliance, interoperability with external systems, transparency and accountability. Head-to-head sensitivity findings show that data and transaction security is the alternative that has the greatest sensitivity value when compared to other alternatives shown in Figure 4. The measurement results show that data and transaction security have the highest sensitivity value compared to other alternatives. These findings help determine the priority scale in decision making. The implementation of head-to-head sensitivity findings regarding data and transaction security not only helps organizations make informed decisions but also reinforces their ability to adapt to an increasingly complex and dynamic security landscape.

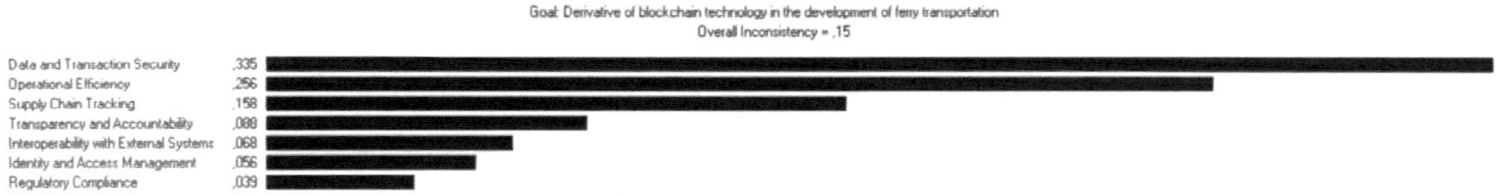

Figure 5. Synthesis based on priority on blockchain technology in ferry transportation

Synthesis results refer to calculating priorities through comparison of indicators in the form of criteria or alternatives based on hierarchy in AHP analysis. Synthetic measurement by combining weights or priority measurement results as consideration in decision making. Priority order of measurement results on alternatives to blockchain technology concept in ferry transportation. Data and transaction security, operational efficiency and supply chain tracking are three superior priorities as alternative measurement results shown in Figure 5. The priority alternative as the main finding is a derivative of blockchain technology in ferry transportation and has the opportunity to be developed in the future. However, four other alternatives such as identity and access management, regulatory compliance, interoperability with external systems, transparency and accountability are derivative parts that also have the potential to be followed up even though they are not yet the main priority in this study. While the priority may currently lie in exploring the main finding related to blockchain integration in ferry transportation, these derivative parts present promising avenues for future development and implementation. By considering and addressing these alternative applications, stakeholders can harness the full potential of blockchain technology to enhance various aspects of ferry transportation operations and management.

The inconsistency ratio in AHP analysis is a measure used to determine how consistent the preferences or assessments given are in pairwise comparisons. This ratio indicates the level of inconsistency in the assessment where higher values indicate more inconsistency. The AHP weighting results in this study with an inconsistency ratio of 15% are considered quite high in the AHP context, which means that the assessments given are not completely consistent and there are several assessments that do not comply with the logic of rationality. Inconsistencies occur in alternatives that have low value, including regulatory compliance, identity and access management, and interoperability with external systems. A high inconsistency ratio indicates that there is error or confusion in the scoring process, or that the scoring does not fully reflect real preferences. Use of Indicator Consistency: Using tools or indicators to help decision makers understand and reduce inconsistencies. Although a lower inconsistency ratio of 10% is preferable, a result with a ratio of 15% does not mean the analysis is invalid. However, it shows that there is room for improvement in the assessment process to achieve more consistent and reliable results.

Furthermore, novelty was obtained by finding three main derivatives of blockchain technology in ferry transportation which can be followed up in developing safety in ferry transportation in the future, including data and transaction security, operational efficiency, and supply chain tracking. Furthermore, the three new findings derived from blockchain technology in ferry transportation need to be followed up by researchers or academics in the future. The three new concepts have a correlation with the development of ferry transportation which is in line with transportation safety. Apart from that, it was also stated that the novelty in this study was applying a mixed method which developed qualitative analysis using Vosviewer and quantitative analysis which optimized AHP as part of the final findings. Researchers and academics in the future can follow up on this novelty by conducting further research to optimize the implementation of blockchain technology in ferry transportation. It is also necessary to develop mathematical models or computer simulations to understand its impact in more depth, conduct case studies of successful implementations, and design an appropriate regulatory framework to facilitate widespread adoption of this technology. By strengthening data and transaction security, increasing operational efficiency, and improving supply chain tracking, blockchain technology has the potential to significantly improve the safety of ferry transportation in the future.

4. Conclusion

The findings obtained are a derivative of blockchain technology in ferry transportation which is important to follow up for the development of science and technology to prioritize transportation safety in the future. The findings of three main topics have the potential to be novel in strengthening knowledge related to blockchain technology in ferry transportation, including data and transaction security, operational efficiency, and supply chain tracking. Four other derivative topics which include identity and access management, regulatory compliance, interoperability with external systems, transparency and accountability are not considered to be priority findings but still have opportunities to develop knowledge and technology for researchers and academics. It is hoped that blockchain technology can bring significant changes in improving safety, efficiency and transparency in ferry transportation. In addition, the results of this research can become a basis for further development in other fields related to transportation and logistics.

5. References

[1] C. Durán, C. Fernández-Campusano, R. Carrasco, and E. Carrillo, "DMLBC: Dependable machine learning for seaports using blockchain technology," *J. King Saud Univ. - Comput. Inf. Sci.*, vol. 36, no. 101918, pp. 1–21, 2024, doi: 10.1016/j.jksuci.2024.101918.

[2] K. Shahzad, Q. Zhang, A. U. Zafar, M. F. Shahzad, and W. Liu, "Consumers' concerns and the role of blockchain technology in mobile food delivery applications," *J. Destin. Mark. Manag.*, vol. 32, no. 100877, 2024, doi: 10.1016/j.jdmm.2024.100877.

[3] M. Javaid, A. Haleem, R. P. Singh, R. Suman, and S. Khan, "A review of Blockchain Technology applications for financial services," *BenchCouncil Trans. Benchmarks, Stand. Eval.*, vol. 2, no. 100073, pp. 1–18, 2022, doi: 10.1016/j.tbench.2022.100073.

[4] M. Ben Farah *et al.*, "A survey on blockchain technology in the maritime industry: Challenges and future perspectives," *Futur. Gener. Comput. Syst.*, vol. 157, pp. 618–637, 2024, doi: 10.1016/j.future.2024.03.046.

[5] G. Kapnissis, G. K. Vaggelas, H. C. Leligou, A. Panos, and M. Doumi, "Blockchain adoption from the Shipping industry: An empirical study," *Marit. Transp. Res.*, vol. 3, no. 100058, pp. 1–13, 2022, doi: 10.1016/j.martra.2022.100058.

[6] Y. Xu, G. Xu, Y. Liu, Y. Liu, and M. Shen, "A survey of the fusion of traditional data security technology

and blockchain," *Expert Syst. with Appl. Available*, vol. 124151, 2024, doi: 10.1016/j.eswa.2024.124151.

[7] Y. Teng, S. Ma, Q. Qian, and G. Wang, "SEIR-diffusion modeling and stability analysis of supply chain finance based on blockchain technology," *Heliyon*, vol. 10, no. e24981, pp. 1–20, 2024, doi: 10.1016/j.heliyon.2024.e24981.

[8] S. Yadav and S. Prakash Singh, "Modelling procurement problems in the environment of blockchain technology," *Comput. Ind. Eng.*, vol. 172, no. 108546, 2022, doi: 10.1016/j.cie.2022.108546.

[9] S. Anomah, B. Ayeboafo, M. Aduamoah, and O. Agyabeng, "Blockchain technology integration in tax policy: Navigating challenges and unlocking opportunities for improving the taxation of Ghana's digital economy," *Sci. African*, vol. 24, no. e02210, pp. 1–12, 2024, doi: 10.1016/j.sciaf.2024.e02210.

[10] P. Giganti, M. Borrello, M. Falcone, and L. Cembalo, "The impact of blockchain technology on enhancing sustainability in the agri-food sector : A scoping review," *J. Clean. Prod.*, vol. 142379, 2024, doi: 10.1016/j.jclepro.2024.142379.

[11] Y. Liu, Z. Li, and L. Huang, "The application of blockchain technology in smart sustainable energy business model," *Energy Reports*, vol. 8, pp. 7063–7070, 2022, doi: 10.1016/j.egyr.2022.05.002.

[12] Y. Mesengiser and N. Miloslavskaya, "Problems of Using Redactable Blockchain Technology," *Procedia Comput. Sci.*, vol. 190, pp. 582–589, 2021, doi: 10.1016/j.procs.2021.06.068.

[13] L. Stockburger, G. Kokosioulis, A. Mukkamala, R. R. Mukkamala, and M. Avital, "Blockchain-enabled decentralized identity management: The case of self-sovereign identity in public transportation," *Blockchain Res. Appl.*, vol. 2, no. 100014, pp. 1–18, 2021, doi: 10.1016/j.bcra.2021.100014.

[14] Z.-H. Yang *et al.*, "An attribute-based access control scheme using blockchain technology for IoT data protection," *High-Confidence Comput.*, vol. 100199, 2024, doi: 10.1016/j.hcc.2024.100199.

[15] F. Yang, H. Yu, C. Wilson, G. Jacoby, and Z. Wu, "Blockchain technology and international countertrade," *J. Int. Financ. Mark. Institutions Money*, vol. 91, no. 101933, 2024, doi: 10.1016/j.intfin.2023.101933.

[16] A. Kharche, S. Badholia, and R. K. Upadhyay, "Implementation of blockchain technology in integrated IoT networks for constructing scalable ITS systems in India," *Blockchain Res. Appl.*, vol. 100188, 2024, doi: 10.1016/j.bcra.2024.100188.

[17] L. Huang and T. Zhou, "How does blockchain technology enhance firm operation and cooperation?," *Int. Rev. Econ. Financ.*, vol. 92, pp. 34–49, 2024, doi: 10.1016/j.iref.2024.01.049.

[18] E. Abad-Segura, A. Infante-Moro, M. D. González-Zamar, and E. López-Meneses, "Influential factors for a secure perception of accounting management with blockchain technology," *J. Open Innov. Technol. Mark. Complex.*, vol. 10, no. 100264, pp. 1–10, 2024, doi: 10.1016/j.joitmc.2024.100264.

[19] X. Yang, M. Liu, J. Wei, and Y. Liu, "Research on investment optimization and coordination of fresh supply chain considering misreporting behavior under blockchain technology," *Heliyon*, vol. 10, no. e26749, pp. 1–17, 2024, doi: 10.1016/j.heliyon.2024.e26749.

[20] P. Jiang *et al.*, "Blockchain technology applications in waste management: Overview, challenges and opportunities," *J. Clean. Prod.*, vol. 421, no. 138466, 2023, doi: 10.1016/j.jclepro.2023.138466.

[21] M. Marzouk, N. Labib, and M. Metawie, "Blockchain technology applications in maintaining heritage buildings," *J. Cult. Herit.*, vol. 67, pp. 62–72, 2024, doi: 10.1016/j.culher.2024.02.005.

[22] B. Wen, Y. Wang, Y. Ding, H. Zheng, B. Qin, and C. Yang, "Security and privacy protection technologies in securing blockchain applications Author," *Inf. Sci. (Ny).*, vol. 645, no. 119322, 2023, doi: 10.1016/j.ins.2023.119322.

[23] A. A. Amponsah, A. F. Adekoya, and B. A. Weyori, "Improving the Financial Security of National Health Insurance using Cloud-Based Blockchain Technology Application," *Int. J. Inf. Manag. Data Insights*, vol. 2, no. 100081, pp. 1–21, 2022, doi: 10.1016/j.jjimei.2022.100081.

[24] Y. Maythu, A. O. J. Kwok, and P. L. Teh, "Blockchain technology diffusion in tourism: Evidence from early enterprise adopters and innovators," *Heliyon*, vol. 10, no. e24675, pp. 1–15, 2024, doi: 10.1016/j.heliyon.2024.e24675.

[25] K. Duan, G. Pang, and Y. Lin, "Exploring the current status and future opportunities of blockchain technology adoption and application in supply chain management," *J. Digit. Econ.*, vol. 2, pp. 244–288, 2023, doi: 10.1016/j.jdec.2024.01.005.

[26] A. G. Gad, D. T. Mosa, L. Abualigah, and A. A. Abohany, "Emerging Trends in Blockchain Technology and Applications: A Review and Outlook," *J. King Saud Univ. - Comput. Inf. Sci.*, vol. 34, no. 9, pp. 6719–6742, 2022, doi: 10.1016/j.jksuci.2022.03.007.

[27] H. Kusumastuti, D. Pranita, M. Viendyasari, M. S. Rasul, and S. Sarjana, "Leveraging Local Value in Post-Smart Tourism Village to Encourage Sustainable Tourism," *Sustainability*, vol. 16, no. 873, pp. 1–26, 2023, doi: 10.3390/su16020873.

[28] D. Pranita, S. Sarjana, B. M. Musthofa, H. Kusumastuti, and M. S. Rasul, "Blockchain Technology to

Enhance Integrated Blue Economy : A Case Study in Strengthening Sustainable Tourism on Smart Islands," *Sustainability*, vol. 15, no. 5342, pp. 1–24, 2023, doi: 10.3390/su15065342.

[29] S. Sarjana, "Green transportation : Development opportunities in support of sustainable transportation," in *E3S Web of Conferences*, 2023, vol. 429 (03003, pp. 1–10, doi: 10.1051/e3sconf/202342903003.

[30] S. Sarjana, "Smart City in Supporting Sustainable Cities," in *2023 10th International Conference on Information Technology, Computer, and Electrical Engineering, ICITACEE 2023*, 2023, pp. 365–370, doi: 10.1109/ICITACEE58587.2023.10277619.

[31] S. Sarjana, "Near Field Communication Studies in Transportation Business," *2023 3rd Int. Conf. Intell. Cybern. Technol. Appl. ICICyTA 2023*, no. 10428896, pp. 266–271, 2023, doi: 10.1109/ICICyTA60173.2023.10428896.

[32] M. Y. N. Attari, A. A. M. Beirami, A. Ala, and E. N. Jami, "Resolving the practical factors in the healthcare system management by considering a combine approach of AHP and ANP methods," *Eval. Program Plann.*, vol. 100, no. 102339, 2023, doi: 10.1016/j.evalprogplan.2023.102339.

[33] J. Nazarian-Jashnabadi, S. Rahnamay Bonab, G. Haseli, H. Tomaskova, and M. Hajiaghaei-Keshteli, "A dynamic expert system to increase patient satisfaction with an integrated approach of system dynamics, ISM, and ANP methods," *Expert Syst. Appl.*, vol. 234, no. 121010, pp. 1–19, 2023, doi: 10.1016/j.eswa.2023.121010.

[34] A. Thilagavathy and S. Mohanaselvi, "T-Spherical fuzzy TOPSIS method based on distance measures and Hamacher Heronian mean averaging aggregation operators and its application to waste management," *Appl. Soft Comput. Vol.*, vol. 162, no. 111868, 2024, doi: 10.1016/j.asoc.2024.111868.

[35] S. Sarjana, "Energy Security Faces Critical Global Attention," *Int. J. Energy Econ. Policy*, vol. 14, no. 1, pp. 37–44, 2024, doi: 10.32479/ijeep.14982.

[36] M. Munir, M. K. Muqorrobin, R. A. M. Yunus, and S. Sarjana, "Ruang Bersama: Monitoring Application for Sexual Harassment in Public Transport," *2023 Int. Conf. Informatics, Multimedia, Cyber Inf. Syst. ICIMCIS 2023*, no. 10349058, pp. 257–262, 2023, doi: 10.1109/ICIMCIS60089.2023.10349058.

[37] G. Guidi, G. Goffo, and A. C. Violante, "Application of the Analytic Hierarchy Process (AHP) method to identify the most suitable approach for managing irradiated graphite," *Nucl. Eng. Technol.*, vol. 56, no. 7, pp. 1–6, 2024, doi: 10.1016/j.net.2024.06.045.

[38] C. Bai, P. Dallasega, G. Orzes, and J. Sarkis, "Industry 4.0 technologies assessment: A sustainability perspective," *Int. J. Prod. Econ.*, vol. 229, no. 107776, pp. 1–15, 2020, doi: 10.1016/j.ijpe.2020.107776.

[39] F. Solano, "Melanins: Skin Pigments and Much More—Types, Structural Models, Biological Functions, and Formation Routes," *New J. Sci.*, no. 498276, pp. 1–28, 2014, doi: 10.1155/2014/498276.

[40] N. N. Misra, Y. Dixit, A. Al-Mallahi, M. S. Bhullar, R. Upadhyay, and A. Martynenko, "IoT, Big Data, and Artificial Intelligence in Agriculture and Food Industry," *IEEE Internet Things J.*, vol. 9, no. 9, pp. 6305–6324, 2022, doi: 10.1109/JIOT.2020.2998584.

[41] Y. Lu, "Artificial intelligence: a survey on evolution, models, applications and future trends," *J. Manag. Anal.*, vol. 6, no. 1, pp. 1–29, 2019, doi: 10.1080/23270012.2019.1570365.

[42] M. Fowlie, M. Reguant, and S. P. Ryan, "Market-Based Emissions Regulation and Industry Dynamics," *J. Polit. Econ.*, vol. 124, no. 1, 2016, doi: 10.3386/w18645.

[43] M. Wazid, A. K. Das, V. Odelu, N. Kumar, M. Conti, and M. Jo, "Design of Secure User Authenticated Key Management Protocol for Generic IoT Networks," *IEEE Internet Things J.*, vol. 5, no. 1, pp. 269–282, 2018, doi: 10.1109/JIOT.2017.2780232.

[44] S. E. Quadir *et al.*, "A Survey on Chip to System Reverse Engineering," *ACM J. Emerg. Technol. Comput. Syst.*, vol. 13, no. 1, pp. 1–34, 2016, doi: 10.1145/2755563.

[45] S. A. Haque, S. M. Aziz, and M. Rahman, "Review of cyber-physical system in healthcare," *Int. J. Distrib. Sens. Networks*, no. 217415, pp. 1–21, 2014, doi: 10.1155/2014/217415.

[46] F. Casino *et al.*, "Blockchain-based food supply chain traceability: a case study in the dairy sector," *Int. J. Prod. Res.*, pp. 1–13, 2020, doi: 10.1080/00207543.2020.1789238.

[47] K. Németh and G. Kereszturi, "Monogenetic volcanism: personal views and discussion," *Int. J. Earth Sci.*, vol. 104, no. 8, pp. 2131–2146, 2015, doi: 10.1007/s00531-015-1243-6.

Evaluation of tol laut in Tahuna and Natuna as underdeveloped, remote, outermost, and border areas

A Kurniawan[1], A A Kharisma[2], K I P Yasadhi[3], S Bahri[4] and A Wirawan[5]

[1,2,3] Transportation Policy Agency, Jalan Medan Merdeka Timur No 5, Jakarta, 10110
[4] Polytechnics of Makassar Merchant Marine, Jalan Tentara Pelajar No. 173 Makassar, 90172
[5] Polytechnics of Land Transportation Indonesia, Jalan Raya Setu No.89 Bekasi,17520

achmad.andhika@kemenhub.go.id

Abstract. Tahuna (North Sulawesi) and Natuna (Riau Islands) are included in the 3TP (underdeveloped, remote, outermost, and border) areas and use sea transportation to support for their economic activities. High inflation and price disparities are economic problems in this area. "Tol Laut" is government initiative designed to distribute essential basic goods, through a subsidized sea transportation system. These routes connect the economic hub in Java with the 3TP areas. The analysis process uses a mix method, quantitative and qualitative analysis. Port Assessment analysis using weighted scoring, Supply Demand Analysis using index consumption demand, and Regulation Review. To evaluate the implementation of Tol Laut, it focuses on aspects of port location, connectivity, port facilities, potential feeder ships, production performance, hinterland, as well as administration and commerce. Tol Laut in Tahuna can fulfill the supply-demand of essential basic goods significant quantities around 30% from local demand and is supported by the regional government's commitment, allowing Tol Laut to reduce price disparities. On the contrary, Tol Laut in Natuna can only the supply 16%, resulting inability to control price disparities effectively. There is an urge to optimize and coordinate regarding Tol Laut by involving agencies for the successful enforcement of regulations.

1. Introduction

Tahuna in North Sulawesi and Natuna in the Riau Islands are two regions included in the 3TP category (underdeveloped, remote, outermost, and border) [1],[2] and [3]. The limited local production capabilities result in meeting community needs through inter-regional trade with sea transportation as the main support. This condition makes economic inequality such as inflation and high price disparities a problem in this area. Tol Laut Program as a strategic government program exists to distribute essential basic goods in the form of special subsidized sea transportation with routes connecting the economic center in Java with the 3TP area [4]. Natuna is served by the T3 route and Tahuna is served by the H-1 and T-5 [5]. The dynamic changes in Tol Laut routes in these two regions will have an impact on the effectiveness of the program, so it needs to be evaluated based on aspects of port location, connectivity, port facilities, liner ship specifications, feeder ship potential, performance of Tol Laut production, hinterland, as well as administration and commercial [6] and [7]. These aspects are studied to produce policy recommendations as considerations for program continuity, port suitability, and efforts to improve services.

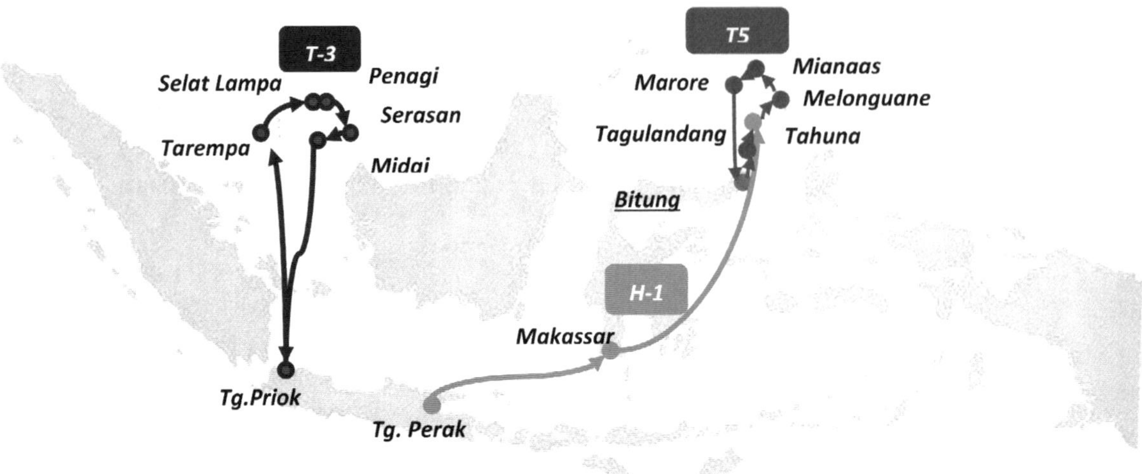

Source : Directorate General Sea Transportation (2023)

Figure 1. Map of Tol Laut's Route in Tahuna (left) and Natuna (Right)

Based on Figure 1 and Table 1 we can see that Tol Laut's route in Natuna and Tahuna using different scheme. Tol Laut in Natuna using direct-multiport scheme with Tanjung Priok Port as homebase, meanwhile in Tahuna using Hub and Spoke scheme with Tanjung Perak and Bitung as homebase hub port then distributed among the islands in Sangihe, Talaud and Siau Tagulandang Biaro Regency [8]. Tol Laut have different route codes, direct routes use code T, while hub and spoke routes use code H.

Table 1. Overview Tol Laut's Route

No	Route Code	Home Base Port	Routes	Ship
1	H-1	Tanjung Perak	Tanjung Perak – 438 – Makassar – 766 - Tahuna – 1071 - Tanjung Perak	Logistik Nusantara 06
2	T-3	Tanjung Priok	Tanjung Priok – 83 - Patimban - 524 – Kijang – 144 – Letung - 36 - Tarempa – 149 – Pulau Laut - 76 - Selat Lampa – 61 – Subi – 47 – Serasan – 85 – Midai - 605 – Patimban – 83 – Tanjung Priok	Logistik Nusantara 04
3	T-5	Bitung	Bitung - 89–ulu Siau/Tagulandang – 64- Tahuna – 84 - Marore – 84 - Miangas – 57 - Marampit – 55 - Lirung / Melangoane– 83 – Mangaran – 171 – Bitung	Kendhaga Nusantara 01

Source : Directorate General Sea Transportation (2023)

2. Methods

The analysis process employs a mixed method approach that amalgamates quantitative and qualitative analyses, the findings of which are expounded upon in a descriptive verification format [9]. Primary and secondary data are gathered through focus group discussions involving representative stakeholders from agencies, operators, and users of Tol Laut services at the Lampa Natuna Strait port and the Nusantara Tahuna port [10]. The stakeholders for these discussions are selected through non-probability sampling, specifically purposive sampling [11]. This method

IOP Conf. Series: Earth and Environmental Science **1423** (2024) 012015 doi:10.1088/1755-1315/1423/1/012015

involves choosing respondents who represent different relevant agencies and possess key responsibilities, functions, and authorities in port planning, development, and operations.

Analysis process started with Port Assessment Analysis to measure the implementation of Tol Laut by using Weighted Scoring based on port performance indicators [12]. We added several new indicators based on local condition like ships specification for liner and feeder carrier, hinterland factor like the availability of users and information technology support, administrative and commercial aspects to identify the local stakeholder's participation and supporting regulations. Next step in analysis is identifying local demand for primary and important goods that are determined by the regulation [13]. Supply potency use production data from local area statistics. Analysis result that shows positive value means that the local production of basic good and important goods is enough to supply the local demand or get some surplus, as vice versa negative value means that there is deficit in that area. The final comparation will show the difference of supply of local demand which distributed by Tol Laut ship and another cargo ships. In order to get another perspective then we use descriptive comparation analysis for regulation which manages Tol Laut programs, in this comparation we use regulation from Indonesian Ministry of Transportation and Ministry of Trade [14][15]. Based on research background and analysis methode then arranged in Research framework as shown in figure 2.

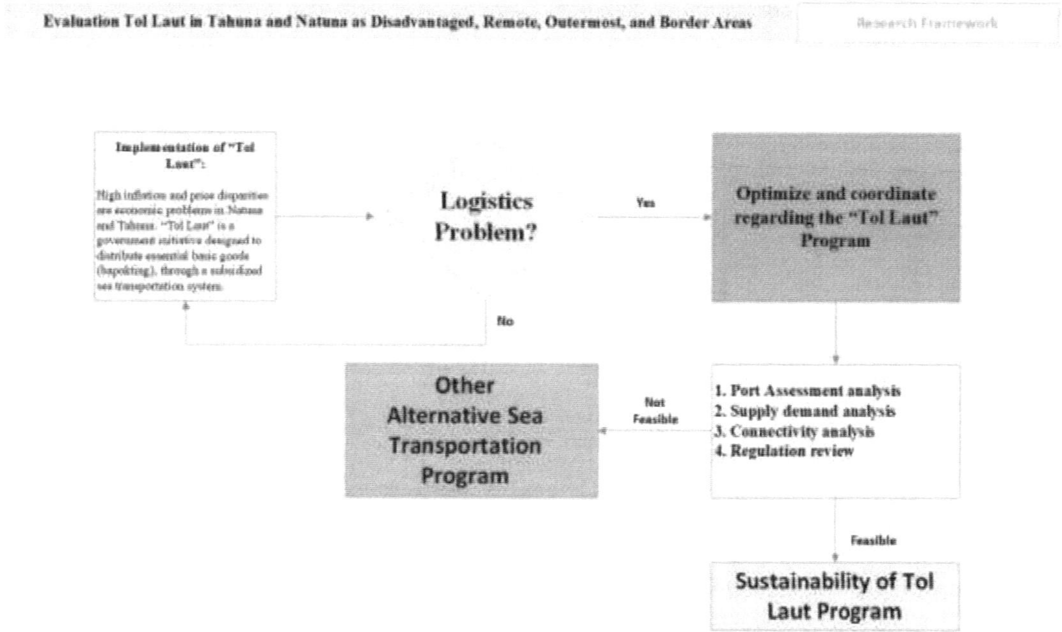

Source : Analysis (2024)
Figure 2. Research framework

3. Findings

Tahuna and Natuna ports are situated in naturally sheltered locations, enabling year-round operations. Nevertheless, connectivity to key economic centers in Java (such as Jakarta and Surabaya) heavily relies on the Tol Laut liner route, while connections to provincial capitals depend on passenger ships, sea pioneers, and ferry pioneers. Additionally, there exist other routes linking to Java's economic hubs, albeit in the form of tramper routes with irregular frequencies. Route accessibility serves as a primary indicator of connectivity and can accommodate requests for new routes from local regional governments [14]. Tahuna is serviced by the H-1 and T-5 routes, whereas Natuna is served by the T3 route. Presently, container ships are utilized for Tol Laut liner

IOP Conf. Series: Earth and Environmental Science **1423** (2024) 012015 doi:10.1088/1755-1315/1423/1/012015

services, whereas for Tol Laut feeder transportation, employing sea pilot vessels is deemed ideal due to the availability of onward routes, smaller cargo volumes, and more economical rates.

3.1 Port Assessment analysis

Port facility support plays a crucial role in enhancing the efficiency of Tol Laut Program, primarily due to the substantial portion of service time per voyage allocated to port-related activities such as loading and unloading, container stuffing and unstuffing, cargo distribution to consignees, and waiting for return cargo from hinterlands [16]. The Natuna port, constrained by limited loading and unloading equipment, notably impacts the efficiency of these operations. With reliance solely on ship cranes and forklifts with a capacity of 1.5 tons, loading, unloading, and cargo consolidation activities are restricted to dockside shifts and manual handling by TKBM (Manpower to Loading and Unloading) porters, consequently elongating Tol Laut Program's turnaround time to up to 4 days. Conversely, the availability of a 28-ton forklift at the Tahuna port facilitates the movement of containers from the dock to the container yard, albeit with TKBM porter assistance. This, coupled with efficient loading and unloading equipment, enables handling of a larger volume of containers simultaneously, thereby reducing Tol Laut Program's turnaround time to a maximum of 2 days at the Tahuna port. Nonetheless, both ports exhibit limitations in warehouse and container yard capacities, unable to accommodate reefer containers despite the substantial potential for significant backloading of fishery products in these regions. The formulation of alternative policies for optimizing the maritime highway program begins with identifying the results of the analysis in the form of a port assessment described as follows figure 3.

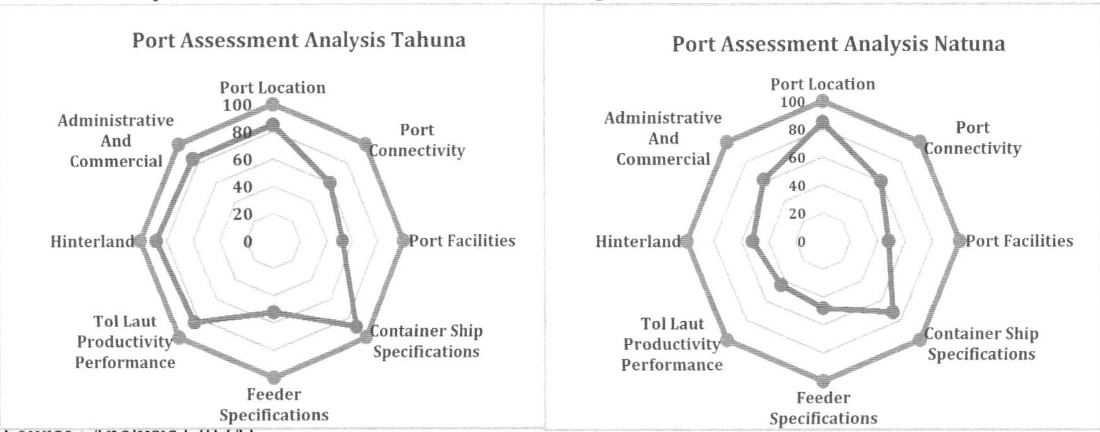

Source : Analysis (2024)

Figure 3. Port Assessment Analysis Tahuna and Natuna

Table 2. Comparison between Tahuna and Natuna

Port	Tahuna	Natuna
Port Location	85	85
Port Connectivity	60	60
Port Facilities	52	48
Tol Laut Productivity Performance	89	72
Container Ship Specifications	52	48
Feeder Specifications	84	44
Hinterland	88	52
Administrative And Commercial	85	62

Source : Analysis (2024)

Initial research of port assessment show the different result between tahuna and natuna. Tol laut impact in Tahuna is better then Natuna. Based on the table 2, assessment results of Tol Laut Program transit ports on H-1, T-3, and T-5, it is evident that there are numerous aspects requiring improvement to optimize Tol Laut Program implementation. While port locations and liner ship specifications generally meet satisfactory standards, particular attention must be directed towards ship carrying capacity, which remains limited and fails to adapt to escalating demands, particularly in areas where Tol Laut activities are well-established, such as Tahuna, Morotai, and Soasio.

The productivity aspect of Tol Laut Program underscores service user satisfaction levels with the program, notably reflected in the utilization of container quotas for unloading and reverse loading. High utility rates are observed in ports proximate to local economic centers and consistently served by Tol Laut Program over many years, including Tahuna, Morotai, Soasio/Tidore, and Buli/Maba. Conversely, ports experiencing frequent route alterations and situated far from economic hubs exhibit lower utility rates. Conversely, aspects receiving poor ratings encompass connectivity, feeder ship potential, inadequate port facilities, insufficient hinterland support, as well as administrative and commercial deficiencies. The connectivity aspect reveals a lack of well-established sea transportation linkages between many ports in the 3TP region and the national economic center, evident from indicators such as direct route availability and ship call frequency.

Cargo distribution sustainability in the 3TP area heavily relies on feeder transportation availability, such as sea pilot ships, ferry pioneer ships, and local shipping. While several feeder routes intersect with Tol Laut Program route, disparities in movement patterns often arise, leading to asynchrony with Tol Laut vessel movements and arrival times. Futhermore, Tol Laut Program cargo distribution process hinges on adequate port facilities aligned with the characteristics of Tol Laut vessels and their cargo. Many Tol Laut transit ports still suffer from minimal availability and capacity of loading and unloading equipment, such as land cranes and forklifts, thus hindering container handling capabilities and necessitating heavy reliance on TKBM porters whose productivity remains relatively low at approximately 5 tonnes/box/hour.

Hinterland, administrative, and commercial support constitute pivotal drivers of Tol Laut-based economic activities in the downstream sector and serve as regulatory mechanisms to mitigate adverse economic practices. Geographical constraints of islands and limited land accessibility contribute to diminished monitoring activity intensity by relevant agencies. The results of this analysis are in line with other research which states that port aspects, especially port facilities and the environment, are supporting factors for Tol Laut distribution [21].

3.2 Supply Demand Analysis

The restriction on the number of reefer containers transportable by Tol Laut vessels poses a significant challenge frequently cited by service users. With a capacity limited to approximately 5 reefer container units per voyage, not all service users are guaranteed access to a reefer container quota per journey [17]. Cargo production serves as a key metric for assessing the efficacy of Tol Laut Program program, whereby service users located in designated ports of call can leverage container quotas to import goods from economic centers. Tol Laut Program production in both regions is outlined Figure 4.

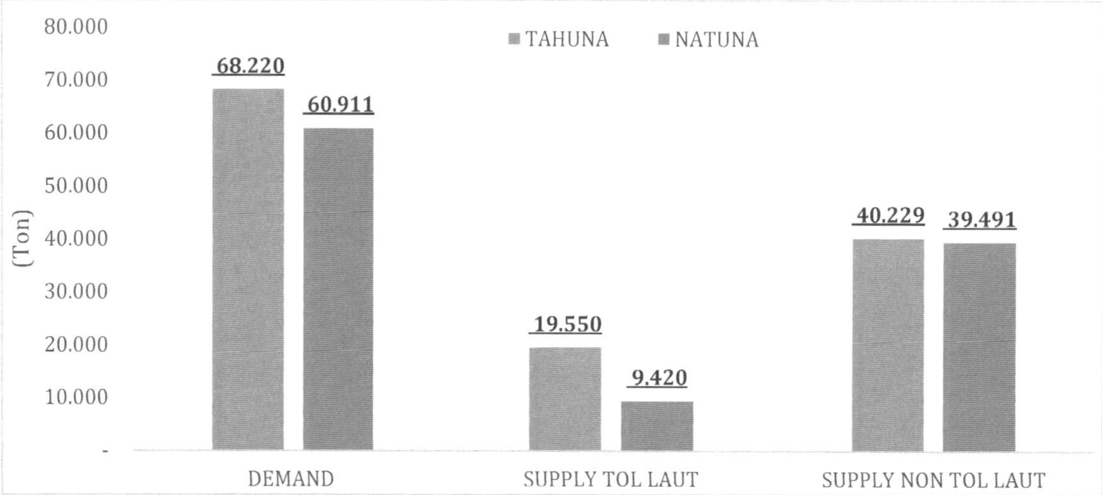

Source : Analysis (2024)
Figure 4. Supply-Demand Tol Laut in Tahuna and Natuna

The fulfillment of supply-demand dynamics for essential basic goods in Tahuna and Natuna based on the quota determination equation formula that outlined in the Chapter 3 of Standard Operating Procedure (SOP) for Reserving Ship Loading Space and Issuing Shipping Instructions for Tol Lauts [15]. The ship capacity is a predetermined factor contingent upon the operator's fleet assets. The potential exists for a reduction in cargo quotas within a region upon the addition of new ports of call. For instance, the T-3 route in Natuna (and Anambas) initially serviced only 2 ports of call, but has now expanded to 9 ports, while the T-5 route caters to just 3 ports. The expansion of service areas along a route could potentially diminish the efficiency of the Tol Laut route [18]. In Tahuna, the Tol Laut Program demonstrates an ability to adequately meet the supply-demand requirements for essential basic goods, bolstered by the proactive involvement of regional governments in monitoring trade activities within their respective jurisdictions. This proactive involvement ensures that Tol Laut cargo plays a role in mitigating price disparities. Conversely, the Tol Laut Program in Natuna has yet to significantly meet the supply-demand for essential basic goods, consequently failing to exert control over price disparities.

The mitigation of price disparities represents the pinnacle objective of Tol Laut Program program. The selling price of essential basic goods to end consumers is subject to market dynamics influenced by various price-forming factors, as elucidated by traders' considerations [19]. Notably, within the SOP Shipping Instructions, an integrity pact stipulates that consignees commit to selling essential basic goods at a reduced rate, thus refraining from exploiting the benefits of the Tol Laut program.

IOP Conf. Series: Earth and Environmental Science **1423** (2024) 012015 doi:10.1088/1755-1315/1423/1/012015

3.3 Regulation Review

Figure 5 shown the implementation of Tol Laut based on main duty of two different government agencies, Ministry of Transportation regulates the transportation meanwhile Ministry of Trade regulates the trade aspects. These two government agencies will be interacted in aspect of supply-demand. The ships and routes that arranged by Ministry of Transportation should accommodate the local demand of primary and important goods that will be distributed by Tol Laut ship to destination area based on local hinterland condition. Ministry of Trade then doing local evaluation to make sure the inbound cargo from Tol Laut are based on local demand in sufficient quantity.

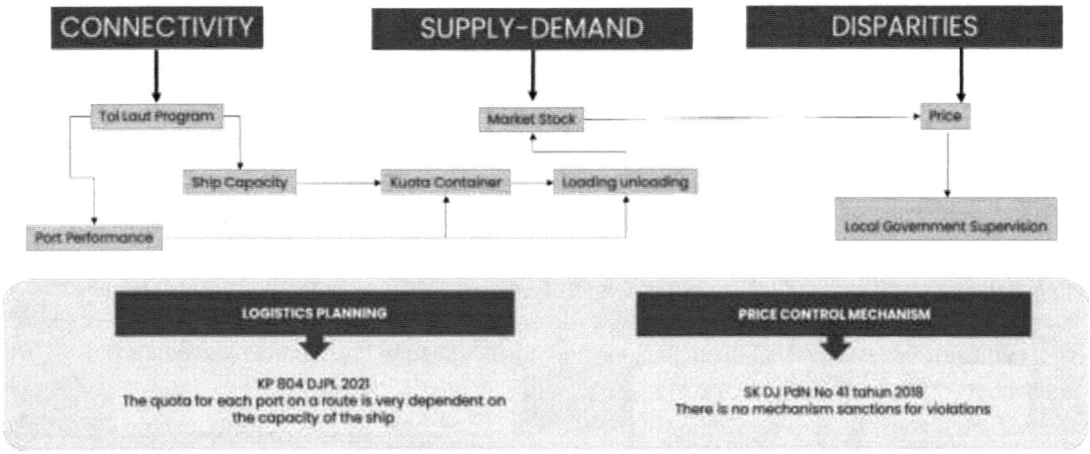

Source : Analysis (2024)
Figure 5. Regulation Review Tol Laut

The Tol Laut cargo requires supervision when reaching the final destination at the consignee or reseller location by an agency that has authority in the trade sector, guided by related regulations [20]. It is common to find violations of the integrity pact where resellers sell Tol Laut essential basic goods higher than the prevailing average price. In the Decree of the Director General of Hubla No. KP.804/DJPL/2021 there is a special section which regulates sanctions for violations within the scope of distribution, especially in port activities, shipping and conformity of documents, while in the Decree of the Director General of Domestic Trade Number 41 of 2018 there is no regulation regarding sanctions. The sanctions mechanism in the Director General of Hubla's Decree No. KP.804/DJPL/2021 cannot be used to follow up on findings related to violations of the integrity pact, because these violations have become the domain of trade activities which are ideally accommodated by regulations related to trade. These results are in accordance with other research which states that the Tol Laut program requires collaboration from various stakeholders [19].

4. Discussion/Conclusion

The Tol Laut program is a government initiative designed to provide essential goods to Underdeveloped, remote, outermost, and border regions through subsidized sea transport containers. An analysis was conducted to evaluate and enhance the Tol Laut Program along routes H-1, T-3, and T-5, using port assessment analysis with weighted scoring to measure effectiveness, supply-demand analysis with index consumption demand, and a review of regulations. This framework considers key aspects like Port Location, Port Connectivity, Port Facilities, Liner Ship Specifications, Feeder Ship Potential, Tol Laut Productivity, Hinterland Support, and administrative and commercial factors. A qualitative analysis was also performed to compare regulations related to transportation and trade.

The port assessment analysis reveals that the Tol Laut program's implementation in Tahuna is more successful than in Natuna. Tahuna Port benefits from comprehensive facilities, leading to higher maritime highway productivity. The region also enjoys strong hinterland support, with a diverse user base including logistics companies and consignees, and proactive regional government involvement in monitoring Tol Laut cargo stocks and prices. In contrast, Natuna suffers from limited port facilities, resulting in low cargo handling productivity. The hinterland has fewer service users, and local government engagement in monitoring prices and shipping is minimal, leading to significant price disparities.

Supply-demand analysis indicates that Tahuna's estimated annual need for essential goods is about 68,000 tons, compared to 61,000 tons in Natuna. Tahuna effectively meets about 30% of its local demand for essential goods, aided by regional government support, which helps reduce price disparities. Conversely, Natuna struggles to meet demand, fulfilling only 16% of local needs.

The regulatory review shows that the Tol Laut program is overseen by two ministries, each with distinct regulations aligned with their core responsibilities. The Ministry of Transportation regulates ships, routes, and ports, while the Ministry of Trade oversees cargo identification and distribution monitoring. These regulations intersect with supply and demand, requiring coordination in ship provision, port designation, and container quota adjustments based on the Ministry of Trade's demand assessments.Tol Laut Program, a strategic government initiative, demonstrates varying achievements across different regions. In the Tahuna area, the supply-demand dynamics of Tol Laut Program are notably significant, contributing to the control of price disparities. Conversely, in Natuna, the impact is less pronounced. The effectiveness of Tol Laut Program is influenced by factors such as the comprehensiveness of port facilities, ship specifications, hinterland support, as well as administrative and commercial considerations. The implementation and oversight of Tol Laut Program program, both upstream and downstream, are governed by regulations established and enforced by multiple agencies.

Ongoing monitoring and evaluation are essential to ensure that Tol Laut Program operates within the designated parameters while facilitating continuous improvements. To optimize the program, several recommendations can be made: enhancing route efficiency through the Hub and Spoke model, increasing loading and unloading capacities, supporting the cold supply chain infrastructure, and revising relevant regulations.

Reference

[1] Kusuma, L.W.N., 2022. Cluster Analysis Of Marine Toll Service Area To Minimize Price Disparity To Inprove The Economy Of The East Indonesian Region In The Era Of The Covid-19 Pandemic. *Journal of Innovation and Applied Technology*, ***8*(2)**, pp.1451-1458.

[2] Gugat, R.M.D., Raga, P., Marpaung, E. and Adety, C.S.S.Z.P., 2022. Load Factor Improvement Analysis of Return Cargo on The Tol Laut Ship Route. *Advances in Transportation and Logistics Research*, **5**, pp.49-69.

[3] Pambudi, Muhammad Aji Luhur, and Tika Handayani. 2023. "Application of Sea Toll As a Means to Maintain the Availability of Goods in the 3T Area (Lagging, Frontier and Outermost)". *Maritime Park Journal of Maritime Technology and Society* **2 (2)**:37-45. https://doi.org/10.62012/mp.v2i2.26360.

[4] Triantoro, W., 2020. Comparative cost analysis of domestic container shipping network: a case study of Indonesian sea-toll concept. *Jurnal Penelitian Transportasi Laut*, **22(1)**, pp.33-46.

[5] Kusuma, L.T.W.N., Hamdala, I., Cahyawati, A.N., Hafiyyan, G.A., Ega Ulaya, H., Damaris, D., Rifqi, M., Azizah, F. and Buana, I.G.S., Sea Toll Services Criteria Cluster Analysis to Minimize Price Disparity Of Indonesian Eastern Areas Amidst The COVID-19 Pandemic.

[6] Kurniawan, Abdy, Dewi Indira Biasane, Rosita Sinaga, Chairul Paotonan, and Samsul Bahri. "Synchronization of ship services areas between ports in the Lampa Strait of Natuna." *In AIP Conference Proceedings*, vol. 2675, **no. 1**. AIP Publishing, 2023.

[7] TLSDP, Puslitbang. (2021). Prosiding Webinar FGD "Hub and Spoke" dalam Menunjang Tol Laut. Pusat Penelitian dan Pengembangan Transportasi Laut, Sungai, Danau dan Penyeberangan.

[8] Direktorat Jenderal Perhubungan Laut. 2023. Penetapan Jaringan Trayek Penyelenggaraan Kewajiban Pelayanan Publik Untuk Angkutan Barang di Laut Tahun Anggaran 2024, Nomor KP.DJPL 693 Tahun 2023. Direktorat Lalu Lintas dan Angkutan Laut.

[9] Sugiyono. (2013). Metode Penelitian Kuantitatif, Kualitatif dan R&D. Bandung: *Alfabeta*.

[10] Hadi, S., Hamzah, M.Z. and Sofilda, E., 2023. Public Private Partnership Development Model (PPP) In The Provision Of Airport Infrastructure In Indonesia. *Ecobestha*, **2**, pp.178-198.

[11] Mabuza, L., Sonnenberg, N. and Marx-Pienaar, N., 2023. Natural versus synthetic dyes: Consumers' understanding of apparel coloration and their willingness to adopt sustainable alternatives. *Resources, Conservation & Recycling Advances*, **18**, p.200146.

[12] De Langen, P., Nidjam, M., & Van der Horst, M. (2007). New indicators to measure port performance. Journal of maritime research, 4(1), 23-36.

[13] JDIH BPK. (n.d.-a). Perpres No. 59 tahun 2020. Database Peraturan | JDIH BPK. https://peraturan.bpk.go.id/Details/136174/perpres-no-59-tahun-2020

[14] Direktorat Jenderal Perhubungan Laut. 2020. Usulan Jaringan Trayek/Rute Baru/Tambahan Kapal Penyelenggaraan Pelayanan Publik Angkutan Barang, Nomor 001/SOP/DITLALA-SD 3/08/2020. Direktorat Lalu Lintas dan Angkutan Laut.

[15] Direktorat Jenderal Perhubungan Laut. 2021. Standar Operasional Prosedur Pemesanan Ruang Muatan Kapal dan Penerbitan Shipping Instruction Dalam Penyelenggaraan Kewajiban Pelayanan Publik Untuk Angkutan Barang Di Laut Dari Dan Ke Daerah Tertinggal, Terpencil, Terluar Dan Perbatasan, Nomor KP.804/DJPL/2021.

[16] Kusuma, Lalu Tri Wijaya Nata, and Fu-Shiang Tseng. "Analysis of the impact of the "sea toll" program for seaports: Resilience and competitiveness." *Applied Sciences 9*, **no. 16** (2019): 3407.

[17] Kurniawan, A. (2023). Evaluasi Program Tol Laut di Natuna "Peningkatan Konektivitas, Pemenuhan Supply-Demand dan Pengendalian Disparitas Harga". Pusat Kebijakan Sarana Transportasi, Badan Kebijakan Transportasi, Kementerian Perhubungan.

[18] Sofiyandi, Y., Kurniawan, Y.R, Yudhistira, M.H. (2013). The Impact of Maritime Logistics Subsidy on Food Prices: Evidence from Indonesia. *Economic Analysis and Policy*, **7**, 1026-1045.

[19] Bhinadi, A. (2012). Struktur Pasar, Distribusi, Dan Pembentukan Harga Beras. *Jurnal Ekonomi dan Studi Pembangunan*. **13(1)**. 24-32.

[20] Direktorat Jenderal Perdagangan Dalam Negeri. 2018. Petunjuk Teknis Pendataan, Pemantauan, Dan Evaluasi Terhadap Jenis, Jumlah Dan Harga Barang Di daerah Yang Termasuk Dalam Program Penyelenggaraan Kewajiban Pelayanan Publik Untuk Angkutan Barang Dari Dan Ke Daerah Tertinggal, Terpencil, Terluar, Dan Perbatasan, Nomor 41 tahun 2018.

[21] Pribadi, O. S., Suartika, I. M., Istiyanto, B., & Jepriadi, K. (2022, November). Analysis of Airport Optimal Location in Papua Province as a Support for Logistics Center Distribution Development. In RSF Conference Series: Engineering and Technology (Vol. 2, No. 2, pp. 282-293).

Information Extraction from Automatic Identification System (AIS) Data for Identifying Shipping Line based on Maritime Mobile Service Identity (MMSI)

M. Ma'ruf[1]*, R. Ruslin[2], Uswatun Hasanah[3], E. Efendi[4], Willdan Aprizal Arifin[5], Dhimas Widhi Handani[6], Gede Bagus Dwi Suasti[6], Farih Syamsudin[5], Johannes Valentino Sinaga[1], Billal Soebhi Eryawan[1], and Ken Cindha Bumi Al-Kautsar[1]

[1]Maritime Logistics Study Program, Universitas Pendidikan Indonesia, Serang, Indonesia.
[2]Akademi Ilmu Pelayaran Nusa Tenggara, Jl. Raya Labuhan Haji, Lombok Timur, Indonesia.
[3]Geophysical Engineering, Universitas Syiah Kuala, Jalan Teuku Nyak Arief, Banda Aceh, Indonesia.
[4]Economics Study Program, UIN Ar Raniry, Jl. Syeikh Abdul Rauf Darussalam, Banda Aceh, Indonesia.
[5]Marine Information System Study Program, Universitas Pendidikan Indonesia, Serang. Indonesia.
[6]The Center of Excellence of Maritime Safety and Marine Installation, Institut Teknologi Sepuluh Nopember, Surabaya, Jawa Timur, Indonesia.

* E-mail: maruf.lk@upi.edu

Abstract. Automatic Identification System (AIS) is a technology used to track the movement of ships through the transmission of ship data using Very High Frequency (VHF) waves. Ships around the world transmit their information, which is then captured by AIS receivers. This AIS data collected and stored at the ground station. This study uses the data on the number of ships categorized based the Maritime Mobile Service Identity (MMSI) in the southern Indonesian Archipelago Sea Route II (ALKI II), Madura strait, and its surroundings during one month period of March 2023. The AIS raw data processed using The Jupyter Notebook to generate the information such as MMSI, longitude, latitude, and navigation status. The information then plotted on QGIS to see the difference pattern of shipping lanes based on the ship's MMSI number. The aim of this research is to demonstrate how AIS data processing and analysis can generate sufficient information for maritime spatial planning, including depicting maritime traffic density and traffic constellations, shipping lanes and navigation status, the hierarchical network of maritime routes, fishing zones, and spatial-temporal interactions between shipping activities. The findings from maritime spatial planning can offer valuable insights for law enforcement at sea and offer guidance for developing port infrastructure based on specific needs through the utilization of AIS data.

1. Introduction

Automatic Identification System (AIS) is a technology that leverages the utilization of big data, which is analyse to transform an information into a knowledge to help humans in a decision-making process. It is crucial to undertake appropriate preprocessing procedures to achieve precise fresh understandings from big data [1] [2]. An example of the use of big data is the processing of AIS data sent from ships into useful information in shipping activities [3] [4]. AIS data is periodically dispatched by ships and collected via AIS ground stations positioned along coastlines or via satellite receptors [5] [6] Nevertheless, due to the vast amount of data used to facilitate data processing, there is a need for an application capable of measuring the efficiency aspects of data storage using designed data filtering algorithms. By implementing data filtering algorithms, efficiency in data storage utilization within the database can be achieved [7] [8].

The AIS data can be used to visualize and identify various aspects such as maritime traffic density, shipping routes, navigation patterns, the hierarchical structure of maritime pathways, mapping fishing zones and activities, spatial temporal interactions among activities, and accident risk indicator evaluation [9] [10] [11] [12] [13]. The AIS Team at the UN Global Platform also analyse AIS data for use in various research fields, such as migration, environmental, maritime and fisheries, as well as economic and trade fields. In addition, that AIS data can be utilized for data processing and analysis, resulting in relevant information related to Maritime Spatial Planning (MSP) [14]. This research focuses on analysis of information obtained from AIS data to determine the number of ships based on MMSI that have been converted and grouped per ship route that often pass through the waters of the Madura Strait and its surroundings during March 2023. The purpose of this paper is to provide valuable insights for law enforcement at sea and offer guidance for developing port infrastructure based on specific needs through the utilization of AIS data.

2. Methodology

Based on Bhatthacharjee [15] AIS data is categorized into four groups: static data, encompassing details about a ship's attributes; dynamic data, which constantly changes owing to the continuous motion of vessels; voyage-specific data associated with the current journey; and short-safety related message

Table 1. AIS data comprised into four information categories

Static	Dynamic	Voyage related	Short-safety related
Information on ship characteristics (MMSI, IMO number, call sign, ship name, type, dimensions, etc.)	Information on ship movements [Ship's position (longitude, latitude), speed over ground (SOG), course over ground (COG), navigation status, etc.]	Information of current voyage (destination, estimated time of arrival, draught, etc.)	A text message without a specific format sent to one or multiple recipients or broadcasted to all stations in the vicinity (buoy missing, iceberg sighting, etc.)

As mentioned in United Nation Statistic Data [16], there are 27 different types of AIS message information, each conveying different types of data. AIS transmits voyage-related information, including details such as vessel location, speed, course, direction, turning speed, destination, and estimated time of arrival. In addition, it includes static details such as vessel name, MMSI ID, message ID, vessel type, and dimensions. Dynamic data, such as the current latitude and longitude of the vessel, is transmitted automatically, depending on the vessel's speed and course. When the vessel is moving, this data is transmitted every 2 to 10 seconds, and when the vessel is at anchor the data is transmitted every 3 minutes.

Table 2. Typical AIS message types

ID	Types	Description
1-3	Position report	(Assigned) Scheduled position report, or response to interrogation
4	Base station report	Position, UTC, date and current slot number of base station
5	Static and voyage related data	Scheduled static and voyage related vessel data report
6-8	Binary related message	Binary communication
10-11	UTC related message	Request/Response-to UTC/date
12-14	Safety-related message	Communication/Acknowledgement/Broadcast safety data
15	Interrogation	Request for special response
21	Aids-to-navigation report	Position and status report for aids-to-navigation
27	Position report for long-range applications	Class A and Class B "SO" shipborne mobile equipment outside base station coverage

In the static data, there is a field known as MMSI (Maritime Mobile Service Identity), this serves as the code identifier on the vessel. The generation of the MMSI code is governed by document ITU-R M.585-8, which specifies that the MMSI code consists of nine digits in the format M1I2D3X4X5X6X7X8X9. The initial three digits represent the vessel's country of origin code. Static information related to shipping is sent every 6 minutes or can be done upon request from the authorities. While the dynamic data transmission interval is determined based on the status and speed of the vessel [17].

AIS navigational status represents a signal transmitted by the vessel, providing real-time information about the vessel's condition. In fact, by examining the navigational status code, you can discern the vessel's current actions. This code can assume various values within the range of 0 to 15. Understanding the navigational status of vessels holds paramount importance in maritime activities. This knowledge serves multiple critical purposes. Firstly, it plays a pivotal role in collision avoidance, allowing vessels to make informed decisions about adjusting their course and speed to prevent accidents and mitigate damage to vessels, property, and the environment. Additionally, displaying the appropriate lights and shapes to signify a vessel's navigational status serves as a universal communication tool, enabling vessels to convey their intentions effectively, especially in situations where vessels are in close proximity or visibility is hindered by adverse weather conditions. Moreover, vessel status information contributes significantly to safety, as it provides insights into a vessel's activities and requirements. For instance, a vessel reported as "aground" or "restricted in its ability to maneuver" may necessitate extra caution or assistance, ensuring the safety of all involved. Finally, adhering to international rules governing the display of signals for navigational status is imperative for legal compliance, as failure to do so can lead to legal repercussions, financial penalties, and an increased risk of accidents and associated damages.

As mentioned by Schumannthat [18]AIS navigational status can be explained as descibe in table below:

Table 3. AIS navigational status

Navigation Status	Description
0	Under way using engine
1	At anchor
2	Not under command
3	Restricted maneuver ability
4	Constrained by her draught
5	Moored
6	Aground
7	Engaged in Fishing
8	Under way sailing
9	Reserved for future amendment of Navigational Status for HSC
10	Reserved for future amendment of Navigational Status for WIG
11	Reserved for future use
12	Reserved for future use
13	Reserved for future use
14	AIS-SART is active
15	Not defined (default)

2.1. Database Processing

The database collection from AIS data analyses using a Jupyter notebook which python has been installed [19]. To perform commands in python, it is necessary to download the required libraries in

IOP Conf. Series: Earth and Environmental Science **1423** (2024) 012016 doi:10.1088/1755-1315/1423/1/012016

the form of pandas and math, the use of pandas serves to enter data files in the form of *maret.csv* which will be data frame to obtain results used in plotting.

Code	
Input	: import pandas as pd import math

Output : The library is active

The data in *march.csv* is still in the form of data frame which that cannot be used for plotting, so it should separate and filter to get complete data that needed in ship plotting. The library input in the data retrieval will be forwarded using the python command which is used to separate the rows. This process is used to obtain rows of data that are suitable for analysis. The data needed in the analysis is MMSI, Longitude, and Latitude. The code used as below:

Code	
Input	: data_ais = df[['MMSI', 'Longitude', 'Latitude']]

Output :

	MMSI	Longitude	Latitude
0	200000000	13.951118	108.567617
1	210718000	112.721900	-7.201667
2	215390000	112.653733	-7.087883
3	312256000	52.178768	-6.947093
4	352401408	51.920315	-4.873445
..	...	...	...
506	525000265	14.004558	108.494688
507	525000360	112.740843	-7.202142
508	525000369	13.919998	108.780310
509	525000653	57.238288	-13.532493
510	525001047	112.729698	-7.207488

Figure 1. Result data after separation

The separated data is added to the ship type using data frames. The use of data frames intended to convert MMSI into ship type without losing the MSSI. The code is below:

IOP Conf. Series: Earth and Environmental Science **1423** (2024) 012016 doi:10.1088/1755-1315/1423/1/012016

Code

Input :
```
ship_types = {
    '00-09': 'Passenger ships',
    '10-19': 'Cargo ships',
    '20-29': 'Tankers',
    '30-39': 'Fishing vessels',
    '40-49': 'High-speed craft',
    '50-59': 'Pilot vessels',
    '60-69': 'Search and rescue vessels',
    '70-79': 'Tugs',
    '80-89': 'Sailing vessels',
    '90-99': 'Other types of vessels'
    }
```

Output :

	MMSI	Longitude	Latitude	Ship_Type
0	200000000	13.951118	108.567617	Passenger ships
1	210718000	112.721900	-7.201667	Unknown ship type
2	215390000	112.653733	-7.087883	Fishing vessels
3	312256000	52.178768	-6.947093	Unknown ship type
4	352401408	51.920315	-4.873445	High-speed craft
..	...	...	...	...
506	525000265	14.004558	108.494688	Passenger ships
507	525000360	112.740843	-7.202142	Passenger ships
508	525000369	13.919998	108.780310	Passenger ships

Figure 2. Result data after convert

After the data is converted, the result is MMSI, Longitude, Latitude, and Ship Type. The data that has been displayed is in accordance with the plotting requirements, then saved using Jupyter Notebook with csv format.

2.2. Information Retrieval from the Database

The information that has been received from MMSI conversion results type of ships. To plot in Jupyter Notebook according to the type of ship must enter an additional library such as matplotlib. The code that inputted for plotting ship types is:

Code

Input :
```
In [1] import numpy as np
       import pandas as pd
       from mpl_toolkits.basemap import Basemap
       import matplotlib.pyplot as plt
```

IOP Conf. Series: Earth and Environmental Science **1423** (2024) 012016 doi:10.1088/1755-1315/1423/1/012016

Output : The library is activeThe data containing MMSI, Longitude, and Latitude will be done plotted to find out the type of the ships. Before that, the coordinates of the Madura Strait should be known to describe the position and colour of the ship. Then, the plotting can be done with code:

Code

Input :
```
fig,ax=plt.subplots(figsize=(8,8))
    m = Basemap(llcrnrlon=minlon,llcrnrlat=minlat,urcrnrlon=maxlon,
    urcrnrlat=maxlat,rsphere=(6378137.00,6356752.3142),
    resolution='l',projection='merc',lat_0=lat0,lon_0=lon0,
    lat_ts = lat1)
    m.drawmapboundary(fill_color='white')
    m.fillcontinents(color='lightgrey',lake_color='white')
        colors = {
              #'tankers': 'royalblue',
              'passenger': 'red',
              #'vessels': 'blue',
              }
    for shiptype, color in colors.items():
              subset = bro_inidata[bro_inidata['Ship_Type'] == shiptype]
          print(f"\nFound {len(subset)} {shiptype} ships.")
              if not subset.empty:
x, y = m(subset['Longitude'].values, subset['Latitude'].values)
              print(f"Plotting {shiptype} ships at coordinates: {list(zip(x, y))}")
          m.scatter(x, y, s=5, c=color, marker='o', alpha=0.7, label=shiptype)
        else:
   print(f"No {shiptype} ships found in the data.")
        plt.legend(loc='lower left')
        plt.title('Ship Types Distribution')
        plt.show()
```
Output :

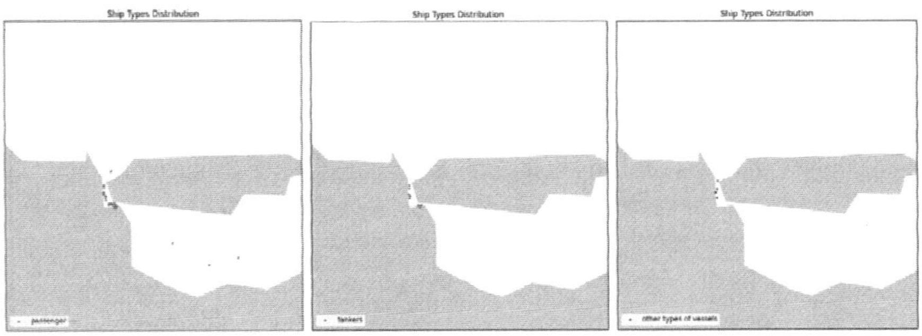

Figure 3. Plotting ship type

3. Result and Discussion

After going through selection and classification process of AISITS database which were obtained in Surabaya area during one month period of March 2023, then the data are analyze as follows:

3.1. Ship classification information based on MMSI

The following figure is the number of ships by type in Surabaya waters during a one-month period in March 2023. the resulting image uses a dataframe that is projected into bar form so that it can be analyzed easily

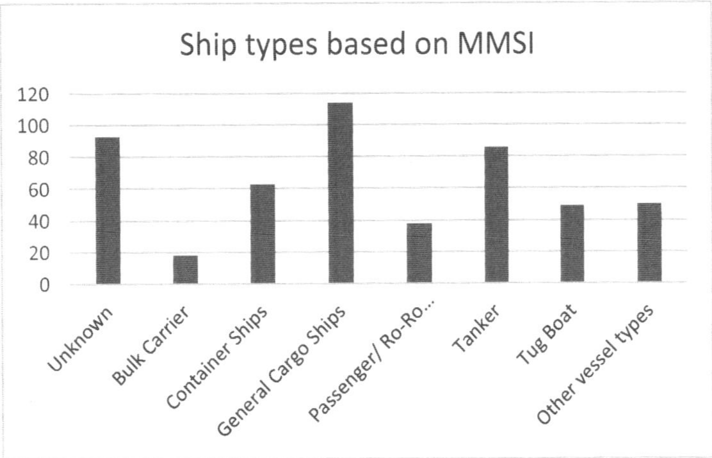

Figure 4. Number of vessels based on its function

3.2. Ship classification information based on navigational status

The following is the navigational status of the ship in Surabaya waters during one month period of March 2023 in percent. In March, many boats sail with engines which means these waters are only used as a shipping lane.

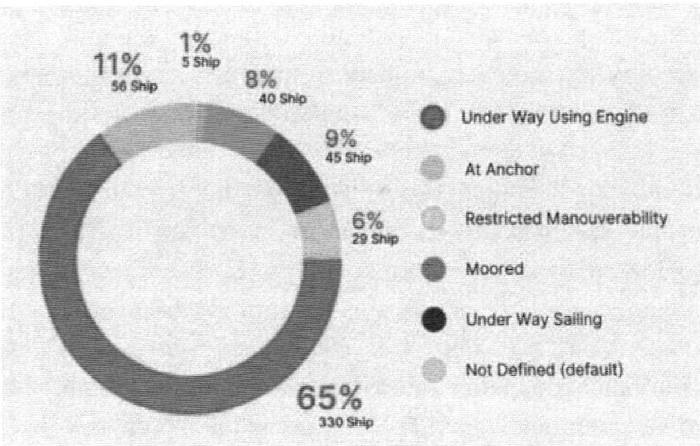

Figure 5. Navigational status in percent

3.3. Shipping Lanes Plotting based on MMSI

After grouping the data by MMSI, it was found that each MMSI had a varying number of log data entries according to the data obtained from the AIS receiver. The data obtained by the AIS receiver was then plotted on a map based on the translated longitude and latitude. Each MMSI is assigned a different colour, resulting in a different visualization. The figure below illustrates the density of vessel traffic over a one-month period in Surabaya and surrounding waters.

Figure 6. Traffic constellation in Madura Strait and its surrounding during one month period of March 2023.

4. Conclusion

Based on the analysis of the data on the AIS receivers stored in the database system, we were able to extract statistical insights about the vessels in the waters of Surabaya's Madura Strait during a one-month period in March 2023. However, prior to acquiring this information, there's a need for initial data processing within the database. One step in this process involves establishing a view or relational table to ensure the completeness of the data obtained. Some challenges in data conversion include coordinates that fall outside the coverage area of Surabaya waters, which are considered as an error in plotting shipping lanes. When processing latitude and longitude data, excel files often identify numerous duplicate entries, it needs a backend application to reduce the number of duplicates for easier data conversion.

Based on the MMSI distribution, The vessel types were dominated by General Cargos, Unknown Ships, and Tankers. The most navigational status founded was under way using engine. The results of shipping lanes plotting, it reveals a traffic constellation and even crossing activities. From those conditions, it required to conduct further analysis for mapping the traffic constellation and potential vessel crossings to reduce the risk of near misses and ship collisions, and ultimately enhancing maritime safety. It is important for port operators to create a priority scale in the development of port infrastructure based on the number of ships passing through Surabaya waters. The priorities are sequentially for general cargo ships, tankers, and container ships to provide optimum port services.

Acknowledgement

Authors sincerely appreciate Mr. Prof. Dr. Ketut Buda Artana, S.T., M.Sc., Mr. Dr. Eng. Dhimas Widhi Handani S.T., M.Sc., and Mr. Gede Bagus Dwi Suasti, S.T., M.M.T., from The Center of Excellence of Maritime Safety and Marine Installation, Institut Teknologi Sepuluh Nopember (PUI KEKAL ITS) for their support regarding AIS data used on this research.

References

[1] D. L. W. S. W. H. J. a. K. X. L. Yang, "How big data enriches maritime research–a critical review of Automatic Identification System (AIS) data applications," *Transport Reviews,* vol. 39, no. 6, pp. 755-773, 2019.

[2] S.-H. M.-I. R. M.-J. O. S.-W. P. a. I.-I. K. Kim, "Estimation of ship operational efficiency from AIS data using big data technology," *International Journal of Naval Architecture and Ocean Engineering,* vol. 12, no. -, pp. 440-454, 2020.

[3] K. D. S. R. W. a. R. M. Bao, "AIS big data framework for maritime safety supervision," *International Conference on Robots & Intelligent System (ICRIS),* Vols. -, no. -, pp. 150-153, 2020.

[4] D. M. S. K. W. a. W. A. Filipiak, "Big data for anomaly detection in maritime surveillance: spatial AIS data analysis for tankers," *Maritime Technical Journal,* vol. 215, no. 4, pp. 5-28, 2018.

[5] J. &. N. R. Nickelson, "Export-Import Value Nowcasting Procedure Using Big Data-AIS and Machine Learning Techniques," *The East Asian Journal of Business Management,* vol. 12, no. 3, pp. 1-12, 2022.

[6] A. N. Skauen, "Ship tracking results from state-of-the-art space-based AIS receiver systems for maritime surveillance," *CEAS Space Journal,* vol. 11, no. 3, pp. 301-316, 2019.

[7] A. M. M. E. P. Depandi Enda, "Rancang Bangun Aplikasi AIS Backend Untuk Pemantauan Lalu Lintas Kapal di Selat Melaka," *Jurnal Inovtek Polbeng Seri Informatika* , vol. 6, no. 2, pp. 284-294, 2021.

[8] N. F. D. A. R. R. P. P. R. H. W. H. Muazam Nugroho, "Information Extraction From Lapan Satellite Ais Database For Ship Classification In The Indonesian Territorial Waters," *Jurnal Teknologi Dirgantara,* pp. 81-92, 2020.

[9] F. M. G. A. A. M. V. a. A. P. Natale, "Mapping fishing effort through AIS data," *PloS one,* vol. 10, no. 6, p. e0130746, 2015.

[10] L. Y. X. Q. W. F. W. a. Z. X. Wu, "Mapping global shipping density from AIS data," *The Journal of Navigation* , vol. 70, no. 1, pp. 67-81, 2017.

[11] R. J. a. A. L. A. ". n. a. a. r. i. Bye, "Maritime navigation accidents and risk indicators: An exploratory statistical analysis using AIS data and accident reports," *Reliability Engineering & System Safety,* vol. 176, no. 1, pp. 174-186, 2018.

[12] A. M.-E. I. a. J. v. H. Dobrkovic, "Maritime pattern extraction and route reconstruction from incomplete AIS data," *International journal of Data science and Analytics,* vol. 5, no. 1, pp. 111-136, 2018.

[13] L. F. G. a. P. K. Du, "Review and analysis of methods for assessing maritime waterway risk based on non-accident critical events detected from AIS data," *Reliability Engineering & System Safety,* vol. 200, no. 1, p. 106933, 2020.

[14] M. D. L. G. F. G. a. B. Q. Le Tixerant, "How can Automatic Identification System (AIS) data be used for maritime spatial planning?," *Ocean & Coastal Management* , vol. 166, no. -, pp. 18-30, 2018.

[15] S. Bhatthacharjee, "What is Automatic Identification System (AIS)- Types And Working (FAQs)," 27 July 2022. [Online]. Available: https://www.marineinsight.com/marine-navigation/automatic-identification-system-ais-integrating-and-identifying-marine-communication-channels/.

[16] UNSD_MM, "Overview of AIS dataset," 05 May 2020. [Online]. Available: https://unstats.un.org/wiki/display/AIS/Overview+of+AIS+dataset.

[17] datalastic, "AIS Navigational Status," -, 12 April 2022. [Online]. Available: https://datalastic.com/blog/ais-navigational-status/. [Accessed 14 05 2023].

[18] S. Schumann, "AIS Navigational Status," - - 2023. [Online]. Available: https://help.fleetmon.com/en/articles/4476744-ais-navigational-status. [Accessed 14 05 2023].

[19] H. S. ,. W. H. M. M. R. Haryadi, "Sistem Penguraian Data Automatic Identification System (AIS) dengan Bahasa Pemograman Python," in *Seminar Nasional Rekayasa dan Teknologi*, Jakarta, 2019.

IOP Conf. Series: Earth and Environmental Science **1423** (2024) 012017 doi:10.1088/1755-1315/1423/1/012017

A Multicriteria Approach for Ship Bunkering System

Fini[1], K B Artana[1], N W D Puspitasari[1], G J Kusuma[1] and D S Wiraguna[1]

[1] Marine Engineering Department, Institut Teknologi Sepuluh Nopember, Indonesia

Finikemenhub@gmail.com

Abstract. This research aims to enhance maritime fuel supply's efficiency, safety, and sustainability by advocating for the standardization of bunkering systems in ports. Through the application of Multi-Criteria Decision Making (MCDM) and the Analytical Hierarchy Process (AHP), this study assesses three primary bunkering methods: Ship to Ship, Truck to Ship, and Onshore to Ship. The criteria for evaluation encompass safety, economic factors, operational efficiency, technical considerations, and additional relevant aspects. Stakeholder feedback underscores the significance of safety (0.281) and economic (0.274) criteria, eclipsing operational (0.274), technical (0.137), and other factors (0.033) in importance. The findings prominently advocate for the Onshore to Ship method as the superior option for standardization in principal ports, marked by a value of 0.480, with Truck to Ship and Ship to Ship methods trailing in preference. Sensitivity analysis reinforces the stability of this outcome, indicating that adjustments in criteria weightings scarcely affect the overall conclusion, thereby affirming the Onshore to Ship method's predominance. This method, characterized by a pipeline connecting the jetty to ships, holds considerable potential for both research and practical applications in port logistics.

1. Introduction

The maritime sector recognizes the substantial role of ship transportation in facilitating large-scale and cost-effective movement of goods, distinguishing it from alternative modes of transportation such as air, land, and rail [1]. Comparative analyses of various transportation modes reveal the prevalent favourability towards maritime transport. This is exemplified by Indonesia's utilization of 90% of its maritime network for international trade. Additionally, UNCTAD data from 2011 underlines the remarkable capacity of ships to carry a total weight of 8,4 billion tons, emphasizing their pivotal contribution to global trade and logistics [2].

Table **1**. Number of Sea Transportation Fleet Applications by Ownership 2018-2022

No	Detail	Amount per Year (Unit)				
		2018	2019	2020	2021	2022
1	National	25.559	32.634	33.770	37.733	19.104
2	Charter	92	51	78	45	37
3	Agent	7.612	8.237	29.642	34.546	11.564
	Total	**33.236**	**40.922**	**63.490**	**72.324**	**30.705**

Amount of ship in Indonesia by perspective ownership has a significant increased every year. According to Operational Statistics Transportation Data in 2022, the average growth naval fleet by ownership from 2018 until 2022, has increased 8,64% as presented in Table 1 [3]. This value is high category, with consider that Indonesia's territorial water and economic condition is still centralised in Java Island. Because of that, ownership of vessels in Indonesia needs sustainability acceleration.

The necessity for fuel supply to ships is directly related to their number. According to International Maritime Organization's (IMO) 2020 fuel oil consumption report, quantities for each type of fuel oil used by vessels over 5.000 GT are detailed. Documented fuel types include Heavy Fuel Oil (HFO), Light Fuel Oil (LFO), Diesel/Gas Oil (MDO), Liquefied Natural Gas (LNG), Methanol, Liquefied Petroleum Gas (LPG), and Ethanol. The reported consumption amounts to 31361496 tons for container ships, 30361496 tons for bulk carriers, 23510528 tons for tankers, and 160267 tons for passenger ships, specifically regarding HFO [4].

Bunkering of vessels can be efficiently executed at dedicated terminals within harbors. Presently, there are three predominant methods of bunkering utilized: truck-to-ship, shore-to-ship, and ship-to-ship [5]. Each of these methods presents its own set of benefits and limitations. The truck-to-ship approach is favored for its enhanced mobility, lower upfront investment, and reduced upkeep expenses, but is hampered by its inability to cater to large-scale fuelling requirements due to the limited capacity of truck tanks. Conversely, the ship-to-ship and shore-to-ship methods, despite necessitating greater initial financial outlay, are capable of servicing the fuelling needs of larger maritime vessels [6].

In 2021, PT. Pelindo Energi Logistik, a branch of PT. Pelindo Marine Service, unveiled the Filling Station On Shore (FSOS) service at the Tanjung Perak Port. The FSOS system facilitates bunkering through pipelines directly at the dock, aiming to streamline the bunkering process at Tanjung Perak in Surabaya. This innovation underscores the importance of tailoring the bunkering system to the specific requirements of each port's terminal. A comprehensive evaluation is essential to identify the optimal conditions for the deployment of such systems, with the overarching goal of enhancing bunkering operations' efficiency and effectiveness. Employing the Multiple Criteria Decision Making method, particularly the Analytical Hierarchy Process, is critical in the optimization of the bunkering system tailored to each port's unique characteristics.

2. Literature Review

2.1. Bunkering System. Bunkering is the process of refueling a ship following certain standardized procedures, such as those outlined in ISO 13739. Several methods are employed in bunkering, namely ship to ship, onshore to ship, and truck to ship, each with its own advantages and disadvantages. [7]

2.2. Ship to Ship. The ship-to-ship (STS) method involves transferring fuel from one ship to another. This process can be carried out at various bunkering terminal locations. The ships involved in the transfer are positioned alongside each other, and the transfer is conducted using equipment such as manifolds, loading arms, or flexible hoses. Several important procedures in STS include checking the conditions between the two ships involved in the transfer, preparing a fuel transfer plan, executing mooring (securing the ships), connecting the cargo hose, cooling down the cargo hose, pipes, and tanks, controlling the transferred cargo, monitoring weather conditions, checking the stratification of fuel product density, checking the stratification of fuel product density. [8]

2.3. Onshore to Ship. This strategy involves the use of bunkering terminals that link onshore fuel storage to ships via pipelines [9]. Suitable for large fuel demands, this method employs flexible hoses for fuel pumping from onshore facilities to the vessel. The construction of a dedicated bunkering terminal is necessitated for employing this strategy [10].

2.4. Truck to Ship. Employing trucks for fuel transport and refueling offers flexibility and is commonly used for immediate, lesser-scale requirements. Despite the benefit of lower initial costs and higher mobility, this method is constrained by the trucks' limited capacity, rendering it viable solely for minor operations. [11].

IOP Conf. Series: Earth and Environmental Science **1423** (2024) 012017 doi:10.1088/1755-1315/1423/1/012017

2.5. Multi Criteria Decision Making

Multi-Criteria Decision Making (MCDM) is a technique for making decisions from several alternative options. The Analytic Hierarchy Process (AHP) simplifies the decision-making problem by structuring hierarchical elements such as goals, criteria, sub-criteria, and alternatives. [12]. This approach can be depicted in a model flow diagram and serves as a research method for resolving decision-making problems [13]. Having multiple criteria necessitates the use of a Decision Support System to facilitate the selection of decision alternatives. The use of the Analytic Hierarchy Process (AHP) is advantageous for making effective decisions.

2.5.1. Decomposition. This stage involves breaking down the overall problem into smaller, manageable parts. Decomposition results in various levels of the problem, making it easier to handle a complex and unstructured issue by breaking it into more structured parts.

2.5.2. Assessment of Criteria and Alternatives. This stage involves pairwise comparisons (weighting) at each hierarchy level based on their relative importance. The level of importance can be adjusted using the chosen analysis method, typically utilizing a 1-9 scale. This method evaluates the relative importance of two components at a given level compared to the level above.

2.5.3. Priority Determination. This stage involves processing to determine the relative weights or priorities of all alternatives. Weights or priorities are calculated through matrix manipulation (pairwise comparison matrix) or by solving matrix equations. The eigenvalues obtained from the pairwise comparison matrix are used to determine priorities.

2.5.4. Consistency Test. This stage involves calculating the consistency of respondents' preferences to ensure they reflect actual consistency. Consistency is assessed with two categories of comparators: if the Consistency Ratio (CR) > 0.1 (10%), it indicates inconsistency, whereas if CR ≤ 0.1 (10%), it is acceptable.

3. Research Methodology

Multi-Criteria Decision Making (MCDM) or Multi-Criteria Decision Analysis (MCDA) is a method to get the best solution with use multiple aspect as a criteria and subcriteria. This approach tackles problems by considering several factors, not just one. Here's how it works: first, relevant criteria for different bunkering options are established. Then, a quantitative method is applied to weigh these criteria and identify the most suitable option.[14]

This research employed the Analytic Hierarchy Process (AHP) method because it effectively integrates both qualitative and quantitative aspects of the criteria and simplifies the weighting of each criterion through expert opinion. Each criterion is associated with specific sub-criteria, with alternatives detailed in Table 2. The AHP model is illustrated in Figure 1. Once the model is established, the weights for criteria and sub-criteria are derived through pairwise comparisons. The Saaty Pairwise scale is utilized for these comparisons. This process yields a pairwise comparison matrix, which is then normalized and computed into a priority matrix using Equations 2 and 3.

$$C_{ij} = \begin{bmatrix} C_{11} & C_{12} & C_{13} & C_{14} \\ C_{21} & C_{22} & C_{23} & C_{24} \\ C_{31} & C_{32} & C_{33} & C_{34} \\ C_{41} & C_{42} & C_{43} & C_{44} \end{bmatrix} \quad ...(1)$$

$$X_{ij} = \frac{C_{ij}}{\sum_{i=1}^{n} C_{ij}} \quad ...(2)$$

IOP Conf. Series: Earth and Environmental Science **1423** (2024) 012017 doi:10.1088/1755-1315/1423/1/012017

$$W_{ij} = \frac{\sum_{f=1}^{n} x_{ij}}{n} \quad ...(3)$$

Where, Cij : Pairwise Matrix
Xij : Normalize Matrix
Wij : Matrix Weighted Priority

This study will include a sensitivity analysis, as the weights assigned to the criteria significantly impact the overall priorities. By performing a "what-if" analysis, also known as sensitivity analysis, we can evaluate how the final results might vary with different criteria weights. This step is essential in our AHP methodology, enabling us to test the robustness of our original decision and pinpoint the key criteria that influenced the initial outcomes.[15] Figure 1 shows hierarchy of process selection bunkering method with criteria definition in Table 2.

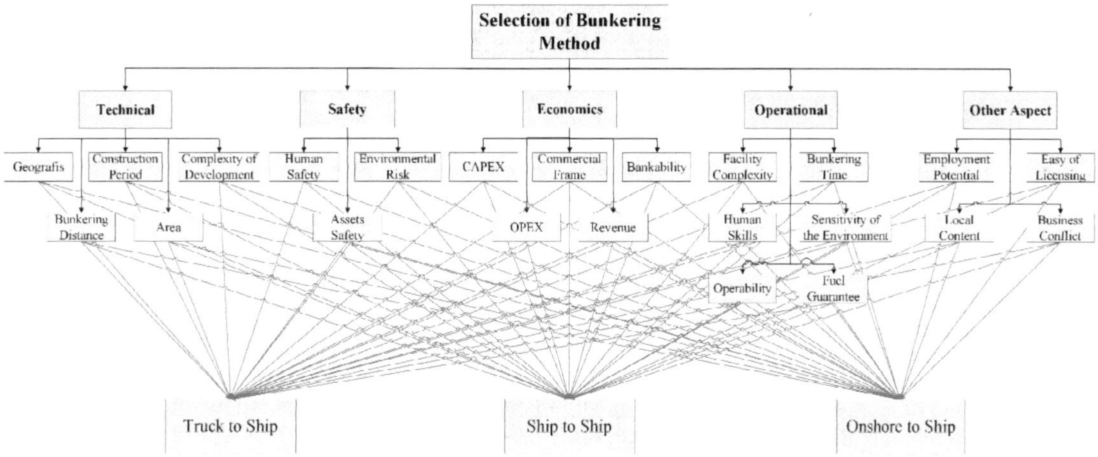

Figure 1. Analytical Hierarchy Process Selection Bunkering Method

Table 2. Criteria Definition

Criteria	Sub-criteria	Definition
Technical	Geographic	Soil structure and minimum sea depth before and after Construction.
	Bunkering Distance	Distance required to move fuel from storage to ship.
	Construction Period	Time for land provision, equipment and construction process for each type of bunkering.
	Area	Area required for provision of each type of bunkering .
	Complexity of Development	Level of complexity and difficulty of construction of each facility
Safety	Human Safety	Human safety during the bunkering process, both crew and human resources.
	Assets Safety	Security of assets in the bunkering process, both vessels and bunkering facilities.
	Environmental Risk	Minimize the occurrence of malfunctions and work procedure errors that impact the environment during the bunkering process.
Economics	CAPEX	Calculation of economic analysis in terms of initial investment.

Criteria	Sub-criteria	Definition
Economics	OPEX	Calculation of economic analysis in terms of bunkering operations.
	Revenue	Comparison between investment required and income to be earned.
	Commercial frame	Potential cooperation with several parties related to supporting facilities and main facilities for each type of bunkering.
	Bankability	Bank interest in lending against investments in each type of bunkering.
Operational	Facility Complexity	Completeness of facilities and infrastructure to support operations in the bunkering system.
	Bunkering Time	Comparison of bunkering time from fuel tank to ship from 3 existing methods STS, TTS, OTS.
	Sensitivity of the Environment	Comparison of operations between 3 bunkering systems whether affected by weather, wind, etc.
	Human Skills	Competencies of workers required to run this type of bunkering.
	Fuel Guarantee	Assurance of the quality and quantity as well as correction of the amount of fuel when refueling.
	Operability	Ease of operation among bunkering system options.
Other Aspect	Employment Potential	Number of jobs generated for surrounding communities from each type of bunkering.
	Easy of Licensing	Ease of regulations used for each type of bunkering during the operational process.
	Business Conflict	Interest in the party establishing the bunkering business.
	Local Content	Ease of sourcing goods domestically for the facility construction process.

4. Result and Discussion

4.1 Focus Group Discussion

Initial phase of bunkering study involved focus group discussion (FGD). FGD is commonly employed as a qualitative approach to deeply understand social issues. This particular FGD aimed to identify criteria and sub-criteria impacting evaluation of bunkering options. Furthermore, FGD sought to collect opinions on weighting assessment among criteria and sub-criteria. Participants included key stakeholders in bunkering ecosystem, such as fuel suppliers, shipping lines, regulators, and facility construction firms with totally number 27 person. Selection of criteria and sub-criteria was informed by qualitative feedback from experts in bunkering field, who provided insights based on real-world conditions. Comprehensive details on each sub-criteria, criteria, and available alternatives, as well as weighting assessment, are elaborated in following sections.

4.2 Analytical Hierarchy Process

Analytical Hierarchy Process (AHP) is carried out with 27 people in the focus group discussion. The number of respondents to be invited for discussion is important consideration. It is generally accepted that minimum eight participant are sufficient [16]. The respondents are from various who impact of facility bunkering, such as otority harbour, crew of ship, figure of regulatory harbour, shipping line, assurance ship companies at major port type. Each respondent fill in a questionnaire to do pairwise comparison. This research has five criteria to choose the type bunkering of main port. Each criteria has different advantages and disadvantages if implemented for bunkering system of vessel. This advantages and disadvantages is shown in **Table** *3*.

Table 3. Advantage and Disadvantage Each Method

Alternatives	Advantages	Disadvantages
Truck To Ship	Cost investment is cheaper than other alternatives, it doesn't need complexity of building, flexibility for the filling location.	It has limited capacity, it depends on distance source tank and the vessel, it depends on infrastructure.
Onshore To Ship	Highest efficiency, It has high of capacity, The process of transfer fuel doesn't depends on weather condition	It has the big investment cost, Building of facility is complex
Ship To Ship	It has high of capacity to transfer fuel.	It dependence on the weather condition, Highest environment risk (oil spill), it has medium complexity of operational

In the AHP assessment process, there are several stages to compare weight values with each other. Weighting of each criterion is conducted by comparing correlation between sub criteria and other sub criteria. Subsequently, scale normalisation is performed to obtain the weighting. Assessment of each sub-criterion to assess existing alternatives is executed by inputting value method by expert respondents using focus group discussion. The following figure 2 is an instance of comparative value conducted between existing sub-criteria.

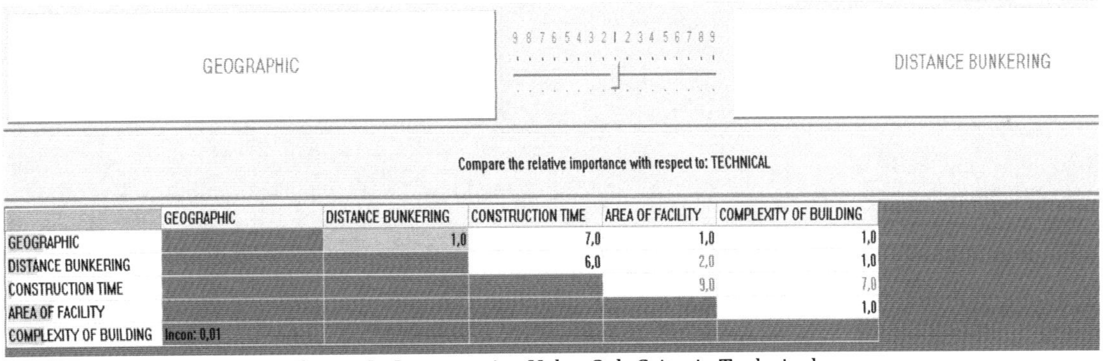

Figure 2. Comparative Value Sub Criteria Technical

From the result of questionnaire this standarization of bunkering, the biggest weight of criteria is Safety with a score 0.281. Meanwhile criteria Other Aspect is the lowest criteria of this questionnaire, with a score 0.033. Furthermore, for the sub criteria with the biggest value is Safety of Human with score 0.615. For each criteria, has the biggest sub criteria. In Safety Criteria, Safety for Human has the biggest value. Then, economical criteria has Revenue for the biggest weight of sub criteria. Next, the biggest weight of Operational criteria is complexity facilities. And the end, for the Other Aspect criteria has Business Conflict for the biggest value of this sub criteria. After all global score input for standarization type of bunkering especially in major port like Tanjung Perak, the biggest alternative score is Onshore to Ship with a value 0.480. So that, Onshore to ship is choosen for the standardization bunkering type main port like Tanjung Perak. The detail global score, alternative with darkest colours mean the score is higher than other alternatives. The overall weight and ranking result is show in **Table** *4*.

Table 4. Weight and Ranking Result of AHP

Criteria	Weight	Sub Criteria	Global Weight	Global Score		
				Truck To Ship	Onshore To Ship	Ship To Ship
Safety	0.281	Safety for Human	0.615	0.297	0.522	0.181
		Safety for Asset	0.319	0.281	0.444	0.276
		Environment Risk	0.066	0.226	0.401	0.373
Economical	0.274	CAPEX	0.194	0.379	0.440	0.181
		OPEX	0.152	0.263	0.544	0.192
		Bankability	0.040	0.275	0.422	0.303
		Commercial Frame	0.040	0.226	0.448	0.326
		Revenue	0.574	0.185	0.540	0.275
Technical	0.137	Geographic	0.237	0.266	0.533	0.202
		Distance of Bunkering	0.202	0.330	0.363	0.307
		Construction Time	0.033	0.445	0.293	0.262
		Area of Facility	0.290	0.218	0.501	0.281
		Complexity of Building	0.237	0.267	0.500	0.233
Operational	0.274	Complexity Facilities	0.473	0.260	0.506	0.233
		Time of Bunkering	0.153	0.249	0.394	0.357
		Sensitivity Environment	0.077	0.246	0.501	0.253
		Operability	0.134	0.249	0.395	0.356
		Human Resource Competence Requirement	0.039	0.173	0.569	0.258
		Fuel Guarantee	0.124	0.295	0.447	0.259
Other Aspect	0.033	Local Content	0.065	0.379	0.408	0.214
		Potential Employment Oppurtinities	0.211	0.210	0.572	0.218
		Ease of Licensing	0.119	0.338	0.321	0.341
		Business Conflict	0.604	0.261	0.456	0.284
		Total Score		0.265	0.480	0.254

The table illustrates that the highest weight within the Human Safety sub-criteria is assigned to the Land to Ship alternative. This result is directly proportional to the chosen alternative. While the Ship to Ship (STS) and Truck to Ship (TTS) methods do not imply a neglect of safety, they involve additional human, asset, and environmental sub-criteria considerations. Perspective analysis from this table indicates that, for almost every sub-criteria, the Onshore to Ship method is the preferred alternative. This preference is attributed to the numerous advantages offered by the Onshore to Ship method, particularly when implemented at a major bunkering port like Tanjung Perak. These advantages align with the characteristics of a primary port, leading respondents, who are relevant workers, to assign the highest value to the Onshore to Ship method over other alternatives.

4.3 Sensitivity Analysis

Sensitivity analysis is utilized to validate decisions or judgments made, by identifying which elements or values are most critical in influencing the outcome. This helps ensure that decisions made are accurate [17]. The methodology of analyzing weight through a modification of ±10% in the weights is adopted to ascertain the resilience or susceptibility of decision-making or outcome results to variations in the criteria weights. In this framework, the influence of altering each criterion's weight by either increasing or decreasing it by 10% is meticulously examined to determine its effect on the ultimate conclusion or decision [18]. Sensitivity analysis is carried out with the change of weight of the criteria to see the effect on alternative selection. The change of the selection at each of the criteria weight changes can be seen through the gradient sensitivity graph. The sensitivity analysis all the criteria and the result is shown in this **Figure** *3*.

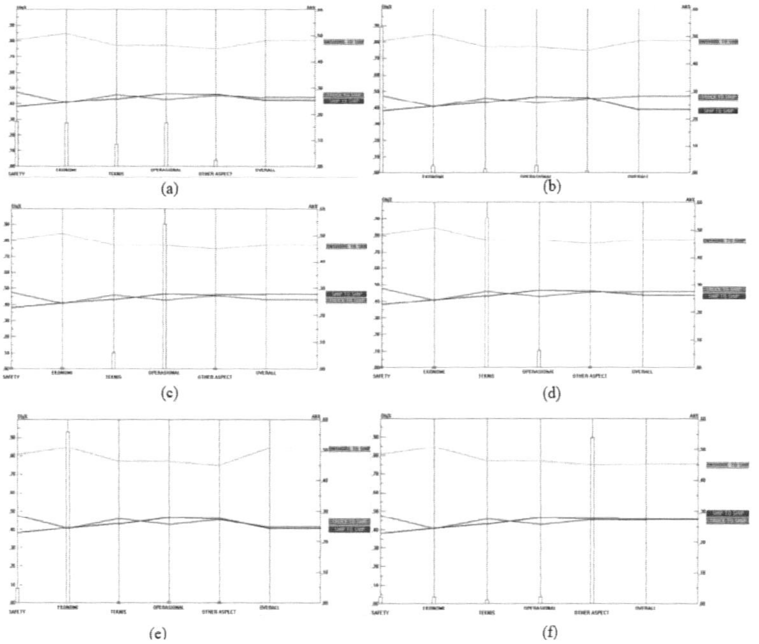

Figure 3. Sensitivity Analysis Result

Sensitivity analysis has function to ensure that miss perception/judgement will not fully change the selection result. Figure 3 shows some result of the sensitivity analysis by implementing different weight for one attribute. Also, the weight automatically changes of other attributes. The original weight in Figure 3 (a), that is shown as Onshore to Ship is selected for standardization type bunkering in main port. Increasing the weight of economy, keeps Onshore to Ship as most preferable type bunkering for main port. It has same result when we increase the weight of operational, safety, technical criteria. All this criteria would then seem insensible. As a presented above of sensitivity analysis, it is unquestionable then Onshore to Ship is choosen for standardization type bunkering at main port.

5. Conclusion

In the conclusion of this paper, the selection of the appropriate bunkering type for the main port is determined using five criteria: Safety, Economical, Technical, Operational, and Other Aspects. The AHP method was employed to weight the criteria and select the best alternative, utilizing the opinions of experts related to bunkering in the port. The results showed that Safety was the top priority with a weight of 0.281, followed by Economical at 0.274, Operational at 0.274, Technical at 0.137, and lastly Other Aspects at 0.033. Based on these criteria, the appropriate bunkering type for standardization at the main port is Onshore to Ship, followed by Truck to Ship, and finally Ship to Ship. Sensitivity analysis demonstrated that changes in the weighting of the criteria did not affect the alternative results, confirming that the optimal choice for main port bunkering is Onshore to Ship. The Onshore to Ship type of bunkering is done through a pipeline installed along the jetty. The pipeline will be connected to a fuel storage tank which will then be piped directly to the ship. This method has not been widely implemented in commercial or specialized ports, so research and application opportunities can be carried out massively.

References

[1] Dong J X, Lee C Y and Song D P 2015 Joint service capacity planning and dynamic container routing in shipping network with uncertain demands Transp. Res. Part B Methodol. 78 404–21
[2] Dijk C M P 2015 Indonesia Maritime Hotspot. Maritime by Holland.
[3] Kementerian Perhubungan 2022 Data Operasional Statistik Perhubungan 2022.
[4] IMO 2021 Energy Efficiency of Ships IMO
[5] Park N K & Park S K 2019 A study on the estimation of facilities in LNG bunkering terminal by Simulation-Busan port case. Journal of Marine Science and Engineering, 7(10).
[6] Peng Y Zhao, X Zuo T, Wan W and Song X 2021 A systematic literature review on port LNG bunkering station. Transportation Research Part D: Transport and Environment, 91.
[7] DMA. (n.d.). North Euroean LNG Infrastructure Projectm full report.
[8] Sigtto 2011 Ship to Ship Transfer Guide for Petroleum, Chemicals and Liquefied Gases. Witherby Seamanship.
[9] EMSA 2017 Guidance on LNG Bunkering to Port Authorities and Administration. 31 January, 430.
[10] Pangestu, S 2017 Penilaian risiko sosial pada kapal penumpang saat melakukan pengisisan bahan bakar LNG, ITS, Surabaya.
[11] Saputro, Galih 2015 Kajian Teknis dan Ekonomis Sistem Bunkering LNG untuk Bahan Bakar di Kapal. (Skripsi Sarjana, Institut Teknologi Sepuluh Nopember)
[12] Saaty, T. L 1993 Pengambilan Keputusan Bagi Para Pemimpin. PT. Bustaman Binaman Presindo. Jakarta.
[13] Zimmer, Sebastian., Labib and Ashraf 2009 Analysis Hierarchi Process dan Expert Choice: Benefit and limitations. ORinsight.22(4) University of portmouth-business school, United Kingdom.p201-22.
[14] P. Sen and J.-B. Yang 1998 Multiple Criteria Decision Support in Engineering Design. Springer London.
[15] E. Mu and M. Pereyra-Rojas 2017 Practical decision making - an introduction to the analytic hierarchy process (ahp) using super decisions.
[16] Taherdoost Hamed 2017 Decision Making Using the Analytic Hierarchy Process (AHP); A Step by Step Approach. International Journal of Economics and Management Systems.
[17] Ivanco M, Hou G, Michaeli J. Sensitivity analysis method to address user disparities in the analytic hierarchy process. Expert Syst Appl. 2017 Dec;90:111–26.
[18] Muanley YY, Son AL, Mada GS, Dethan NKF. Analisis Sensitivitas Dalam Metode Analytic Hierarchy Process dan Pengaruhnya Terhadap Urutan Prioritas Pada Pemilihan Smartphone Android. VARIANSI J Stat Its Appl Teach Res. 2022 Dec 15;4(3):173–90.

Acknowledgments

The author would like to extend appreciation to PT Pelindo Energi Logistik, Port Authority, District Navigating, shipping line, and Pertamina for all support during the study, such as support to join foccuss group discussion for standardization type bunkering, and tobe a respondent for this research.

IOP Conf. Series: Earth and Environmental Science **1423** (2024) 012018 doi:10.1088/1755-1315/1423/1/012018

A System Dynamics Approach for Optimizing Shipping Delivery Strategies under Demand Uncertainty: A Case Study of the Fertilizer Industry

Y C Katon[1*], K B Artana[1] and I N Pujawan[1]

[1] Department of Technology Management, Institute Technology of Sepuluh Nopember, Surabaya, Indonesia

*E-mail: yokekaton@gmail.com

Abstract. The subsidized fertilizer industry faces significant supply chain costs, especially in logistics. Various specific issues require attention and resolution concerning the distribution of fertilizers. These challenges include high shipping costs, uncertain demand, limited funds, and a mismatch between requests and deliveries, which change according to need and are included in the entire supply chain. The supply chain for subsidized fertilizer is designed to be flexible and efficient, allowing for quick adjustments to changes and disruptions in plans and operations. There are complex interactions among the variables in the supply chain, wherein changes in one variable can affect others, leading to dynamic and complex situations. Due to its complexity, the research will utilize the dynamic system method to create an optimal delivery strategy. Two methods will be used: clustering for delivery location and system dynamics. Causal loop diagrams will determine the relationship between variables that influence the system, and stock and flow diagrams will be used to determine delivery routes. The output of this research compares the total cost, responsiveness, and reliability by considering two delivery scenarios: single-trip and milk-run. The results of this research successfully calculated the three outcomes and determined whether to use single-trip or milk-run for the delivery.

1. Introduction

World fertilizer consumption reached 125 million tonnes in 2019, and the global demand is rising [1]. It is important to intensify agriculture with fertilizers to meet the food requirements of the growing global population [2], [3]. Indonesian government regulates the procurement and distribution of subsidized fertilizers to farmer groups based on five principles: right type, right quantity, right price, right place, and right time to ensure national food security [4]. By this concept, the role of the supply chain is crucial in ensuring the accurate and timely distribution of fertilizers [5], [6].

National Medium-Term Development Plan (RPJMN) 2020-2024 outlines the direction of food and agricultural development through the following measures: (1) improving the quality of food consumption, safety, fortification and bio fortification; (2) increasing the availability of agricultural food and seafood sustainably to maintain supply stability and prices of necessities; (3) increasing agricultural sustainability productivity (HR) and market certainty; (4) increasing

IOP Conf. Series: Earth and Environmental Science **1423** (2024) 012018 doi:10.1088/1755-1315/1423/1/012018

productivity, sustainability of agricultural resources and digitalization of agriculture; and (5) improving the governance of the national food system [7]. In the 2020 State Budget, the Ministry of Finance implemented fertilizer subsidies to boost food production. These subsidies positively impact farmers by increasing their capital, fostering the development of fertilizer markets, promoting technology adoption, boosting productivity, and improving farming income. [8].

Definitive Plan for Group Needs (RDKK) is a document used in the agricultural sector to detail the specific needs of farmer groups for a particular planting season, aimed to improve subsidized fertilizer distribution effectiveness and efficiency. In Indonesia, RDKK serves as the basis for the distribution of subsidized fertilizers to ensure effectiveness and efficiency in the process [9], [10]. In the RDKK system, the amount of fertilizer needed in each region is predicted to help calculate the volume and cost. In 2021, 16.6 million farmers requested 23 million tons of fertilizer, which would cost 67 trillion rupiah [11]. However, the 2021 budget for subsidized fertilizer was only 25.3 trillion rupiah for 8.2 million tons, creating a significant gap between supply and demand. This same issue occurred in the previous year due to budget constraints. In 2019, Indonesia imported 6.13 million tons of fertilizer from over 19 countries [12]. Local producers (In Indonesia, fertilizer factory are in Sumatera, Java and Kalimantan) must compete with cheaper imported fertilizers, impacting the cost of goods sold for the agricultural sector [13]. The cost of distributing fertilizer is closely tied to the government subsidy budget. Indonesia's fertilizer industry relies on imported raw materials from different countries and needs to distribute its products across the country [14]. Keeping fertilizer prices competitive and affordable depends largely on managing logistics costs.

The fertilizer industry's supply chain process begins with the supply of raw materials from suppliers. The port is responsible for unloading and transporting the raw materials to warehouses, where the factory processes them into bulk or packaged fertilizer. When it comes to subsidized fertilizer, the quantity of fertilizer produced is influenced by the annual subsidized fertilizer budget. As mentioned earlier, this budget fluctuates according to the manufacturer's allocation from the National Budget. The quantity of fertilizer produced will vary based on the demand from farmers, which is directly impacted by the budget. This will affect the availability of fertilizer in the warehouses and will significantly impact supply, subsequently affecting the distribution of fertilizers. The Bulk fertilizer is distributed to consumers (exports) and local Distribution Centers (DC) warehouses via sea routes, while packaged fertilizer is distributed to distributors via sea and land routes. While loading the fertilizer onto the ship, the availability of the heavy machine for loading and unloading also has significance to the distribution cost. It depends on the number of products being loaded into the ship.

The distribution of goods via sea routes is crucial for areas outside Java, such as Kalimantan, Sumatra, and Sulawesi. The distribution processes is costly. Companies must overcome the main challenge of high logistics costs, which reached 1.7 trillion rupiah in 2019. These high logistics costs pose a significant challenge that companies and the government must address. Distribution costs for fertilizer from factory warehouses to DC/GP warehouses outside Java totalled 731.5 billion rupiah, accounting for 43% of total distribution costs, despite covering only 40% of the product quantity. This is a significant problem in distribution cost efficiency. There is potential for efficiency improvements, considering the various shipping methods available (bulk, packaged, container).

Fertilizer distribution often encounters uncertainty in demand, supply, and distribution processes [15]. One primary reason is the limitation of the National Budget allocated for subsidized fertilizer. This budget constraint directly affects the supply, causing it to be limited. As a result, fertilizer availability is often insufficient to meet the farmers' needs. Furthermore, the demand for fertilizer is continuously rearranged and adjusted based on the specific requirements of the farmers, which can vary significantly due to changing agricultural conditions, crop types, and seasonal variations. These adjustments create additional layers of uncertainty in ensuring timely and adequate distribution of fertilizers to meet the actual demand. The changing demand will affect the number of distributed fertilizers to the farmer. The different numbers will affect the needs of heavy machinery for loading and unloading products and also affect the time. The fertilizer will be transported using sea transportation or land transportation, or they will be used on a single or multi-trip basis. System Dynamics help manage uncertainty and model how variations affect supply chain performance and interactions with the external environment, such as government policy. Changes in government policy through the Ministry of Agriculture impact prices and supplies of fertilizer, especially for subsidized fertilizer, and Indonesian government regulations regarding fertilizer subsidies affect fertilizer prices for sustainable agricultural development [16]. Research indicates that the cost of fertilizer subsidies is relatively small compared to the government purchase price for rice, which is harmful to farmers [17]. Due to the complexity and variability affecting fertilizer distribution, supply and demand quantities can vary over time. System Dynamics can handle changes in policy and market conditions that affect fertilizer distribution [18], such as changes in government policy or market conditions. This research focuses on measuring performance in cost efficiency, timeliness, and appropriateness of the amount of fertilizer distributed [19].

The System Dynamics model aims to achieve cost efficiency, timeliness, and appropriateness of fertilizer distribution in Indonesia. This research is based on government policy, emphasizing flexibility in meeting fertilizer needs. The model is designed to achieve several goals simultaneously, requiring harmony in these aspects to ensure the optimal amount of fertilizer is provided to farmers in Indonesia. It involves an extended supply chain with third parties in the fertilizer distribution chain.

In a study by Camur, et al. [20], it was suggested that the distribution of fertilizer in Indonesia can be analysed through System Dynamics simulations. The distribution of fertilizer is constantly changing and is impacted by government regulations, which necessitate adjustments to production and delivery. System Dynamics have the capability to depict changes in the system and generate signals for decision models, providing feedback for system dynamics [21]. Due to the importance of fertilizer in Indonesia, this research focuses on analysing the distribution of fertilizer in Indonesia using a System Dynamics approach. The objective is to identify the most efficient distribution plan that minimizes costs while ensuring the timely delivery of the correct quantity. This approach is anticipated to result in an increased allocation of subsidized fertilizer to farmers. This research are conducted as below: literature review to explain the concept of System Dynamics in Fertilizer Supply Chain, then Research Methodology and then our approach to explain further is using case study of System Dynamics in fertilizer delivery and Data Analysis, and Result and Discussion lastly is conclusion.

This study has the following structure: In Chapter 2, previous research on the system dynamics of the fertilizer supply chain is explained. Chapter 3 presents the conceptual model, which outlines the research flow of the study. Chapter 4 discusses the case study used, while Chapter 5 presents the results and discussion. Finally, Chapter 6 serves as the conclusion. This research delves into the complexities of the delivery strategies approach for subsidized fertilizer. The system dynamics approach will elaborate on the necessity of a flexible delivery system to reduce costs.

2. System Dynamics in Fertilizer Supply Chains

Various studies have analysed the complexity of the fertilizer supply chain. Xuan Xi and Yulin Zhang studied the competition model between traditional and green agriculture in the agricultural product supply chain. They found that production adjustments affect stability and profitability [22]. Helder Zavale et al. identified supply and demand constraints in the fertilizer value chain in Mozambique. They suggested an enhanced subsidy program could strengthen the supply chain [23]. Improving input subsidy programs through vouchers and effective targeting can increase farmer demand for fertilizer and strengthen supply chains. Dmitry Ivanov discussed the issue of complexity in supply chains, highlighting various aspects and applications of complexity [24]. Jianliang Peng proposed a method to dynamically and automatically adjust supply chain optimization strategies to adapt to dynamic environments [25].

In addition to the aforementioned research, a study conducted in West Lombok Regency found that the distribution of subsidized fertilizer from distributors to farmers was adequate, with 83.0% of the distribution met the criteria of being in the right place, at the right time, in the correct quantity, and at the right price [26]. Another study explored the potential use of blockchain technology in fertilizer supply chain management. The study found that this technology can significantly improve the fertilizer supply chain's transparency, traceability, and efficiency [27]. Another study focusing on organic fertilizer utilization and soil carbon sequestration found that increasing subsidies can increase organic fertilizer utilization. However, excessive subsidies may not be conducive to long-term sustainable agricultural development [22]. Another study in Indonesia revealed that government policy, supervision, distribution reliability, and innovation factors all play an essential role in implementing the subsidized fertilizer supply chain [28]. From the studies on fertilizer and agro-industries, it was discovered that the supply chain system is complex and involves many uncertainties. This complexity is dynamic and flexible. Therefore, similar problems must consider system dynamics.

The concept of System Dynamics refers to a system's ability to adapt and learn from new information in a constantly changing environment. This involves integrating new knowledge into existing knowledge and making decisions based on real-time insights. System Dynamics are applied in various domains, such as big data classification and stability of renewable energy distribution networks. In supply chain management, System Dynamics are utilized to model the relationship between variables and the decision-making process [29]. Innovative contract technology employs dynamic monitoring and analysis methods to optimize applications by identifying inefficient behaviour [30]. These systems offer increased flexibility and control, making organizations more responsive and agile in competitive environments [31]. Dynamic supply chain management systems can pinpoint causality factors from historical order data, reducing variability [32].

The main challenge of System Dynamics is their increasing complexity, which makes them difficult to represent and control [33]. Supply chains are interconnected entities that continually evolve. To address disaster risks and environmental uncertainty, supply chains must shift from a static to a dynamic perspective [34]. System Dynamics can increase operational efficiency, reduce costs, and improve collaboration. The growing complexity of supply chain systems, driven by changing customer preferences and globalization, demands a rigorous decision-making framework [35]. Control theory has been applied in supply chain management to understand and manage this complexity [36]. Supply chain adaptability is crucial, and System Dynamics enable the automatic adjustment of optimization strategies to adapt to dynamic environments [37]. By using System Dynamics, the complexity of the fertilizer supply chain can be better understood and managed, improving decision-making and performance.

System Dynamics analysis in the subsidized fertilizer supply chain helps understand decision-making, identify value chain constraints, and evaluate the impact of complexity on supply chain behaviour [38]. This approach also allows the development of inventory-capacity management models that optimize material flows and reduce costs [39] and systemic risk management models that consider the complex interactions between risks in the food supply chain [40]. A System Dynamics can minimize variability in order lead times and fill rates [41]. In addition, the System Dynamics approach helps manage supply chains sustainably by describing the dynamics of the supply chain's impact on the environment and achieving balance and stability [42]. The dynamic integration of analytical processing into supply chain business processes results in more automated and efficient operations [43]. Analysing the subsidized fertilizer supply chain improves decision-making for increased productivity and sustainability in the industry.

3. Conceptual Model

To elaborate the research, a conceptual model is designed to explain the proposed model. The conceptual model (Figure 1) was created to clarify the research plans and stages. This research consists of five steps arranged comprehensively. Step 1 is the data collectiveness, such as DC location, demand, and supply capacity. The second step is the Clustering stage, which groups delivery areas according to the distance between regions. Step 3 is the Scenario test, which consists of a single run and a milk run, which can be separated or working together. Step 4 is verifying and validating the model using a structural scoring test and data comparison. Step 5 is scenario testing and drawing conclusions and analysis. To explain further, we conducted a case study, as shown in the Figure 1.

Figure 1 shows two scenarios, which are the main discussion. The first scenario (Figure 2) involves each ship transferring to one warehouse destination for each activity. This is an alternative scenario, aiming to carry out responsible and quick transfers, according to the tonnage required at each destination. This approach can also simplify coordination and port operational activities. Meanwhile, the second scenario (Figure 3) involves a milk-run scheme, where one ship will serve multiple warehouse destination ports. To facilitate this, all destination ports will be grouped into clusters based on their geographical proximity, allowing for efficient coverage by a single shipping trip. This research utilizes the K-Means Clustering method to create these clusters based on the closest distance, taking into account request requirements and the location of the request origin. The decision to use either the first or second scenario is influenced by the volume of requests from the destination warehouse, the Distribution Center (DC), or Buffer Warehouse (GP). The dynamic nature of demand will determine the scenario chosen.

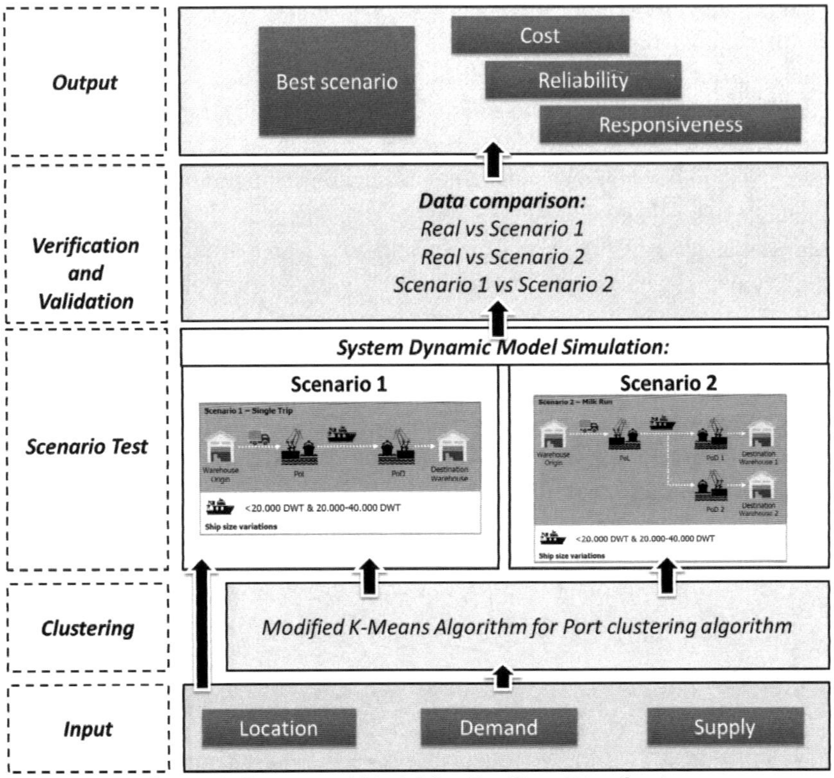

Figure 1. Conceptual Model for Fertilizer Distribution

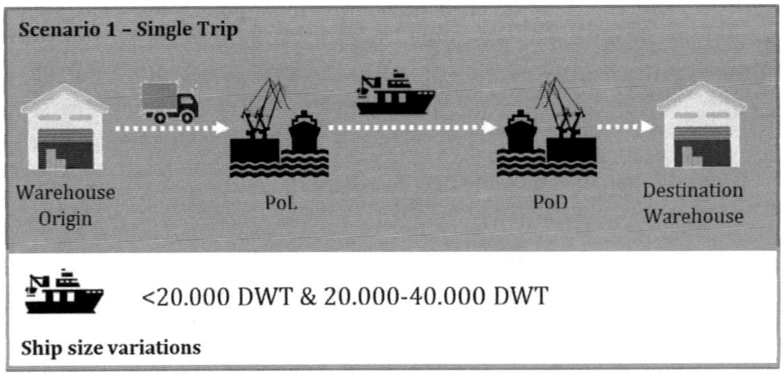

Figure 2. Experimental Design for Scenario 1

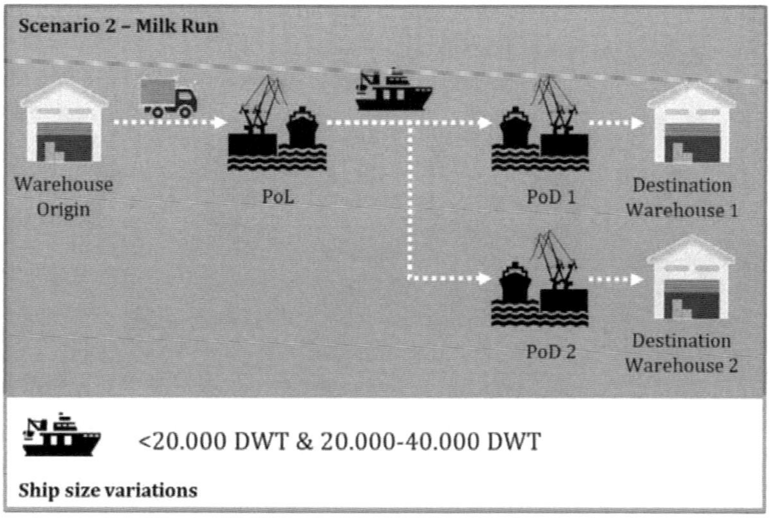

Figure 3. Experimental Design for Scenario 2

4. Case Study of Flexible Delivery System in Fertilizer Supply Chain

At this case study of fertilizer delivery system, there are two main steps. The first step involves clustering to group destination ports for shipping. Meanwhile, the second stage consists of creating a System Dynamics algorithm. This algorithm aims to analyse and calculate optimal cost, responsiveness, and reliability results in the entire delivery process.

4.1 Demand Clustering

Clustering is carried out to group port locations that have the same or adjacent variables so that they can be used to group fertilizer shipments. This clustering uses the K-Means Algorithm method, which is an unsupervised learning method used to group data points that have similarities by creating clusters or groups [44]. The modified k-means algorithm utilized in this study ensures that the clustered area is accessible for the shipping route with the certain capacity (shipping capacity). Closer ports are clustered and depicted as the accessible route. The distance between ports is calculated using the Euclidean theory. It is followed by a verification process that assesses the capacity and potential locations for ships to navigate through and the accessibility for ships delivering bags or bulk shipments to ports within a specific area.

This case study specifically applies the Distribution Center (DC) or Buffer Warehouse (GP) method to the island of Sumatra with a k number of four, which accommodates the demand. There are three DC (DC Lampung, DC Medan, and DC Padang) which has high demand average and five GP (GP Bengkulu, GP Dumai, GP Lhoksemawe, GP Pangkal Pinang, GP Palembang) throughout the island of Sumatra. Sumatra has relatively high demand than other location in Indonesia because the agricultural sector played the significance sector in Sumatra [45].

The data demand for subsidized fertilizer, which can be seen in Table 1, is the data coming from the demand pattern from National Fertilizer Company data from 2019-2020. Each delivery location (GP and DC) has its own minimum and maximum demand amounts, as shown in Table 1 which we calculated based on the annual demand from each location during the same period

(2019-2020). The number of demands will be different because the requirements of each location are not identical.

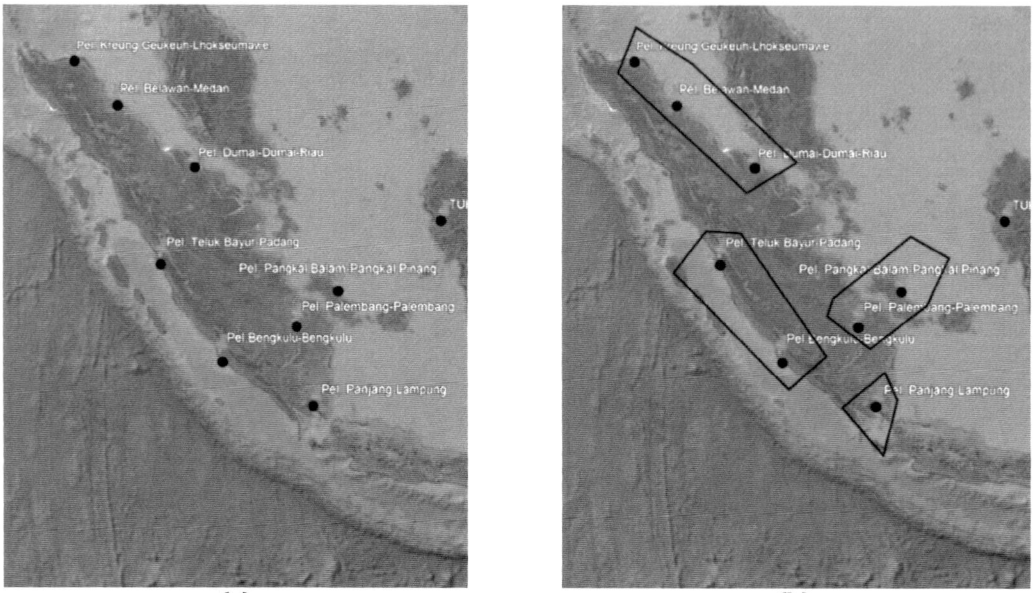

(a) (b)

Figure 4. Location of DC/GP in Sumatra: (a) DC/GP Location before clustering and (b) after clustering

Table 1. Average demand from DC/GP in Sumatera

From	To	*Demand* Min (Ton)	Average *Demand* (Ton)	*Demand* Max (Ton)
DC Gresik	DC Lampung	1,800	6,850	11,900
DC Gresik	DC Medan	1,800	7,600	13,400
DC Gresik	DC Padang	2,000	5,100	8,200
DC Gresik	GP Bengkulu	2,400	4,100	5,800
DC Gresik	GP Dumai	1,800	3,150	4,500
DC Gresik	GP Lhoksemawe	3,000	3,750	4,500
DC Gresik	GP Pangkal Pinang	1,900	2,775	3,650
DC Gresik	GP Palembang	1,800	3,150	4,500

4.2 Application of Scenario into System Dynamics

To produce the expected results from the case study, first create a description of the activities that will later be used to create stock and flow in the research. The written activities are based on events that are typical of each scenario. The description of the activities can be seen in the Table 2 as follows:

Table 2. Activities for two scenarios in the subsidized fertilizer distribution activities

Activities	Scenario 1	Scenario 2
The process of loading fertilizer from PoL to ships using trucks and shore cranes at the Loading Port	The capacity loaded is the requested capacity per PoD location	The capacity loaded is the requested capacity per cluster
Ship activities when sailing from PoL to PoD with full load	The ship sails from PoL to one PoD location point in full condition	Ships sail from PoL to PoD points that are in one cluster
Unloading activities when the ship is at PoD using shore cranes and trucks	The unloaded capacity is the demand capacity per PoD location	The unloaded capacity is the demand capacity per PoD location
Ensure that the ship returns to the port of origin	The ship sails from one PoD location point to PoL	The ship sails from the last nth PoD of a cluster route to the PoL

*PoD = Port of Discharge; PoL = Port of Loading

By using the activities shown in the Table 2, then we conduct the experiment by making causal loops and Stock and flow as shown in the Appendix A and Appendix B. Causal loop are used to understand the relationships between different elements. When creating a causal loop, we consider various company assets such as human resources (labour), heavy machinery like conveyors and cranes, and transport vehicles such as trucks and ships. We also consider uncertainty, which can begin with farmers' requests, leading to changes in financial policy, influencing the budget for subsidized fertilizer production, and affecting the amount of fertilizer produced and sent to farmers. Delivery time is influenced by response speed, which contributes to a measure of responsiveness. All expenses incurred are totalled to determine the overall cost. The causal loop illustrates the relationships between these factors and yields three outcomes: reliability, responsiveness, and total cost. Stock and flow is conducted to understand the system, identifying variables, specifying causal relationships, and designing diagrams. The purpose of these diagrams is to visually represent the system, conduct simulations on real systems, and aid in decision-making for existing systems.

For the case study, it is assumed that Demand Min (Tons) reflects low demand (not during the planting period), Average Demand (Tons) indicates regular demand (in standard times), and Demand Max (Tons) reflects high demand (during the planting period) (Table 1). The ship capacity used is a 3-hold bulk carrier with a hold capacity of 1800 tons, with a total capacity of one ship of 5400 tons. These demands are then calculated through System Dynamics simulations. The two distribution scenarios used are single trip and milk run, both of which are the basis for determining trips with varying numbers of requests. It is assumed that because they are in a time region that has similar seasons, the points in the Sumatra area in the first stage of the numerical test have a uniform pattern. Therefore, the tested demand patterns include low, regular, and high demand without considering random variations between demand points. The result is presented in the Table 3 and will be explained in the result and discussion.

Table 3. Result for each scenario (Single Run and Mixed run)

Scenario	Number of Fleet	Capacity total (tons)	Time (hr)	Time (days)	Total Cost (Rupiah)	Average Comparison between Load Capacity and Ship Capacity (%)
Single Run	8	16500	1192.57	49.69	15,377,149,840.42	35.42%
Mixed Run (Single run + milk run combine)	5	16500	691.98	28.83	9,172,091,635.96	61%

5. Result and Discussion

The required data is processed by considering dynamic changes in demand between requests at each Distributor Center (DC) and Buffer Warehouse (GP) in Sumatra. Numerical tests were conducted twice. The initial experiment was conducted under Low Demand Conditions, using a single trip. Meanwhile, the second experiment used a mixed run (a single trip and a milk run as needed). Table 3 shows the results of this research. The results of the two experiments showed that the second experiment had lower costs and response times than the first experiment, which used a single trip only. This was attributed to the fact that the capacity sent was less than 50% of the ship's maximum capacity when using a single trip. However, the shipping costs for the same route remained similar. It was observed that when the delivery was maximized, the costs incurred per capacity unit decreased. It is possible because the ship is maximized according to ship capacity. The total number of ships used also varies. With the number of destinations being 8, as in Table 1, experiment 1 requires eight ships according to the destination point, while experiment 2 requires five ships because some locations are reached using milk-run.

The selection of the method for the second experiment depended on the results of the clustering of shipping areas, ensuring that the total shipping capacity did not exceed the ship's capacity. The results of the two experiments showed that the second experiment had lower costs and response times, saving 59.6% compared to the first experiment, which used only the single-trip approach. This was attributed to the fact that when using a single trip, the capacity of fertilizer sent was less than 50% of the ship's maximum capacity. However, the shipping costs did not significantly differ for the same route. When delivery was maximized, the costs incurred per capacity decreased.

The experimental results which is the result is shown in Table 3, the case studies at the initial stage indicate that the decision to use the single trip or milk run method is based on delivery capacity. The company can decide whether to use scenarios 1 (single trip) or 2 (milk run) based on ordering and delivery simultaneously. Total shipping capacity is calculated based on the ship's carrying capacity. If the ship's carrying capacity for one trip can be filled by orders from two or more destinations (DC or GP), then the milk run method is used. This enables companies to make decisions that align with the needs in the field.

6. Conclusions

This research uses a System Dynamics approach to analyse fertilizer distribution. The conclusions drawn are as follows. First, there is a cause-and-effect relationship between supply and demand in fertilizer distribution, influenced by the decisions of the Ministry of Agriculture through the Minister of Agriculture. Companies need to implement dynamic simulations based on changes in demand and subsidy allocation. Second, the total distribution costs are calculated from loading, unloading, ship travel, and vehicles. Experiments show that using maximum vessel capacity with mixed (single and milk run) shipping routes can reduce costs and response times. Third, the best distribution scenario is selected based on fertilizer capacity and demand conditions, with a dynamic process to minimize response time and costs. Fourth, the relationship between the Ministry of Agriculture's decisions and subsidized fertilizer production will determine the company's reliability in meeting annual demand, with improved System Dynamics modelling to observe demand patterns dynamically. Future research needs to elaborate on the possibility of different planting periods and consider the data from other regions. The distribution pattern will also affect the total cost and shipping needs. Future research must also address contemporary government policy issues, such as significantly subsidized fertilizer. The need to measure the needs of other areas in Indonesia is also considered.

References

[1] M. Sane, M. Hajek, C. Nwaogu, and R. C. Purwestri, "Subsidy as An Economic Instrument for Environmental Protection: A Case of Global Fertilizer Use," *Sustainability*, vol. 13, no. 16, p. 9408, Aug. 2021, doi: 10.3390/su13169408.

[2] C. Di Bene, R. Francaviglia, R. Farina, J. Álvaro-Fuentes, and R. Zornoza, "Agricultural Diversification," *Agriculture*, vol. 12, no. 3, p. 369, Mar. 2022, doi: 10.3390/agriculture12030369.

[3] A. Ickowitz, B. Powell, D. Rowland, A. Jones, and T. Sunderland, "Agricultural intensification, dietary diversity, and markets in the global food security narrative," *Global Food Security*, vol. 20, pp. 9–16, Mar. 2019, doi: 10.1016/j.gfs.2018.11.002.

[4] Rahmanta, A. Prasetyo, and Iskandar Muda, "It's True The Effectiveness Of The Fertilizer Subsidy Policy And Its Effects On The Income Of The Farmers ? (Case In Indonesia)," *International Journal of Scientific & Technology Research*, vol. 8, no. 6, pp. 99–102, 2019.

[5] T. Li *et al.*, "Exploring optimal nitrogen management practices within site-specific ecological and socioeconomic conditions," *Journal of Cleaner Production*, vol. 241, p. 118295, Dec. 2019, doi: 10.1016/j.jclepro.2019.118295.

[6] A. Dayanti, Mgs. A. Firdaus, A. Meiriza, D. R. Indah, and A. Heryanto, "Implementation of Distribution Requirement Planning in the Subsidized Fertilizer Management System," in *Proceedings of the Sriwijaya International Conference on Information Technology and Its Applications (SICONIAN 2019)*, Palembang, Indonesia: Atlantis Press, 2020. doi: 10.2991/aisr.k.200424.082.

[7] Bappenas, "Lampiran Pidato Kenegaraan Presiden Republik Indonesia Dalam Rangka HUT Ke-74 Proklamasi Kemerdekaan Republik Indonesia," Badan Perencanaan Pembangunan Nasional, Jakarta, Aug. 2019.

[8] W. R. Susila, "Kebijakan Subsidi Pupuk: Ditinjau Kembali," *Jurnal Penelitian dan Pengembangan Pertanian*, vol. 29, no. 2, pp. 43–49, 2010.

[9] Asti Ananda, Jumiati, and Akbar, "Kinerja Kelembagaan Terhadap Distribusi Pupuk Bersubsidi (Studi Kasus pada Gapoktan Sepakat di Desa Pattinoang Kecamatan Galesong Kabupaten Takalar)," *KaliAgri*, vol. 4, no. 1, pp. 19–26, Jun. 2023, doi: 10.56869/kaliagri.v4i1.470.

[10] S. A. Sembiring, H. Sihaloho, J. Hutauruk, M. F. Nainggolan, and P. Sagala, "Implementation of Subsidized Fertilizer Policy on Farmer Groups Based on the Regulation of the Minister of Agriculture Number 41 of 2021 in Percut Sei Tuan District, Deli Serdang Regency, North Sumatra Province, Indonesia," *theijbm*, Jan. 2023, doi: 10.24940/theijbm/2023/v11/i1/BM2301-012.

[11] Direktorat Penyusunan APBN, "Pokok - Pokok APBN 2020 Akselerasi Daya Saing Melalui Inovasi dan Penguatan Kualitas Sumber Daya Manusia," Kementrian Keuangan, Jakarta, 2020.

[12] T. Perdana, H. N. Utami, A. H. Sadeli, and F. R. Hermiatin, "Small-Scale Farmer Preferences: Exploring the Gap Towards Product Attributes of Local and Imported Fertilizers," *International Journal on Advanced Science, Engineering and Information Technology*, vol. 9, no. 2, p. 502, Apr. 2019, doi: 10.18517/ijaseit.9.2.7702.

[13] Kementrian Pertanian, "Permentan Nomor 1 Tahun 2012: Tentang Komponen Harga Pokok Penjualan Pupuk Bersubsidi Untuk Sektor Pertanian," Kementerian Pertanian Republik Indonesia, Jakarta, 2012.

[14] Badan Pusat Statistik (BPS), "Perkembangan Ekspor dan Impor Indonesia Juni 2020," Badan Pusan Statistik, Jakarta, 2020.

[15] T. Oliviaputie and K. Sa'diyah, "Analisa Ekonomi Pra-rancangan pabrik Kimia Pembuatan Pupuk Organik Cair dari Sekam Padi Kapasitas 8000 Ton/Tahun," *Distilat: J. Tekn. Sep*, vol. 8, no. 3, pp. 646–653, May 2023, doi: 10.33795/distilat.v8i3.495.
[16] G. Conway and E. Barbier, "Pricing Policy and Sustainability in Indonesia," in *The Earthscan Reader in Sustainable Development*, 1st ed., London: Routledge, 2023, pp. 151–157. doi: 10.4324/9781003403432-20.
[17] H. Ashari *et al.*, "Policy urgency and development of the highest retail price (HRP) of subsidised fertilizer," *E3S Web Conf.*, vol. 316, p. 02037, 2021, doi: 10.1051/e3sconf/202131602037.
[18] H. Baturohmah, "Pemodelan Sistem Dinamik dalam Peningkatan Profitabilitas Produksi Menggunakan Ventana Simulation," *Jurnal Restikom : Riset Teknik Informatika dan Komputer*, vol. 5, no. 1, pp. 64–72, Apr. 2023.
[19] Kementrian Perindustrian, "Undang-undang (UU) Nomor 11 Tahun 2020 tentang Cipta Kerja." Nov. 02, 2020.
[20] M. C. Camur, C.-Y. Tseng, A. E. Thanos, C. C. White, W. Yund, and E. Iakovou, "An Integrated System Dynamics and Discrete Event Supply Chain Simulation Framework for Supply Chain Resilience with Non-Stationary Pandemic Demand," 2023, doi: 10.48550/ARXIV.2305.00086.
[21] W. N. Tanjung, S. Hidayat, and S. Azmiyati, "Simulasi Sistem Untuk Meningkatkan Kinerja Rantai Pasok," *SST*, vol. 3, no. 4, p. 171, Dec. 2017, doi: 10.36722/sst.v3i4.231.
[22] Y. Jiang, K. Li, S. Chen, X. Fu, S. Feng, and Z. Zhuang, "A sustainable agricultural supply chain considering substituting organic manure for chemical fertilizer," *Sustainable Production and Consumption*, vol. 29, pp. 432–446, Jan. 2022, doi: 10.1016/j.spc.2021.10.025.
[23] H. Zavale, G. Matchaya, D. Vilissa, C. Nhemachena, S. Nhlengethwa, and D. Wilson, "Dynamics of the Fertilizer Value Chain in Mozambique," *Sustainability*, vol. 12, no. 11, p. 4691, Jun. 2020, doi: 10.3390/su12114691.
[24] D. Ivanov, "Entropy-Based Supply Chain Structural Complexity Analysis," in *Structural Dynamics and Resilience in Supply Chain Risk Management*, vol. 265, in International Series in Operations Research & Management Science, vol. 265. , Cham: Springer International Publishing, 2018, pp. 275–292. doi: 10.1007/978-3-319-69305-7_9.
[25] J. Peng, "Research on the construction of supply chain dynamic system based on adaptive control method," in *2010 International Conference on Future Information Technology and Management Engineering*, Changzhou, China: IEEE, Oct. 2010, pp. 368–370. doi: 10.1109/FITME.2010.5656285.
[26] Suprianto, "Effectiveness of Subsidized Fertilizer Distribution from Distributors through Farmers' Kiosks to the Farmers' Level (Case Study in West Lombok District)," *IJSSERS*, vol. 02, no. 10, Oct. 2022, doi: 10.55677/ijssers/V02I10Y2022-09.
[27] P. Jangid, S. Maurya, A. Kushwah, S. Tiwari, and A. Shaikh, "Fertilizer Supply Chain Management using Blockchain," *IJSRCSEIT*, pp. 457–462, Apr. 2023, doi: 10.32628/CSEIT2390256.
[28] A. N. Sahim, N. K. Mat, and E. Sudarman, "The Power of Innovation, Distribution and Supervision Factor in Improving Performance of Supply Chain Management of Subsidized Fertilizer in Indonesia," *International Journal of Supply Chain Management*, Feb. 2018.
[29] S. Barrad, R. Valverde, and S. Gagnon, "The Application of System Dynamics for a Sustainable Procurement Operation," in *Innovative Solutions for Sustainable Supply Chains*, H. Qudrat-Ullah, Ed., in Understanding Complex Systems. , Cham: Springer International Publishing, 2018, pp. 179–196. doi: 10.1007/978-3-319-94322-0_7.
[30] Y. Ding, C. Wang, Q. Zhong, H. Li, J. Tan, and J. Li, "Function-Level Dynamic Monitoring and Analysis System for Smart Contract," *IEEE Access*, vol. 8, pp. 229161–229172, 2020, doi: 10.1109/ACCESS.2020.3046005.
[31] I. Lebhar, A. Dadda, and L. Ezzine, "Artificial Intelligence Applications in the Global Supply Chain: Benefits and Challenges," in *International Conference on Advanced Intelligent Systems for Sustainable Development*, vol. 712, J. Kacprzyk, M. Ezziyyani, and V. E. Balas, Eds., in Lecture Notes in Networks and Systems, vol. 712. , Cham: Springer Nature Switzerland, 2023, pp. 282–295. doi: 10.1007/978-3-031-35251-5_27.
[32] H. Delavar, H. Gilani, and H. Sahebi, "A system dynamics approach to measure the effect of information sharing on manufacturing/remanufacturing systems' performance," *International Journal of Computer Integrated Manufacturing*, vol. 35, no. 8, pp. 859–872, Aug. 2022, doi: 10.1080/0951192X.2022.2027015.
[33] D. N. Nya and H. Abouaissa, "An Efficient Framework for Tactical Management in Supply Chain Systems," in *2022 14th International Colloquium of Logistics and Supply Chain Management (LOGISTIQUA)*, EL JADIDA, Morocco: IEEE, May 2022, pp. 1–6. doi: 10.1109/LOGISTIQUA55056.2022.9938109.
[34] A. Chaouni Benabdellah, I. Bouhaddou, and A. Benghabrit, "Supply Chain Challenges with Complex Adaptive System Perspective," in *Trends and Advances in Information Systems and Technologies*, vol. 745, Á. Rocha, H. Adeli, L. P. Reis, and S. Costanzo, Eds., in Advances in Intelligent Systems and Computing, vol. 745. , Cham: Springer International Publishing, 2018, pp. 1081–1093. doi: 10.1007/978-3-319-77703-0_105.
[35] E. Ekinci, S. K. Mangla, Y. Kazancoglu, P. R. S. Sarma, M. D. Sezer, and M. Ozbiltekin-Pala, "Resilience and complexity measurement for energy efficient global supply chains in disruptive events," *Technological Forecasting and Social Change*, vol. 179, p. 121634, Jun. 2022, doi: 10.1016/j.techfore.2022.121634.
[36] C. Bell, R. Higgs, S. Vickers, S. Toncinich, and T. Haslett, "Using Systems Modelling to Understand the Dynamics of Supply Chains," p. 467685 Bytes, 2017, doi: 10.4225/03/5934DA85B7D91.

[37] M. Abbasi and L. Varga, "Steering supply chains from a complex systems perspective," *EJMS*, vol. 27, no. 1, pp. 5–38, Sep. 2022, doi: 10.1108/EJMS-04-2021-0030.

[38] X. Xi and Y. Zhang, "Complexity analysis of production, fertilizer-saving level, and emission reduction efforts decisions in a two-parallel agricultural product supply chain," *Chaos, Solitons & Fractals*, vol. 152, p. 111358, Nov. 2021, doi: 10.1016/j.chaos.2021.111358.

[39] H. Taboada, Y. A. Davizón, J. F. Espíritu, and J. Sánchez-Leal, "Mathematical Modeling and Optimal Control for a Class of Dynamic Supply Chain: A Systems Theory Approach," *Applied Sciences*, vol. 12, no. 11, p. 5347, May 2022, doi: 10.3390/app12115347.

[40] M. Azizsafaei, A. Hosseinian-Far, R. Khandan, D. Sarwar, and A. Daneshkhah, "Assessing Risks in Dairy Supply Chain Systems: A System Dynamics Approach," *Systems*, vol. 10, no. 4, p. 114, Aug. 2022, doi: 10.3390/systems10040114.

[41] T. K. Nadella, L. Ming, M. Pop, P. K. Chavva, and Y. Xin, "Dynamic supply chain management systems and methods," Apr. 2018.

[42] G. Wang and A. Gunasekaran, "Modeling and analysis of sustainable supply chain dynamics," *Ann Oper Res*, vol. 250, no. 2, pp. 521–536, Mar. 2017, doi: 10.1007/s10479-015-1860-2.

[43] T. Fukui, "A systems approach to big data technology applied to supply chain," in *2016 IEEE International Conference on Big Data (Big Data)*, Washington DC,USA: IEEE, Dec. 2016, pp. 3732–3736. doi: 10.1109/BigData.2016.7841041.

[44] J. Wang and Y. Zhao, "Time Series K-Nearest Neighbors Classifier Based on Fast Dynamic Time Warping," in *2021 IEEE International Conference on Artificial Intelligence and Computer Applications (ICAICA)*, Dalian, China: IEEE, Jun. 2021, pp. 751–754. doi: 10.1109/ICAICA52286.2021.9497898.

[45] Y. M. Manik, N. E. Prasetyo, S. Sulistyo, and M. Waluyo, "Analysis of the Dynamis of Labor Absorption in West Sumatra Province in 2018-2020," *JPIS*, vol. 32, no. 2, pp. 226–247, Dec. 2023, doi: 10.23917/jpis.v32i2.20367.

Appendix A: Causal Loop for study case using output such as Reliability, Responsiveness, and Total Costs

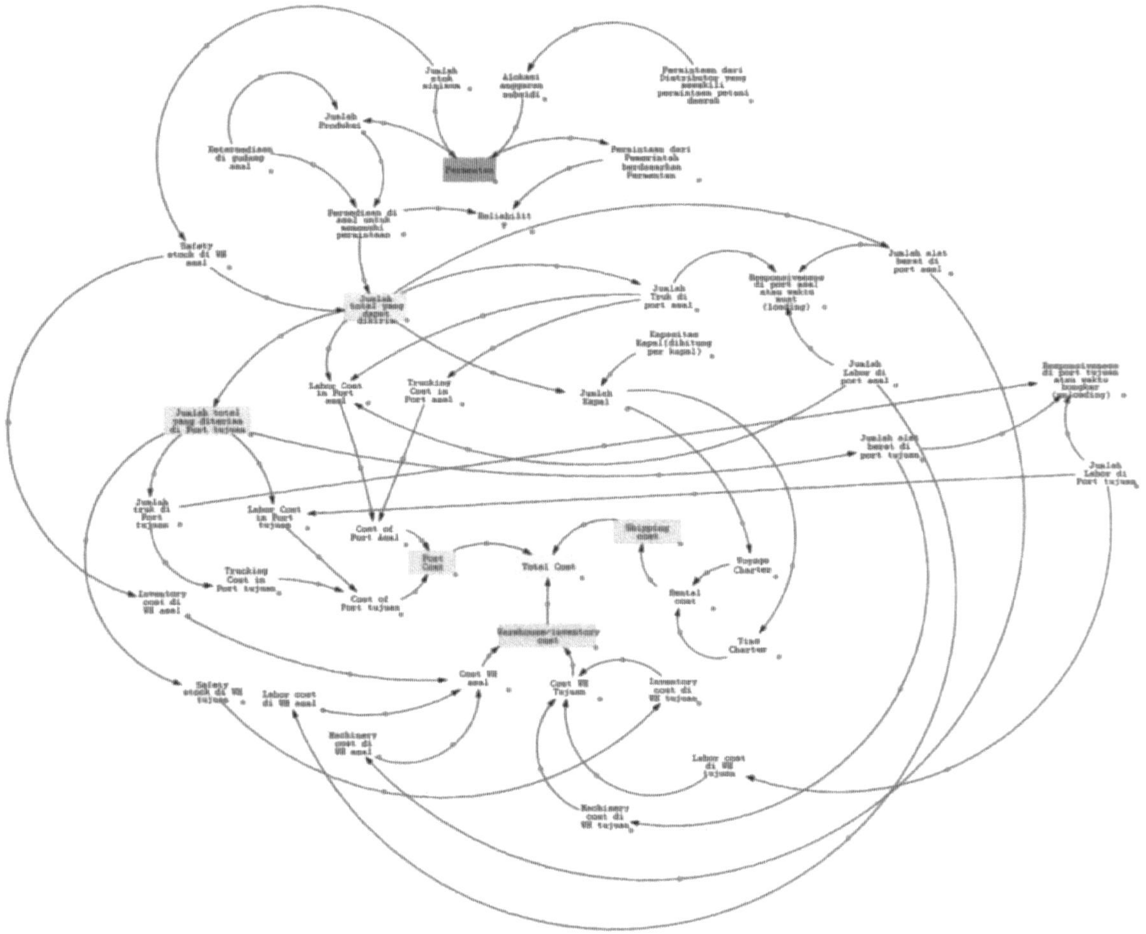

IOP Conf. Series: Earth and Environmental Science **1423** (2024) 012018 doi:10.1088/1755-1315/1423/1/012018

Appendix B: Stock and Flow diagram for Formulating Total Cost which consists of: (a) Loading cost in PoL; (b) Shipping cost; (c) Warehouse Cost; (d) Unloading cost in PoD

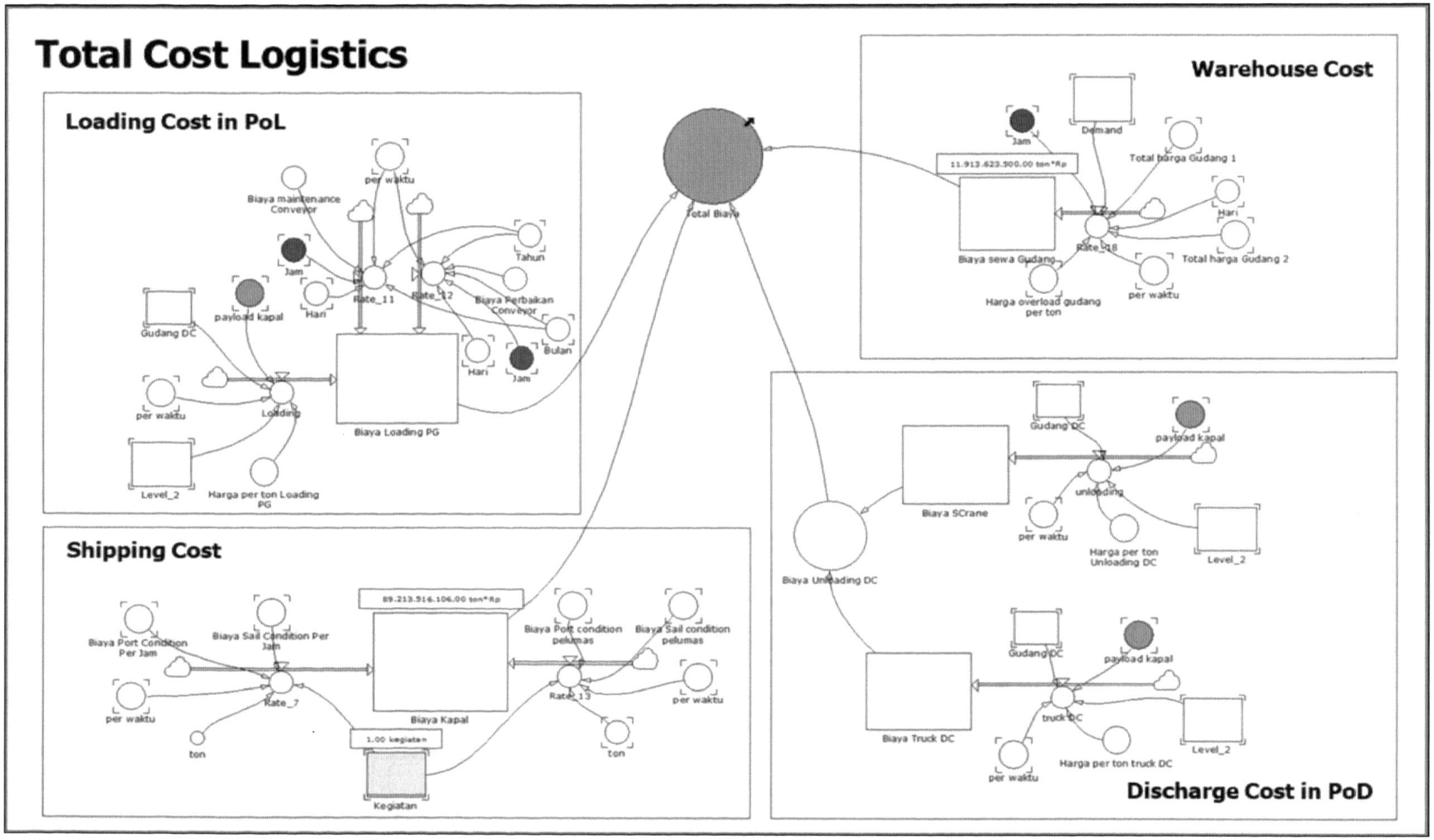

Real-Time Load Monitoring System for Tuna Fishing Ships: A Solution to Combat Illegal Transshipment

E S Koenhardono[1*], S P Giafi[1], A Kurniawan[1], I R Kusuma[1], S Sarwito[1], and J Prananda[1]

[1] Department of Marine Engineering, Institut Teknologi Sepuluh Nopember, Surabaya, Indonesia

*E-mail: eddy_koen@its.ac.id

Abstract. Illegal fishing activities are the biggest challenge for Indonesia in protecting its territorial waters from illegal fish transhipment. To address this issue, a real-time load measurement system is needed on ships. Real-time measurement devices can also replace land-based measurement functions, thus improving the quality of catch during the transfer of tuna from the ship to the cold storage. The method used in this research is testing, which requires real data to support and simulate the implementation of the study. The research involves a prototype that utilizes a load cell as a sensor for detecting the mass of fish, and NodeMCU as a microcontroller programmed with Arduino IDE language. The prototype is integrated with Wi-Fi to send data to a MySQL database. The sensor circuit consists of four load cells, a junction box, and an HX711 amplifier. The results of the sensor experiment at a 12-degree tilt condition, which indicates the roll or trim condition of the ship, show a decrease in sensor accuracy and relatively unstable results compared to normal conditions. The average accuracy rate of sensor readings under normal conditions is 99.42%, while the average accuracy rate of sensor readings at a 12-degree tilt is 96.98%..

1. Introduction

Tuna fish is a pelagic marine fish belonging to the tribe Thunnini and is considered a premium food, thus having high commercial value. Unlike most fish with white flesh, tuna meat ranges from pink to dark red due to its high myoglobin content. Indonesia has significant tuna fish potential, as its waters cover both the Indian Ocean and the Pacific Ocean. Currently, Indonesia's tuna production accounts for approximately 16 percent of the world's total tuna production, which reaches 7.7 million metric tons. Despite Indonesia's substantial tuna production, it remains unstable (fluctuating) annually, as evidenced in Figure 1 [1].

The instability in tuna production is caused by illegal, unreported, and unregulated (IUU) fishing activities that are prevalent in Indonesian waters [1], [2]. The government's policy of sinking ships aims to combat IUU fishing activities. However, Indonesia's large tuna potential still attracts foreign fishermen to continue illegal fishing using different methods, such as transhipment. Transhipment involves transferring catch from fishing vessels to carrier vessels at sea [3], [4]. If carried out legally, transhipment can streamline fishing operations and streamline operational costs for fishing vessels. Without proper oversight, transhipment can facilitate the movement of illegally caught products into the global supply chain.

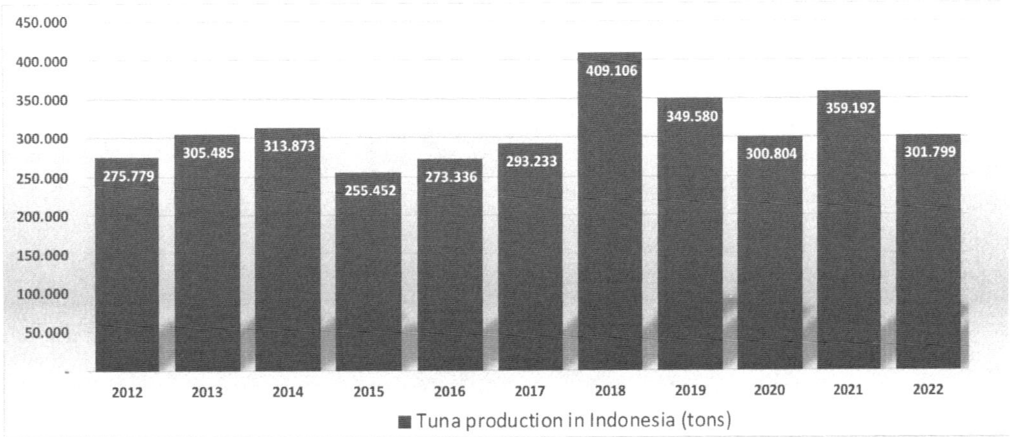

Figure 1. Graph of tuna production in Indonesia (tons) [1].

In order to reduce losses to the state and society, the Indonesian Government issued Ministerial Regulation No. 58/PERMEN-KP/2014 of 2014 concerning the Prohibition of Illegal Transhipment as a result of improving the regulation No. 30/PERMEN_KP/2012 of 2012 [5], [6]. However, in reality, this regulation applies to all transhipment activities, resulting in a decrease in the number of fish catch production due to a decrease in the number of trips, an increase in operational costs and a decrease in the quality of the catch [7], [8]. Therefore, the Indonesian Government issued Ministerial Regulation No. 11/PERMEN-KP/2020 of 2020 and No. 58/PERMEN-KP/2020 concerning the regulation of the legal transhipment process [9], [10].

Therefore, this paper aims to design and test a real-time cargo monitoring system for tuna fishing vessels, which can be monitored by the relevant authorities on a laboratory scale, as mentioned in regulation of 58/PERMEN-KP/2020. The presence of a real-time cargo monitoring system and VMS on tuna fishing vessels and tuna fish carrier vessels is expected to prevent illegal transhipment activities at sea. Thereby stabilizing Indonesia's tuna production according to its potential. Additionally, this real-time cargo monitoring system can replace land-based measurements as currently practiced, thus maintaining the quality of tuna catches better.

2. Methodology

2.1 Problem Definition and Objectives

The lifting of the ban on transhipment of fish in the middle of the sea will result in the emergence of illegal transhipments again. Monitoring the cargo and VMS installed on tuna fishing vessels and tuna collecting vessels and their position online can make it easier for interested parties to know the movement of fishing results from one vessel to another until landing at the port at any time.[11] This combination of monitoring will be able to prevent the occurrence of Illegal, Unreported, and Unregulated Fishing practices, so that the Measurable Fishing policy can be achieved. Monitoring, evaluation, and data collection activities on fisheries results that run well can prevent overfishing.

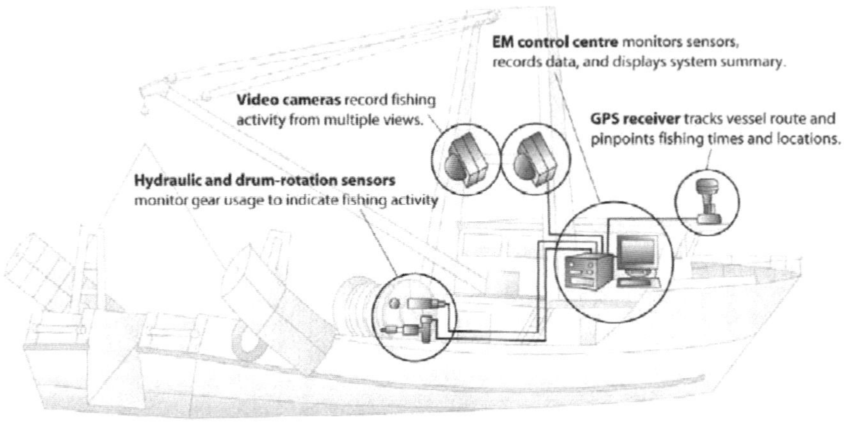

Figure 2. Stylized schematic of a trawl vessel outfitted with an EM system.

2.2 Literature Review

According to Regulation No. 05/Permen-KP/2007 regarding the implementation of the fisheries Vessel Monitoring System (VMS), VMS is part of monitoring and control measures to address issues related to illegal fishing. Referring to Regulation No. 42/Permen-KP/2015, fishing vessels in Indonesia with a size above 30 GT are required to install VMS transmitters, and vessel owners gain access to their vessel's movements through the VMS website. In 2022, the Ministry of Maritime Affairs and Fisheries (KKP) also launched the Integrated Maritime Intelligent Platform, which oversees a VMS-based vessel monitoring system that supports sustainable fishing practices and ensures a secure fish supply.

The Food and Agriculture Organization (FAO) has conducted trials of electronic monitoring (EM) on fishing vessels as a cost-effective complement to catch monitoring programs in the fisheries sector, as seen as on Figure 2.[12], [13] The EM system consists of various activity sensors and cameras placed on the vessel to remotely record fishing activities and catch results. Electronic monitoring has the potential to be an effective tool in future fisheries monitoring, especially when integrated into existing monitoring programs. The EM in fisheries has several benefits:

- Cost-Efficiency: EM systems, which consist of various activity sensors and cameras positioned on vessels to remotely record fishing activity and catches, have emerged as a cost-efficient supplement to existing catch monitoring programs[14], [15].
- Representative Coverage: EM has the potential to provide more representative coverage of the fleet than any observer program[14], [15].
- Enhanced Registration: EM enhances the registration of fishing activity and location[14], [15].
- Data Collection: EM is a tool used to collect fishing data including the number of fish that are caught, fishing effort (e.g., number of hours or days spent fishing), and bycatch[16], [17]. These data support and improve stock assessments and ensure that catch limits are sustainable in the long term.
- Transparency: EM allows vessel owners or fishing companies to monitor catches and activities on their vessels to ensure their legality[15].

IOP Conf. Series: Earth and Environmental Science **1423** (2024) 012019 doi:10.1088/1755-1315/1423/1/012019

- Compliance: EM can incentivize better compliance and discard reduction[14]. It helps to document conformity with conservation and management measures and international obligations[15].

Despite these advantages, the fishing managers and industry are often reluctant to its uptake due to concerns such as intrusion of privacy, liability, and costs[14]. However, better understanding of these concerns and exploration of EM benefits, such as increased traceability, sustainability claims, and market access, may enhance implementation on a larger scale[14].

2.3 Components of Real-Time Load Measurement System

2.3.1 Load Cell

A load cell is one type of sensor used to convert mechanical force into electrical signals through changes in resistance that occur in a strain gauge due to pressure in the form of deformation. The output signal of a load cell is only in millivolts, requiring a differential amplifier before use[18]. Load cells are used in electromechanical devices (transducers) to convert force into electrical signals. This indirect conversion procedure is divided into two steps. The applied force can be sensed by deforming the strain gauge with a corresponding mechanical structure. The strain gauge is a small electrical device that exhibits changes in resistance when subjected to strain. Subsequently, the strain gauge converts the slight resistance changes into electrical signals using an electrical circuit known as the Wheatstone bridge[19].

2.3.2 NodeMCU

NodeMCU is an electronic board based on the ESP8266 chip with the ability to perform microcontroller functions and also connect to the internet (Wi-Fi), as seen in Figure 3. It has several I/O pins, allowing it to be developed into a monitoring or controlling application for IoT projects. NodeMCU is an open-source IoT platform that integrates the ESP8266 chip into a board with various features similar to a microcontroller. NodeMCU has Wi-Fi access capability and a communication chip in the form of USB to serial. Therefore, programming only requires a USB data cable. The main source of NodeMCU is the ESP8266, particularly the ESP-12 series, which includes ESP-12E. NodeMCU ESP8266 requires a power supply of around 3.3V and has three WiFi

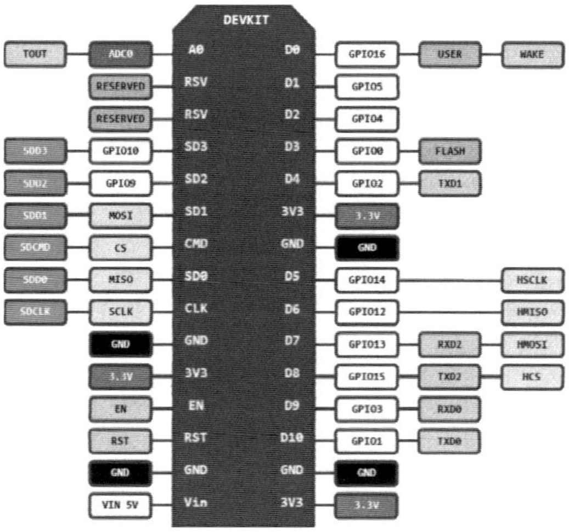

Figure 3. NodeMCU Pin Designation[20]

modes: Station, Access Point, and both. This module can operate independently without using any other microcontroller because it already has the necessary equipment like a microcontroller[20].

2.3.3 HX711

The HX711 module functions as an amplifier for the weight sensor (load cell) readings. The output voltage from the bridge circuit is typically in millivolts, so it needs to be amplified by the amplifier device. The HX711 is a precision 24-bit analog-to-digital converter (ADC) integrated component designed for digital weighing sensors in industrial control applications connected to bridge sensors. The working principle of the HX711 is to convert the measurable changes in resistance into voltage values through the existing circuitry. Communication with the microcontroller is done through TTL232. The advantages of the HX711 include its simple structure, ease of use, stable and reliable results, high sensitivity, and ability to measure changes quickly[21].

2.4 Sensor Functionality Test

Calibration of sensor functionality on the fish hold load measuring prototype was carried out to obtain measurement accuracy. The calibrated load mass was used as a guideline to obtain the level of mass sensor accuracy. The calibration value refers to the adjustment made to ensure that the load cell or load sensor readings are in accordance with the actual value. The readings from the load cell were recorded and compared to the reference value. The difference between the actual reading and the reference value was used to calculate the calibration factor, as shown in Table 1.

In Table 1 of the sensor functionality test, each sensor worked well, displaying the measured mass. However, there was a difference in the reading results of load cells 2 and 4 compared to the actual mass. This was caused by the different lengths of the load cell cables 2 and 4, due to the extension of the cable to reach the input terminal junction box. The length of the cable can affect the load cell reading results because the cable has internal resistance, which causes a voltage drop as the cable length increases. This voltage change can cause variations in the load cell output signal, which ultimately results in errors in reading the load or force applied to the load cell. Therefore, if the calibration value used is adjusted to the difference in voltage of each load cell, the results displayed will be more accurate.

2.5 Web Server Test

Testing the web server is necessary to ensure that the device can successfully send the readings to the web server. The Arduino IDE program used for database transmission should be configured so that the NodeMCU operates in client or station mode (STA) to connect to the available Wi-Fi network and send the sensor readings to the MySQL database. The SSID and password in the Arduino IDE program should be entered correctly, representing the Wi-Fi network to be used. The IP address or host address used as the database should be entered in the program to send the sensor readings to the designated IP address.

To reduce the number of mass values stored in the database, a logic is implemented such that if the constant mass value persists for more than 4 readings, only one value is displayed on the

Table 1. Sensor Functionality Result

Real mass (kg)	Measured Mass from Each Load Cell (kg)			
	Load Cell 1	Load Cell 2	Load Cell 3	Load Cell 4
2	1.9	2	1.9	2

web server for the specified time interval until a change in the mass value occurs again, as seen on Figure 4. In the Arduino IDE program itself, the successfully transmitted sensor value will indicate the targeted IP address.

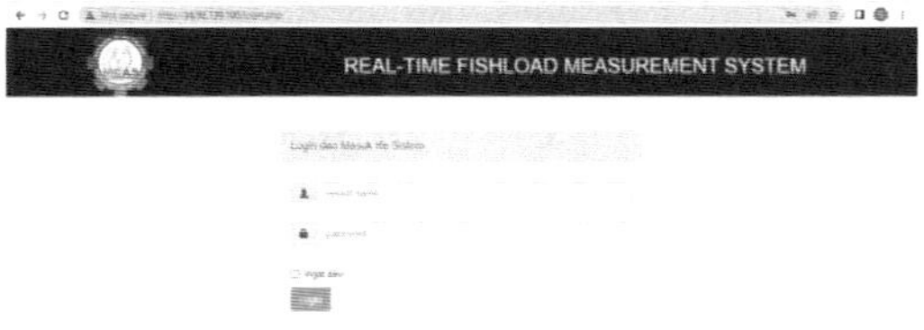

a. First Page of Web Server

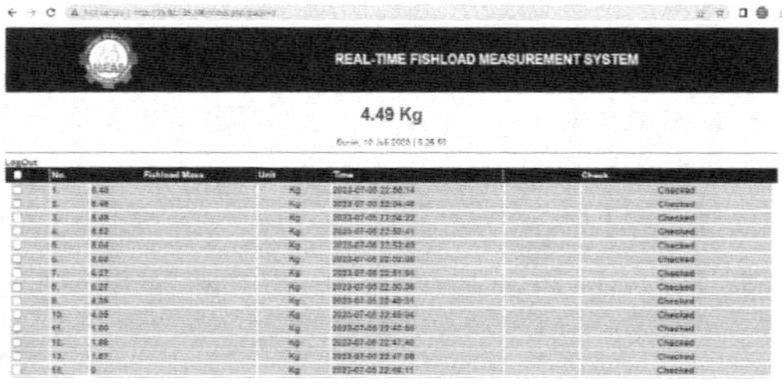

b. Second page of web server

Figure 4. Web Server

2.6 Performance Test

2.6.1 1st Scenario

The test in the first scenario is a test of the accuracy of the sensor components when the ship is in an even keel or flat condition, as a representation of calm sea conditions. The fish hold here is

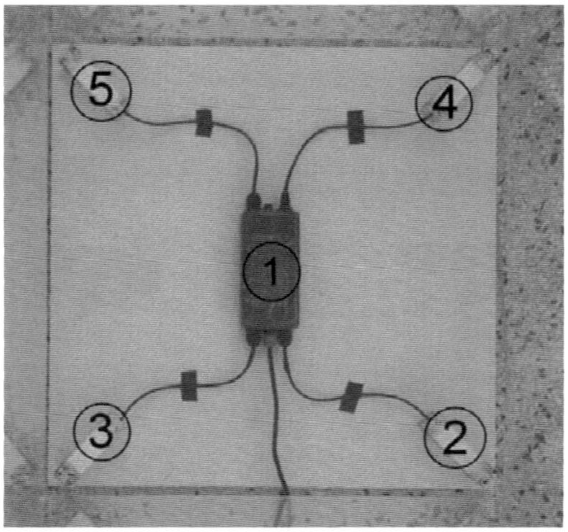

Figure 5. Placement of load measuring sensors in five positions for performance testing in scenario 1 and 2

Table 2. The first scenario performance test result

Real mass (kg)	Measured Mass (kg)				
	1st position	2nd position	3rd position	4th position	5th position
2	2	2,05	1,93	2,04	1,99
3	3,01	3,08	2,91	3,09	3,02
Accuracy	99,8%	97,4%	96,8%	97,5%	99,4%
Average			98,2%		

assumed to be a mechanically cooled hold, using the IQF method to cool the fish. In this test, the sensor is placed under the hatch model at five locations, as in Figure 5 which produces the measurement results in Table 2.

2.6.2 2nd Scenario

The second test scenario illustrates the accuracy testing of the mass sensor when the ship is rolling in wavy sea conditions. The test was carried out at a fish hatch model inclination of 12 degrees for 5 minutes at five mass sensor placement positions, as shown in Figures 6. The measurement results with a 12-degree tilt position are shown in Table 3.

3. Result and Discussion

The design of an accurate fish load measuring device that can be connected to a real-time network combines a load cell, HX711 and NodeMCU. A load cell is a sensor that can detect pressure or load applied to it. When a load is applied, the electrical resistance of the load cell changes, and this information is used to determine the correct weight. A load cell has four wires: red (VCC), black (GND), data, and clock. The HX711 is an ADC converter board that amplifies the signal from the load cell so that it can be read by the Arduino. The HX711 reads information from the load cell, converts the resistance change into a voltage, and sends it to the Arduino for processing. NodeMCU is an ESP8266-based microcontroller module that is used to connect the device to a WiFi network. In this project, the NodeMCU functions as a connector between the load cell, HX711, and an OLED display or other device.

Figure 6. Mass sensor accuracy testing when the ship experiences a 12 degree roll

Table 3. 2nd scenario performance test result

Real mass (kg)	Measured Mass (kg)				
	1st position	2nd position	3rd position	4th position	5th position
2	1,94	2	1,89	1,99	1,93
3	2,90	2,98	2,9	2,82	2,89
Accuracy	96,8%	99,7%	95,6%	96,8%	96,4%
Average			97,1%		

In the sensor functionality test, each sensor worked well displaying the measured mass, but the reading results of load cell sensors 2 and 4 differed from the actual mass. This is due to the difference in the length of the load cell cable to the input terminal junction box on load cells 2 and 4. The length of the cable can affect the reading results of the load cell because the cable has internal resistance which causes a voltage drop along with the length of the cable used. This voltage change can cause a change in the output signal from the load cell, which can ultimately result in errors in reading the load or force applied to the load cell. Therefore, if the calibration value used is adjusted to the difference in voltage of each load cell, the results shown will be more accurate. In each scenario, the mass value is displayed in real time on the ship's cargo monitoring web server, including when the load cell detects 0 kilograms or no load. Thus, to indicate the transfer of fish cargo to foreign vessels illegally, it can be seen from the significant decrease in mass at certain times when the ship is in the middle of the sea. A comparison of the level of accuracy in the experiment can be seen in Table 2, and Table 3.

In the first scenario, placing the load in the middle of the hatch model tends to be more accurate compared to placing the load in the corners of the model. This proves that uneven load distribution can affect load cell readings because load cells are usually designed to provide a linear response to loads that are applied evenly over the entire sensor surface. If observed through the potentiometer resistance value of each load cell, the potentiometer resistance also affects the mass reading results. This proves that the amount of resistance on the potentiometer influences the output value of the load cell, where if the resistance of the potentiometer is small, the load reading at that corner will be greater, and vice versa. Potentiometers on load cells are usually used to set the zero or reference point when no load is applied. The position of the potentiometer determines how the load cell responds and provides its output when there is no load acting on the sensor. When the potentiometer experiences a zero shift in the positive direction, the load cell output value when there is no load will increase from the previous value. To achieve a higher zero shift, the potentiometer must make the resistance relationship in the load cell circuit change to become smaller. If the load cell output value at zero shifts to a higher value, then when adding a load, the output value will be higher than the value that should be according to the applied load. On the other hand, if the potentiometer experiences a zero shift in the negative direction, the load cell output value will decrease and the potentiometer will be more resistive or have greater resistance.

The results of the sensor detection mass data in the second scenario show different mass values for each load cell. Apart from the impact of load distribution, differences in cable length, and potentiometer resistance, this can also be caused by the influence of tilt. When the load cell is not in the correct position or not vertical, the applied force does not travel only along the vertical axis of the sensor. Part of the force will travel along the horizontal axis, causing undesirable force components to appear in the reading. This can result in errors in the actual load reading, because the load cell is only supposed to respond to vertical forces or in accordance with the axis of application.

4. Conclusions

The design of a Real-Time Load Monitoring System for Tuna Fishing Ships on a laboratory scale, which has three main components, namely the load cell, NodeMCU, and HX711, has been successfully created and tested. The testing was carried out using 3 scenarios:

- Scenario 1 is the testing of a flat fish hold box representing the condition of tuna fishing ships in an even keel state, with good results.

- Scenario 2 is the testing of a tilted fish hold box at 12 degrees in four conditions (tilting right-left and front-back) to represent the condition of the ship experiencing rolling and trimming movements. In this test, the sensor device did not work well, so the measurement results showed different numbers.

Furthermore, the target system for measuring cargo can work in real-time as indicated by measurement results that can already be displayed on a website on a computer monitor that can be accessed by interested parties if an internet network is available.

References

[1] Baihaqi Yahya Nursya'ban, Fajar Adhany, Hafizd Razaan Zaidan, and Renny Kurniawati, "Daya Saing Ekspor Ikan Tuna Indonesia di Pasar Global," *Jurnal Publikasi Ekonomi dan Akuntansi*, vol. 4, no. 1, pp. 177–187, Jan. 2024, doi: 10.51903/jupea.v4i1.2705.

[2] M. Imron, R. Yusfiandayani, and M. S. Baskoro, "Productivity and Production Tuna by Tuna Longline Based on PPN Pelabuhan Ratu," *Jurnal Teknologi Perikanan dan Kelautan*, vol. 10, no. 2, pp. 173–181, 2019.

[3] C. V. Yunita, "Praktik Illegal Transhipment di Laut Lepas Berdasarkan Hukum Laut Internasional," *Mimbar Hukum - Fakultas Hukum Universitas Gadjah Mada*, vol. 31, no. 1, p. 75, 2019, doi: 10.22146/jmh.35718.

[4] Y. H. Sihombing, "Optimalisasi Hukum Laut Nasional untuk Pengembangan Potensi Sumber Daya Perikanan di Indonesia," *Jurnal Hukum Lingkungan Indonesia*, vol. 3, no. 2, pp. 97–123, 2017, doi: 10.38011/jhli.v3i2.43.

[5] K. K. dan Perikanan, *Moratorium Perizinan Usaha Perikanan Tangkap, Alih Muatan (Transhipment) di Laut, dan Penggunaan Nahkoda dan Anak Buah Kapal Asing*. Indonesia: BN 2014 (1789): 5 hlm., 2014.

[6] K. K. dan Perikanan, *Usaha Perikanan Tangkap di Wilayah Pengelolaan Perikanan Negara Republik Indonesia*. Indonesia, 2012.

[7] Restu Isroq Badahi, Mercy Patanda, and Hendrawan Syafrie, "Dampak Peraturan Pelarangan Transhipment Terhadap Aktivitas Kapal Longline di Pelabuhan Nizam Zachman Jakarta," *Jurnal Satya Minabahari*, vol. 4, no. 2, pp. 133–137, 2019.

[8] A. Hudayana and T. N. Utami, "Analysis of Transhipment Policy on Long Line Fisheries Vessels Business Performance," *Economic and Social of Fisheries and Marine*, vol. 005, no. 01, pp. 78–89, Oct. 2017, doi: 10.21776/ub.ecsofim.2017.005.01.08.

[9] K. K. dan Perikanan, *Pencabutan Peraturan Menteri Kelautan dan Perikanan Nomor 58/PERMEN-KP/2014*. 2020.

[10] P. M. K. dan Perikanan, *Mencabut Peraturan Menteri Kelautan dan Perikanan Nomor 12/MEN/2012; 30/MEN/2012; 57/PERMEN-KP/2014 dst*. Indonesia, 2020.

[11] M. Bachtiar, "Penerapan Vessel Monitoring System untuk Kapal-Kapal Ikan," *Jurnal Ilmiah Teknik Mesin Cylinder*, vol. 8, no. 2, 2022.

[12] H. Murua, J. Ruiz, A. Justel-Rubio, and V. Restrepo, "Minimum Standards for Electronic Monitoring Systems in Tropical Tuna Purse Seine and Longline Fisheries," Washington, DC, Apr. 2022.

[13] K. Stobberup *et al.*, "Electronic monitoring in tuna fisheries: strengthening monitoring and compliance in the context of two developing states," Rome, 664, 2021.

[14] L. Weiser, "Overview Electronic Monitoring: A Key Tool for Global Fisheries How governments and RFMOs can better monitor high-seas fleets," 2019. Accessed: Jul. 22,

2024. [Online]. Available: https://www.pewtrusts.org/en/research-and-analysis/issue-briefs/2019/09/electronic-monitoring-a-key-tool-for-global-fisheries

[15] A. T. M. van Helmond *et al.*, "Electronic monitoring in fisheries: Lessons from global experiences and future opportunities," *Fish and Fisheries*, vol. 21, no. 1, pp. 162–189, Jan. 2020, doi: 10.1111/faf.12425.

[16] E. Gilman, V. De Ramón Castejón, E. Loganimoce, and M. Chaloupka, "Capability of a pilot fisheries electronic monitoring system to meet scientific and compliance monitoring objectives," *Mar Policy*, vol. 113, p. 103792, Mar. 2020, doi: 10.1016/j.marpol.2019.103792.

[17] T. J. Emery *et al.*, "The use of electronic monitoring within tuna longline fisheries: implications for international data collection, analysis and reporting," *Rev Fish Biol Fish*, vol. 28, no. 4, pp. 887–907, Dec. 2018, doi: 10.1007/s11160-018-9533-2.

[18] S. Sumardi and I. Iswahyudi, "Prototype Sistem Pemantau Batas Beban Pada Kapal Penumpang," *Jurnal Teknik Elektro dan Komputasi (ELKOM)*, vol. 1, no. 2, pp. 64–74, Dec. 2019, doi: 10.32528/elkom.v1i2.3085.

[19] O. S. Al-Dahiree *et al.*, "Design and Shape Optimization of Strain Gauge Load Cell for Axial Force Measurement for Test Benches," *Sensors*, vol. 22, no. 19, p. 7508, Oct. 2022, doi: 10.3390/s22197508.

[20] T. Suryana, "Automation and Remote Control of Electronic Equipment Using the Internet with NODEMCU ESP8266 Interface and Apache MYSQL Web Server," 2021.

[21] D. Y. Widagdo, "Recording System of Weighing Results Using Load Cell Sensors Through a Database Based on Arduino Uno," *Jurnal Jartel: Jurnal Jaringan Telekomunikasi*, vol. 10, no. 1, pp. 13–19, Mar. 2020, doi: 10.33795/jartel.v10i1.161.

IOP Conf. Series: Earth and Environmental Science **1423** (2024) 012020 doi:10.1088/1755-1315/1423/1/012020

Techno-economic and supply chain model of a small-scale LNG for BMPP in Sambelia

H N Abdillah[1*], K B Artana[2,3] and M I Zakaria[2]

[1] Shipbuilding Engineering Department, Politeknik Perkapalan Negeri Surabaya, Indonesia
[2] Department of Marine Engineering, ITS, Surabaya, Indonesia
[3] Center of Excellence of Maritime Safety and Marine Installation (PUI-KEKAL), Faculty of Marine Technology, ITS, Surabaya, Indonesia

*E-mail: hayyabdillah@ppns.ac.id

Abstract. This paper addresses a method for creating the supply chain model of liquefied natural gas (LNG) to a barge-mounted power plant (BMPP) of 30 MW in Sambelia. This paper conducts the supply chain using a landing craft tank (LCT) ship as the LNG carrier, equipped with a specific principal dimension, and an ISO tank as the LNG storage facility. The use of LCT has the advantage of being able to sail in coastal areas and can be converted using existing ships. This comprehensive techno-economic assessment consists of calculating the LNG demands for the BMPP, the arrangement of ISO containerized tank stacking in the LCT, and identifying the LNG terminal facilities. Moreover, the research points for a future LNG economy are highlighted; cost-effective production and utilization solutions for enhancing the feasibility of LNG. The Landing Craft Tank (LCT), a vessel with a load capacity of 32 container tanks (20 ft) and a volume of 20.38 m^3, distributes LNG to BMPP Sambelia. The economic analysis of LNG distribution patterns from Benoa LNG Terminal to BMPP 30 MW Sambelia, using LCT ships with FSRU (Floating Storage and Regasification Unit) receiver terminals, reveals that at a margin of $2.2, the most optimal option has an IRR of 11.6%, a payback period of 5.7, and an ROI of 17.41%.

1. Introduction

Indonesia has natural gas reserves of up to 142.72 TSCF (trillion standard cubic feet), with the explanation that 100.36 TSCF is a proven reserve and 42.36 TSCF is a potential reserve. Indonesia ranks sixth globally in terms of liquefaction capacity and utilization and seventh in terms of LNG exports [1]. Natural gas could be converted into a liquid phase by cooling it to approximately -160 °C at atmospheric pressure and its volume to 1:600 from its initial condition. Indonesia produces twice as much natural gas as consumption, but it is still not sufficient for domestic needs [2]. The use of natural gas in the form of LNG is an alternative solution to the problem of overcoming the power crisis. Supporting facilities, including transport ships and facilities for production, processing, and distribution, can facilitate this utilization. Floating power plants, also known as barge-mounted power plants (BMPP), can utilize LNG fuel, a technology that enables power plants to run on both gas and diesel.

The LNG supply chain can be interpreted as a natural gas network that extends from the gas field to the liquidation plant, turning the natural gas phase into liquid. The LNG transport then distributes these fluids to users. The LNG supply chain, often referred to as the LNG supply chain, has four interdependent segments: production, liquidation, transportation, and storage, with or without regasification [2]. This study examines the supply chain from the perspective of transportation and storage, including the regasification process. The research designs BMPP technology to supply electricity in some remote areas located near the sea. Because of its geographic location and required infrastructure, Sambelia is a potential model location. The location is close to Eastern Lombok PLTU, and the electricity needs in the Sambelia region can accommodate a total of 30 MW [3].

Natural gas, as a fuel, can reduce exhaust emissions by 25–30%, SOx and solid particles by 100%, and NOx by 90% [4]. Therefore, based on economic studies, this research will explore the optimal supply chain design for LNG to the BMPP as well as assess the necessary operations for LNG fuel use at the BMPP. The LCT serves as a vessel that transports various types of cargo, including passengers, vehicles, and containers [5]. This research used the LCT to transport ISO containerized tanks filled with LNG. The use of LCT has the advantage of being able to sail in coastal areas and can be converted using existing ships. Containerization is the process of transporting goods between modes using intermodal containers (also known as shipping containers or ISO containers). The ISO container tank can be used for loading and unloading, stacking, long-distance transportation, and transitioning from one mode of transport to another. In this research on LNG supply chain planning, LCT ships will transport LNG using an ISO containerized tank with a design that matches the characteristics of LNG. The container tank has a volume specification of 20.38 m^3, a weight of 34.000 kg, a holding time of 44 days, and a working pressure of 6.6 bar [6]. Table 1 shows the conversion factor used in this study.

2. LNG terminal identification

There are four LNG refineries in Indonesia: Donggi Senoro (Central Sulawesi), Bontang (East Kalimantan), Arun (Aceh), and Tangguh (West Papua). However, because gas production has decreased, the refinery in Arun has changed its function to become a regasification terminal. As a result, there are only three refineries available: the Bontang, Tangguh, and Donggi refineries [7]. The closest refineries for BMPP destinations are in Bontang and Donggi Senoro; however, there is an LNG receiving facility near the BMPP 30 MW Sambelia location, namely FSRU Karunia Dewata. Based on a capacity of 30 MW, LNG demand analysis at BMPP Sambelia is 6 mmcfd per day, with a total LNG demand per year of 95,396.4 m^3. Details of the calculation of gas consumption can be seen in Table 2.

Table 1. Conversion factors

	Tons LNG	mmcm[a] LNG	mmcm Gas	mmcf[b] Gas	MMBtu[c]
Tons LNG		2 – 2.222	0.0013	0,0459	53.38
Cubic meter LNG	0.45 – 0.5		5.85 x 10^{-4}	0,0207	24.02
mmcm gas	769.2	1,700		35.31	41,100
mmcf gas	21.78	48	0.0283		1,200
MMBtu	0.0187	0.0416	2.44 x10^{-5}	8.601 x 10^{-4}	

[a] mmcm : million cubic meters.
[b] mmcf : million cubic feets.
[ac] MMBtu : million metric British thermal units.

Table 2. LNG demand at BMPP Sambelia

Information	Unit	Value
Barge Mounted Power Plant Capacity	Megawatt (MW)	30
Natural Gas Demand per day	mmscfd	6
LNG Demand per day	tons/day	130.68
LNG Demand per day	m^3/day	261.36
LNG Demand per week	m^3/week	1,829.52
LNG Demand per year	m^3/year	95,396.40

[a] mmcfd : million standard cubic feet per day.

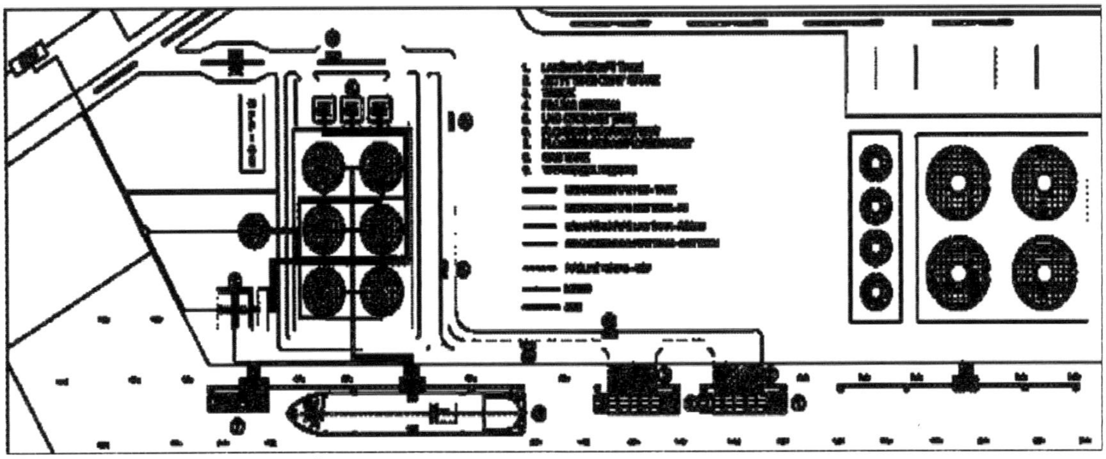

Figure 1. Layout LNG terminal

The LNG terminal complies with the LNG demand for BMPP in Sambelia by planning a filling station in the reclamation area of Benoa. This filling station has the potential to transfer LNG into the ISO containerized tank, because the infrastructure and and facilities are sufficient. The adjustments are made based on the activities carried out for the loading and unloading process of ISO containerized tanks, as well as the time required from the loading and unloading process to the LNG filling station. Layout of LNG terminal shown in Figure 1. The pump at the filling station has a capacity of 90.6 m^3/hour, so that 32 containers for containerized tanks have a capacity of 20.36 m^3, which only takes about 7.25 hours to fill. Meanwhile, filling 24 container tanks takes around 5.44 hours.

3. Methodology

Figure 2 illustrates the methodological steps used in this study. It is divided into technical and economic aspects, technical aspects consist of data collecting; calculation of LNG; arrangement of container ISO tank; the BMPP and FSRU arrangement; and determination of the supply chain models. The economical aspects consist of the CAPEX and OPEX data; calculation of PP, IRR, NPV and ROI. The data collection includes information on power plant capacity, the location of the power plant, the distance matrix, the LNG carrier, support facilities, as well as CAPEX and OPEX costs. LNG requirements are calculated within one year. Economical aspects based on the distribution scheme comply with ESDM Minister Regulation No. 8 Year 2020 and ESDM Regulation

IOP Conf. Series: Earth and Environmental Science **1423** (2024) 012020 doi:10.1088/1755-1315/1423/1/012020

91K/12/MEM/2020 by considering CAPEX, OPEX, revenue, PP, IRR, NPV, and ROI. It is economically feasible if the result of the payback period calculation is less than 10 years and the ROI is greater than the regulation of the bank (between 16% and 20%) [8]. The economical aspects determine the most optimal supply chain model.

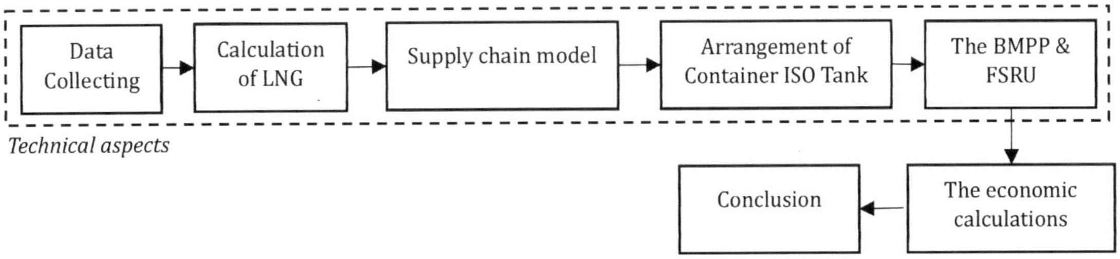

Figure 2. Methodological framework

4. Result and discussion

4.1 Arrangement of containerized ISO tank

The selection of container tanks used is based on standard ISO 1496 Part 3, which will require that the ISO containerized tank be built to withstand a superimposed load of 192,000 kg [9]. The ISO Containerized Tank measures 6.058 x 2.438 x 2.591 meters in length, width, and height, consecutively with a gross weight of 34,000 kgs (34 tons). By considering the ship load capacity, which is 1500 DWT, and the area of the LCT ship deck covering 655 m^2, it is necessary to check the full load weight and full deck area. The ISO container tank is placed on the deck with the lashing point. The lashing system consists of the turn buckle, the lushing bar, and the extension hook. The main dimension of LCT, LOA 64m, LPP 59.7m, B 14 m, T 4.5m, and Vs 9 knots. The size of this LCT is suitable for demand and is capable of sailing in coastal areas, depending on the geographical location.

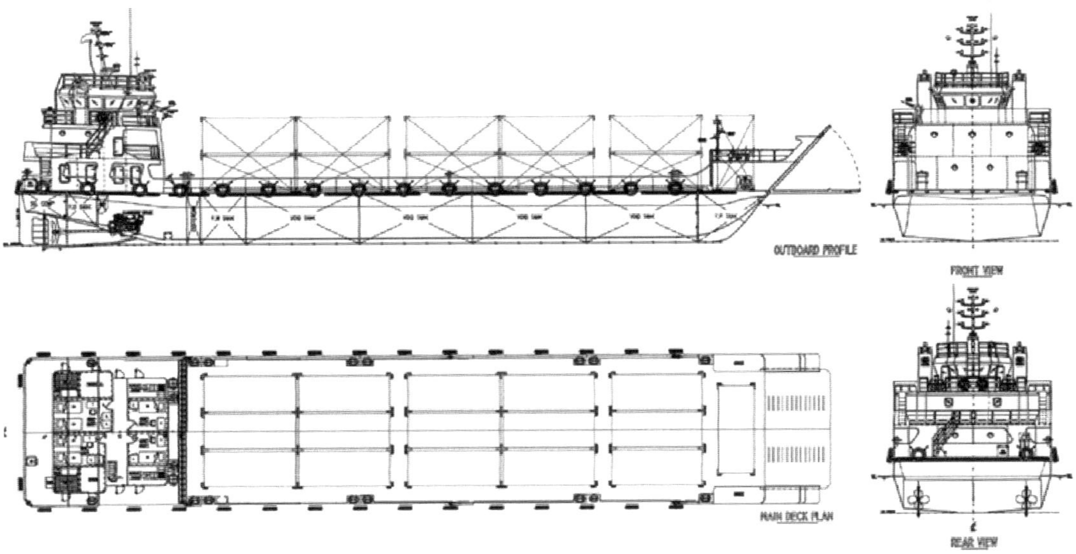

Figure 3. LCT arrangement

4.2 The BMPP & FSRU arrangement

Figure 4 shows the arrangement of the BMPP. The dimensions of the BMPP for 30 MW Sambelia are LOA 54 m, B 27.4 m, H 6.5 m, and T 4.7 m. There is a buffer tank that is sufficient to accommodate the safety stock of LNG for BMPP when it is not supplied by the LCT ship. BMPP compartments are arranged with double bottoms and bulkheads, followed by the selection of a buffer tank and the determination of regasification facilities [10]. The study utilized a type C LNG tank, with a pressure of 8 bar, a capacity 600 m^3, a design BOG rate 0.08%, a length 23 m and a tank diameter 7 m as the LNG buffer tank. This choice ensures a longer storage period for BOG, which is unaffected by sloshing because of its cylindrical geometry [11]. The determination of the LNG tank volume is calculated as follows (Equation 1). This calculation requires the buffer tank to load the LNG. The buffer tank can accommodate 22 container tanks, each with a capacity of 20.38 m^3, resulting in a total volume of 448.36 m^3 of LNG for the safety stock. Then, Equation (1) is used to calculate the minimum volume of the buffer tank for selection..

$$LL = FL \ x \ \frac{\rho R}{\rho L} \tag{1}$$

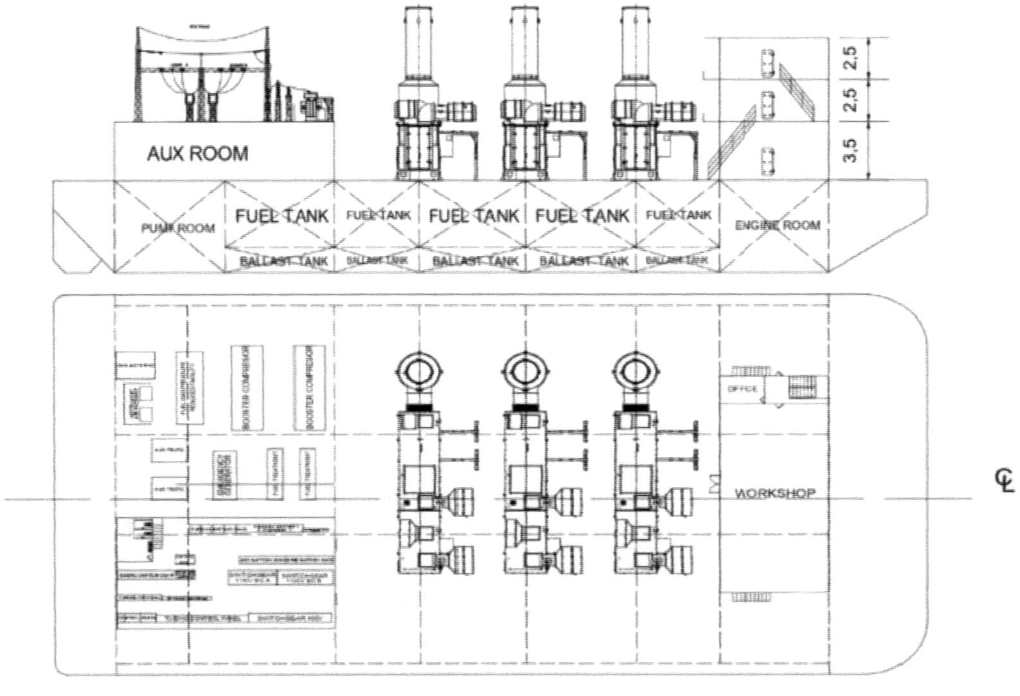

Figure 4. BMPP arrangement

The BOG normal rate value in units of Nm^3/hour determines the choice of compressor capacity [12]. The BOG normal rate represents the conversion value of LNG to gas, which is calculated based on the tank design and then multiplied by the total volume of LNG storage. These facilities include cranes, buffer tanks, container yards, regasification, and accommodation. The FSRU BMPP (30 MW) Sambelia main dimensions are the LOA of 35 m, B of 22 m, and H of 5.5 m. Figure 5 displays the FSRU 30 MW arrangement.

IOP Conf. Series: Earth and Environmental Science **1423** (2024) 012020 doi:10.1088/1755-1315/1423/1/012020

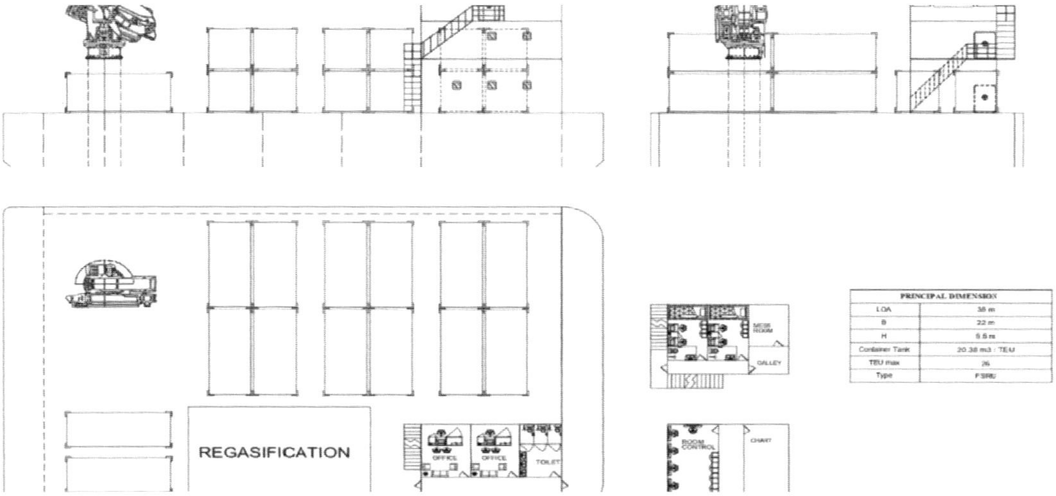

4.3 Supply chain model

Based on political views, the objectives for small-scale LNG in Indonesia are to enhance national energy security, increase the electrification ratio in the eastern part of the country, reduce fuel subsidy costs, and increase the proportion of gas for domestic use. The supply chain model is affected by the market, infrastructure, and capability. The eastern part of Indonesia dominates the gas market and has the potential to create multiplier effects. The main challenge of the LNG supply chain in Indonesia is the small and distributed demand, so the hub and spokes model can be implemented to address it. The other aspects require increased skill, experience, and technology to ensure sustainability.

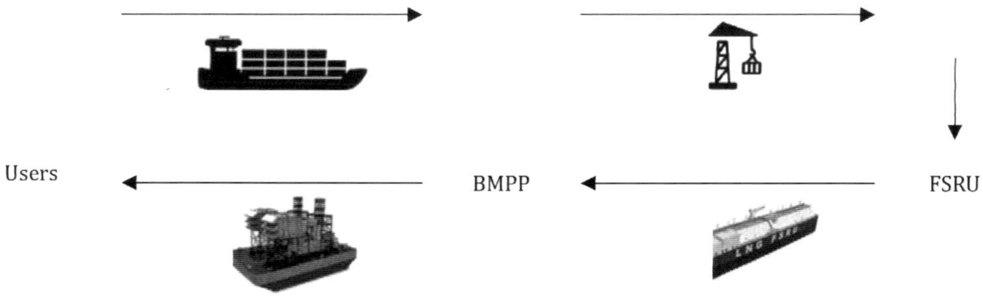

Figure 6 illustrates the LNG supply chain from the terminal to the end users. An LCT ship distributes LNG from Benoa, an LNG terminal, to the FSRU. The FSRU, as the storage unit, supplied LNG to the BMPP Sambelia. One roundtrip from Benoa to Sambelia takes five days. The roundtrip calculation for BMPP Sambelia 30 MW a year totals 182 roundtrips. LNG consumes 261 m^3/day.

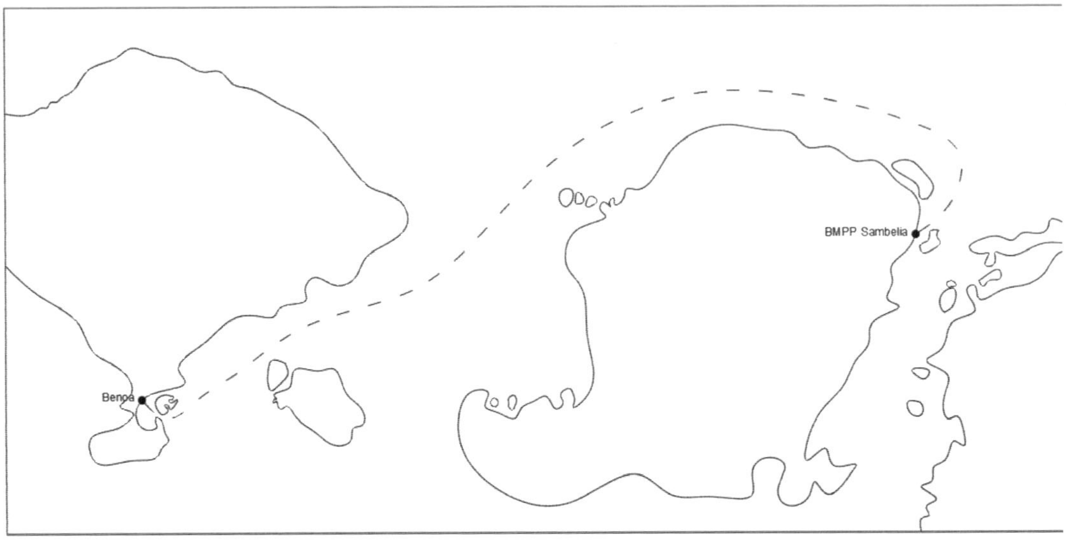

Figure 7. LNG distribution route

The distance between the LNG terminal and the BMPP Sambelia is about 118.3 Nm. Figure 7 illustrates the route from the LNG terminal to the Sambelia BMPP. The number of voyages that can be calculated by adding up the time available for one year divided by the total time needed to perform one voyage. The spare time in a year must be provided for maintenance, surveys to maintain the condition of the ship, and docking [13]. After obtaining the data, it can be ascertained that the supply chain used is sufficient to supply BMPP 30MW Sambelia with 1 voyage for 2 days carrying 24 container tanks at a time. Due to the demand for BMPP at 30 MW per day, it only requires 12 ISO containerized tanks. This development supply chain model is expected to meet the challenges of the small-scale LNG concept. The small-scale LNG concept is to reach "stranded" LNG markets, which means too far from the existing pipeline system and not economically feasible for pipeline extension or new pipeline construction [14]. Demand volume, geographical profile, and existing infrastructure are the main considerations in choosing the right match technologies [15].

4.4 The economic calculations

This research used two variables to determine investment: capital expenditure (CAPEX) and operational expenditure (OPEX). CAPEX is an investment cost incurred at the beginning to carry out the distribution of LNG supplies because the LCT ship, ISO Container Tank, and FSRU are the responsibility of the distributor to ensure the distribution pattern in this research is carried out. This investment cost will have an impact on the actual distribution of LNG supplies [16]. This analysis is based on the unit cost of delivery to the receiving terminals in both BMPPs. For the Sambelia BMPP 30 MW supply chain, a single LCT ship serves as a backup in case other ships arrive for maintenance, necessitating the use of six LCT vessels.

To create a distribution cycle in the Sambelia BMPP 30 MW supply chain, each ship must have its own 24 ISO containerized tanks and 24 additional ISO containerized tanks. For the FSRU, it requires 2 units of investment, with each BMPP accounting for 1 unit, as the BMPP itself lacks an area for LNG storage and regasification. Operating expenses (OPEX) are all costs incurred to carry out operations for a certain period, usually one year. The transportation costs in this study, which also include crew, port, and insurance expenditures, describe the distribution pattern for the ship operating costs. The proceeds from the LNG selling price, also known as the margin or the difference between the selling price and the buying price of LNG, result in income, also commonly referred to as revenue.

In this distribution, the sales margin increased by $2.2, $2.3, and $2.4 for scenario 1, and by $3.0, $3.1, and $3.2 per 1 mmbtu of LNG for scenario 2. These two scenarios are developed based on LNG margin at plant gate to be implemented in economic analysis. The BMPPs consume 6,323,913 mmbtu/year for scenario 1 and 5,533,619 mmbtu/year for scenario 2, multiplied by the selling price of LNG. Table 3 and Table 4 show the income per year from a predetermined margin. The price of LNG at LNG Terminal A aligns with the Minister Regulation of ESDM Number 8 of 2020 and the Minister Regulation of ESDM 91K/12/MEM/2020, which stipulates a 10% ICP + midstream. The regulation stipulates that the upstream LNG tariff, known as Bontang LNG, is 10% ICP, or the price of Indonesian crude oil, which begins at approximately 47 USD in January 2021, resulting in an LNG price of $4.7 per mmbtu. The estimate for the midstream tariff at LNG Terminal A is x per mmbtu. Therefore, the total cost of buying LNG is $4.7 + x.

Table 3. Scenario 1 – economical aspect

		Margin $ 2.2		Margin $ 2.3		Margin $ 2.4
Purchase Price	$	4.7 + *x*	$	4.7 + *x*	$	4.7 + *x*
Margin	$	2.20	$	2.30	$	2.40
Selling Price	$	10.70	$	10.80	$	10.90
Annual Revenue	$	13.912.609	$	14.545.000	$	15.177.392
NPV	$	9.994.904	$	12.847.671	$	15.700.439
IRR		11.60%		13.20 %		14.77 %
PP		5.74		5.38		5.06
ROI		17.41%		18.58 %		19.75 %

Table 4. Skenario 2 – economical aspect

		Margin $ 3.0		Margin $ 3.1		Margin $ 3.2
Purchase Price	$	4.7 + *x*	$	4.7 + *x*	$	4.7 + *x*
Margin	$	3.00	$	3.10	$	3.20
Selling Price	$	7.7 + *x*	$	7.8 + *x*	$	7.9 + *x*
Annual Revenue	$	16,600,859	$	17,154,221	$	17,707,583
NPV	$	-2,091,170	$	1,865,084	$	5,821,339
IRR		7.70 %		9.01 %		10.26 %
PP		9.07		8.30		7.65
ROI		9.05 %		9.92 %		10.78 %

Bank interest at this time is the ratio of the interest paid on loans over a certain period; bank interest is usually expressed as a percentage. In this distribution, OJK uses 8.4% of the total average interest for investment loans from 30 banks in Indonesia. Meanwhile, it can interpret the payback period (PP) as the time it takes to recoup the investment through the benefits of a planned project. The Internal Rate of Return (IRR) is used to calculate the interest rate when the NPV value is equal to 0. IRR is useful for finding out at what interest rate some fixed investments are profitable. NPV is commonly used to calculate the return on investment, whether the investment is profitable or not. This method converts all cash flows to present value (P) and adds them up, thereby reflecting the net value of all cash flows that occur during the planning period. If the NPV value is greater than 0, the investment can be said to be profitable. When the NPV value is zero, the investment yields a return that matches the expenses. Finally, the realization of the project is impossible if the NPV value falls below 0. The results of an economic study conducted in the form of payback.

Period, IRR, and NPV, with sales-purchase margins of \$2.7, \$2.8 and \$2.9 for scenario 1 and for scenario 2 the sales margin is varied with a margin of \$3.0, \$3.1 and \$3.2 for 20 years contract can be seen in Table 3 and Table 4. From the results of the economic study that has been shown in the table above, the margins that meet the requirements for this research are in the form of PP < contract years, IRR > Bank Interest and ROI more than 10%, which is at a margin of \$2.9 for scenario 1 where the margin has an IRR of 10.24%, then PP at 7.62 years and ROI of 10.71%. For scenario 2 is at a margin of \$3.2 where the margin has an IRR of 10.26%, then PP at 7.65 years and ROI of 10.78%.

5. Conclusion

This techno-economic review evaluated the feasibility of the BMPP distribution scenario in Sambelia. To meet the 30 MW electricity supply requirement, four LCT vessels with 32 tanks are required. The gas purchase price at the Benoa Terminal complies with ESDM Minister Regulation No. 8 of 2020 and ESDM Ministerial Regulation 91 K/12/MEM/2020, which stipulate a 10% ICP or equivalent to \$4.7. The terminal LNG midstream cost of Benoa, which ranges from \$3.8 to \$8.5, includes this price. The Landing Craft Tank (LCT), a vessel with a load capacity of 32 container tanks (20 ft) and a volume of 20.38 m3, distributes LNG to BMPP Sambelia. The economic analysis of LNG distribution patterns from Benoa LNG Terminal to BMPP 30 MW Sambelia, using LCT ships with FSRU receiver terminals, reveals that at a margin of \$2.2, the most optimal option has an IRR of 11.6%, a payback period of 5.7, and an ROI of 17.41%.

References

[1] IGU . (2020). 2020 World LNG Report – IGU. Retrieved from: https://igu.org/resources/2020-world-lng-report/.

[2] Putra, M. A. S. D. (2016). *Desain Rantai Pasok Gas Alam Cair (LNG) untuk Kebutuhan Pembangkit Listrik di Indonesia Bagian Timur. Jurnal Teknik ITS* 5(2), 12-19.

[3] Putra, H. R. L. (2019). Societal Risk Assessment On 60MW Barge Mounted Power Plant In South East Sulawesi. *Jurnal Teknik ITS,* 45-58.

[4] Siahaya, Y. (2014). Benefits of Use LNG as Material Main Engine Fuel Ship. *J. Pen. Transla* 16(3), 87–92.

[5] Abdillah, H.N., Artana, K.B., Ariana, I.M., and Masroeri, A.A. (2018, July 9-11), *Technical and Economical Analysis of LCT (Landing Craft Tank) Vessel conversion into Non-Conventional LNG (Liquid Natural Gas) Carrier*. Proceedings of Mastic, Bali, Indonesia.

[6] Soegiono dan Artana, K. B. (2006). Transportasi LNG Indonesia. Surabaya: Airlangga University Press.

[7] Yoo, B. (2017). Economic Assessment of Liquefied Natural Gas (LNG) as a Marine Fuel for CO2 Carriers Compared to Marine Gas Oil (MGO). *Journal of Energy* (121), 772-780.

[8] Pertamina. (2014). Annual Report – Inspiring Indonesia to the World. Jakarta : Pertamina.

[9] Abramowski, T., Cepowski, T., & Zvolensky, P. (2018). Determination of Regression Formulas for Key Design Characteristics of Container Ships at Preliminary Design Stage. *New Trends Prod. Eng.* (1), 247-257.

[10] Gronhaug, R., & Christiansen, M. (2009). *Supply Chain Optimization for the Liquefied Natural Gas Business*. Proceeding of Innovations in Distribution Logistics, Heidelberg, Berlin.

[11] Yao, Y., Guo, Z., Duan, M., Li, Z., & Liu, B. (2015). The Structure Design of Type-C Independent Tank on LNG Ship. *Material Science and Engineering* (5), 73-82.

[12] Pettersson, F., Bittante , A., & Saxen, H. (2018). Optimization of a Small-Scale LNG Supply Chain. *Journal of Energy* (148), 79-89.

[13] Raj, R., Ghandeharium, S., Kumar, A., Geng, J., & Linwei, M. (2016). A Techno-Economic Study of Shipping LNG to The Asia-Pacific from Western Canada by LNG Carrier. *Journal of Natural Gas Science*

and Engineering (34), 979-992.
[14] Budiyanto, M., Riadi, A., Buana, I., & Kurnia, G. (2020). Study on The LNG Distribution to Mobile Power Plants Utilizing Small-Scale LNG Carriers. *Journal of Heliyon* (6), 38-45.
[15] Thomson, H., Cobette, J., & Winebrake, J. (2015). Natural Gas as a Marine Fuel. *Journal of Energy Policy* (87), 153-167.
[16] Uemura, T., & Ishigami, K. (2018). *LNG Supply Chain Infrastructure Configuration - Formulating Policy Options for Promoting Natural Gas Utilization in the East Asia Summit Region Volume II: Supply Side Analysis*. Jakarta: ERIA.

IOP Conf. Series: Earth and Environmental Science **1423** (2024) 012021 doi:10.1088/1755-1315/1423/1/012021

An Optimization Model for the Offshore Well Intervention Campaign: Literature Review

Sri Konsep Harum Wicaksono[1*], Ahmad Rusdiansyah[2] and Haryo Dwito Armono[3]

[1] School of Interdisciplinary Management and Technology, Institut Teknologi Sepuluh Nopember, Surabaya, Indonesia
[2] Department of Industrial and Systems Engineering, Institut Teknologi Sepuluh Nopember, Surabaya, Indonesia
[3] Department of Ocean Engineering, Institut Teknologi Sepuluh Nopember, Surabaya, Indonesia

*E-mail: konsep.wicaksono@gmail.com

Abstract. In the offshore oil industry, well-intervention barges provide crucial support for the production enhancement of offshore platforms. These specially designed ships contain tools and equipment to perform perforation activities that keep the wells active. However, with the relevance and practical operations, operators meet two principal challenges: 1) limited barge availability to serve a set of platforms and 2) each platform exhibits specific job requirements (different number of wells, duration, production gain, cost, time windows). This short literature review provides an overview of the Well Intervention Campaign Problem (WLICP) adopting the model of Vehicle Routing Problem considering Profits (VRPP). We propose four carefully designed questions to answer. It relates to 1) problem structure 2) modelling approach 3) solution technique and 4) optimization scenario. According to the result, WLICP exhibits a rigorous project management structure. Its structure contains the breakdown of n-sub-project representing VRP nodes. Combinatorial optimization helps barges find routes and schedules to maximize profit.

1. Introduction

As the population and economy grow, global energy demands are rising. However, oil production is declining over time due to decreasing reserves from onshore fields, especially from developed areas [1], [2]. Amidst the development of the onshore field, offshore field development has risen significantly over the past decade. The main cause of this expansion is oil scarcity onshore around the world thus pushing forward the development to the frontiers. Offshore reserves, which have a proven rate of just 30%, make significant contribution of the world's oil reserves (nearly 30% share). This suggests that offshore oil exploitation holds significant potential [3]. This leads to oil operators and services focusing on the offshore area to explore the near future.

In the offshore oil industry, the well intervention (WLI) barge provides crucial support for the production enhancement operations of offshore platforms [4]. It takes the role of providing well intervention means such as electric lines, slicklines, snubbing units, and coiled tubing when the wells in the respective platforms start to decline [5]. These specially designed ships contain

IOP Conf. Series: Earth and Environmental Science **1423** (2024) 012021 doi:10.1088/1755-1315/1423/1/012021

tools, equipment, and personnel to perform well intervention activities such as perforations, water shut-offs, data acquisition and other related activities to keep the wells in the platforms delivering continuous production [6]. Figure 1 shows a snapshot of the well intervention barge that comprises of crane, accommodation area, and equipment storage. During the operation, this vessel approaches the platform, connects a gangway between the barge and platform, thus ensuring personnel and equipment movements.

1.1 Well intervention campaign problem

In the context of a dry wellhead, an offshore platform is a home for producing wells. There are various types and sizes of offshore production platforms installed in the oil field. However, typically, a platform contains 10-20 wells for tripod and quadric-pod type, considering the operator's production level to achieve. As the OPEX (operation expenditure) of offshore platforms is expensive, operators attempt to maximize oil production by assigning a barge to work in certain wells in the platform. The high number of platform and well brings complexity for operators because they need to plan and prioritize certain well and platform to be visited by the barge to perform the intervention, once the individual well's production decrease. It is even more challenging to plan since the respective well in the platform has unique intervention gain and job duration due to its subsurface complexities (such as reservoir properties). Moreover, operators have limited well intervention barge to serve due to its high operation cost.

The vessel is an accommodation crane/work-barge purpose build to performs and fulfill the following duties:
- Well intervention (perforation, zone shut-off)
- Accommodation/floatel services

Gross Tonnage	**5107**
Summer Deadweight (t)	**5300**
Length Overall, LOA (m)	**85**
Beam (m)	**24**

Figure 1. Well intervention barge contains crane, accommodation, and intervention tools/ equipment (courtesy Logindo, 2023)

1.2 Barge Routing and Scheduling

Based in the cost structure of well intervention activities, operation duration of a barge is the main factor of budget spending. Job characteristic or intervention from individual well shares the uncertainty of barge duration. Thus, it also impacts most of the barge operation time that implies to cost [7]. The expenses associated with these operations, including the costs of the barges needed to carry them out, make this research is important and anticipated to give significant contribution of managing barge scheduling that impacts to costs. Barges are a costly resource, as the charter rates and the fuel costs for operating the barges are expensive [8]. Efficient use of the barges is therefore of great interest, both for cost-effectiveness and high delivery. Research is needed to assess the problem structure and also identify the best methodology for generating optimal solutions to offshore vehicle routing problems. To date, there has been minimal research in the area of well intervention scheduling.

The primary objective of this paper is to identify appropriate techniques for addressing profitability in vehicle routing problems/scheduling that impacts on cost and to evaluate their performance. To achieve this aim, the following specific objectives have been outlined: 1) to construct a problem into a structured model, and 2) to identify suitable techniques/models which can be used for solving the profitability in VRP.

2. Methodology

VRP in the context of well intervention barge planning optimization is relatively underrepresented in the research literature both in the scope of optimization studies and petroleum management. However, there are some exceptions, primarily involving simulation studies [9]. There is some additional optimization-related research, although it addresses slightly different practical scenarios. Current studies on optimization in well intervention campaigns have been categorized into specific clusters using search descriptors such as area, algorithm, constraints, and models. The main sources of the literature are extracted from Science Direct, ResearchGate, Web of Science, and Springer.

2.1 Research Questions

In this paper, we propose an optimization model for planning and scheduling context related to well-intervention campaigns, which we call the Well Intervention Campaign Problem (WLICP). Despite the extensive use of optimization in the petroleum industry, literature specifically addressing optimization in well-intervention planning is limited [10]. The authors build research questions and objectives related to the WLICP.

RQ1: How to construct the well intervention campaign into a structured model?

Motivation: This research question aims to create a structured model to address the complexity of well intervention campaigns and to identify an appropriate technique or model for solving profitability issues in vehicle routing problems (VRP).

RQ2: Which VRP models are most suitable and efficient for addressing offshore barge planning and scheduling challenges?

Motivation: This research question aims to identify a the most optimum outcome-based VRP model for addressing problems related to maximizing profitability. The findings could be valuable for future research in developing solutions for other typical offshore routing issues, such as well plug and abandonment campaigns.

2.2 Knowledge Gap and Research Direction

It can be reviewed from the current landscape, that there are research gaps and knowledge needed in the field of well-intervention planning optimization [11]. On top of the cluster respectively, the authors provide research directions in Figure 2. Thus, the research direction will be: 1) alternative/studies in developing countries/regions, 2) how to implement planning optimization for well intervention campaigns, 3) modeling of additional uncertainties: traveling time and work duration, 4) examine the role of routing management to model optimization in multi-platforms layout, and 5) Study prioritization for routing & scheduling.

The field service routing problem is an important problem in the modern industry [12]. Yet the research done on the real-world problem is very limited, specifically for offshore well intervention campaigns [13][14]. This paper aims to deliver a scientific contribution based on the gap analysis and new methodology as presented in Figure 3. This research focuses on a tactical time horizon for planning and scheduling well interventions across multiple offshore wells, where production systems are situated on dedicated platforms. In this context, a well intervention involves assigning barges to perform well intervention activities on a group of producing wells or

IOP Conf. Series: Earth and Environmental Science **1423** (2024) 012021 doi:10.1088/1755-1315/1423/1/012021

platform. The wells have dry wellheads means manifold and control systems are in the platform, so all well interventions cannot be performed unmanned.

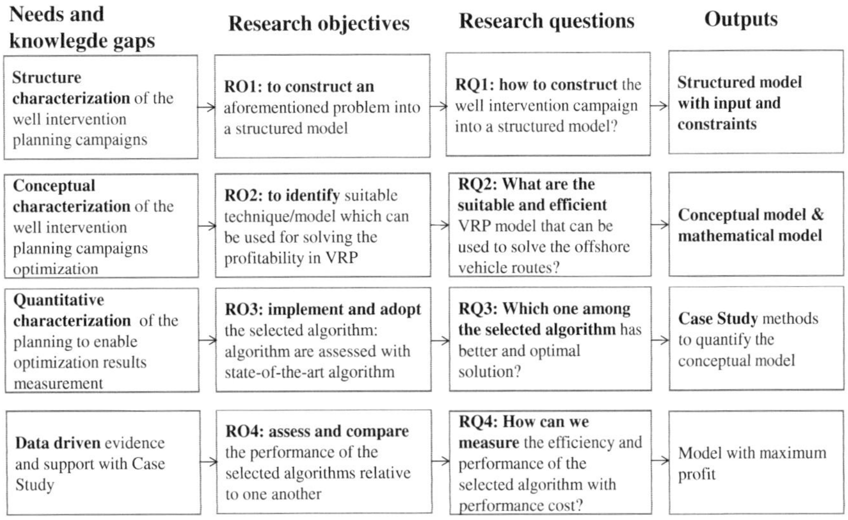

Figure 2. The relationship between research gap, purpose, and research question

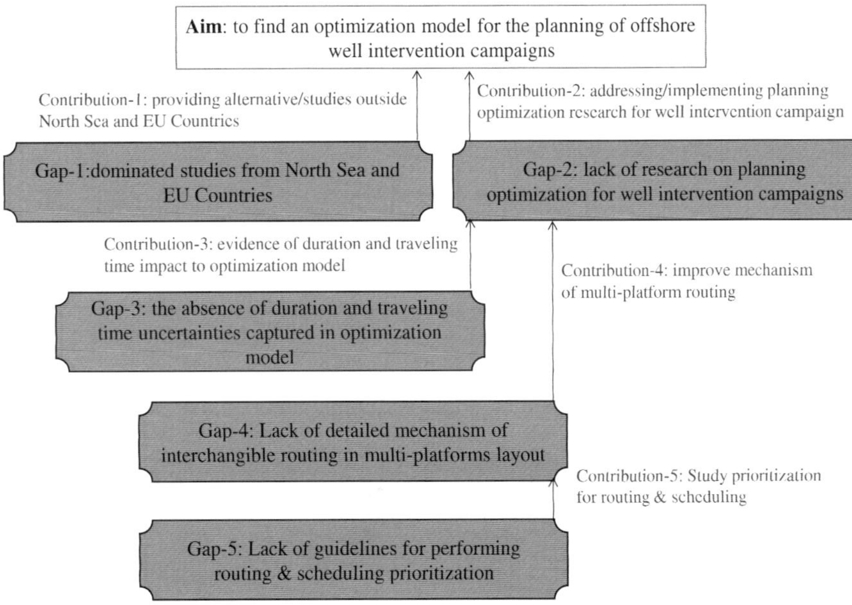

Figure 3. Overview of research gap addressed and scientific contribution.

2.3 Vehicle Routing Problem (VRP) Typology

The VRP is a well-established and advanced method for optimizing logistics [12]. This method attempts to find the nearest and optimized route for a given customer located in a specific area with a set of vehicles [15]. The VRP problem exhibits more complex constraints and conditions, making this method has been extensively developed [16]. Figure 4 shows the typology of many types of VRP and the position of the research (dashed light red).

In the context of the well intervention campaign problem (WLICP), there are combinations of VRP such as VRP-Profit, VRP-time windows, and VRP-multiple depots. Despite the typology, each VRP contains its constituent elements such as path/network, customer, depots, vehicle, and side constraints.

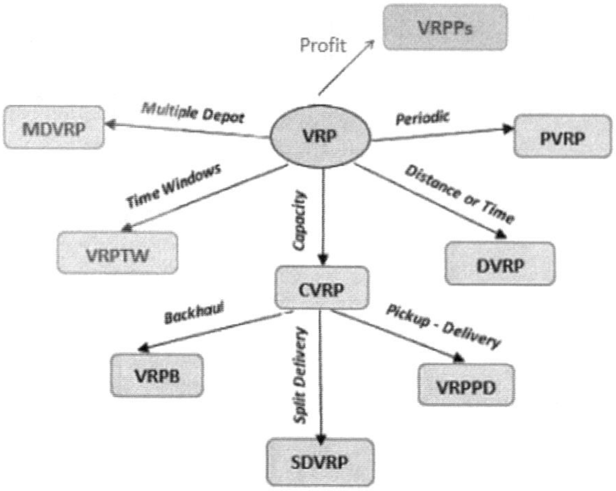

Figure 4. Different Variants of VRP (modified after [17])

3. Results and Discussions

3.1 Constituent Elements of Vehicle Routing Problem

3.1.1 Path/Network

Path/network is the key element of VRP. It represents the movement path from one point to another where the routing goes [18][19]. In the context of VRP, the network becomes the connection between vertex (in this case, point of departure and arrival), while the arch represents the route of the vehicle. In WLICP, network means the connection between platform A to platform B, platform B to platform C, etc. The path may be irregular and non-pattern but the traveling distance/duration between those points is considered the nearest (distance term) or fastest (duration term). WLICP describes road networks as the distance between platforms, whatever the number, and expressed as matrix distance. This paper utilizes matrix distance in days (duration) instead of meterage due to barge contractual duration.

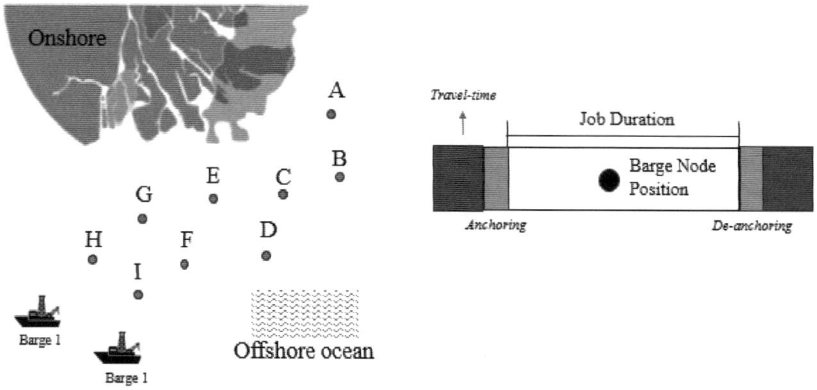

Figure 5. Schematic of path/network characterization of WLICP [20]

3.1.2 Customer

In the classical VRP, costumer means the position of service demand and a vertex of network connection. WLICP describes the customer as a platform to visit where it represents a set of jobs to be done for a certain duration. Figure 5 demonstrates the example of customer A (Platform A)

IOP Conf. Series: Earth and Environmental Science **1423** (2024) 012021 doi:10.1088/1755-1315/1423/1/012021

with a set of sub-projects. In platform A, there is a project A to be done in wells A1, A2, and A3. Each of the wells has a unique job duration, type, gain, cost, and profit. In aggregate, Project A consists of three sub-projects from three different wells. The project is then considered as customer 1. In the expansion of the study, the customer comprises several projects (B, C, D, etc.) from several platforms, not a single point of objective.

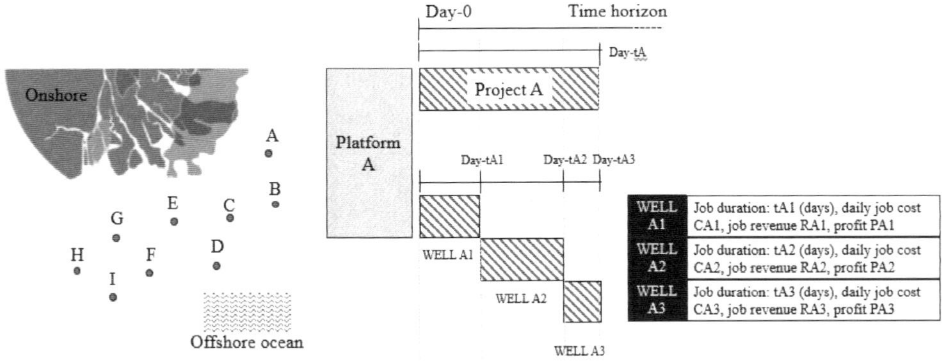

Figure 6. A schematic characterization of well intervention campaign problem (WLICP) in platform A that contains job/project for 3 wells.

3.1.3 Depot

For any logistic study, the depot takes a role at a point where the vehicle starts to depart and end its journey [21]. It can be critical in classical VRP because it is included in the route to be optimized. However, in WLICP, the depot is less important because the vessel isn't required to back to the depot and the starting point and end point (if any) could be different. Thus, this paper introduces the absence (or flexibility) of depot requirements for the routing and scheduling optimization.

3.1.4 Vehicle

In any context, the vehicle becomes the principal element of the VRP study [22]. It ensures that the supply is where the customers need it and has a certain capacity. The vehicle has specific characteristics such as capacity, type, rental cost, and size. WLICP employs multiple vehicles, however, they have the same capacity to do jobs such as perforation, water shut-off, and data acquisition. There is no operation limitation of this vehicle (24 hours) and no supply issue. In WLICP, it is called a WLI barge. Figure 6 explains the typical time sequencing of the WLI barge including travel time between vertex (costumer), anchoring, main job, and de-anchoring.

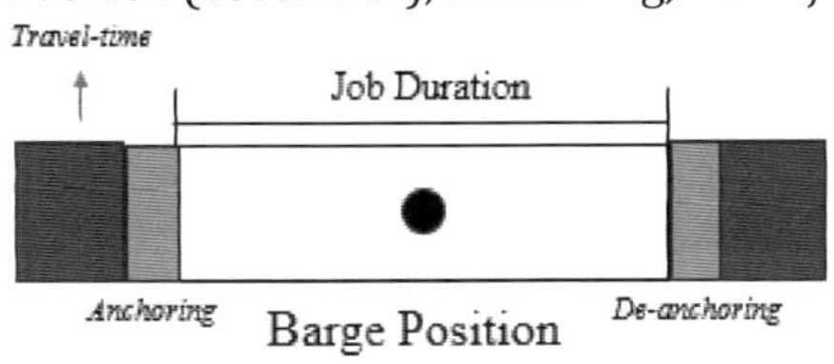

Figure 7. A Typical vehicle time window to visit the platforms that contains travel time from origin to destination, anchoring, main job, de-anchoring, and travel time to next destination.

3.1.5 Side Constraint

In WLICP, the main side constraint is the time window of platform operation. There is a specific time when the platform cannot be visited by WLI barge such as during drilling rig and construction works. Operators provide a particular time of platform window operation as a basis of the WLI barge to visit.

3.2 Model 1: WLICP Project Node (Full Duration)

In this model, Project A represents a node in the VRP. As seen in Figure 8, Project A consists of several sub-projects (called A1, A2, and A3) with specific duration. It explains that the total duration of the project equals to summation of all sub-projects. Thus, in Model 1 (Full Duration), the barge should finish all the jobs in Project A from day 0 to day tA. Similar to Platform A, the next visit to Platform B performs Project B which comprises two sub-projects (called B1 and B2). The time duration in Project B is the summation of two sub-projects. Barge only will move out after all jobs related to Project B finish.

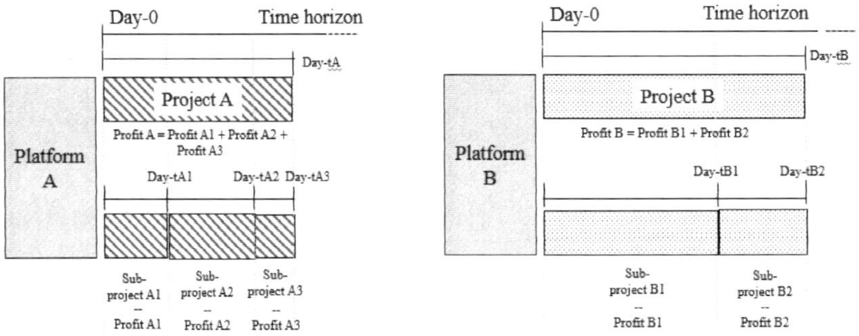

Figure 8. Schematic of WLICP full duration model

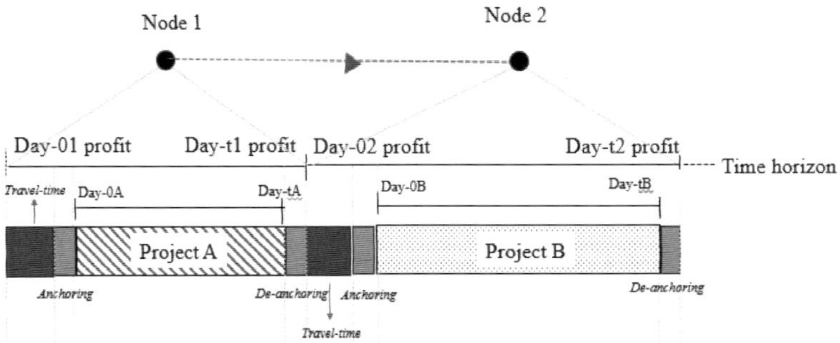

Figure 9. Routing schematic of barge movement full duration

In terms of cost and gain, Project A and Project B exhibit full project gain and cost. Thus, operators should prioritize barge planning on a project basis to gain the maximum profit. Figure 9 shows the routing schematic of the barge from Node 1 to Node 2 and exhibits pre and post-job sequences such as traveling time, anchoring, and de-anchoring. Through optimization and other respected VRP algorithms, it achieves WLICP barge planning in Figure 10 in a certain time horizon. It explains that, for example, Project A to Project G are selected to be visited, while due to profit evaluation Project H and Project I are excluded in the selected time horizon.

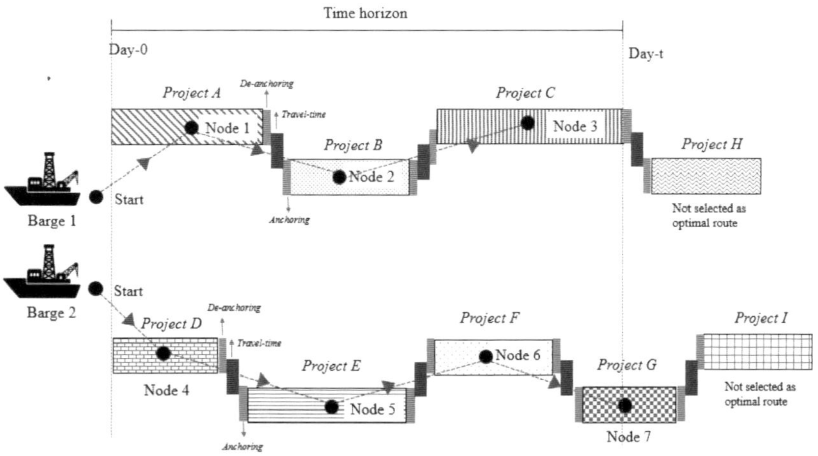

Figure 10. Schematic of WLICP barge planning full duration

3.3 Model 2: WLICP Sub-Project Node (Partial Duration)

Different from Model 1, this Model (or called Model 2) refers to a sub-project basis. As seen in Figure 8, Project A consists of several sub-projects (called A1, A2, and A3) with specific duration. In Model 2, the barge should not finish all the jobs in Project A from day 0 to day tA, but it can choose which sub-project to select. Thus, the time duration is quite different and is parameter to be included in the optimization evaluation.

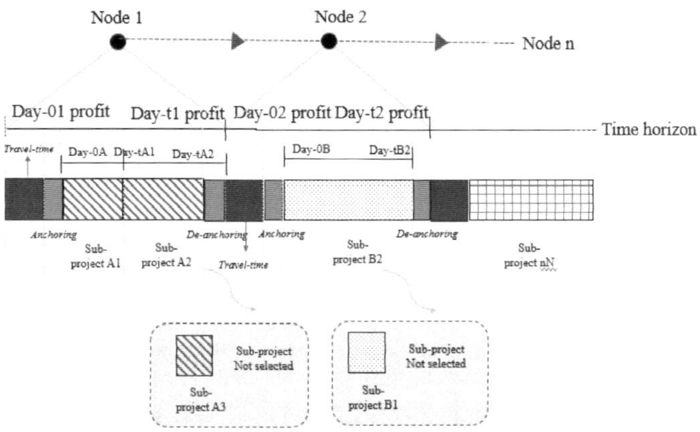

Figure 11. Routing schematic of barge movement partial duration

Barge only will move out after selected jobs related to Project A finish. But the drawback of this approach is more traveling time and non-producing time such as anchoring and de-anchoring. Different from Model 1, it can be seen in Figure 11 that sub-project A3 and sub-project B1 are excluded. Moreover, Figure 12 shows that more movement of barge is anticipated and the inclusion of particular sub-projects I and H previously less profitable, are now being included and push other less profitable sub-projects. Each project aims to find optimum profit solution for a given real case.

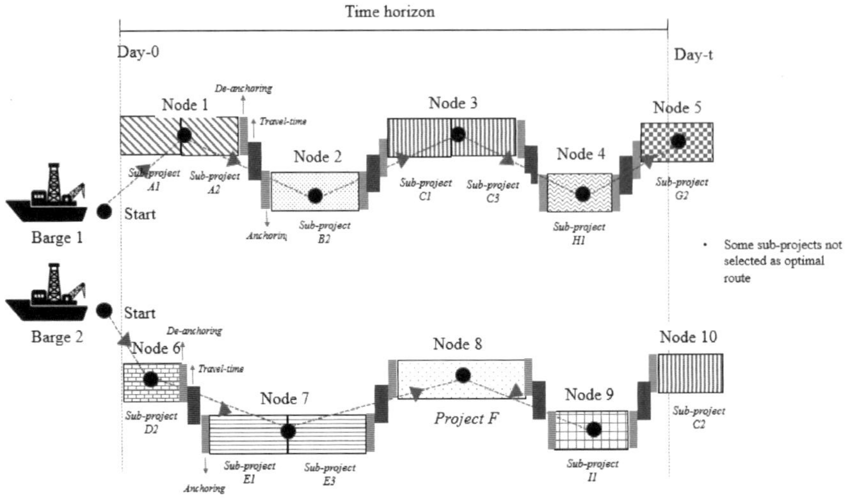

Figure 12. Schematic of WLICP barge planning partial duration

4. Conclusions

Finding the most optimized routes and schedules for barges to visit preferred platforms may accomplish profit maximization. Although optimization has been widely utilized in the petroleum industry, there is limited literature on its application in well-intervention planning. The findings indicate that the Well Intervention Campaign Problem (WLICP) demonstrates a robust project management framework. Its structure contains the breakdown of n-sub-project representing VRP nodes. Combinatorial optimization helps barges find routes and schedules to maximize profit. There are two models proposed for WLICP: 1) Model 1 (Full Duration, Project Basis) and 2) Model 2 (Partial Duration, Sub-project Basis). Model 2 hypothetically demonstrates the more profit delivered compared to Model 1. Future study suggests enrichment of the optimization problem such as a) clustering based on distance, b) commercial approach considering fluid price as the function of time, and c) risk-based scheduling to deliver the most optimum solution.

References

[1] Board, M., Board, O.S., and Council, N.R. (2003). Oil in the sea III: inputs, fates, and effects. National 786 Academies Press786 Academies Press

[2] Hughes, L., and Rudolph, J. (2011). Future world oil production: growth, plateau, or peak? Current 842 Opinion in Environmental Sustainability, 3, 225-234.

[3] Sandrea, I., and Sandrea, R. (2007). Growth expected in global offshore crude oil supply. Oil & Gas 890 Journal, 105, 34-34.

[4] Bakker, S., Vrålstad, T., & Tomasgard, A. (2019). An optimization model for the planning of offshore plug and abandonment campaigns. Journal of Petroleum Science and Engineering, 180 , 369–379. doi:10.1016/j. petrol.2019.05.042.

[5] Almuallim, M. Ullah, S. Z., Ruzhnikov, A. Silva, P. Al Shaikh, J., Agarwal, R. Al-Herz, M. A Breakthrough Solution to Set ESP Packer at Higher Inclination by a Creative Combination of Re-Designing Slick-Line Tool String, Completion String and Well Trajectory Saving up to 30% of Completion Time in a Middle East Depleted Oil Field. 2023. International Petroleum Technology Conference held in Bangkok, Thailand, 1 - 3 March 2023

[6] Dolgui, A., D. Ivanov, and B. Sokolov. 2018. "Ripple Effect in the Supply Chain: An Analysis and Recent Literature." International Journal of Production Research 56 (1–2): 414–430. doi: 10.1080/00207543.2017.1387680

[7] T. Ferg, H.-J. Lund, D. Mueller, M. Myhre, A. Larsen, P. Andersen, G. Lende, C. Hudson, C. Prestegaard, D. Field Novel approach to more effective plug and abandonment cementing techniques Society of Petroleum Engineers - Arctic and Extreme Environments Conference and Exhibition, 1 (2011), pp. 1-14

[8] Aarlott, M.M. 2016. Cost Analysis of Plug and Abandonment Operations on the Norwegian Continental Shelf. Master thesis, Norwegian Universityof Science and Technology (2016)

[9] Baldacci, R., Hadjiconstantinou, E., Mingozzi, A. 2004. An Exact Algorithm for The Capacitated Vehicle Route Problem. Operations Research 52 [5], S. 723–738.

[10] Board, M., Board, O.S., Council, N.R., 2003. Oil In the Sea III: Inputs, Fates, And Effects. National Academies Press.

[11] Christiansen, M. Fagerholt, K. Ronen, D. 2004. Ship Routing and Scheduling: Status and Perspectives. Transportation Science Vol. 38, No. 1, February 2004, pp. 1–18

[12] Dantzig, G. and Ramser, J. 1959. The Truck Dispatching Problem. Management Science, 6, 80-91.

[13] Haugen, S., & Vinnem, J. E. 2015. Perspectives on risk and the unforeseen. Reliability Engineering and System Safety, 137, 1–5.

[14] Hughes, L., Rudolph, J., 2011. Future World Oil Production: Growth, Plateau, Or Peak? Curr. Opin. Environ. Sustain. 3, 225–234.

[15] Ivanov, D., Sokolov, B., Käschel, J. 2010. A multi-structural framework for adaptive supply chain planning and operations with structure dynamics considerations. Eur J Oper Res 200(2): 409–20

[16] Laporte, G., 1992. The Vehicle Routing Problem: An Overview of Exact and Approximate Algorithms. European Journal of Operational Research, Volume 59(3), pp. 345–358

[17] Kaja, S. C. A New Approach for Solving the Disruption in Vehicle Routing Problem During the Delivery A Comparative Analysis of VRP Meta-Heuristics. Faculty of Computing, Blekinge Institute of Technology, 371 79 Karlskrona, Sweden.

[18] Liu, R., Xie, X., Augusto V. and Rodriguez, R., Heuristic Algorithms for a Vehicle Routing Problem with Simultaneous Delivery and Pickup and Time Windows in Home Health Care, European Journal of Operational Research, 230 (3), 2013, pp. 475-486.

[19] Mandal, P. Thakur, M. 2024. Higher-order moments in portfolio selection problems: A comprehensive literature review. Expert Systems with Applications Volume 238, Part A, 15 March 2024

[20] Purwanto, E. Y., Herawati, S., Rau, I.T., Anggoro, J., Suwito, E. 2017. Perforation Portfolio Management in Tunu Mature Giant Gas Field by Implementation of Intensive Well Review and Its Application in Production Forecast Improvement and Optimization of Well Intervention Planning. SPE-187023-MS. Proceeding the SPE/IATMI Asia pacific Oil & Gas Conference and Exhibition, Jakarta, Indonesia, October.

[21] Pollaris, H., Braekers, K., Caris, A., Janssens, G. K., & Limbourg, S. 2015. Vehicle Routing Problems with Loading Constraints: State-Of-The-Art and Future Directions. OR Spectrum, 37(2), 297–330.

[22] Toth, P., & Vigo, D. 2014. Vehicle Routing: Problems, Methods, and Applications. Society for Industrial and Applied Mathematics.

IOP Conf. Series: Earth and Environmental Science **1423** (2024) 012022 doi:10.1088/1755-1315/1423/1/012022

Optimization of disaster management vessel networks in integrated disaster management governance in Indonesia using genetic algorithm

Erwin Rasyid[1*], Ketut Buda Artana[2], Kriyo Sambodho[2]

[1]Interdisciplinary School of Management and Technology Institute of Technology of Sepuluh Nopember (ITS) Surabaya Indonesia
[2]Faculty of Marine Technology, Institute of Technology of Sepuluh Nopember (ITS) Surabaya Indonesia

*erwinrasjid1970@gmail.com

Abstract. Indonesia is highly susceptible to natural disasters due to its location at the convergence of three major tectonic plates: the Indo-Australian, Eurasian, and Pacific plates. These disasters significantly impact the nation's economic development. This study focuses on natural hazards such as earthquakes, tsunamis, volcanic eruptions, landslides, floods, and flash floods. Currently, maritime disaster management primarily involves ships from the Indonesian National Armed Forces (TNI). However, this often leads to mismatches between disaster-prone areas and the locations of available ships, resulting in delayed emergency response times and higher disaster management costs. In addition to TNI ships, the government has vessels from the National Search and Rescue Agency (BASARNAS) and the Indonesian Maritime Security Agency (BAKAMLA) that could be utilized for disaster management. There are also situations where certain response functions cannot be effectively performed by ships, leading to inadequate assistance for disaster victims and potentially increasing fatalities. This research aims to optimize the use of existing ships for disaster management by considering operational costs and coverage areas through the application of the Genetic Algorithm (GA) method. GA is used to identify the optimal combination of ships that minimize costs and maximize coverage areas. The findings of this study provide several ship combinations suitable for responding to natural disasters in Indonesia, which can serve as valuable input for the government in enhancing disaster management efforts.

1. Introduction

Geographically, the Indonesian archipelago is located between the continents of Asia and Australia, and between the Indian Ocean and the Pacific Ocean. Certain regions of Indonesia are especially prone to natural disasters due to its position at the convergence of two young mountain ranges: the Pacific Circumference, which consists of folded mountains surrounding the Pacific Ocean, and the Mediterranean Circumference, which stretches from the Atlas Mountains in North Africa to Nicobar and into Indonesia[13].

Additionally, Indonesia lies at the junction of three major tectonic plates: the Indo-Australian, Eurasian, and Pacific plates. This positioning places Indonesia within the world's ring of fire, a region that frequently experiences earthquakes and volcanic eruptions around the Pacific Ocean basin. According to the Directorate of Volcanology and Geological Hazard Mitigation (DVMBG) of the Department of Energy and Mineral Resources, there are 28 areas in Indonesia identified as prone to earthquakes and tsunamis. These areas include NAD, North Sumatra, West Sumatra, Bengkulu, Lampung, Banten, Central Java and southern Yogyakarta, southern East Java, Bali, NTB and NTT, North Sulawesi, Central Sulawesi, South Sulawesi, North Maluku, South Maluku, Biak, Yapen, Fak-Fak in Papua, and Balikpapan in East Kalimantan. This study focuses on natural hazards, specifically five types: earthquakes, tsunamis, volcanic eruptions, landslides, floods, and flash floods. These types of disasters are highlighted due to their rapid onset and significant impact on both the environment and human lives, necessitating immediate response.

In determining the type of transportation for delivering aid, factors such as speed, capacity, cost, and availability of access points are considered to ensure that logistical assistance reaches disaster victims quickly and efficiently. The speed of aid delivery can be influenced by various factors including the distance between the disaster site and the source of aid, the availability of adequate transportation infrastructure, weather conditions, and security. In disaster situations, effective cooperation and coordination among parties such as the government, disaster relief organizations, and local communities are essential to ensure that assistance is promptly received by refugees and victims. However, these factors can change in urgent situations, potentially hampering aid delivery. To expedite aid delivery in such cases, several actions can be taken, including prioritizing safe and adequate transportation routes, strengthening cooperation with relevant organizations and parties, and ensuring that available aid resources are always ready for deployment to the disaster site. Various types of transportation can be used to send logistical assistance to disaster victims, depending on the disaster location and the availability of access routes. Common modes of transportation for delivering aid include ships, aircraft, cars, helicopters, unmanned aircraft, and trains.

This study focuses specifically on sea transportation, namely ships. Ships are effective for sending logistical assistance to disaster sites located in coastal areas or islands. These ships can come from various sources, including the government, disaster organizations, non-governmental organizations, and private companies. In some cases, ships belonging to the National Army or Navy can be used to deliver logistics and aid to disaster sites, especially those difficult to reach by land. International organizations such as the World Food Programme (WFP) also have their own fleet of logistics vessels used to deliver aid and logistics to disaster sites worldwide. These organizations work with local governments and agencies to ensure that aid and logistics reach refugees and disaster victims. Ships used for delivering relief logistics to disaster sites must meet safety requirements and have sufficient cargo capacity to ensure aid and logistics are received safely and promptly. The government is responsible for ensuring that logistics and aid reach disaster victims and refugees, and government vessels are one way to fulfill this responsibility.

Regarding aid delivery by ship, the selection of the vessel is based on several considerations: load capacity, speed, operational feasibility, availability, accessibility, security, and infrastructure. These considerations ensure that the vessels used to deliver logistics to disaster victims meet the requirement that aid reaches victims quickly, effectively, and safely. In this study, the disaster area is divided into six sectors, each covering a specific area. Each sector has multiple ships designated to cover it in the event of a disaster. Additionally, each sector has a logistics warehouse ready to send supplies to affected areas.

The purpose of this research is to identify the optimal combination of ships for delivering aid to disaster victims, ensuring that assistance is provided swiftly and at minimal cost. Several similar studies have been conducted, such as Yen-Hung Lin's research in 2011 [16]. This research utilizes a tour-based, multi-objective logistics model for optimal scheduling of essential item deliveries in disaster relief operations. The model accounts for multi-item, multi-vehicle, multi-period, and separate delivery scenarios in a disaster context. The study addresses the problem with three main objective functions:

1. Minimizing Unmet Demand: The first objective function is a penalty function designed to minimize the unmet demand after the relief operation within a given period. This function consists of two parts: the total accrued penalty cost, which aggregates the penalty costs for various levels of severe delays during the planning period, and the penalty cost incurred at the end of the planning period. The goal is to minimize the total unmet demand, especially for high-priority items..
2. Minimizing Total Travel Time: The second objective function aims to minimize the total travel time of all vehicles involved in the operation. The goal is to maximize vehicle utilization during working hours.
3. Balancing Satisfaction Levels: The third objective function seeks to minimize the disparity in satisfaction levels between clusters. Satisfaction level is defined as the ratio of requested demand to the actual amount delivered. The aim is to ensure balanced service levels across different clusters.

In another study, Akwafuo conducted research on disaster logistics distribution using genetic algorithms. Akwafuo's study formulated the problem of vehicle and logistics routing mathematically and used optimization algorithms to determine the location of depots. The study further explores algorithms for route construction and improvement, including evolution, artificial intelligence-based algorithms, and artificial immune systems. The algorithms are categorized into three groups:

1. Facility Location Allocation Problem
2. Routing Problem
3. Integrated Approach

Yong Ye proposed a sequential approach to humanitarian logistics in natural disasters based on Bayesian group information updates (GIU). The model with GIU was revised into a single objective model, and a matrix coding-based genetic algorithm was developed to solve the revised model. These studies illustrate the diverse approaches and methodologies applied in the field of disaster logistics, each contributing valuable insights and tools for improving the efficiency and effectiveness of relief operations.

2. Methodology

The main objective of this research is to design the distribution of ships in numbers and types that can be used in disaster management and find operational patterns that will guarantee short emergency response times, minimum operational costs, and low investment in providing ships using the Genetic Algorithm method. The way the genetic algorithm works is to evaluate solutions [6] [11]. The process will evaluate each population by calculating the fitness value for each chromosome and evaluating it until a stopping criterion is met. If the stopping criteria are not met, a new generation will be formed by repeating step 2, as shown in Figure 1. Some commonly used stopping criteria include:

- Stops after several consecutive generations and the maximum fitness value does not change.
- If no higher fitness value is obtained in the next n generations, stop.

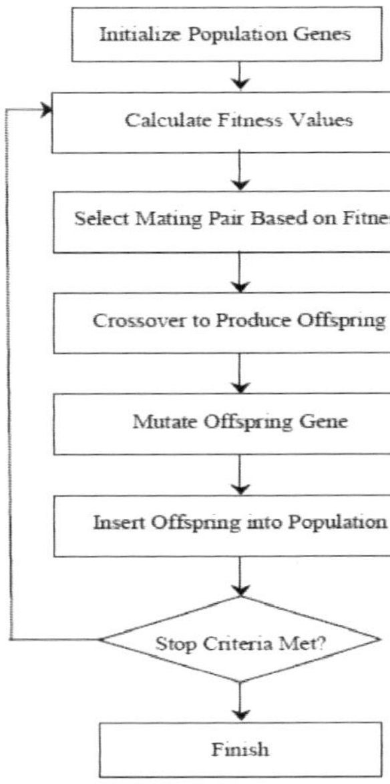

Figure 1. Research Steps in Ship Selection

Based on the calculation of fitness values, a combination of 20 selected solutions is selected to initialize the initial population:

$$Fitness = \sum_{i=1} \frac{CA_i \times N.K_i}{Ls_i} \leq 5\%Ls_n \quad (1)$$

The following stages must be carried out to design and apply the Genetic Algorithm method [1] [5]:

1. Defining the Individual: An individual represents one possible solution to the problem.
2. Defining Fitness Value: This measures the quality of an individual or how good the obtained solution is.
3. Determining the Initial Population Generation Process: This is typically done using random generation methods such as a random walk.
4. Determining the Selection Process: Identify the selection process to be used.
5. Determining the Crossover and Mutation Processes: Define the crossover process and gene mutations to be used.

a. Understanding the Individual

Each individual is a potential solution. A group of individuals forms two types of solutions [7]:

1. Chromosome: Represents the basic genetic information (genotype).
2. Phenotype: Represents the model of a chromosome.

A chromosome is composed of genes, where each gene is a candidate solution. Each solution corresponds to a gene on a chromosome. One solution is represented by one chromosome, distinct from others, ensuring that a chromosome contains information about the solution it represents [15].

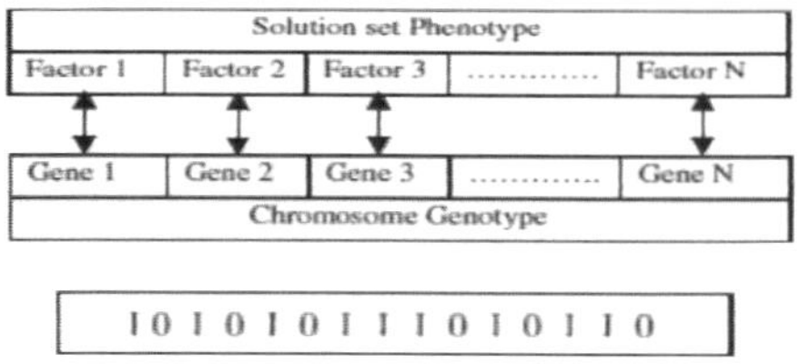

Figure 2. Chromosome and Phenotype Representation [13]

b. Understanding Genes

Genes are the fundamental units of instructions in a genetic algorithm. A chromosome is composed of a series of genes. While genes can describe solutions to problems, they are not the solutions themselves. A gene consists of a bit string represented in binary, with values ranging from a lower limit to an upper limit. This range can be divided into intervals, which are indicated by the bit string in the gene. A bit string of length n can represent the values 0 to $(2^n - 1)$ [8]. The binary gene representation can seeing as in Figure 3.

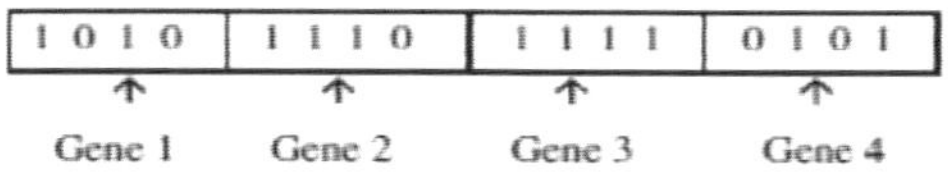

Figure 3. Binary Gene Representation [13]

c. Definition of Fitness

The fitness function in multi-criteria optimization can be challenging to determine, particularly when a solution is better for one criterion but worse for another. The main issue is defining what constitutes an optimal solution rather than implementing a Genetic Algorithm (GA) to find it. If a fitness function derived from combining various criteria can produce good results, then these criteria can be consistently combined.

d. Understanding Population

A population in a genetic algorithm is a collection of individuals, each representing a potential solution to the problem. These individuals are evaluated based on their phenotype parameters and other relevant information available during the search process. This evaluation determines the fitness of each individual, which then guides the selection, crossover, and mutation processes, evolving the population towards optimal solutions. Two critical aspects of the population in genetic algorithms are the initial generation population and the population size. [10].

The population size in a genetic algorithm depends on the complexity of the problem. During initialization, the population size can be either random or predetermined [12]. Ideally, the initial population should have a diverse gene pool to explore a wide range of candidate solutions [14]. To achieve this, the initial population is often chosen at random, although heuristic approaches can sometimes be used to optimize the initial conditions. This can result in a high fitness value for the population, helping the genetic algorithm to find solutions more quickly [1].

However, maintaining a large gene pool can be challenging. If the population lacks sufficient diversity, the algorithm will explore only a small portion of the solution space and may fail to find the optimal solution. The size of the population also raises other considerations. A larger population makes it easier to explore the search environment thoroughly. Computation can be considered complete when individuals have minimal differences and further improvements are unlikely through mutation [9].

In summary, while large populations are beneficial for exploring the search space and ensuring diverse solutions, they also require more time, computational resources, and memory. Balancing population size is crucial for the efficiency and effectiveness of the genetic algorithm.

3. Results and Discussion

The steps that must taking to complete the optimization of ship optimization that will using in disaster management in Indonesia with genetic algorithms.

a. Defining Individuals.

During the initial initialization, the best 20 individuals will selecting to carry out the next several stages of GA. The initialization process refers to the problem of how to choose the best combination of best disaster management vessels in each sector by referring to the data below.

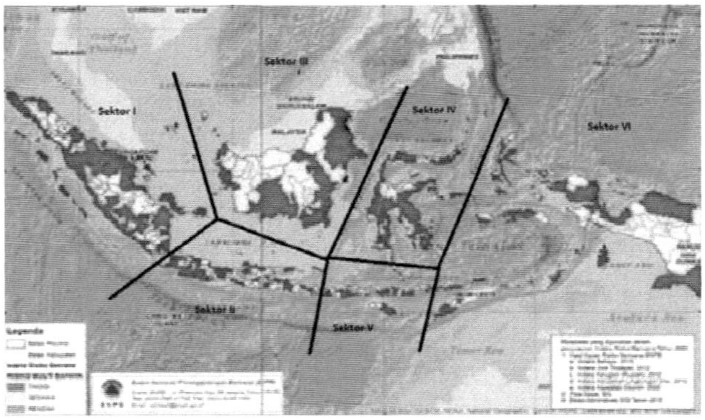

Figure 4. Proposed Region/Sector Division

Based on Figure 4 above, it is possible to determine how much of the total area must be covered by ships delivering logistical supplies to disaster victims. Therefore, the formula for sector area is as follows:

$$\text{Total Area} = \text{Sector I Area} + \text{Sector II Area} + \text{Sector III Area} + \text{Sector IV Area} + \text{Sector V Area} + \text{Sector VI Area} \quad (2)$$

Initialize the initial population by selecting 20 combinations of solutions that selecting according to the calculation of their fitness values. Therefore, the individual initialization results (solutions) given by the genetic algorithm are as follows:

K1	141310151310140212120804120407110511070801121109151406120305050809130404081009020612030907041300

K1 is the selected individual who meets the fitness value

NO	Populasi
K1	141310151310140212120804120407110511070801121109151406120305050809130404081009020612030907041300
K2	011502131012061010150000140611080314140513150306130213090108080111011113051301140403100209070508
K3	140309060815140708020709120903050504071315021115150405060311151309030310030509080206141407091305
K4	120613031002060409051106101110140307101013050312120709140113080606060702050712050409060704120414
K5	130805110705091207070303111403110010020711070604100901111400150308081515031004130111130406031211
K6	111112080507020905101111090110041413101513101401121208041203071106110608011211101514061204060408
K7	091304050710090207120208070413010015021311130610101501011406110804141405131504071301130902080800
K8	110112130513011405041001090705091402090608151407080208091208030506050613150211151504050604111513
K9	090304100204090803061414070913061105130310030503090511070911101403081010130503121206091402140706
K10	060607030407120504090506041205141308051108050912070704041114031101100107110707051009011114011502
K11	080815150209041302121303060312111110120806080109051011120901100414130915131014021212080412040711
K12	061107080012111015140512040605080913040508100802071203090603130101150112111306101014010114061008
K13	041414051315040713021309020808011000121305130014050410020806050914030905081514070701080912090205
K14	060507131501111515040506041115130803041003050807030614140609130612061202100306040904110710111013
K15	030810101304031212070814021408060605070305071204040906070411051413080410080509120707040411140210
K16	011002071107070510090011140115030808151503100312021213040602121111111207060802090409111209010903
K17	141310151310140212120804120407110511070801121109151406120305050809130404081009020612030907041300
K18	011502131012061010150000140611080314140513150306130213090108080111011113051301140403100209070508
K19	140309060815140708020709120903050504071315021115150405060311151309030310030509080206141407091305
K20	120613031002060409051106101110140307101013050312120709140113080606060702050712050409060704120414

Figure 5. Results of initialization of 20 initial individuals (proposed solution from GA)

b. Defining Fitness Values

The fitness value is the objective function of optimization in a genetic algorithm. For the problem of optimizing the combination of disaster management ships with types, the type of ship that has been determined with the constraint value is the area of each sector.

$$Fitness = \sum_{i=1}^{8} \frac{\mathrm{CA_i} \times \mathrm{N.K_i}}{\mathrm{Ls_i}} \leq 5\%\mathrm{Ls_n} \tag{3}$$

c. Determining the Selection Process

Selection using to select which individuals will selecting for the crossbreeding and mutation process. Selection using to get good parent candidates. "Good parents are likely to produce good offspring." Individuals with higher fitness values have a greater chance of being selected. The first step taken in this selection is searching for fitness values. This fitness value will be used in the subsequent selection phase. Each individual in the selection container will receive a reproduction probability that depends on its own objective value against the objective values of all individuals in the selection container. There are several selection methods, but in this simulation I chose to use the Roulett Wheel model. Based on the results of the Roulette Wheel selection algorithm above, the following model obtaining:

K17	141310151310140212120804120407110511070801121109151406120305050809130404081009020612030907041300
K3	140309060815140708020709120903050504071315021115150405060311151309030310030509080206141407091305
K14	060507131501111515040506041115130803041003050807030614140609130612061202100306040904110710111013
K9	090304100204090803061414070913061105130310030503090511070911101403081010130503121206091402140706
K2	011502131012061010150000140611080314140513150306130213090108080111011113051301140403100209070508
K15	030810101304031212070814021408060605070305071204040906070411051413080410080509120707040411140210
K6	111112080507020905101111090110041413101513101401121208041203071106110608011211101514061204060408
K14	060507131501111515040506041115130803041003050807030614140609130612061202100306040904110710111013
K20	120613031002060409051106101110140307101013050312120709140113080606060702050712050409060704120414
K20	120613031002060409051106101110140307101013050312120709140113080606060702050712050409060704120414
K10	060607030407120504090506041205141308051108050912070704041114031101100107110707051009011114011502
K10	060607030407120504090506041205141308051108050912070704041114031101100107110707051009011114011502
K11	080815150209041302121303060312111110120806080109051011120901100414130915131014021212080412040711
K18	011502131012061010150000140611080314140513150306130213090108080111011113051301140403100209070508
K7	091304050710090207120208070413010015021311130610101501011406110804141405131504071301130902080800
K9	090304100204090803061414070913061105130310030503090511070911101403081010130503121206091402140706
K4	120613031002060409051106101110140307101013050312120709140113080606060702050712050409060704120414
K8	110112130513011405041001090705091402090608151407080208091208030506050613150211151504050604111513
K7	091304050710090207120208070413010015021311130610101501011406110804141405131504071301130902080800
K1	141310151310140212120804120407110511070801121109151406120305050809130404081009020612030907041300

Figure 6. Selection results using a roulette wheel machine of *20 initial individual*

IOP Conf. Series: Earth and Environmental Science **1423** (2024) 012022 doi:10.1088/1755-1315/1423/1/012022

d. Determining the Crossover Process

The principle of crossover involves performing operations (such as exchange or arithmetic) on corresponding genes from two parents to produce new individuals. The crossover process is applied to each individual with a specified crossover probability. After completing the selection process, the next step is the crossover process, where two parents are involved in forming new chromosomes. The following are the hybridization results from the multi-point hybridization model.

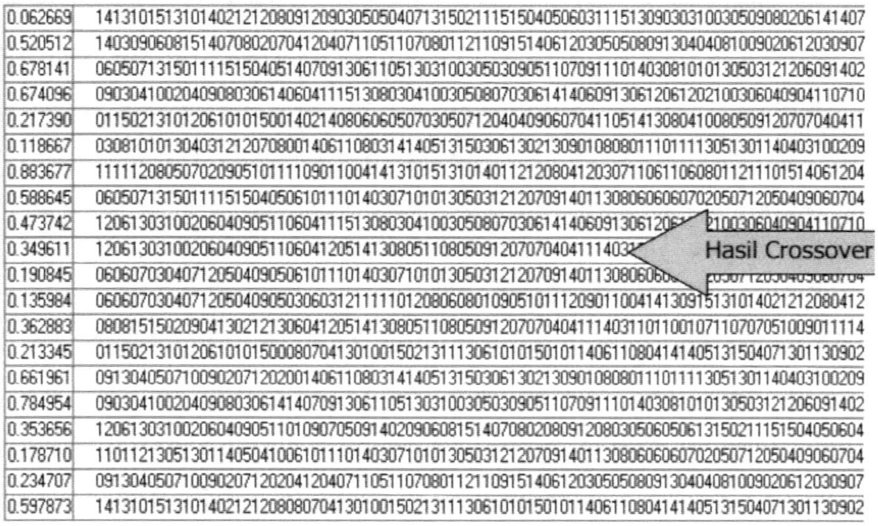

0.062669	14131015131014021212080912090305050407131502111515040506031115130903031003050908020614 1407
0.520512	14030906081514070802070412040711051107080112110915140612030505080913040408100902061203 0907
0.678141	06050713150111151504051407091306110513031003050309051107091110140308101013050312120609 1402
0.674096	09030410020409080306140604111513080304100305080703061414060913061206120210030604090411 0710
0.217390	01150213101206101015001402140806060507030507120404090607041105141308041008050912070704 0411
0.118667	03081010130403121207080014061108031414051315030613021309010808011101111305130114040310 0209
0.883677	11111208050702090510111109011004141310151310140112120804120307110611060801121110151406 1204
0.588645	06050713150111151504050610111014030710101305031212070914011308060606070205071205040906 0704
0.473742	1206130310020604090511060411151308030410030508070306141406091306120[illegible]210030604090411 0710
0.349611	1206130310020604090511060412051413080511080509120707040411140[illegible]
0.190845	06060703040712050409050610111014030710101305031212070914011308060[illegible]
0.135984	06060703040712050409050306031211111012080608010905101112090110041413091513101402121208 0412
0.362883	08081515020904130212130604120514130805110805091207070404111403110110010711070705100901 1114
0.213345	01150213101206101015000807041301001502131113061010150101140611080414140513150407130113 0902
0.661961	09130405071009020712020014061108031414051315030613021309010808011101111305130114040310 0209
0.784954	09030410020409080306141407091306110513031003050309051107091110140308101013050312120609 1402
0.353656	12061303100206040905110109070509140209060815140708020809120803050605061315021115150405 0604
0.178710	11011213051301140504100610111014030710101305031212070914011308060606070205071205040906 0704
0.234707	09130405071009020712020412040711051107080112110915140612030505080913040408100902061203 0907
0.597873	14131015131014021212080807041301001502131113061010150101140611080414140513150407130113 0902

Figure 7. The result of crossing 2 parents to form chromosomes new

e. Carrying out the Mutation process

After the cross-breeding process carrying out, the next step is the mutation process by applying the random-relocation method, by selecting one non-zero gene and zeroing it, then the value placing on the gene that has a value of zero. This mutation does not occur often, but it must happen because the probability of mutation greatly affects the determination of the new individual.

K17	1413@01513101402121208#91#09030505040713150#@115150405060#1115130#0303100305090802#6141407#91305
K3	1403090608151407080207041204071105110708011211091514#61203#505#8091304040810090206@2#3090#041300
K14	#60507131#0111151504051407091#06110#1303100##503#9051107091110140308101#1305#3121206091402@#0706
K9	09##0410020409080306140604111#1308030410030#0807#306141#0609130#1206@202100306040#041@0710111013
K2	01150213101206101015001402140#060605#70305071204040#0607041105#4130#0410080509120707040411140210
K15	#30810101304#312120##8#0140#110803141#05@3150306@302130901080#011@0111@30#13011404031002#90705##
K6	111112#8050702#9051#1@11090110041413101#131014011212080#1203071106110608011211101514061204060408
K14	06050713150111151504050610111014030710101#05031212070914011308#60#0607020507120504#90607#41#04@#
K20	@20613031002060#09051106041115130803041003050807030614@4#609@#061206@202100[illegible]409041107@011101[illegible]
K20	1206@303100206040905@106041205141308051108050#@20707040411140#1101@0010[illegible]
K10	60607##0407@2050409#50610@11014#307101013050312@2070#@4011#080606060702[illegible]
K10	0606070304#712#5040905030#0#12@11110120#060#010905101112090110041413091513@014[illegible]12120804120407@1
K11	#808151502090413021##3060#120514130805@@0#05091#0707040411140#@101#00107110707051#09011114011502
K18	1150#131012#610@01#000807##130@00150213111306@01015#10114##110804@414051315040###011#09#2080#0
K7	091#040507100902071202001#0611#8031414051315#30613021#09010808011@0@111#051301140403100#090705#8
K9	09030410020#0908030614140#09130611051###1003050309051@070911@0140#0810@0130503121206091402140706
K4	1206130310#2060409051101#907050#1#020906081514#7080#0809120803050605061315#21115150405#604111513
K8	@10@12130513011405041006101110140307101013050#121207091401@30806060607020507120504090#070412041#
K7	0913040507@009020#120204120407110511070#011#1109151406120305050809130404081009020612030907041300
K1	14131015@31014021212080807#41301001502131113#61010150101140#1108#41#14051315040713011309020#0800

Figure 8. Gene mutation results using a random relocation model

f. Next is the replacement process, part of the crossover process to carry out the process of exchanging the corresponding genes from the two parents to produce a new individual.

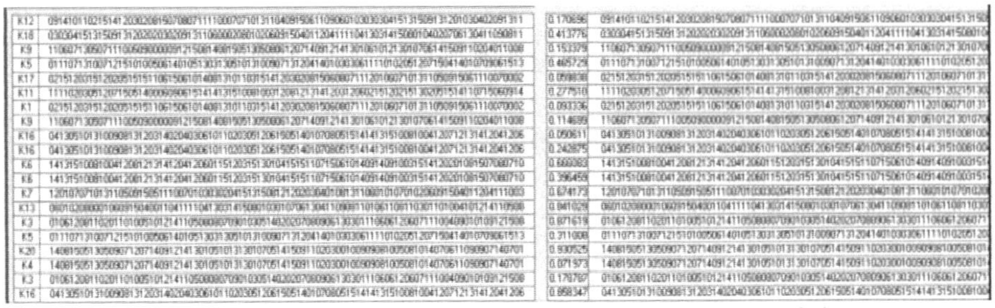

Figure 9. Replacement results after the selection process

g. The next process is the new population, which is a new population generated after several stages of selection, crossover, mutation and replacement processes, so the next stage of GA is to determine the new population.

NO	Populasi
K1	110612081102011010051012141105080807090103051402020708090613030111060612060711100409010109121508
K2	010607130507111005090000091215081408150513050806120714091214130106101213010706141509110204011008
K3	030107071013111050915051110070103030204151315081212020304010813110601010701020609150401120411003
K4	120304151315091312020203020913110600020801020609150401120411110413031415080104020706130411090811
K5	110107131007121510100506140105130313051013100907131204140103030611110102051207150414010709061513
K6	011607130507111005090000091215081408150513050806120714091214130106101213010706141509110204011008
K7	140102080001060915040011041111041303141508010301070613041109081101061108110301101004101214110508
K8	060815051305090712071409121413010510131301070514150911020300100909081005081014070611090907140701
K9	111305101310090813120314020403061011020305120615051401070805151414131510081004120712131412041206
K10	041102030512071505140006090615141413151008100312081213141203120602151202151302051514110715060814
K11	140815051305090712071409121413010510131301070514150911020300100909081005081014070611090907140701
K12	140107131007121510100506140105130313051013100907131204140103030611110102051207150414010709061513
K13	011815051305090712071409121413010510131301070514150911020300100909081005081014070611090907140701
K14	041305101310090813120314020403061011020305120615051401070805151414131510081004120712131412041206
K15	041607130507111005090000091215081408150513050806120714091214130106101213010706141509110204011008
K16	110305101310090813120314020403061011020305120615051401070805151414131510081004120712131412041206
K17	041305101310090813120314020403061011020305120615051401070805151414131510081004120712131412041206
K18	141304151315091312020203020913110600020801020609150401120411110413031415080104020706130411090811
K19	030304151315091312020203020913110600020801020609150401120411110413031415080104020706130411090811
K20	030315100810041208121314120412060115120315130104151511071506101409140910031514120200108150708071 0

Figure 10. New population results after all GA stages have been carried out

This process will evaluate each population by calculating the fitness value of each chromosome and evaluating it until the stopping criteria meeting. For this GA simulation, the researchers determined the stopping criteria based on the number of iterations that the GA machine would perform. If the number of iterations has reached a predetermined value, the system will stop. After completing the specified number of iterations, the GA engine stops and provides recommendations for the 5 (five) best solutions based on the fitness values.

Table 1. Five best solutions recommended by GA

Best solution	Sector: 1					Sector: 2					Sector: 3				
GA's proposal	HOS	TKR	PAT	PCG	LSH	HOS	TKR	PAT	PCG	LSH	HOS	TKR	PAT	PCG	LSH
1	1	1	1	4	1	1	1	1	2	1	1	1	1	2	1
2	1	1	1	1	1	2	1	1	2	1	1	1	2	1	1
3	1	1	1	1	1	1	1	2	2	1	1	1	2	1	1
4	1	1	1	1	1	1	1	1	2	1	1	1	2	1	1
5	1	1	1	1	1	1	1	1	1	1	1	1	2	2	1

Sector: 4					Sector: 5					Sector: 6				
HOS	TKR	PAT	PCG	LSH	HOS	TKR	PAT	PCG	LSH	HOS	TKR	PAT	PCG	LSH
1	2	2	2	2	1	1	1	1	2	1	1	2	3	2
1	1	1	1	1	1	1	1	1	2	1	1	1	5	1
1	2	3	1	2	1	1	1	1	2	1	1	1	4	2
1	1	2	1	1	1	2	1	1	2	1	1	1	3	1
1	2	1	1	1	1	1	1	3	2	1	1	1	3	1

Table 2. The function and suitability of the ship in disaster management are as follows:

Ship Function	Ship Type				
	Hospital	Tanker	Patrol	Passenger and Cargo ships	LCT / LCU
Victim Evacuation	√	√	√	√	√
Distribution Logistics	√	√		√	√
Shore Landing and Relief Distribution					√
Heavy Equipment Distribution				√	√
Hospital	√		√		
Personnel transportation	√				
Transportation of reconstruction materials		√		√	√
Temporary Community Shelter				√	
Clean Water Supply		√			
Fuel Supply		√			
Monitoring and Early Detection			√		

4. Conclusion

This research indicates that the Genetic Algorithm effectively identifies the best combination of ships for disaster management in Indonesia, taking into account operational costs and coverage areas. The chosen vessel combinations are tailored to meet the specific needs of different sectors, leading to a more efficient disaster response. The study suggests enhancing these results further by incorporating qualitative factors such as ship maneuverability and water conditions.

References

[1] Arabali, Amirsaman, Mahmoud Ghofrani, Mehdi Etezadi-Amoli, M. Sami Fadali, and Ya-hia Baghzouz. "Genetic-algorithm-based optimization approach for energy management." IEEE Transactions on Power Delivery 28, no. 1 (2012): 162-170.

[2] Badan Naional Penanggulangan Bencana, 2011, Jurnal Penanggulangan Bencana, Volume 2 no 1 Juni tahun 2011, ISSN 2087636X, Jakarta.

[3] Badan Naional Penanggulangan Bencana 2021. Peraturan Badan Nasional Penanggulangan Bencana -no-5-Tahun-2020.

[4] Badan Naional Penanggulangan Bencana, 2020. Indeks Risiko Bencana Indonesia.

[5] Goldberg DE, Holland JH, Genetic algorithms and machine learning, 1988.

[6] Holland, J. H., Genetic algorithms. Scholarpedia, (2012), 2 10, Volume 7, p. 1482.

IOP Conf. Series: Earth and Environmental Science **1423** (2024) 012022 doi:10.1088/1755-1315/1423/1/012022

[7] Kramer, Olive, Genetic algorithm essentials, Vol. 679. Springer, 2017.
[8] Mathew K, Tabassum M, A genetic algorithm analysis towards optimization solutions, International Journal of Digital Information and Wireless Communications (IJDIWC), 2014 Jan 1, 4(1), 124-42.
[9] Mirjalili, Seyedali, Genetic algorithm, In Evolutionary algorithms and neural networks, pp. 43-55. Springer, Cham, 2019.
[10] Mirjalili S, Dong JS, Sadiq AS, Faris H. Genetic algorithm: Theory, literature review, and application in image reconstruction. In Nature-Inspired Optimizers 2020 (pp. 69-85). Springer, Cham.
[11] S. R, R. T, A review of selection methods in genetic algorithm, Int. j. of eng. Sc. and tech., 2011 May, 3(5), 3792-7.
[12] Sampson E. Akwafuo, Armin R. Mikler, Fariba A. Irany, Optimization Models for Emergency Response and Post Disaster Delivery Logistics: A Review of Current Approaches, International Journal of Engineering Technologies and Management Research August 2020, Vol 7(08), 35 – 49 2020
[13] Sivanandam, S. N., and S. N. Deepa. "Genetic algorithm optimization problems." In Introduction to genetic algorithms, pp. 165-209. Springer, Berlin, Heidelberg, 2008.
[14] Surono, 2010. Perspektif Analisis Resiko Bencana dari Aspek Geologis dan Pengembangannya; Workshop Pengembangan Analisis Resiko Bencana untuk Pembangunan Nasional; Hotel Millenium 25 Oktober 2010, Badan Naional Penanggulangan Bencana Jakarta.
[15] Yang, Jinhui, Chunguo Wu, Heow Pueh Lee, and Yanchun Liang. "Solving traveling salesman problems using generalized chromosome genetic algorithm." Progress in Natural Science 18, no. 7 (2008): 887- 892.
[16] Yen-Hung Lin, Rajan Batta, Peter Rogerson, Alan Blatt, Marie Flanigan, A logistics model for emergency supply of critical items in the aftermath of a disaster, Socio-Economic Planning Sciences Volume 45, Issue 42011, 132-145.

IOP Conf. Series: Earth and Environmental Science **1423** (2024) 012023 doi:10.1088/1755-1315/1423/1/012023

Utilization of sandblasting waste as additives to enhance mechanical performance of concrete mortar

Moh Najib Rizal [1]*, Wahyuniarsih Sutrisno[2] and Kiki Dwi Wulandari[2,3]

[1] Enviromental Engineering, Institut Teknologi Sepuluh Nopember, Surabaya, Indonesia
[2] Civil Engineering, Institut Teknologi Sepuluh Nopember, Surabaya, Indonesia
[3] Politeknik Perkapalan Negeri Surabaya, Surabaya, Indonesia

*E-mail: muhammadnajibrizal01@gmail.com

Abstract. This research explains about the utilization of sandblasting waste, a byproduct of the shipping industry's blasting process, as an additive in concrete mortar. The blasting process involves spraying silica sand to remove impurities from hull layers, resulting in waste that contains heavy metals like Fe2+, Mn4+, and Cr3+. Given its hazardous nature, alternative disposal methods are necessary to reduce landfill accumulation. By pre-treating this silica sand through cleaning and grinding, a new material, Ground Sandblasting Waste (GSW), with a particle size of approximately 390.4 nm and 94.75% SiO2 content, was produced. Concrete mortar specimens with dimensions 5 x 5 x 5 cm were prepared with GSW added in 1%-5% by weight of cement. Furthermore, this research also explores the combination of GSW with another additive named silica fume with a percentage of 6%. The compressive strength tests showed that adding 2% GSW and GSW-SF combinations, with proportions of 2% and 6% from the cement weight, greatly enhanced performance of mechanical the concrete mortar. This research highlights the potential for using pre-treated sandblasting waste to improve concrete performance and manage hazardous waste effectively in the shipping industry.

1. Introduction

In the last decade, dry abrasive blasting has become more efficient in removing rust, dirt, and old paints from ship surfaces because it is easy to use and low cost [1]. It aims to maintain and repair the ship to be more durable. High velocity abrasive streams are pushed using some principle blasting like air pressure, water pressure, and centrifugal force [2]. Silica sand is one of the media abrasive blasting is commonly used [3]. 200 – 2000 tons of abrasive media are required for 10,000 m2 of ship surface area, while 0.3 – 1kg/m2 abrasive blasting media (ABM) waste is formed [2]. The solid waste generated is categorized as hazardous waste according to the Government Regulation of the Republic of Indonesia due to its high heavy metal content, raising environmental and health concerns [4].

Based on the annual report of the Ministry of Transportation of the Republic of Indonesia in 2018, the fleet of ships from 2005 to 2016 was 6041 to 24646 units [5]. As the shipping industry in Indonesia experiences rapid growth, the demand for abrasive blasting media (ABM) utilities is expected to increase correspondingly, leading to a proportional rise in ABM waste generation. To address the issue of ABM waste, which is currently mitigated primarily through disposal, it is crucial to implement strategies that significantly reduce its accumulation in the environment. Maximizing ABM waste reduction aims to protect the environment and conserve natural resources [6]. A common method to limit the environmental dispersion of ABM waste involves applying a resin coating to control dust [7]. An alternative approach with the potential for reducing both the generation and the environmental content of heavy metal ABM waste is its repurposing as a construction material.

Silica sand, which is used as ABM, is usually used in many sectors, such as the construction sector. Research conducted by Malathy et al. mentioned that silica sand can be used as a substitute for fine aggregate to enhance the concrete properties. Adding silica powder to replace cement with 30% - 40% in concrete will enhance the bond of the chemical properties, i.e., CSH to C2SH, which leads to an enhancement in the concrete compressive strength [8]. The physical properties of silica sand support its use as a construction material, such as a fine aggregate and cement substitute. However, with the times, the use of silica sand is increasingly widespread and heading towards advanced materials, namely making it an additive material with a smaller size than cement to nano-sized materials [9]. One of the most familiar examples is the use of silica fume as an additive material that significantly improves the mechanical and concrete durability [10].

Silica is used more frequently in concrete production to enhance material properties and optimize performance [1]. A high concrete density is achieved by filling the empty spaces between relatively large particles, such as cement particles, with nanomaterials fine dust particles that can be reactive or nonreactive [11]. In addition to the increased density, the porosity will decrease because additive silica sand will fill the voids formed [12]. Additionally, silica additives are recognized for enhancing concrete's resistance to chemical attack [13]. Adding silica sand will also increase the workability of fresh concrete [2]. Silica sand properties and characteristics are not only mechanical but also increase the durability of concrete, so the concrete will be more adaptive when applied to extreme environments.

The most popular additive in concrete is silica fume. The concentration of SiO2 in silica fume and ABM waste is similar to about >90% [14] and 93.02% [1]. However, the fundamental difference is that silica fume is a by-product while ABM waste is hazardous waste, so ABM waste has the potential to be utilized. On the other hand, the effectiveness of combining nano - silica with silica fume surpasses that of using nano - silica alone, whereas combining commercial nano silica did not show any notable impact [11]. Silica fume has improvement of concrete quality can be attributed to the continuous CSH formation bonds by pozzolanic reaction of reactive silica in silica fume [15].

The research primary objective is to reduce of amount of hazardous ABM waste generated and explore its potential for reuse. A worthy opportunity for construction companies is presented by utilizing ABM waste as a substitute for construction materials. This approach can enhance the mechanical and durability of concrete and other building products while reducing natural resource extraction. Previous studies have investigated using ABM waste as a substitute for fine aggregate, leveraging its similarity to silica sand in raw material composition and physical appearance post-blasting. Given its high silica content, ABM waste can potentially serve as a pozzolanic material, either as a cement substitute or a concrete additive. This research will focus on applying ABM waste as a concrete additive. Initially, the ABM waste, which resembles sand in its untreated form, will undergo a pre-treatment process by mechanical grinding using the milling machine. The grinding process conducted by PT Nanotech Indonesia Global aims to achieve a finer average particle size of the ABM waste. Additionally, mechanical properties tests on concrete mortar are performed to assess the impact of incorporating silica sand into the concrete.

Table 1 Chemical Oxide Content of Cement

Compound	Concentration (%)
Al_2O_3	1,9
SiO_2	9,25
P_2O5	0,45
SO_3	1,5
K_2O	0,64
CaO	79,65
TiO_2	0,41
V_2O_5	0,02
MnO	0,078
Fe_2O_3	5,765
CuO	0,094
ZnO	0,02
Rb_2O	0,009
SrO	0,079
ZrO_2	0,03
BaO	0,1

2. Experimental Program

2.1 Materials

2.1.1 Cement

Portland composite cement is used in this research. The chemical composition shown in the Table 1. The cement composition meets the PCC Chemical Composition requirements based on ASTM C150 [16] which is the Portland Cement Standard Specification.

2.1.2 Ground Sandblasting Waste (GSW)

ABM waste was sourced from one of the largest shipyard industries located in Lamongan, East Java, Indonesia. A visual representation of the untreated ABM waste is provided in Figure 1 The raw ABM Waste is pretreated, which involves a screening and washing process. The screening process uses a number 4 sieve to separate large particles. Subsequently, the washing process is carried out to reduce mud content and other impurities that cannot be removed through sieving. The chemical composition of the effluent before and after preliminary treatment by washing is shown in the Table 2 according to the result of the research by Sutrisno et al, 2024 [1].

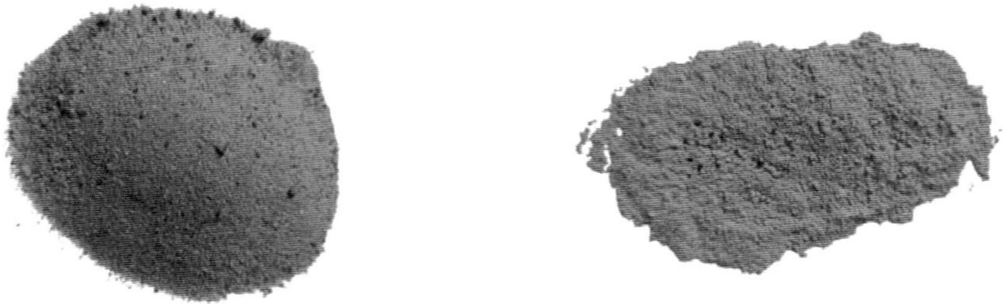

Before **After**

Figure 1 Before and After treatment of ABM Waste.

Table 2 Chemical Oxide Composition of ABM Waste

Chemical Oxide Composition	ABM Waste (%)	ABM Waste after Washing (%)
SiO_2	93.02	94.75
Al_2O_3	1.02	0.87
Fe_2O_3	2.00	1.26
CaO	0.63	0.52
MgO	0.43	0.43
Na_2O	0.24	0.2
K_2O	0.39	0.05
TiO_2	0.07	0.09
MnO_2	0.02	0.01
Cr_2O_3	0.01	<0.01
SO_3	0.07	0.06

The process of reducing the size of the sandblasting waste to a smaller particle size is carried out at PT Nanotech Indonesia Global Tbk. This procedure is performed to increase the reactivity of the silica content, so it can be used as an additive material. The analysis result before treatment shows that the silica sand has a large size of about 1-3 mm, but after initial pre-treatment, the average particle size of silica sand is 641.1nm according to the particle size analysis result. Pre-treatment process later continued with mechanical grinding and the result shows that the sandblasting waste average particle size becomes 390.4 nm or 0.39 μm. The average silica nanoparticle size of 262.1 nm or 0.262 μm has a compressive strength of 31.21 MPa which is 28.4% greater than the control variable [1].

Table 3. Chemical Properties of Silica Fume (SF)

Compound	Concentration (%)
Al_2O_3	0,6
SiO_2	93,7
P_2O_5	0,975
SO_3	0,855
K_2O	1,15
CaO	2,125
Fe_2O_3	0,28
CuO	0,097
ZnO	0,0285
As_2O_3	0,0095
Re_2O_7	0,04

2.1.3 Silica Fume (SF)

Silica Fume (SF) was used in this research obtained from a local source in Indonesia. SF Chemical Oxide Content is shown in Table 3. The chemical properties test is conducted to verify the suitability of silica fume (SF) for use, specifically by examining its SiO2 content. Typically, the SiO2 content in silica fume ranges from 85% to 98%. The SiO2 content for SF used in this research is 93.7% based on Table 3.

2.1.4 Sand

Local natural sand that has been sieved with no. 4 sieve with a maximum sieve opening size of 4.75 mm. The sand is ideally conditioned to be SSD or surface dry.

2.2 Mixture Design

The mixing properties of mortar were determined based on ASTM C-311 [17]. The mortar mixtures were designed using varying GSW and SF as cement additives. The composition of GSW as an additive is 1% - 5%. The next variation is to find out the influence of using a combination of

GSW and SF additives in mortar. Variations in the combination of GSW and SF are planned based on literature studies. Sutrisno et al., 2024 mentioned that the optimum addition of powdered silica sand to mortar is 2% by weight of binder. Then, the addition of SF is based on the research of Yunchao et al., 2021 [18], which also examines nano-silica and silica fume combinations as a cement additive. The research result shows that using 6% SF and 2% or 3% nano-silica by weight of the binder is recommended to be used as supplementary cementitious materials. Therefore, in this research, the same combination is used by adding 2% GSW and 6% SF as additives. The complete composition can be seen in Table 4.

Table 4 Mixture Design

Sample Name	Cement (Kg/m3)	GSW (% addition)	SF (%Addition)	Fine Aggregate (kg/m3)	w/c
Normal	666,67	0%	0%	1833,33	0,484
GSW1		1%	0%		
GSW2		2%	0%		
GSW3		3%	0%		
GSW4		4%	0%		
GSW5		5%	0%		
SF6		0%	6%		
GSW2+SF6		2%	6%		

The method of making mortar samples was based on ASTM C305 [19]. The first step is homogenizing the cement and additives for 30 seconds. Then, the fine aggregate that has been prepared at ideal conditions is slowly introduced at a low speed for 30 seconds as a dry mix process. Then, water is added at a medium speed for 30 seconds. Then let the mixer stand for 1.5 minutes and mix again at medium speed for another 1 minute.

2.3 Methodology

2.3.1 Compressive Strength and Strength Activity Index

Compressive strength is an important measurement to know the ability of concrete with vertical load pressure to the mortar to determine the ultimate pressure load. Mortar testing was conducted based on ASTM C109. Compressive strength per surface area is calculated based on Eq (1). P is the ultimate pressure load in Kg, Ton, and KN, then A is mortar surface area in m2 and In^2.

$$f'c = \frac{P}{A} \tag{1}$$

Strength Activity Index (SAI) is an equation to compare the compressive strength between variable addition of mortar and control mix of mortar. Variable Strength Activity Index (SAI) refers to the ratio between the strength of concrete mortar control variables and other variables. The calculation of SAI is based on ASTM C311 [17], which can be seen in Eq (2). SAI calculation has (A) as the compressive strength of variable addition and (B) as control mix compressive strength.

$$SAI = \frac{A}{B} x \ 100\% \quad (2)$$

3. Results and Discussion

The development strength over a 7 to 28-day curing for concrete mortar incorporating various compositions of ground sandblasting waste (GSW) and silica fume (SF) was investigated. The compressive strengths recorded for concrete with 1%, 2%, 3%, 4%, and 5% GSW were 32.46 MPa to 43.9 MPa, 33.85 MPa to 47.09 MPa, 29.37 MPa to 37.42 MPa, 28.26 MPa to 36.77 MPa, and 25.88 MPa to 39.7 MPa, respectively. Furthermore, the strength of concrete with 6% SF ranged from 34.2 MPa to 39.7 MP, while a mixture of 2% GSW and 6% SF exhibited the highest strength, ranging from 42.60 MPa to 49.50 MPa. These results demonstrate that the concrete strength increases with longer curing durations. Additionally, the strength improvement with adding 1% to 5% GSW is observable, though not significantly pronounced at the 7-day mark. The combined use of GSW and SF, especially in formulations with 2% GSW and 6% SF, showed a significant increase in compressive strength which makes the best potential for implementation

The compressive strength increases with the 7 – 28 days curing age for each variation of GSW (1% to 5%) is 35.26%, 39.12%, 27.42%, 30.12%, and 35.36%, respectively. Furthermore, for the additions of 6% SF and the combination of 2% GSW with 6% SF, the increases were 13.85% and 13.94%, respectively. The pattern of compressive strength increase from 7 to 28 days is relatively consistent across all variations of GSW and SF addition. The highest development strength of 7 – 28 days curing age was retrieved in the 2% GSW variation, which showed a 39.12% increase.

Notably, concrete using a combination of 2% GSW and 6% SF as the admixture exhibited a minor percentage increase in strength from 7 to 28 days and demonstrated significant early strength development. At 7-day, the 2% GSW and 6% SF mixture had already achieved a compressive strength of 42.6 MPa, significantly higher than the 26.93 MPa recorded for normal concrete. This condition indicates that combining 2% GSW and 6% SF enhances early development strength and substantially increases concrete mortar compressive strength (MPa).

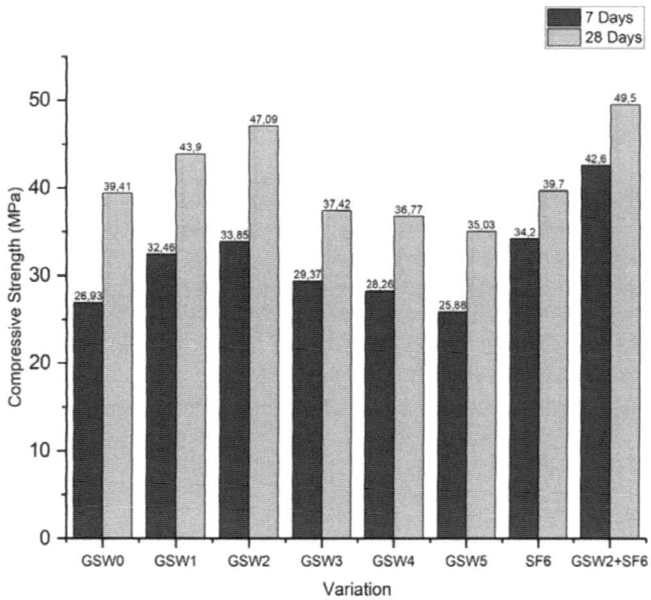

Figure 2 GSW and SF Additives Compressive (MPa) Strength of concrete mortar

Figure 2 illustrates that GSW 2% exhibits the optimal performance among all GSW additions at 7 and 28 days. However, when compared to SF 6% at 7-day, the compressive strength of GSW is slightly lower by 1.03%. Yet, at 28 days, GSW 2% surpasses SF 6% by 18.63%. Comparing GSW 2% to GSW 2% + SF 6% at 7 days, the former shows a 25.84% lower compressive strength. However, at 28 days, GSW 2% demonstrates only a slightly lower difference than GSW 2% + SF 6% by 5.11%.

Further analysis indicates that compressive strength of GSW 2% significantly increases compared to normal mortar at both 7 and 28 curing days, with enhancements of 25.69% and 19.5%, respectively. Similarly, compressive strength adding of SF 6% increases at 7 and 28 curing days compared to normal mortar, with an increase of 26.99% and 0.74%, respectively. Moreover, the combination of GSW 2% + SF 6% shows remarkable enhancements compared to normal mortar at 7 and 28 days, with increases of 58.18% and 25.61%, respectively.

The test result presented that there is an increase with the addition of GSW and SF additives. GSW and SF used in this research contains a large amount of Silicon Oxide (SiO2). This SiO2 will react with lime (Ca (OH2)) in cement. The reaction between SiO2 in the additive and Ca (OH)2 in the cement will form a calcium silica hydrate (CSH) gel which is the main component of the hydrothermal cement solid matrix [20]. The reaction will improve the physical or other properties of the concrete.

Furthermore, the basic concept of adding nanomaterials using GSW and SF is filling pores in concrete mortar [21]. Filling pores with nanoparticles will increase the density of concrete mortar [11] and directly correlate with concrete compressive strength [22]. Nanoparticles such as GSW and SF with petite particle sizes can penetrate and reduce porosity which increase the load pressure resistance of concrete mortar. In addition, this increase in density also contributes to the reduction of micro-cracking, which is usually the starting point of failure of concrete structures. Thus, using GSW and SF in concrete mortar mixes effectively improves concrete's mechanical performance, which often leads to enhanced durability. However, The excessive addition of more than 2% of GSW into the concrete will also reduce compressive strength because the GSW has higher water demand because of its nanoparticle size [1]. This condition occurred in the 4%, 5% GSW variation at 7 days curing age and 3%, 4%, 5% at 28 days curing age.

This study also compared the addition of SF 6% to the mortar with normal concrete without additives. Previous research indicates that adding SF to concrete can increase the initial compressive strength by about 30-50% [23]. In this study, SF showed an increase at 7 days of curing of 26.99% compared to normal mortar. Compared to 2% GSW which showed 25.69% increase than normal mortar, SF showed slightly higher increase. However, at 28 days curing age, while SF 6% showed a 16.08% increase from the 7 days curing age, GSW 2% exhibited a much higher increase of 39.12% over the same period. This discrepancy highlights the fast-reacting nature of SF, which leads to a substantial increase at the age of 7 days but diminishes in comparison to GSW 2% at 28 days of curing.

Furthermore, the research also focused on combining GSW 2% and SF 6% additives. The results demonstrated that compressive strength at 7-day curing increased by 58.18% compared to normal mortar, surpassing the individual additive variations, GSW 2% and SF 6%. This combination enhanced initial compressive strength and showed significant strength development at 28 days of curing, with a 16.2% increase over the 7-day of curing strength. This suggests that the combination of GSW 2% + SF 6% has a more effective impact on compressive strength development in concrete mortars than using the additives separately. This result aligned with the research performed by Yunchao et al, 2021 [18], which addition of SF6% has optimum compressive strength when combined with nano-silica up to 1 - 3%. These findings emphasize the importance of using a combination of additives to achieve optimal concrete strength improvement in the short and long term.

Statistical analysis is also needed to determine the trendline of the effect of the addition of GSW as a concrete mortar filler. Analysis was carried out on the impact of 1% - 5% addition on concrete mortar ultimate compressive strength, which showed on Figure 3.

Based on the 3rd order polynomial trendline graph above, the R-Square value is 0.8419. Based on Chin, 1998 [24] that the R-Square value is declared strong if it exceeds the R-Square of 0.67. This means that 84.19% of the variable data for the addition of GSW filler affects the results illustrated with the 3rd order polynomial equation of strength mortar.

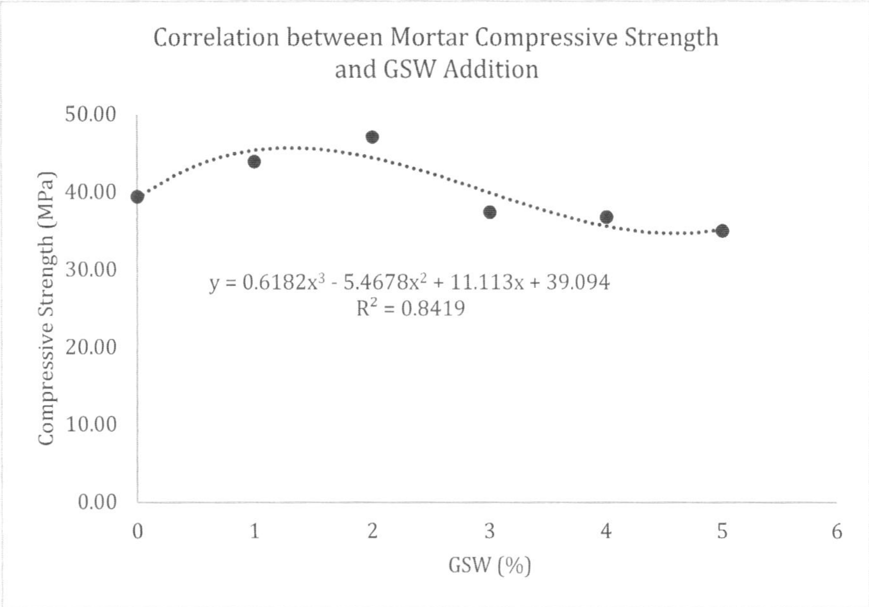

Figure 3 Correlation Graphic between Mortar Compressive Strength and GSW addition

Although this research did not utilize ABM Waste in large quantities, scaling it up to an industrial level could optimize the utilization of ABM Waste for concrete additives, thereby improving hazardous waste management in the shipping industry. Moreover, cost reductions can be achieved by utilizing ABM Waste in high-ultra compressive strength concrete. Previous research states that the concrete mixture of 3% nano silica and 6% silica fume has a compressive strength of 68.86 MPa, 21.99% greater than the normal variable[18]. This means that using the same mix, even with additional ingredients to the normal concrete mix, can produce concrete with higher compressive strength.

4. Conclusion

This research aims to utilize ABM waste as a concrete mortar additive that the result can be concluded:

1. This study explains about ABM waste into GSW, which is rich in SiO2 compounds, can enhance concrete mortar compressive strength. GSW is required only in small quantities as an additive. The highest compressive strength is shown by adding 2% of GSW.
2. Silica fume has significant development strength of concrete mortar at 7-day curing age or early age which is 26.99% compared with normal.

3. The combination of GSW2% and SF6% is the most optimum variation in this study for concrete mortar filler. The strength development can reach 58.18% and 16.2% for 7- and 28-day curing ages.
4. This research is expected to support the management of hazardous waste into additive materials for mortar and concrete to promote a circular economy in the shipping industry.

Acknowledgment

The authors would like to the Centre for Higher Education Funding (BPPT) from The Ministry of Education, Culture, Research, and Technology and the Indonesian Endowment Fund for Education (LPDP) from The Ministry of Finance of Indonesia. Furthermore, gratitude to PT DOK Pantai Lamongan for providing the sandblasting waste material used in this research.

References

[1] Sutrisno W, Wulandari K D, Abidin M Z and Rizal M N 2024 Properties of Concrete Mortar Incorporating Recycle Pulverized Sandblasting Waste as Additives *Heliyon* **10**
[2] Qi C, Weinell C E, Dam-Johansen K and Wu H 2021 A Review of Blasting Waste Generation and Management in The Ship Repair Industry *J Environ Manage* **300**
[3] Flynn M R and Susi P 2004 A Review of Engineering Control Technology for Exposures Generated During Abrasive Blasting Operations *J Occup Environ Hyg* **1** 680–7
[4] Madl A K, Donovan E P, Gaffney S H, McKinley M A, Moody E C, Henshaw J L and Paustenbach D J 2008 State of The Science Review of The Occupational Health Hazards of Crystalline Silica in Abrasive Blasting Operations and Related Requirements for Respiratory Protection *J Toxicol Environ Health B Crit Rev* **11** 548–608
[5] Alison J, Siahaan D L and Sugiharti E 2021 Evaluasi Implementasi Kebijakan Asas Cabotage dalam Perkembangan Industri Pelayaran Nasional *Jurnal Manajemen Transportasi & Logistik* **08** 1–10
[6] Sukandar and Wildani N 2010 *Studi Awal Pemanfaatan Limbah Sandblasting Sebagai Koagulan Preliminary Study of Sandblasting Waste Utilization as a Coagulant* vol 16
[7] Qomariah Q, Sugiharti S and Riyanto S 2020 The Utilization of Sandblasting Sand Waste for Mortar and Normal Concrete *The 1st Annual Technology, Applied Science and Engineering Conference* vol 732 (Institute of Physics Publishing) pp 1–7
[8] Taylor H F W (Harry F W) 1997 *Cement Chemistry* (London: Thomas Telford Publishing)
[9] Wang C, Zhang M, Wang Q, Dai J, Luo T, Pei W, Melnikov A and Zhang Z 2022 Research on The Influencing Mechanism of Nano-Silica on Concrete Performances Based on Multi-Scale Experiments and Micro-Scale Numerical Simulation *Constr Build Mater* **318** 1–9
[10] Panjehpour M, Ali A and Demirboga R 2011 A Review for Characterization of Silica Fume and Its Effects on Concrete Properties *International Journal of Sustainable Construction Engineering Technology* **2** 1–7
[11] Setiati N R 2017 The Potential Use of Silica Sand as Nanomaterials for Mortar *the 3rd International Conference on Construction and Building Engineering (ICONBUILD)* vol 1903 (American Institute of Physics Inc.) pp 1–7
[12] Binici H and Aksogan O 2018 Durability of Concrete Made with Natural Granular Granite, Silica Sand and Powders of Waste Marble and Basalt as Fine Aggregate *Journal of Building Engineering* **19** 109–21
[13] Hamada H M, Abed F, Binti Katman H Y, Humada A M, Al Jawahery M S, Majdi A, Yousif S T and Thomas B S 2023 Effect of Silica Fume on The Properties of Sustainable Cement Concrete *Journal of Materials Research and Technology* **24** 8887–908
[14] Du J, Meng W, Khayat K H, Bao Y, Guo P, Lyu Z, Abu-obeidah A, Nassif H and Wang H 2021 New Development of Ultra-High-Performance Concrete (UHPC) *Compos B Eng* **224**
[15] Wu Z, Khayat K H and Shi C 2019 Changes in Rheology and Mechanical Properties of Ultra-high Performance Concrete with Silica Fume Content *Cem Concr Res* **123** 1–11
[16] ASTM C 150 2007 Standard Specification for Portland Cement 1–8
[17] ASTM C-311 2009 Standard Test Methods for Sampling and Testing Fly Ash or Natural Pozzolans for Use in Portland-Cement Concrete 1–10
[18] Yunchao T, Zheng C, Wanhui F, Yumei N, Cong L and Jieming C 2021 Combined Effects of Nano-silica and Silica Fume on The Mechanical Behavior of Recycled Aggregate Concrete *Nanotechnol Rev* **10** 819–38

[19] ASTM C-305 2009 Standard Practice for Mechanical Mixing of Hydraulic Cement Pastes and Mortars of Plastic Consistency 1–3

[20] Pachideh G, Gholhaki M and Ketabdari H 2020 Effect of Pozzolanic Wastes on Mechanical Properties, Durability and Microstructure of The Cementitious Mortars *Journal of Building Engineering* **29** 1–11

[21] Schmidt M 2004 Influence of the packing density of fine particles on structure, strength and durability of UHPC *International Symposium on Ultra High Performance Concrete* 313–23

[22] Zaid O, Ahmad J, Siddique M S, Aslam F, Alabduljabbar H and Khedher K M 2021 A Step Towards Sustainable Glass Fiber Reinforced Concrete Utilizing Silica Fume and Waste Coconut Shell Aggregate *Sci Rep* **11** 1–15

[23] Mazloom M, Ramezanianpour A A and Brooks J J 2004 Effect of Silica Fume on Mechanical Properties of High-Strength Concrete *Cem Concr Compos* **26** 347–57

[24] Chin W W 1998 *The Partial Least Squares Approach to Structural Equation Modeling*

IOP Conf. Series: Earth and Environmental Science **1423** (2024) 012024 doi:10.1088/1755-1315/1423/1/012024

Air quality monitoring and assessment based on AIS data

A Maulidi[1,4,5*], K Sambodho[2,4], A A B Dinariyana[3,4] and A M Ilmah[5]

[1] Doctoral Program of Ocean Engineering, ITS Surabaya, Indonesia
[2] Department of Ocean Engineering, ITS Surabaya, Indonesia
[3] Department of Marine Engineering, ITS Surabaya, Indonesia
[4] The Center for Excellence of Maritime Safety and Marine Installation, ITS Surabaya, Indonesia
[5] Politeknik Negeri Madura, Indonesia

*Email: amd.2201@gmail.com

Abstract. Indonesia is an archipelagic country, where sea transportation is highly needed to connect the islands among Indonesian regions. Indonesian waters are also passed by international shipping passing through the Indonesia Archipelagic Sea Lane. Number of ships tends to increase by increasing the economic activities in Indonesia. The increasing number of ships will potentially cause the increase of air pollution in Indonesia waters. AIS consortium has been established since 2020 focusing on conducting join research and development on several issues that can be solved using Automatic Identification System (AIS) data gathered from AIS devices that have been installed in several universities as members of the consortium. One potential use of AIS data is the estimation of capacity and spreading of air pollution caused by shipping activities in Indonesia waters. As AIS can provide ship's navigation status time-to-time, the estimation of fuel consumption as well as the emission can be estimated in real time. For highly accusation fuel oil consumption prediction is done by using the Random Forest Model and ship emissions including CO_2, NO_2, SO_2, PM2.5, and PM10 are estimated based on emission factors from ships. This analysis is mapping the contribution of air pollution based on ship type in the Tanjung Perak port area. Later, it is expected to become the basis for Indonesia's maritime operation policy in order to monitor and control the emission contributed by the ships in Indonesian waters.

1. Introduction

Indonesia has a strategic position in the international trade, recorded 40% of world trade through Indonesian waters. This is an opportunity as well as a challenge for Indonesia to be able to manage the sea well. Both in maritime infrastructure, ship security and environmental safety must be managed properly. As known that climate change and global warming are significantly influenced by increased marine transportation. In 2020, the International Maritime Organization (IMO) stated that the shipping industry and marine transportation was contribute for around 2.89% of greenhouse gas effect in 2018 [1]. That increasing need for sea shipping, emissions from this sector continue to increase. A big challenge for the marine transportation and industry which must can decrease air pollution and keep maintaining its important role in national and international trading. Various parties continue to make efforts to reduce gas emissions by making ships more environmentally friendly. IMO reported greenhouse gas effect, that marine transportation and industry currently produces 10^9 tons of gas emissions annually [2]. IMO proposes a new regulation that lower the upper limit for ship fuel sulphur dioxide from 3.5% to 0.5% based on IMO 2020 regulation. On the International Convention for the Prevention of Pollution from Ships (MARPOL) is the main international convention covering prevention of pollution of the marine environment by ships from operational or accidental causes. To decrease emissions sulphur dioxide by 70% compared to 2020 levels [3].

Indonesia has many ports with heavy ship traffic which is the main source of air emissions in the port area and its surroundings. This ship traffic produces air emissions in the form of carbon dioxide, carbon monoxide, nitrogen dioxide, sulphur dioxide and fine particles. European countries and the United States strictly implement emission control areas (ECAs) requiring levels of sulphur dioxide below 0.1% in nearshore areas. Indonesia adopts emission control limits set by MARPOL. Thus, efforts need to be made to estimate air emissions originating from ships by utilizing AIS data, which provides real-time travel information. Every ship that sails emit toxic air pollutant compounds both from its main engine and from its auxiliary engines. Kim & Eom [4] wrote, amount of fuel oil consumption and air emissions can be modelled based on cruise status, such as the operating level of the main engine and auxiliary engines.

In general, the level of air emissions can be influenced by the amount of fuel consumed when the engine is operating. The level of fuel consumption varies greatly based on the engine operating mechanism on the ship, whether sailing, manoeuvring, anchoring, or loading and unloading. AIS can provide machine operating data sent broadcast. Therefore, AIS data can be used to estimate exhaust gas which includes information about the navigation status, speed over ground, longitude, latitude and ships heading [5]. Currently the AISITS Consortium consists of 13 universities, very concerned about maritime safety and pollution especially regarding water and air pollution. By utilizing AIS data collected from Remote Base Stations (RBS) installed around the Madura Strait, previous research has obtained a Navigation Safety Platform Concept through AISITS [6], by integrating ship databases and readings Automatic Identification System (AIS) data in real time [7], followed by development of real time monitoring software prototype based on AIS data [8].

Each ship has different principal dimensions such as Length Over All (LOA), Width, Gross Tonnage (GT), Depth and Draft. Likewise, the engines used on each ship. This causes uncertainty data to predict air pollution from ship activities. To overcome this, Random Forest can provide high accuracy. This algorithm is also able to handle outliers and missing values and helps identify the main factors that contribute for air pollution. By flexibility for regression tasks, Random Forest provides stable and consistent predictions, making it an effective tool in managing and planning environmental policies regarding ship activities. This study can be used to map air emissions caused by ship activities by type. Air emission mapping is obtained from shipping activity data from real-time AIS data based on the ship's navigation status. The hypothesis is that air emissions in the area around the port are highly impacted by the number of ship traffics, which has an impact on air quality in the port area. The method used to estimate this air emission on ship traffic density. To validate the proposed method, AIS data is taken within a monthly span of time included in the Tanjung Perak Port area.

2. Methods

2.1. Remote Base Station (RBS) AIS Connections

The technology used is based on the Internet of Things (IoT) which connects RBS with the Data Server [9]. The RBS consists of a VHF antenna, AIS Responder, AIS Dispatcher and Gateway device. A node-to-node Virtual Private Network (VPN) configuration is used to encrypt and ensure the security of AIS data sent to the Data Server [10]. The IoT system works by utilizing programming algorithms, where each command argument can produce interactions between connected machines automatically without human intervention and without being limited by any distance. Input will be processed according to the algorithm that has been prepared and will continue until it produces the required information [11] [12]. RBS as a client, installs the AIS Dispatcher utility which functions to forward AIS data via a UDP stream connection, TCP connection or serial interface. **Figure 1** shows the data flow originating from the AIS receiver, the AIS Dispatcher functions to check and delete duplicate NMEA messages, calculate overall data and detailed statistics, down sampling the NMEA stream, send the NMEA stream via UDP to the IP Data Server.

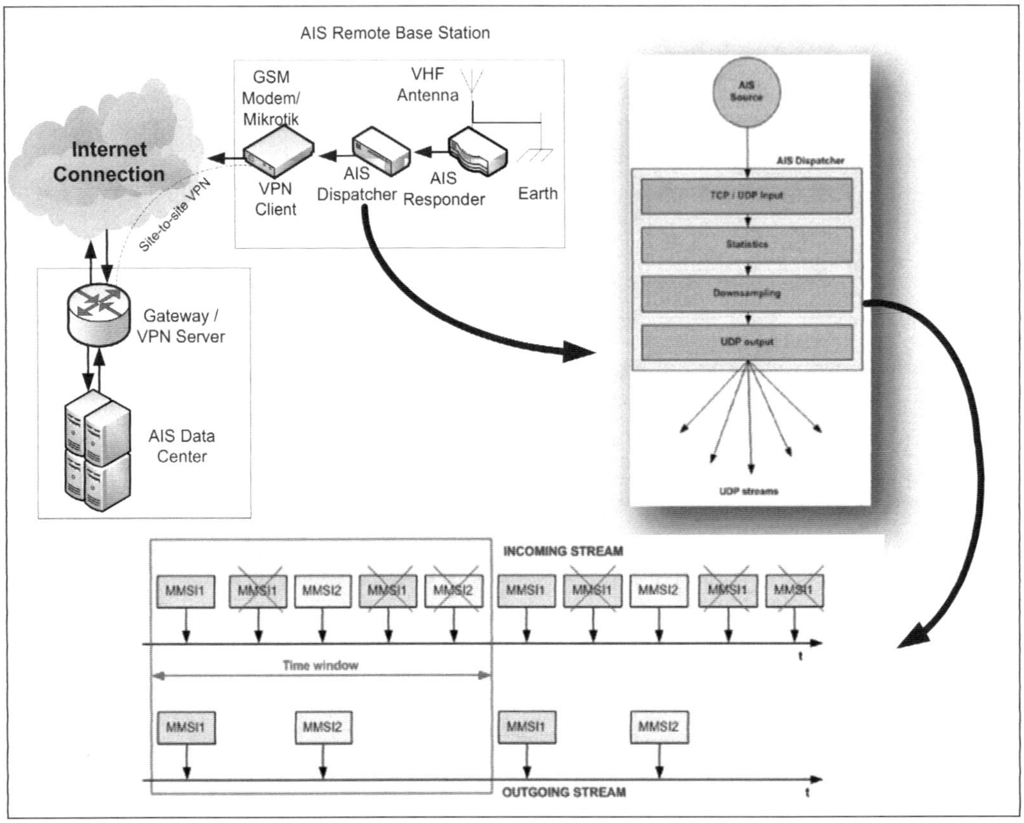

Figure 1 AIS Data Connectivity

2.2. *Data Collection*

The first step in conducting this research is collecting the data needed to conduct the research. The research data used in this research is in the AIS NMEA Message format. This data format is used to encode (convert AIS raw data format) into encoded data (plain data). **Figure 2** is the raw data packet received by the AIS receiver. The AIS NMEA Message Format consists of two types which can be seen at the beginning of the data packet [13].

!AIVDM,1,1,,A,14eG;o@034o8sd<L9i:a;WF>062D,0*7D

Figure 2 AIS raw data format

This research focuses on data received from other ships, because in fact all the data received came from other ships captured by RBS Poltera, ITS and PPNS in the Madura Strait area. Data collection was carried out in the period January - March 2024. In general, the AIS data format sent from ships equipped with an AIS system is type NMEA 0183. An example of the NMEA 0183 data format is as follows [14]:

1. !AIVDM: data received from another ship
2. !AIVDO: information originating from the ship itself

Table 1 is an explanation of AIS NMEA data packages received from a recipient.

Table 1 AIS NMEA Description

Message	Description
!AIVDM	NMEA message type
1	Number of sentences (some messages require more than one)
1	Sentence number (1 except multi-sentence messages)
	The blank is the Sequential Message ID (for multi-sentence messages)
A	AIS channel (A or B)
14eG;...	AIS data encoded using AIS-ASCII6
0*	End of data
7D	NMEA Checksum

The AIS data received by RBS is then sent to the Data Server which can be downloaded using the pre-installed MongoDB database. The MongoDB database interface appears as shown in the **Figure 3** below. Creating a new connection is done by entering the user ID and server IP address in the URI field. Save and connecting the server.
The standard connection string format looks like:

mongodb://[username:password@]host1[:port1][,host2]][/[database][?options]]

Figure 3 Mongo URI Connection

If the connection is successful, the required settings menu will appear as in **Figure 4** below. The setting steps are 1. Selecting the database provided by the server, 2. Selecting the data table to be downloaded, 3. Data export options consisting of export query or export. full collection, 4. The export collection option allows you to download the entire data package or just some of the data needed for assessments.

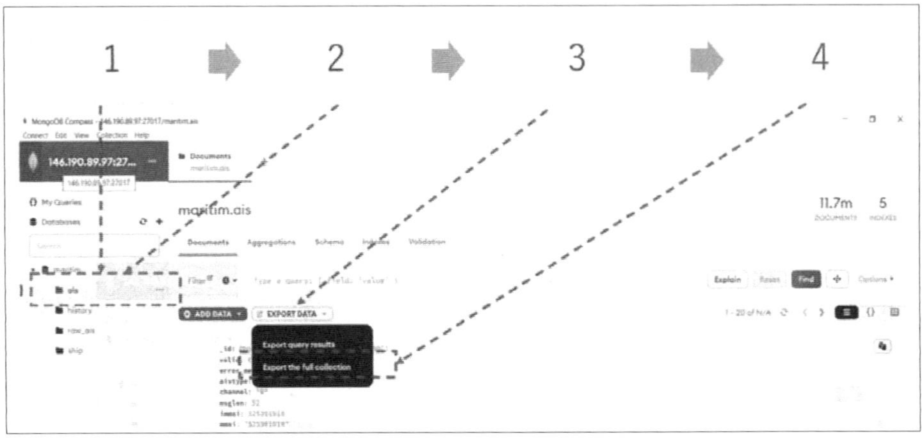

Figure 4 Database Selection

When the ship's engine is operating, it will determine the amount of fuel consumed and will correlate significantly with the level of air emissions produced. This fuel consumption will vary depending on the operating conditions of the engine, especially on the ship's sailing mode, be it cruising, port manoeuvring or anchoring. Engine operating conditions are obtained from the navigation status stated in the AIS data. Therefore, air emission predictions are carried out by utilizing ship AIS data which includes information about the navigation status, speed over ground, longitude, latitude, and ships heading. The Export Collection option allows you to select whatever data is needed as a basis for calculations. This research uses speed data, ship operating modes to estimate air emissions from a ship even in a port area. Next in **Figure 5** is 5. The output data selected for calculating air emissions, 6. The expected output file type. Here, to make data processing easier, you can select the CSV file type, 7. Select the folder to save the output file, and 8. Click Export to carry out the data downloading process.

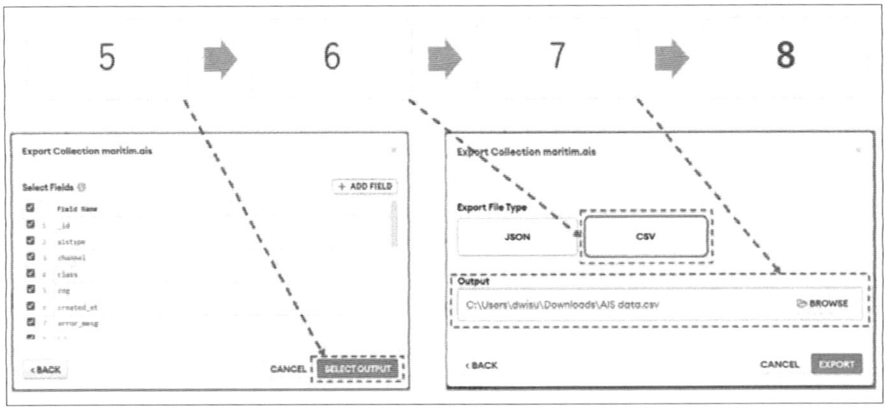

Figure 5 Data Selection and File Type

2.3. *Air Quality Index*

The standard used to measure air quality is called the Air Quality Index. A mention of the air quality index for each country is different, namely the Air Pollution Standard Index (ISPU) in Indonesia, Comparative Air Pollutants (APB) in UK, the Air Quality Index (AQI) in Beijing, the Pollutant Standards Index (PSI) in Singapore, the Comprehensive Air Quality Index (CSI) in South Korea, the Common Air Quality Index (CAQI) in Europe's and others. However, the provisions regarding air emission pollutants are divided into particles and gases. Particle materials include (PM10, PM 2.5) to gaseous materials (CO, NO_2, SO_2, O_3). It is known that this material can have a very bad impact on humans, so that if are exposed to this pollutant it can cause acute respiratory infections, respiratory problems, gas poisoning, etc. [15].

Decree of the Environmental Impact Control Agency (BAPEDAL) Number: KEP-107/Kabapedal/11/1997 about Standard Index Limits for Air Pollutants (ISPU) become the standard for measuring air quality in Indonesia. Table 2 shows the standards and limits for air pollutant levels based on their categories [16] .

Table 2 Standard Index Limits for Air Pollutants (in SI Units) [17]

Air Pollution Standard Index	PM.10 (μg/m³)	CO (μg/m³)	O_3 (μg/m³)	NO (μg/m³)	SO_2 (μg/m³)
50	50	80	5	120	(2)
100	150	365	10	235	(2)
200	350	800	17	400	1130
300	400	1600	34	800	2260
400	500	2100	46	1000	3000
500	600	2620	57.5	1200	3750

Table 2 above is used as a threshold for air emissions with calculation results for each type of ship.

2.4. *Research Definition*

Researchers propose a measurement of the effect of ship traffic density on air quality in the Tanjung Perak. The main research problem how to get calculate amount air emissions resulting from ship traffic density in the Tanjung Perak port area. Our initial hypothesis was that the density of ship traffic in an area would have an impact on the level of air pollution in the port area. The amount of air emissions caused by ship traffic density can be determined using Equations (1) and (2) below.

$$\sum_{i=1}^{n} air\ emission(i) = \sum_{i=1}^{n} air\ factor\ (i) x\ fuel\ oil\ consumtion\ (i) \qquad (1)$$

$$fuel\ oil\ consumtion\ (i) = Power\ (i,t) x\ Specific\ Fuel\ Oil\ Consumtion\ )Power\ (t)) \qquad (2)$$

AIS data has an important role in calculating air emissions, location and speed of ships when operating. Identify the ship as i, and the time as t. Each fuel oil has different emission levels depending on the amount of fuel consumed at a certain time (t). SFOC is a lot of fuel oil consumption engine to become mechanical energy in a certain time period. The SFOC value of a machine can be expressed in g/kWh or kg/kWh (grams or kilograms per kilowatt-hour). Research data collection focuses on the RBS which coverage the Perak port area and around the Madura Strait. **Figure 6** is the research location. Tanjung Perak Port is the centre for transit and loading and unloading of manufacturing and health industry products transported by Tanker ships and containers transported by Cargo ships.

IOP Conf. Series: Earth and Environmental Science **1423** (2024) 012024 doi:10.1088/1755-1315/1423/1/012024

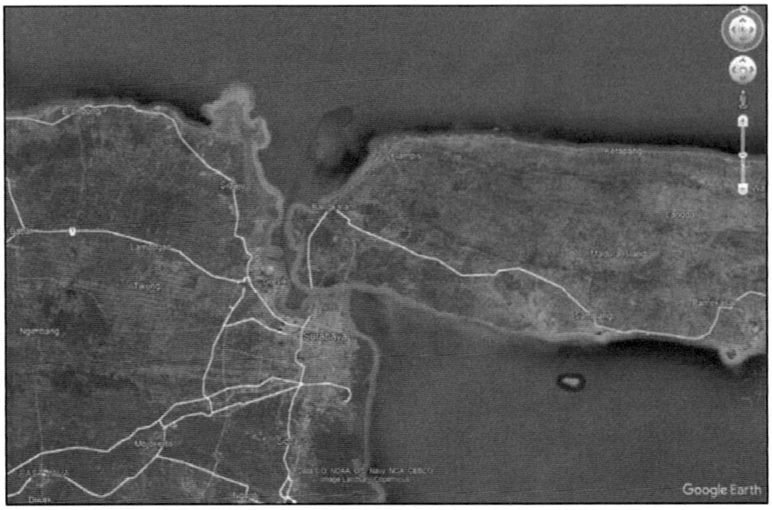

Figure 6 Research location.

Correlation analysis of ship traffic density with air emissions caused by ship traffic uses AIS data which contains travel records, coordinates (Long, Lat), (SOG) speed over ground, (COG) course over ground, and destination. Its collecting from the Remote Base Station (RBS) around Tanjung Perak Port, Surabaya. The data provides the average value of air emissions per hour and is categorized into emissions of particulate matter, carbon, nitrogen, and sulphur. The data taken place from March 1–28, 2024.

3. Result and Discussion

3.1. Ship Navigation Status from AIS data

Important data from AIS is the ship's navigation status in all conditions. This status is important because as a basis for calculating fuel oil consumption it can be calculated based on the navigation status. **Figure 7** shows a simple illustration of a ship's journey, starting from the arrival of the ship and ending with leaving the port.

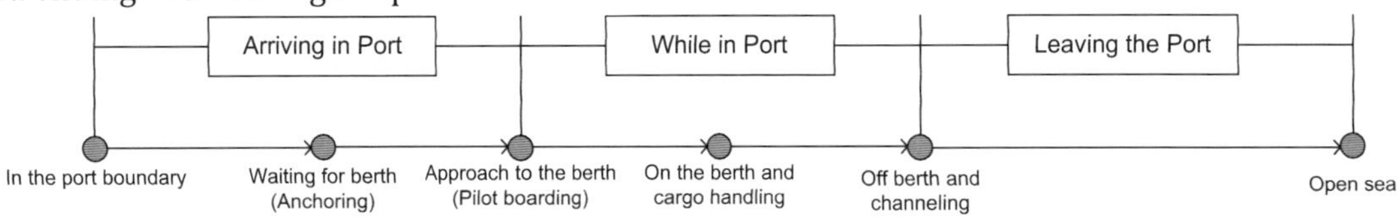

Figure 7 Ship activities at the port.

The ship docking process will begin with the ship being detected by the AIS receiving device at the port. In this condition the ship can be guided directly to the port until it reaches the berth or needs to wait a while until the previous ship has finished. The size of the ship determines whether the ship can enter directly into the berth or still requires being guided by tugboats. The large size of the ship and the narrow port channel require tugboats to reach the berth and be moored safely. After the ship has been successfully moored safely, the loading and unloading process of the ship must be carried out using special equipment and labour that already exists on the ship or can also use equipment available at the port. Loading and demolition activities will take place as long as the ship is at the port and will be carried out sequentially. After the loading and unloading process is complete, the ship will prepare to leave the harbour. Tugboats and harbour pilots are used to ensure the ship can exit the harbour safely. At this stage the ship's navigation tools and the captain's communication skills with local pilots are very important. After the ship can leave the port safely, the ship can then continue to its next destination.

For merchant ships it can be categorized into six assigned categories: 'arriving', 'anchoring', 'departing', 'sailing (before arriving)', and 'sailing (passing)'. Support vessels, that help commercial vessels enter and leave port safely, only have the categories: 'engine use' and 'engine stoppage'. In the previous paragraph, a lot of engine FOC (fuel oil consumption) can be determined by the time operation and amount engine installed on the ship, with different ship navigation statuses the fuel oil consumption will be different. In general, main engine (M/E) in the ship is installed at least two engines which functions as a propulsion and the generator engine (G/Es) which functions as a producer of electrical power. Meanwhile, calculations of FOC must be differentiated based on the use of these machines and the ship's navigation status.

3.2. Specific Fuel Oil Consumption (SFOC)

Ship engines require fuel oil expressed in grams/kilogram. It shows the amount required for the engine to produce power in given time. This term is known as SFOC, the amount is expressed in g/kWh or kg/kWh. Each engine has a different SFOC value based on the provisions of the engine manufacturer which includes SFOC charts for high efficiency turbochargers and values in MCR [18].

The company displays this fuel oil usage estimation model in the form of an SFOC curve as shown in **Figure 8** below. In this curve, it is known that horizontal line x shows the value of engine power (%), while the vertical line y shows the SFOC level. By varying the changed SFOC value to the SFOC value provided in the MCR (maximum continuous rating), the SFOC associated with engine power can be known. The SFOC value in MCR is denoted by SFOCmcr, and this study uses SFOCmcr as in **Table 3**.

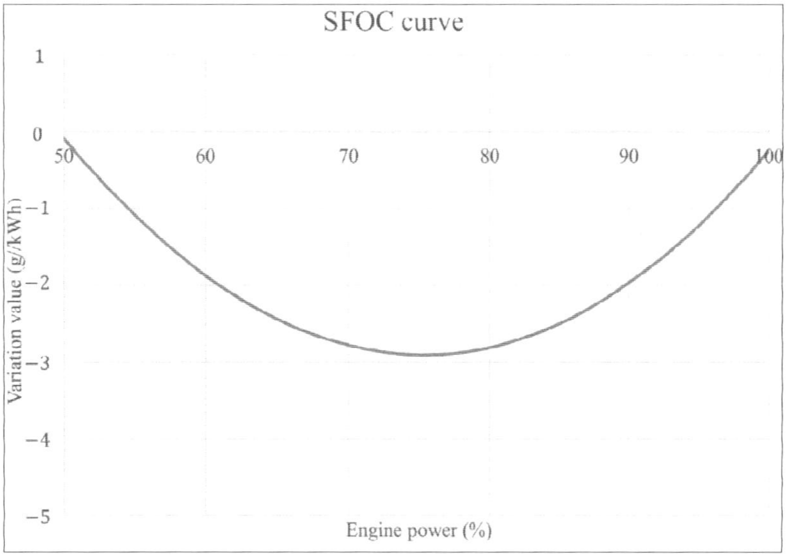

Figure 8 Estimated SFOC curve

Calculating of engine power using maximum and operating speed of each ship, showed in Equation (3), where maximum speed collected from the online database [13] and operating speed collected from AIS data. Data collection was carried out online through web crawling techniques, searched based on identical vessel codes such as IMO, MMSI, or Call Sign numbers. A total of 92%

of the data was recovered, and the maximum speed of the lost vessel was identified from the collected historical AIS data.

$$engine\ power\ (\%) = \left(\frac{operating\ speed}{maximum\ speed}\right)^3 \tag{3}$$

Changes in ship engine power determine the SFOC level, changes in value will be added to $SFOC_{mcr}$ (Equation (4)). The SFOC curve is the basis for calculating changes in FOC consumption values estimated by the ship engine estimation model.

$$SFOC\ (unit\ consumption)\ =\ SFOC\_mcr\ +\ variation\ value \tag{4}$$

The last, to get the FOC (Fuel Oil Consumption) value, the value will be like getting the operating time in the SFOC calculation (Equation (5)).

$$fuel\ oil\ consumption\ =\ SFOC\ operating\ time\ kW\ (t). \tag{5}$$

3.3. Fuel Oil Consumption by Ships Navigation Status

If the ship's main engine is in use and its status is recorded as 'Sailing (passing)', 'Sailing (before arriving)', 'Arriving', or 'Departing', Equation (4) is used to calculate operating time and Equation (5) is used to determine SFOC. Meanwhile, the time and speed from AIS data is used to calculate fuel usage for main engine. The generator (G/Es) will use fuel constantly except in certain circumstances such as being at anchor, leaving port or in emergency conditions. However, data regarding G/Es fuel consumption in emergencies and their electrical power requirements is a challenge due to limited data. G/Es fuel oil consumption estimates using the general FOC G/E consumption charts provided by Generator Source [19]. For ships 'Sailing (arriving)', 'Sailing (passing by)', 'Anchoring', or 'Berth' it assumed that the G/Es constantly consumes approximately 25% of its capacity fuel and 75% of its capacity in the 'Departing', 'Arriving', or 'Sailing (before arrival)'.

3.4. Air Emission Calculation

Calculating air emissions from ships, specific emission factors are adjusted for the type of fuel used and the pollutants produced as shown in **Table 3**. This emission coefficient functions as a metric that provides an illustration of the ship's fuel oil consumption level in tons and to determine the amount of pollutants used. generated based on the type of fuel used.

Table 3 Emission factors by fuel type (kg-pollutant/ton-fuel) [20]

Fuel Type	NO_X	SO_X	CO	PM-10	PM-2.5
MGO (Marine Gas Oil)	78.5	20 x %Sulphur	7.4	1.5	1.4
HFO (Heavy Fuel Oil)	79.3			6.2	5.6

3.5. Estimated Air Emissions from Ships

AIS data used to estimate ship air emissions are presented in **Table 4**. This table presents a comparison of the average fuel oil requirements for each ship. The highest average FOC in metric tons is for pleasure craft, but this data is not proportional to the number of these ships. From the observation data conducted that tanker and cargo ships recorded the largest number of visits, namely 149 and 123 ships. This is because Tanjung Perak Port is close to the Gresik, Sidoarjo and Ngoro Mojokerto areas, which has an impact on the number of tankers. However, even though the number of visit data is less than Cargo ships, Tanker ships have an average FOC four times higher than Cargo ships. This shows that tankers play an important role in ship traffic and air emissions in the Tanjung Perak port area.

Table 4 An Average ship's FOC by ship type

Ship Type	AIS Data	Number of Ships	Average FOC (M/T)
Tanker	4,470	149	2.67
Cargo	3,230	123	0.64
Passenger	4,043	15	4.07
Roro	2,123	16	1.20
Pleasure Craft	1,739	15	20.31
Boat Patrol/Warship	1,332	22	6.82
Other Type	431	13	2.67

Figure 9 shows that the highest FOC value is on Tanker ships, this data is two times higher than on Cargo ships. This data shows that the overall emissions produced by tankers greatly determine the air quality in the Tanjung Perak area. **Figure 10** shows comparitive of two types of ships with SO_X, NO_X, CO, PM-10 and PM-2.5 values. As shown in **Figure 10**, Tankers emit more than double the amount of air pollutants many compared to Cargo ships. Data in March shows that NOX emissions from Tanker ships exceed 65,000 kg, while Cargo ships show NO_X emissions of less than 30,000 kg.

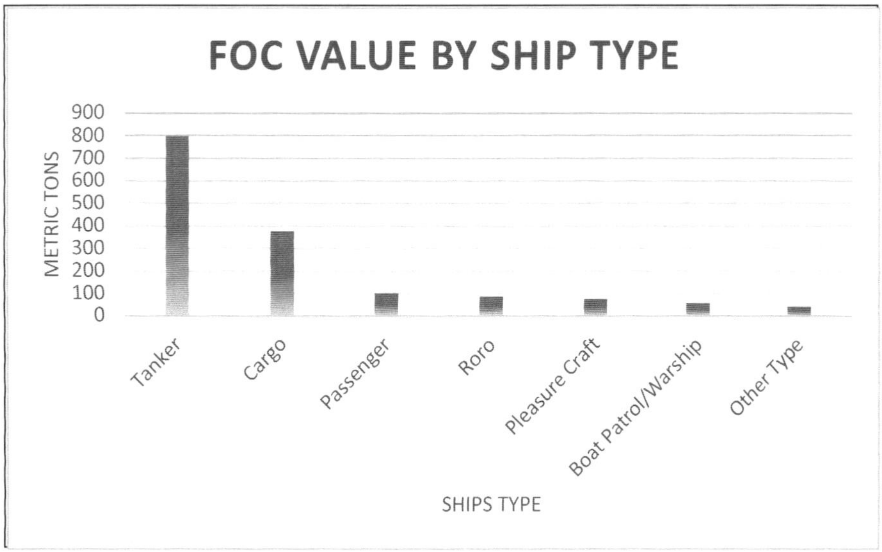

Figure 9 FOC Value by Ships Type

Comparison of significant air emissions between tankers and cargo ships. From the five emissions data, tanker ships have higher emissions than cargo ships, namely CO 184%, NOx 220%, SOx 207%, PM10 129%, PM2.5 184%. The average emissions were 185%, this shows that the number and activity of tankers were greater that month.

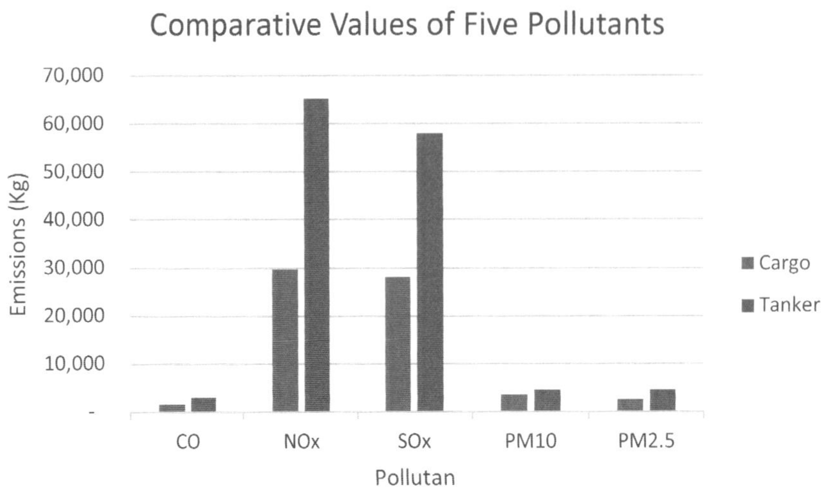

Figure 10 Comparative Values of Five Pollutants

4. Conclusion

AIS provides important data that can be used as a basis for estimating fuel use such as speed, engine type, ship status. In addition, we can estimate the value of air pollutant emissions caused by ships in an area. In this study it is in the Tanjung Perak Port area in Surabaya. By looking at the navigation status of the ship received by the AIS receiving device, we estimate the consumption of ship oil fuel based on its status. The level of specific fuel consumption and a sustainable maximum rating is also estimated for ships whose engine specifications are not known using the random method for stabilities and resource efficiencies. The results of the calculation of fuel consumption can be used to calculate the emissions of Sulphur Dioxide (SO_{X}), Nitrogen Dioxide NO_{X}, Carbon Monoxide (CO), Particulate Matter (PM-10) and Particulate Matter-2.5 (PM-2.5) by usage fuel type and emission factors.

Based on the discussion above, we concluded that there was a significant influence between the amount of ship traffic and the level of fuel consumption of oil on the increase in air emissions from ships operating in Tanjung Perak Port. Tanker ships have 185% higher emissions than cargo ships. However, the results of this calculation are only based on reading from AIS recipients' data. Our next stage, this result needs to be compared with data on air pollutant measurements on land.

Acknowledgments

We would like to thank the Financing Centre Higher Education (BPPT) Ministry of Education and Culture for all the financial support that has been provided, All Members of the AISITS Research Consortium, PUI-KEKAL ITS, Director of the Madura State Polytechnic (POLTERA), Promoter and Co-Promoter who patiently guided us, Poltera Colleagues, All Committee Maritime Safety International Conference 2024 (MASTIC 2024) which has approved and published the results of this research, and my beloved family always supports.

References

[1]. Joung TH, Kang SG, Lee JK, Ahn J. The IMO initial strategy for reducing Greenhouse Gas(GHG) emissions, and its follow-up actions towards 2050. J Int Marit Safety, Environ Aff Shipp. 2020;4(1):1–7. (CrossRef)

[2]. Serra P, Fancello G. Towards the IMO's GHG Goals: A Critical Overview of the Perspectives and Challenges of the Main Options for Decarbonizing International Shipping. Sustainability [Internet]. 2020 Apr 16 [cited 2024 May 26];12(8):3220. Available from: https://www.mdpi.com/2071-1050/12/8/3220 (CrossRef)

[3]. IMO. IMO. 2020. IMO 2020 sulphur limit implementation - carriage ban enters into force. Available from: https://www.imo.org/en/MediaCentre/PressBriefings/pages/03-1-March-carriage-ban-.aspx

[4]. Kim SW, Eom JO. Ship Carbon Intensity Indicator Assessment via Just-in-Time Arrival Algorithm Based on Real-Time Data: Case Study of Pusan New International Port. Sustain. 2023;15(18). (CrossRef)

[5]. Huang L, Wen Y, Zhang Y, Zhou C, Zhang F, Yang T. Dynamic calculation of ship exhaust emissions based on real-time AIS data. Transp Res Part D Transp Environ. 2020;80. (CrossRef)

[6]. Ketut Buda Artana, AAB Dinariyana, I Made Ariana, Dhimas Widhi Handani, Fadilla Indrayuni Prastyasari, Emmy Pratiwi. Milestone For E-Navigation: A Concept for Navigational Safety Platform by Means of AISITS. Pros Politek ILMU PELAYARAN MAKASSAR. 2020;1(4). (CrossRef)

[7]. Maulidi A, Pitana T, Artana KB, Dinariyana AAB, Zaman MB, Masroeri AA, et al. Database Integration Model for Automatic Identification System and Shipping Database In Real Time Traffic Monitoring. IPTEK J Proc Ser. 2015;0(1). (CrossRef)

[8]. Dinariyana AAB, Artana KB, Sambodho K, Kristianto D. Pengembangan Prototype Software Real Time Monitoring Berbasis Data Automatic Identification System (AIS). IPTEK J Proc Ser. 2018;0(1). (CrossRef)

[9]. Maulidi A, Irmiyana T. The development of internet-of-things (IoT)-based ship monitoring system in madura strait. IOP Conf Ser Mater Sci Eng. 2021;1052(1). (CrossRef)

[10]. Maulidi A, Abdullah M, Handani DW. Virtual private network (VPN) model for AIS real time monitoring. In: IOP Conference Series: Earth and Environmental Science. 2022. (CrossRef)

[11]. Maulidi A. Disain Sistem Navigasi Automatic Identification System (AIS) Transceiver Berbasis Mini Computer Pada Kapal Nelayan Tradisional Di Madura. INOVTEK POLBENG. 2019;9(1). (CrossRef)

[12]. Maulidi A. SISTEM PENERIMA (RECEIVER) AUTOMATIC IDENTIFICATION SYSTEM (AIS) BERBASIS MINI COMPUTER PADA KAPAL NELAYAN TRADISIONAL DI MADURA. INOVTEK POLBENG. 2019;9(2). (CrossRef)

[13]. FleetMon. FleetMon. [cited 2024 Jan 19]. Marinetraffic Data Services. Available from: https://www.fleetmon.com/

[14]. Association NME. National Marine Electronics Association (NMEA). [cited 2024 Jan 22]. NMEA 0183 Manufacturer Mnemonic Codes. Available from: https://www.nmea.org/nmea-0183.html

[15]. Hermawan A, Hananto M, Lasut D. PENINGKATAN INDEKS STANDAR PENCEMARAN UDARA (ISPU) DAN KEJADIAN GANGGUAN SALURAN PERNAPASAN DI KOTA PEKANBARU. J Ekol Kesehat. 2016;15(2). (CrossRef)

[16]. Aini N, Ruktiari R, Pratama MR, Buana AF. Sistem Prediksi Tingkat Pencemaran Polusi Udara dengan Algoritma Naïve Bayes di Kota Makassar. Pros Semin Nas Komun dan Inform. 2019;3:83–90. (CrossRef)

[17]. Badan Pengendalian Dampak Lingkungan. Keputusan Kepala Badan Pengendalian Dampak Lingkungan No: KEP-107/KABAPEDAL/11/1997. 1999;1–23.

[18]. Kim KS, Roh M Il. Iso 15016:2015-based method for estimating the fuel oil consumption of a ship. J Mar Sci Eng. 2020;8(10). (CrossRef)

[19]. Source G. Generator Source. [cited 2024 Feb 17]. Approximate Diesel Fuel Consumption Chart. Available from: https://www.generatorsource.com/Diesel_Fuel_Consumption.aspx

[20]. Seol S, Jung S, Seong M, Lim J, Yeo S, Jin H. Estimation of Air Pollutant Emissions from Ships and Their Contributions in Korea. J Korean Soc Atmos Environ. 2021;37(2). (CrossRef)

IOP Conf. Series: Earth and Environmental Science **1423** (2024) 012025 doi:10.1088/1755-1315/1423/1/012025

Risk Analysis of Coral Reef Damage Area due to Grounding Incidents in Karimunjawa National Park

F I Prastyasari[1], AAB Dinariyana Dwi Putranta[1] and E B Z Arifah[2]

[1] Department of Marine Engineering, Institut Teknologi Sepuluh Nopember, Indonesia
[2] Department of Operation and Management of Maritime System, Hochschule Wismar, Germany

fadilla@its.ac.id

Abstract. The designation of Karimunjawa as a National Park in 1999 does not prevent any shipping activities performed by barges, passenger ship, LCT, or even tanker from entering Karimunjawa waters. Such condition can cause damages to coral reefs if the ship run aground. In addition, the possibility of coral reefs contamination due to tourists activities such as diving that requires vessels to transport them. Based on the historical data from 2017 until 2023, fourteen cases of vessel grounding in Karimunjawa were recorded. This study comprises of hazard identification, frequency analysis, severity analysis, and risk representation. The simulation result of IWRAP Mk2 shows that the grounding incident in Karimunjawa is 3.736 incidents/year. Meanwhile the projected coral reefs area that affected by the grounding incident reached 0.55% from the total coverage. Based on Risk Management Australia AS/NZS 4360 (2004), the results showed that support vessels have high risk of grounding, followed by passenger ships with a moderate risk level. Tankers and other types of vessels were found to have a low risk of grounding.

1. Introduction

The Karimunjawa Islands, which located in Jepara Regency, Central Java, are one of the famous marine tourism destinations in Indonesia. This island is well known for snorkeling or diving activities due to its beautiful underwater scenery dominated by coral reefs which covers approximately 713.11 hectares (Greenpeace Indonesia, 2018). The island consists of a land area of 1,500 hectares and surrounding waters of 110,000 hectares. It is home to coral reefs, mangroves, and nearly 400 species of marine fauna, including 242 species of ornamental fish [1].

The International Union for Conservation of Nature (IUCN) defines a national park as a natural area or area on land and/or sea to protect the ecological integrity of one and/or more ecosystems so that they can be maintained for present and future generations. In line with that, the Decree of the Minister of Forestry and Plantations No. 78/Kpts-II/1999 on February 22nd, 1999 stated that the Karimunjawa Nature Reserve and its surrounding water were designated as a National Park [2] which is geographically located at 5°40'39"- 5°55'00" S and 110°05' 57"- 110°31' 15" E. The national park is defined according to the Law of Republic of Indonesia No. 5 of 1990 about Conservation of Natural Resources and Ecosystems, as a Natural Conservation Area which has native ecosystems, managed by a zoning system [3], for which the zoning system for Karimunjawa Islands is presented in Figure 1. Based on that, the allowable activities in the protection zone and cultivation zone are activities that can protect the living natural resources and their respective ecosystems. That means there should be no activity performed that could damage living natural resources and the protected ecosystems [4].

Based on statistical documents from the Karimunjawa National Park Office, damaging coral reefs is against the Law of Republic of Indonesia No. 5 of 1990 about the Conservation of Natural Resources and Ecosystems. In addition, the occurrence of ship grounding incident is also categorized as a situation that against the Law of Republic of Indonesia No. 32 of 2009 concerning Protection and Management of Environment. One of the objectives of this law is to protect the area of Indonesia from pollution and damage to the environment as implied by Article 3. According to the statistical documents from Karimunjawa National Park Office, it stated that during 2017 to 2021, there have always been cases of coral reefs damages due to ship groundings. The cases that occurred were followed up by taking legal action in accordance with the statutory regulations. Based on monitoring in the observation area, the percentage of live coral cover reached 42%, abiotic 30%, dead coral 26%, algae 1%, and other biota 1% [4].

Damage to coral reefs occurred years ago and persists until today due to illegal activities such as barges that sailing close to the Karimunjawa marine protection area. This reason becomes the background of conducting risk assessment to see the risk level currently bears by the area as well as the options that can be applied to preserve the nature in Karimunjawa.

2. Case Study: Karimunjawa Marine National Park

2.1 Marine Traffic in Karimunjawa

Decree of the Minister of Transportation of the Republic of Indonesia no. KM 19 of 2019 on January 25, 2019, has been established to determine the shipping lanes, routeing systems, traffic procedures, and anchorage areas according to their interests in the shipping channels entering Karimunjawa Port [5]. The minister decree stated that ships with a gross tonnage (GT) of 500 GT

Figure 1. Karimunjawa Islands Marine National Park zonation map [website karimunjawa]

or more are required to use ship pilotage services. In prohibition section it is also stated that cargo/container ships are prohibited from entering shipping lanes with an under-keel clearance (UKC) of less than 10% of draft, except with the permission of port authorities and barges with UKC of less than 5% of the vessel draft [5].

The density of shipping routes around the Karimunjawa National Park does not show a heavy traffic in nearshore areas. However, the traffic at the south and north part of Karimunjawa islands shows a great density that approximately spans from 100 to 407,000 trips/0.15sq. km/year as shown in the traffic density from marine traffic website on June 11th, 2024 in Figure 2 (a). A number of tankers, whom activities are forbidden in the Karimunjawa waters, were also found approximately 3 nautical miles to the Menjangan Besar Island as depicted in Figure 2 (b). In addition, as shown in Figure 3 that in 2022, the highest number of ships were passenger ships with a total of 1102 ships. Although tugboats experienced the most incidents of ship grounding, it is in the third position of ship with a total of 456 ships. While the number of other types of ships such as cargo ships is 835, LCT is 54, tanker is 358, and 19 ships for towing ship.

2.2 Environmental Condition

As mentioned in Chapter 1 that Karimunjawa is well-known for its natural beauty, one of which is the coral reefs. Figure 4 shows the coral reefs area where it is colored pink. The figure displays that almost all area in the near shore area is covered by coral reefs. In detail, a study about mapping the coral coverage conducted by Satya et al. [7] has shown that tourism activities significantly reduce the area covered by the coral reefs, as experienced by the Menjangan Kecil Island.

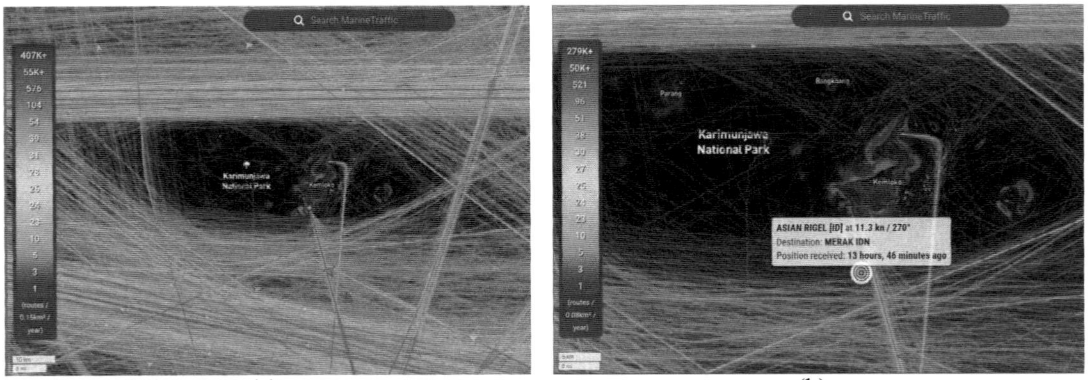

(a) (b)

Figure 2. Marine traffic in surrounding water of Karimunjawa Islands [6].

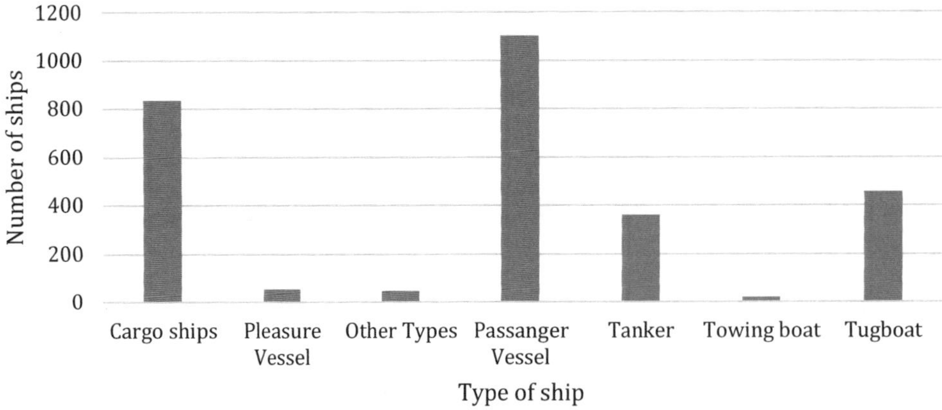

Figure 3. The number of ships sailing nearby Karimunjawa Islands.

In addition, the seabed profile in the proximity of Karimunjawa Islands is relatively shallow as affected by corals and the sandy condition [7] as shown in Figure 5 below. The dark blue area indicates that those are the sandy area whereas the water depth around it is less than 5 meters. This condition would be very much dangerous not only to the environment in terms of coral reefs coverage but also to the ship structure.

2.3 Accidents and Incidents History

In addition to the marine traffic and environmental condition above, several historical accidents have been recorded over 5 years in Karimunjawa. It shows that the coral reefs in Karimunjawa National Park greatly affected by the grounding incidents. Based on this record, ship grounding is considered as the potential hazard that can bring damage to the marine environment around Karimunjawa.

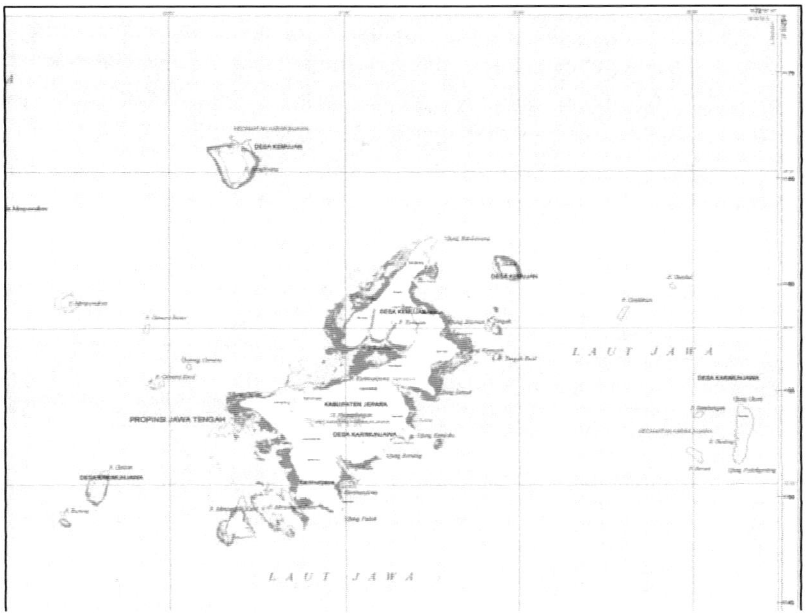

Figure 4. Coral reef map of Karimunjawa Islands.

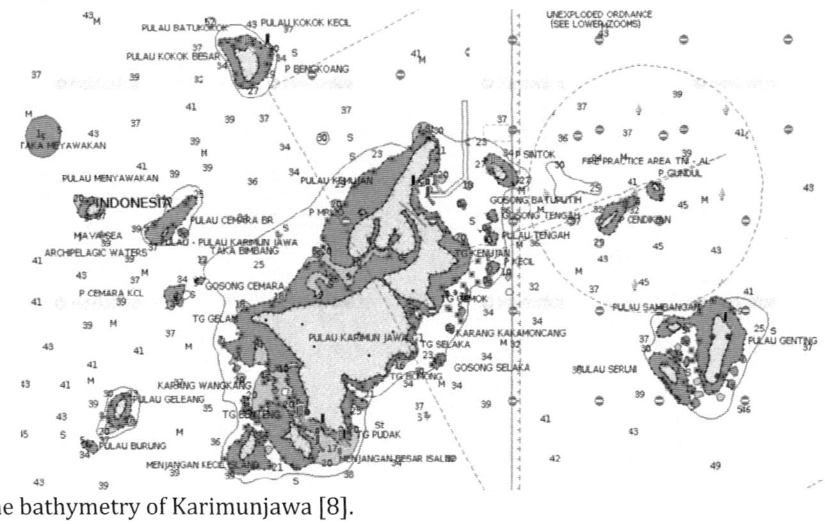

Figure 5. The bathymetry of Karimunjawa [8].

3. Risk Analysis

The quantitative risk assessment (QRA) is proposed to better show the risk level of hazards that have been identified in the previous step. At least three stages of QRA are required, those are frequency analysis, severity analysis, and risk representation. The risk representation could be in the form of risk matrix, f-N curve, or any other visualization.

3.1 Hazard Identification.

The objective of this step is to provide a list of hazards that may occur to the facilities or area being analyzed and can lead to an accident or incident. Any possible hazards are identified through group discussion or other structured method. All hazards associated with past occurrences and possible future hazards could be identified with historical data as well as a discussion with experts. There are many methods for hazard identification that can be utilized in this step, for example, Failure Mode and Effect Analysis (FMEA), Hazard and Operability Study (HAZOP), Fault Tree Analysis (FTA) [9], What-if analysis [10], fishbone or Ishikawa diagram [11], and checklist. All possible hazards that appear in Karimunjawa caused by the marine traffic shall be then screened by a standard method to conclude major potential hazards. In this study, the hazard identification process was done by performing desktop study on the historical data of incidents in Karimunjawa together with the damage caused by each within several years as shown in Table 1. Furthermore, the cause of grounding incidents based on available reports are due to marine traffic, human resources, vessel internal factors, and environmental factors.

Table 1. Accidents and Incidents Recorded in Karimunjawa from 2017-2022

No	Cases	Details
1	Vesselgrounding in Karang Kapal Indonor (January 10, 2017)	Location 2 miles from Karang kapal Indonor and ship engine is damaged by coral reef. Reports have not been handled.
2	Damaging coral reefs accidentally due to a ship grounding. (January 14, 2017)	Location in the waters of Cilik Island.
3	Damaging coral reefs accidentally due to a ship grounding. (February 10, 2017)	Location in the waters of the Tengah Island.
4	Vesselgrounding in Tengah Island (February 11, 2017)	Not reported.
5	Damaging coral reefs accidentally due to a ship grounding. (June 13, 2017)	TNKJ Fisheries Traditional Zone.
6	Vesselgrounding in Tengah Island. (February 10, 2017)	Extent of damage to coral reefs 1.420,32 m^2 ; Coral cover 53,20%; Claim Area 755,61 m^2.
7	Vesselgrounding in Cilik Island. (January 14, 2017)	Area of damage to coral reefs 267,22 m^2; Coral cover 41,8%; Claim area 111,69 m^2.
8	Vesselgrounding in Lake Waters.	Area of damage to coral reefs 52,46 m^2; Coral cover 35,75%; Claim Area 24,09 m^2.
9	Vesselgrounding in Gosong Seloka Waters.	Area of damage to coral reefs 184,45 m^2; Coral cover 49,86%; Claim Area 90,38 m^2.
10	Vesselgrounding in Tanjung Gelang Waters. (September 7, 2018)	Area of damage to coral reefs 7,11 m^2; Absolute coral cover.
11	Vesselgrounding in Gosong Seloka Waters. (2020)	Area of damage : 645,95 m^2
12	Vesselgrounding in Legon Ipik Waters. (2020)	Area of damage: 4.239,21 m^2
13	Vesselgrounding in Menjangan Besar Waters (2021)	Not reported.
14	Vesselgrounding in Tengah Island. (2021)	Not reported.
15	Vesselgrounding (September 22,2022)	Not reported.
16	Vesselgrounding in Gosong Seloka Island (2023)	Not reported.

3.2 Frequency Analysis

Frequency analysis is aimed to estimating how often the hazard occurs in a year. The hazards were selected in the previous step and then be calculated by using the IALA Waterway Risk Assessment Program (IWRAP) Mk2. This tool helps to model the marine traffic in an area where it possesses a risk of ship collisions, allisions and groundings. In IWRAP, the traffic can be modelled by using legs which represent the passage of the ship. Each leg must be completed with the information regarding traffic, i.e. traffic volume and composition, the lateral distribution, as well as its geometry. In addition, the bathymetry of the waterways can also be the input to the model. IWRAP then calculates the frequency of collisions and groundings per year based on the specified model. However, IWRAP does not calculate the severity of collisions and groundings incident. In the case of representing the frequency level, this paper used the frequency index as shown in Table 1 [12].

3.3 Consequences Analysis

In the consequence analysis, the impact of the ship aground that causes damages to the marine biota, especially coral reefs, is estimated. In calculating the affected area, the estimated area was calculated by knowing the area of the waterplane as displayed in Figure 4. A simulation is carried out to model the stopping distance after grounding incidents happens, so the affected area can be calculated as depicted in Figure 5. The result of the analysis is categorized based on the consequence index in the Table 3. Furthermore, the area of waterplane is estimated based on the following equations:

C_{wp} U type sections $= (0.778 \times Cb)+0.248$ (1)

C_{wp} V type sections $= (0.743 \times Cb)+0.297$ (2)

C_{wp} normal sections $= (1+2Cb)/3$ (3)

A_{wp} (Area of the Waterplane) $= C_{wp} \times L \times B$ (4)

Table 2. Frequency index [12].

Index	Frequency	Description	
1	Very unlikely	Less than once per 1000 years	< 1/1000
2	Remote	Once per 100-1000 years	< 1/100
3	Occasional	Once per 10-100 years	< 1/10
4	Probable	Once per 1-10 years	≤ 1
5	Frequent	More than once per year	> 1

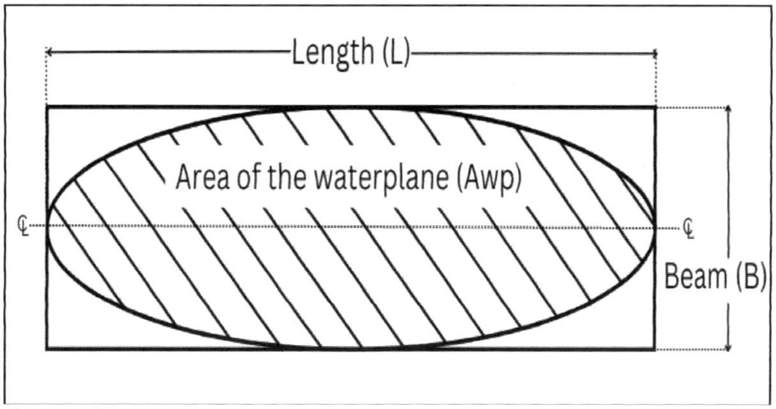

Figure 6. Area of waterplane

Figure 7. Affected area due to groundings

Table 3. Consequence index

Index	Consequence Categories	Total Reef Area Affected
1	Insignificant	0-10%
2	Minor	11-30%
3	Moderate	31-50%
4	Major	51-75%
5	Catastrophic	76-100%

3.4 Risk Representation

The risk representation uses standard from AS/NZS 4360:2004 [13] to define the risk based on the frequency and the consequence. Risk matrix is categorized into four levels as shown in Table 4. Those levels are low, moderate, high, and extreme. This standard provides a generic guide on managing risk. It is not case specific and hence can be used to represent the risk of ship groundings in Karimunjawa Islands.

4. Case Study: Ship Groundings in Karimunjawa Islands

Ship grounding occurs when a vessel makes contact with the seabed either because the water depth is too shallow or there is an obstacle coming from the seabed. A soft seabed might not pose a great danger to the hull and bottom of the ship. However, if the condition of the seabed is rocky or there are coral reefs, there is a risk of severe structural damage to the ship hull and lead the sea water to enter the ship hull and compromise the ship stability. This incident can cause the vessel to list or even sinking when there is no immediate action taken. Two types of grounding incidents are identified based on the circumstances of the accident [14]:

1. Power-driven grounding, and
2. Drifting grounding.

4.1 Types of Groundings

4.1.1 Powered Groundings

Powered groundings happen mostly due to the lack of navigational skills during the sailing time [15]. Hence, the vessel must be sailing using her own main engines and propeller. This type of grounding is the most common one which can be caused by human errors such as error in

Table 4. Grounding scenario categories

Category	Grounding Scenarios
I	This category is for grounding incident when the ship is following the usual waterway a normal speed. It primarily caused by human error but also might be caused b unexpected problems related to the propulsion/steering system that happen around th fixed marine structure or the ground.
II	Ships that cannot change their course at a desired turning point near the obstacle.
III	Ships that carry out evasive actions close to the obstacle and consequently run agroun or collide with the object.
IV	Scenario other than Cat. I, II and III, for example ships that sail out of their course due t propulsion failure.

Table 5. Risk matrix based on AS/NZS 4360:2004.

	Consequences				
Frequency	Insignificant	Minor	Moderate	Major	Catastrophic
Almost certain	High	High	Extreme	Extreme	Extreme
Likely	Moderate	High	High	Extreme	Extreme
Occasionally	Low	Moderate	High	Extreme	Extreme
Unlikely	Low	Low	Moderate	High	Extreme
Rare	Low	Low	Moderate	High	High

making a passage plan, insufficient bridge resource management, vessel traffic service operator lack of skills [16], etc. Table 1 below summarized the power-driven grounding categories based on IWRAP Mk2 working document.

According to the description above, the expected number of groundings in the scenarios in category I (N_I) and category II (N_2) are [17]:

$$N_I = \sum_{Ship\ class,i} P_{c,i} Q_i \int_{Z_{min}}^{Z_{max}} f_i(z) d \tag{5}$$

$$N_{II} = \sum_{Ship\ class,i} P_{c,i} Q_i \exp\left(-\frac{d}{a_i}\right) \int_{Z_{min}}^{Z_{max}} f_i(z) d \tag{6}$$

where,

a_1 : Average distance between position checks by the navigator.
d : Distance from the obstacle to the bend in the navigation route varying with the lateral position,
s, of the ship.
i : Index for ship class, categorized after vessel type and dead weight or length.
$f_i(z)$: Probability density function for the ship traffic.
N_I : Expected number of category I grounding events per year.
N_{II} : Expected number of category II grounding events per year.
$P_{c,i}$: Causation probability, i.e. ratio between ships grounding and ships on a grounding course.
Q_i : Number of ships in class i passing a cross section of the routeper year.
z : Coordinate in the direction perpendicular to the route
z_{min}, z_{max}: Transverse coordinates for an obstacle.

IOP Conf. Series: Earth and Environmental Science **1423** (2024) 012025 doi:10.1088/1755-1315/1423/1/012025

4.1.2 Drifting Groundings

There are two main causes for a ship to be not under command, they are rudder stuck and main engine loss of power. The record shows that the typical number of blackout for a given ship is around 0.1 to 2 blackouts annually. This study used the blackout frequencies shown in Table 6 for the base values [17].

4.2 Risk Analysis of Ship Grounding

In modeling the location of the ship grounding, the author maps based on two scenarios. The first is based on historical data of the ship grounding location from 2017 to 2021 which is recorded in the Karimunjawa National Park Office document. The second is based on areas that have the potential to cause a grounding according to the results of the ship grounding frequency generated by the application IWRAP Mk2. In addition, to determining the grounding area in calculating consequences, the author also mapped several types of ship heading positions when grounded based on the track that usually passed the vessel as can be seen in Figure 8.

4.2.1 Frequency of Ship Grounding

In analysing the frequency of ship grounding with IWRAP Mk2, firstly the water depth in the surrounding area was entered to the software. The water depth data was gathered from the nautical chart from the Indonesian Naval Hydro Oceanographic Center (Pushidrosal) as well as the Navionics website. Using the equation (5) and (6) to estimate the number of grounding events per year, Table 7 shows the results of the frequency analysis. It shows that the powered grounding has a higher frequency compared to the drifting grounding with the total of 2.55 events per year and 1.07E-02 events per year, respectively.

4.2.2 Consequence of Ship Grounding

The damaged coral area due to a ship grounding was modelled in a simulation software as shown in Figure 10. The equations (1) until (4) were used to estimate the area of waterplane that affect the damage on the coral reefs. The results of the estimated damage area based on each vessel type are given in Table 8.

Table 6. Blackout frequencies from IWRAP

Vessel type	Annual Frequency	Hourly Frequency
Passenger / Ro-Ro	0.1 per year	1.15E-05 per hour
Other vessels	0.75 per year	8.56E-05 per hour

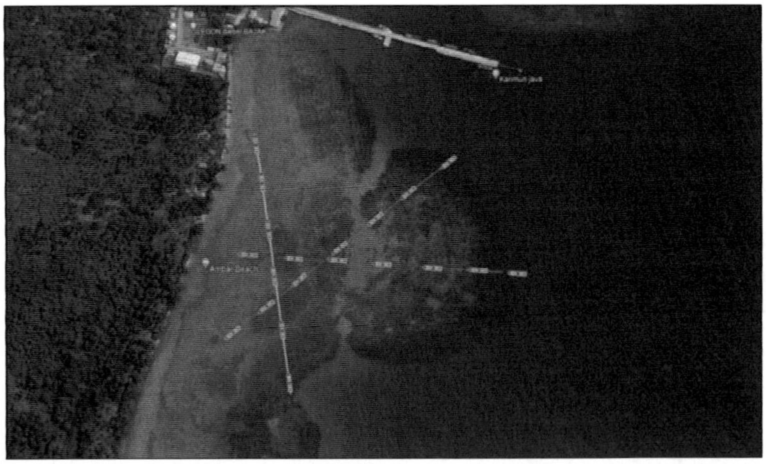

Figure 8. Grounding scenario in Karimunjawa Islands

IOP Conf. Series: Earth and Environmental Science **1423** (2024) 012025 doi:10.1088/1755-1315/1423/1/012025

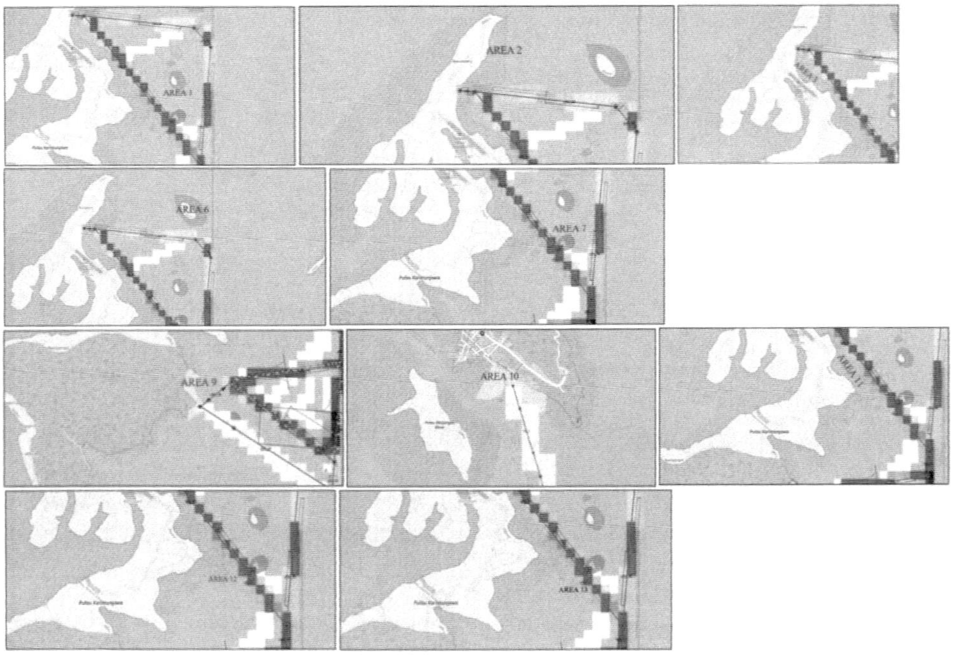

Figure 9. The modelled area for frequency analysis

Table 7. Results of grounding frequency analysis

	Powered Grounding			Drifting Grounding		
Area	Passenger ship	Support ship	Other ship	Passenger ship	Support ship	Other ship
1				6.1E-04		9.8E-03
2				1.3E-03		2.1E-02
3	2.0E-02		4.7E-01	2.3E-03		2.3E-03
6	4.1E-03		1.2E-01			
7	2.2E-02		7.0E-01			
9	3.1E-02		9.7E-01			
10		3.0E-02			3.9E-04	
11	3.0E-02		1.7E-01	3.7E-03		6.0E-02
12				2.7E-04		4.2E-03
13				5.3E-04		7.8E-03
14						1.6E-04
Total	9.9E-02	3.0E-02	2.42E-00	8.8E-03	3.9E-04	9.8E-02

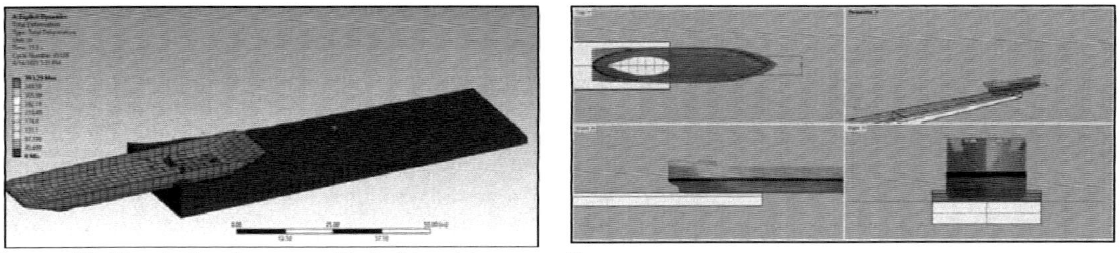

Figure 10. Simulation for area affected by ship grounding

Table 8. Affected area due to ship grounding

Vessel Type	Affected Area (m²)
Passenger	774.58
Tanker	0.011
LCT	309.28
Barge	6912.76
Total	**7996.63**

4.2.3 Risk Representation of Ship Grounding

The frequency and consequence of each type of ship have been calculated. It was found that the affected area of coral reefs is in the first category of consequence, namely 1-10% of the affected coral reef area of the entire coral reef area in Karimunjawa, which approximately 7000 hectares. Table 9 below shows the risk matrix based on AS/NZS 4360:2004 for each type of ship. Based on this table, the results show that the risk of ship grounding due to the activities of support ship (i.e. tugboats and barges) has the highest risk which is high level. The risk due to activities of passenger ships is in the moderate level. Whilst tanker and other types of ship are in the low level of risk.

Table 9. Risk representation due to ship grounding for each vessel type

	Insignificant	Minor	Moderate	Major	Catastrophic
Almost Certain	SUPPORT				
Likely	PASSENGER				
Occasionally	OTHER				
Unlikely					
Rare	TANKER				

Low Medium High Extreme

4.3 Recommended Mitigations

The result of the risk matrix shows that the risk of support vessel grounding incidents to the coral reefs in Karimunjawa is high. That means some mitigations action should be recommended to decrease the risk. As shown in the Table 9 that a high level of risk is caused by the high grounding frequency, so this study focused on how to push the number of support vessel grounding.

At the moment, Karimunjawa VTS have not installed AIS receiver to monitor the movement of ships in the vicinity. In addition, radar and radio are also needed as to communicate directly with every ship that is not activating their AIS at several points in Karimunjawa. To further maximize protection in Karimunjawa conservation area, the deployment of patrol boats needs to be held more frequently.

5 Conclusion

This paper has an objective to find the risk due to ship grounding in Karimunjawa Islands which is a marine national park. The risk was represented based on AS/NZS 4360:2004 Standard for each type of ships that pass nearby Karimunjawa Islands. The results showed that support vessels have high risk of grounding, followed by passenger ships with a moderate risk level. Tankers and other types of vessels were found to have a low risk of grounding. These findings highlight the need for targeted risk mitigation strategies, with a particular focus on improving navigational practices and safety protocols for support vessels operating in the area. Further studies to decide the most effective and efficient risk control options are needed by means of cost-benefit analysis.

References

[1] Indonesia-Tourism.com, "The Paradise of Java: Karimunjawa Islands," Indonesia-Tourism. [Online]. Available: https://www.indonesia-tourism.com/blog/the-paradise-of-java-karimunjawa-islands/

[2] Menteri Kehutanan, *Surat Keputusan Menteri Kehutanan No.78/Kpts-II/1999*. 1999.

[3] Republik Indonesia, *Conservation of Natural Resources and Ecosystems*, vol. Law Number 5 of 1990. 1990.

[4] Taman Nasional Karimunjawa, "Statistik Balai TN Karimunjawa Tahun 2021," Taman Nasional. Karimunjawa, 2021.

[5] Ministry of Transportation of Republic of Indonesia, *Establishment of Shipping Lanes, Route System, Traffic Procedures, and Ship Berthing Areas According to Their Needs in the Karimun Jawa Port Approach Channel*. 2019.

[6] Marine Traffic, "Live Map." Accessed: Aug. 01, 2019. [Online]. Available: https://www.marinetraffic.com/en/ais/home/centerx:115.8/centery:-8.4/zoom:11

[7] E. D. Satya *et al.*, "Mapping coral cover using Sentinel-2A in Karimunjawa, Indonesia," *Biodiversitas*, vol. 24, no. 2, Feb. 2023, doi: 10.13057/biodiv/d240219.

[8] Navionics, "Chart Viewer." Accessed: Jun. 12, 2024. [Online]. Available: https://webapp.navionics.com/?lang=en#boating@9&key=zkqb%40%7DodaT

[9] H. EP Marpaung, N. Suranta, and D. Ferdinal Frengki Simarmata, "Analysis of Formal Safety Assessment (FSA) for Ship Accident Risk Assessment: A Case Study of Work Accident MV Mandarin Ocean at Tanjung Priok Main Port," *ARRUS J. Soc. Sci. Humanit.*, vol. 3, no. 4, pp. 469–478, Aug. 2023, doi: 10.35877/soshum1945.

[10] F. I. Prastyasari, A. A. B. Dinariyana, T. F. Nugroho, and A. S. Sukmana, "Formal Safety Assessment for determining sensitive area due to traffic of tourism vessels at Lombok Strait," *IOP Conf. Ser. Earth Environ. Sci.*, vol. 1166, no. 1, p. 012052, May 2023, doi: 10.1088/1755-1315/1166/1/012052.

[11] T. Luo, C. Wu, and L. Duan, "Fishbone diagram and risk matrix analysis method and its application in safety assessment of natural gas spherical tank," *J. Clean. Prod.*, vol. 174, pp. 296–304, Feb. 2018, doi: 10.1016/j.jclepro.2017.10.334.

[12] M. Zaman, A. Santoso, E. Kobayashi, and ..., "Formal safety assessment (FSA) for analysis of ship collision using AIS data," *... J. Mar. ...*, no. Query date: 2022-10-14 07:30:50, 2015, [Online]. Available: https://yadda.icm.edu.pl/baztech/element/bwmeta1.element.baztech-c3b6e905-1c74-46b0-beca-63899f4918ef

[13] Joint Standards Australia/Standards New Zealand Committee, *AS/NZS 4360:2004 Risk Management*, 2004.

[14] M. Jurdziński, "Causes of Ships Groundings in Terms of Integrated Navigation Model," *Annu. Navig.*, vol. 24, Dec. 2017, doi: 10.1515/aon-2017-0009.

[15] S. Galić, Z. Lušić, S. Mladenović, and A. Gudelj, "A Chronological Overview of Scientific Research on Ship Grounding Frequency Estimation Models," *J. Mar. Sci. Eng.*, vol. 10, no. 2, 2022, doi: 10.3390/jmse10020207.

[16] "Grounding of general cargo vessel BBC Marmara," Marine Accident Investigation Branch, Scotland, Serious Marine Casualty REPORT NO 8/2023, Nov. 2023.

[17] P. Friis-Hansen, "Basic Modelling Principles for Prediction of Collision and Grounding Frequencies," *Tech. Univ. Den.*, p. 59, Mar. 2008.

Development of Measurement Tool for Energy Efficiency Operational Indicator (EEOI)

H Prastowo[1], T Pitana[1], K B Artana[1], I M Ariana[1], A T Pramana[1,2]

[1] Department of Marine Engineering, Institut Teknologi Sepuluh Nopember, Surabaya, 60111, Indonesia
[2] Department of Maritime Studies, Wismar University of Applied Sciences, 23966 Wismar, Germany

*E-mail: h_prastowo@its.ac.id

Abstract. The Energy Efficiency Operational Index (EEOI) is a value that represents the energy efficiency of a ship's operations in relation to CO2 emissions. Improving energy efficiency is one of the important ways to achieve cleaner shipping. Statistical methods are used to predict a number of parameters that determine the value of the EEOI. The aim of this research is an EEOI measurement tool. The analysis was carried out with 100 container ships. The Ship Traffic Emission Model (STEAM) and Multiple Linear Regression (MLR) methods were used in this research. STEAM modeling is based on estimated fuel oil consumption and MLR generated from STEAM EEOI results. This modeling also considers the following parameters overall length, width, draft, depth, service speed, container capacity, average speed of last trip, average draft of last trip, installed main engine power, and travel time. Furthermore, EEOI measurement tools were created in the Android-based version which represent EEOI values based on the ship's last trip per voyage.

1. Introduction

The contribution of the maritime sector to greenhouse gases increased from 2.4% to 3% and this is the IMO's estimate from 2007 to 2050. The projections of the amount of greenhouse gases emitted by the maritime sector from 1.6% to 4.1% of global CO_2 emissions from fuel combustion [1]. Maritime transport is an important avenue for the development and growth of the national economy. There are many ships operating and transporting goods all over the world, this type of transport brings revenue and other benefits to each country. To achieve the rate of environmental pollution and further develop green shipping, ship operations play an important role. In fact, maritime transport contributes the most environmentally friendly when it only emits 2.7% of greenhouse gas (GHG) emissions and for marine pollution, it only emits 12% of total pollution [2].

In 1973, the IMO established the Marine Environment Protection Committee (MEPC) to regulate marine pollution from the shipping sector. In addition to the MEPC, the International Convention for the Prevention of Pollution from Ships (MARPOL) was also inaugurated in the same year for the same reason. The MARPOL Convention addresses a number of marine pollution issues caused by the shipping industry, namely marine pollution from oil spills, noxious liquid substances and other harmful substances carried, sewage, waste and water pollution from ships.

In 1997, MARPOL Annex VI was introduced. The purpose is to prevent the air pollution that came from the shipping sector and gave limitation for the ship's exhaust gas emissions. Then in 2011, the new chapter of MARPOL Annex VI was established. The purpose is to prevent GHG emissions. IMO has made several guidelines to reduce GHG emissions. These guidelines are the Energy Efficiency Design Index (EEDI), the Ship Energy Efficiency Management Plan (SEEMP), and the Energy Efficiency Operational Indicator (EEOI).

The processes of the Ship Energy Efficiency Management Plan (SEEMP) are divided into four stages: planning, implementation, monitoring, self-assessment and improvement. The processes are discussed and described together. Each process is taken from the SEEMP (MEPC) guidance. The main objective of the SEEMP is to establish a mechanism for the shipping company and/or the ship to improve the energy efficiency of the ship and its operations [2].

EEOI is a tool for determining the CO_2 gas emission to the environment per transport or voyage work. So, it means that it could represent the actual voyage efficiency of ship operation. It also could be used to evaluate fleet management for better emissions control. Fuel oil consumption of a ship is one of the critical parameters which effected the energy efficiencies based on the operational condition. This parameter also could be used for controlling the greenhouse gasses (GHG) emission. Nowadays, the shipping sector is strengthening the regulation about GHG emission and preparing the green shipping with less emission emitted. Minimizing fuel oil consumption is one of the solutions for better emission control soon. So, reducing energy consumption equalize to reducing emissions. Nowadays, the operating condition to achieve the low value of EEOI by cargo management is the trim optimization. This method will be affecting not only the ship resistance but also fuel oil consumption. Trim optimization is a relatively new concept recommended by the International Maritime Organization, which requires no ship structural modification and can be attained simply by ballast water management and load distribution [3]. Therefore, the objective of this research is to propose the impact of cargo, distance, power and speed of ship on ship operations on EEOI value, thereby creating an EEOI measurement tool.

2. Methodology

2.1 Fuel Oil Consumption Estimation STEAM Method

Based on data-driven, Ship Traffic Emission Model (STEAM) methods applied. J.-P. Jalkanen proposes this method. Automatic Identification System (AIS) data is used for presenting this estimation method. The predictions of fuel consumption were compared with the actual values obtained from the shipowners. For instance, the annual fuel consumption (neglecting wave effects) is within 5% of the reported values for the RoPax ship in 2007 [4].

Other input data of the STEAM model is extracted from the AIS data center, J.-P. Jalkanen acquires from the Helsinki Commission (HELCOM). Then the technical detail of the ships is taken from ship register data. The specific fuel oil consumption (SFOC) of 200g/kWh is used for analysis as a template value, and this value is taken from the MAN Diesel and Turbo Guide without any correction. This model could be applied universally if the AIS global coverage is achieved. The calculation for the fuel estimation begins with the power P calculation,

$$P_{transient} = CF + CR + CA + CAA\left(\frac{1}{2}V^3S\right)\frac{1}{\varepsilon_0} \tag{1}$$

Where,

CF : Frictional Resistance

CR	: Residual Resistance
CA	: Appendage Resistance
CAA	: Air Resistance
V	: Design Speed
ε_0	: Propulsive Coefficient
S	: Wetted Surface Area

These parameters cannot be found in the available database, so the solution for this is to simplify the Equation (1) by assuming that CF, CR, CA, CAA, S and ε_0 are the ship specific constants. Then, in that case, it can be written only as.

$$P_{transient} = \frac{kV_{transient}^3}{0.514^3} \tag{2}$$

Where k is defined as:

$$k = 0.514^3 * \frac{\varepsilon_p * P_{installed}}{(V_{design} + V_{safety})} \tag{3}$$

Where,

P_transient	: Instantaneous power for calculation
k	: The simplified quantity of Equation (1)
ε_p	: Main engine load at the maximum continuous rating of the main engine
P_installed	: Total installed power (kW) of the main engines
V_design	: Design speed (m/s)
V_safety	: Safety margin (m/s)

2.2 Fuel Oil to CO_2 Conversion Factor

Fuel oil consumption calculation is done using the STEAM method proposed by J.-P. Jalkanen, then came to CO_2 calculation using the conversion table from the IMO guidelines for EEOI. Each type of fuel has it's own carbon chain the chemical properties, so the value from one to another slightly different. The conversion factor is used to calculate the amount of CO_2 released from combustion for a specific volume of fuel oil burned. Fuel mass to CO_2 mass conversion factors (CF) is released by IMO in the guidelines of EEOI, see Table 1.

2.3 Energy Efficiency Operational Indicator

The Marine Environment Protection Committee in 2009, agreed to establish guidelines for the voluntary use of the Ship Energy Efficiency Operational Indicator (EEOI)in the Annex. Ship Energy Efficiency Management Plan is an operational measure to create a mechanism to improve the ship's energy efficiency. According to IMO [5], even though it is voluntary, the approach of applying SEEMP into a ship or fleet could assist ship owners, ship operators, and other parties that are concerned in the evaluation of the performance of their ship or fleet. SEEMP is designed to be implemented for all ships. The EEOI calculation was based on guidelines for voluntary use of the ship's EEOI. Detailed from the formula given by MEPC.1/Circ.684, EEOI is defined as the ratio of the mass of CO_2 emitted per unit of transport work as shown in Equation (4) and (5).

Table 1. Fuel Mass to CO_2 Mass Conversion Factors (C_f) [6]

Type of Fuel	Reference	Carbon Content	C_F (ton-CO_2/ton-fuel)
Diesel/Gas Oil	ISO 8217 Grades DMX through DMC	0.875	3.206000
Light Fuel Oil	ISO 8217 Grades RMA through RMD	0.860	3.151040
Heavy Fuel Oil	ISO 8217 Grades RME through RMK	0.850	3.114400
Liquified Petroleum Gas (LPG)	Propane Butane	0.819 0.827	3.000000 3.030000
Liquified Natural Gas (LNG)		0.750	2.750000

$$EEOI = \frac{\Sigma_j FC_j \times C_{Fj}}{m_{cargo} \times D} \tag{4}$$

$$Average\ EEOI = \frac{\Sigma_i \Sigma_j \times (FC_{ij} \times C_{Fj})}{\Sigma_i (m_{cargo} \times D)} \tag{5}$$

Where,

j : fuel type
i : voyage number
FC_{ij} : the mass of consumed fuel j at voyage i
C_{Fj} : fuel mass to a CO_2 conversion factor of fuel j
m_{cargo} : cargo carried, or work is done (tonnes, TEU, passengers) or gross tonnage of passenger ships
D : distance traveled

2.4 Multiple Linear Regression

It is using GNU Octave to determining the function to be used for android based apps. This function will generate the estimation value based on the variables that have been chosen—then using the term of Multiple Linear Regression (MLR) to determine the estimation value. Data preparation also must be held to match the MLR requirements.

From the fundamental equation of multiple linear regression, then came to choosing the variables for the function. All of the variables must be easy to acquire, and the trendline of each variable must showing a linear line regarding the output value, which is the EEOI value. So by that meaning, the variables that have been chosen are,

EEOI : y/output
LOA : x_1
B : x_2
H : x_3
T : x_4

Vs : x_5
Container Capacity : x_6
Avg Speed : x_7
Avg Draught : x_8
Power Installed : x_9
Travel Time : x_{10}

For x_0 is equal to 1. Then using the GNU Octave, and based on the data that has been prepared, the value of θ as a result.

Table 2. θ Result from GNU Octave.

θ	VALUE
0	0.0003963798
1	0.0000005049
2	0.0000036758
3	-0.0000011131
4	-0.0000038011
5	-0.0000283217
6	-0.0000000284
7	0.0000124749
8	-0.0000033527
9	0.0000000028
10	-0.0000000290

3. Data Analysis and Discussion

This research is using 100 container ship data for calculation and analysis. The online AIS database acquires those data from FleetMon and the ship's register. The ship data collected includes length of overall, breadth, height, draught, speed, power of main engine and auxiliary engine, fuel type, load capacity, voyage data, distance traveled, and travel time.

3.1 Fuel Oil Consumption Estimation STEAM Method

The estimated value of fuel oil consumption is mainly aimed at the main engine that is installed to the ships on the Maximum Continous Rate (MCR). The fuel oil consumption is estimated using the STEAM method's, that proposed by J.-P. Jalkanen. The absolute percentage error is used to determine the range of error from estimation value to the actual data. The actual data using the MV. Meratus Benoa noon report. This report contains essential voyage data, including the bunkering data, which is showing the amount of fuel oil consumption per day, see Figure 1. Then calculate the data with the time of voyage to get the actual fuel oil consumption. This method, as stated by J.-P. Jalkanen, only have 7% of differences between the actual value, so this error result with 14% of error is quite close to those statements.

3.2. Energy Efficiency Operational Indicator

To obtain the estimated energy performance index (EEOI) result, this result is based on the value obtained from EEOI STEAM. It was then using Multiple Linear Regression to generate a new model for estimating the EEOI value. The result of the MAPE is 26% from the MLR equation to the STEAM estimation method. The EEOI value of the two methods has differences; the STEAM method and the MLR has each factor that affects the value to the actual EEOI, see Figure 2. The value of the EEOI has been calculated from the noon report of the MV Meratus Benoa. From the noon report, it determined the actual value of the EEOI. The actual value will be used for determining the reliability of the two methods.

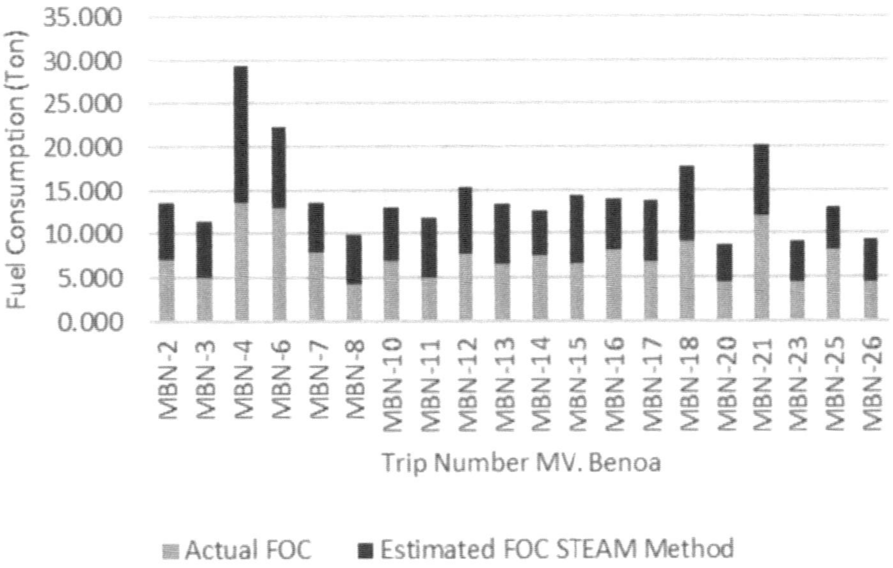

Figure 1. Mean Absolute Percentage Error Actual FOC & Estimated FOC

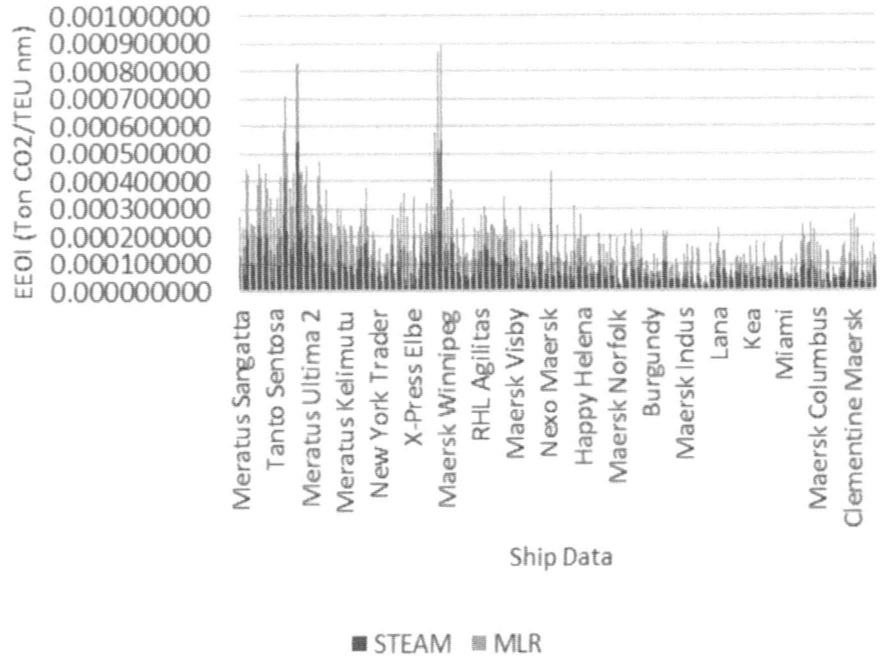

Figure 2. Mean Absolute Percentage Error EEOI STEAM Method & Multiple Linear Regression EEOI

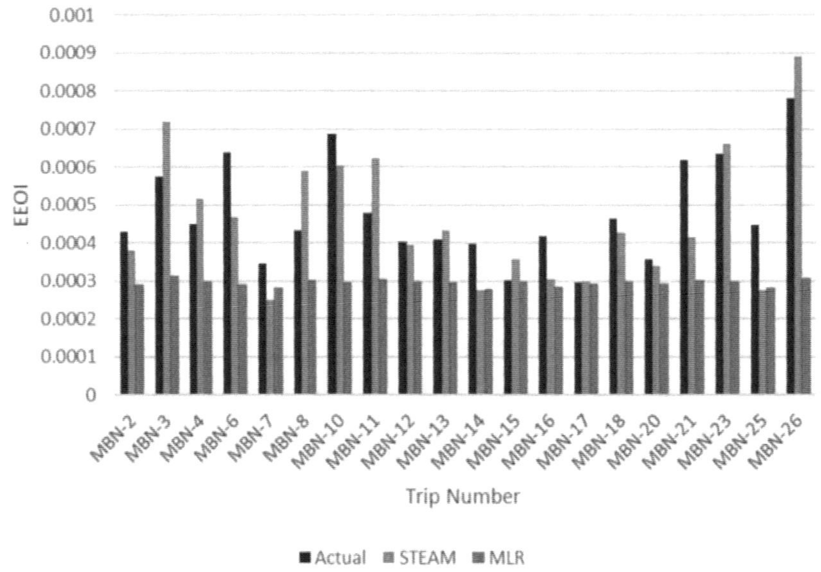

Figure 3. Actual EEOI value, STEAM Method EEOI, & Multiple Linear Regression EEOI

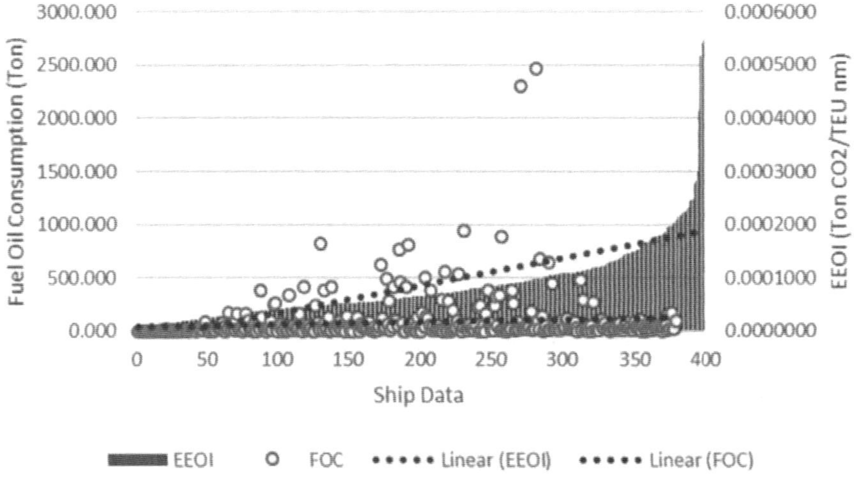

Figure 4. EEOI & Fuel Oil Consumption

Figure 3. shows that the STEAM method is more reliable or relevant than the MLR EEOI value because the variables used for calculation are too scrambled, so the error is massive. The equation for calculating EEOI from the MLR is also using the value from the STEAM Method. It is not directly using the actual value of EEOI due to the minimum data available.

3.3. EEOI Value with the Operational Parameters

Analysis EEOI of the 100 ships data is done with different voyage and different container capacity. Therefore, factors that influence the result of EEOI could be determined.

a) Figure 4. shows the relation between EEOI and fuel oil consumption of the 100 ships that have been acquired before. The value of EEOI is sorted from the lowest to the highest. As shown in the graph, the trend line of fuel oil consumption has a positive gradient and increases insignificantly. The highest fuel oil consumption of a ship is done by Axel Maersk, with 2467.9 Tons of fuel, and it has 0.0001074 Ton CO_2/TEU nm EEOI value. The highest EEOI value of a ship is done by Meratus Gorontalo, with 0.0005485 Ton CO_2/TEU nm, and consumed 102.03 Tons of fuel oil.

b) Figure 5. represents the relation between EEOI and the average speed of the 100 ships. The trend line of average speed shows a slight increase with a positive gradient. The graph shows that the average speed of the ship does have a slight increase correlated with the value of EEOI that increases dramatically. The highest average speed of a ship is done by Munich Maersk, with 20.21 knots, and the EEOI value is 0.0000919 Ton CO_2/TEU nm. The highest EEOI value of a ship is done by Meratus Gorontalo, with 0.0005485 Ton CO_2/TEU nm, and sailed with an average speed of 12.46 knots.

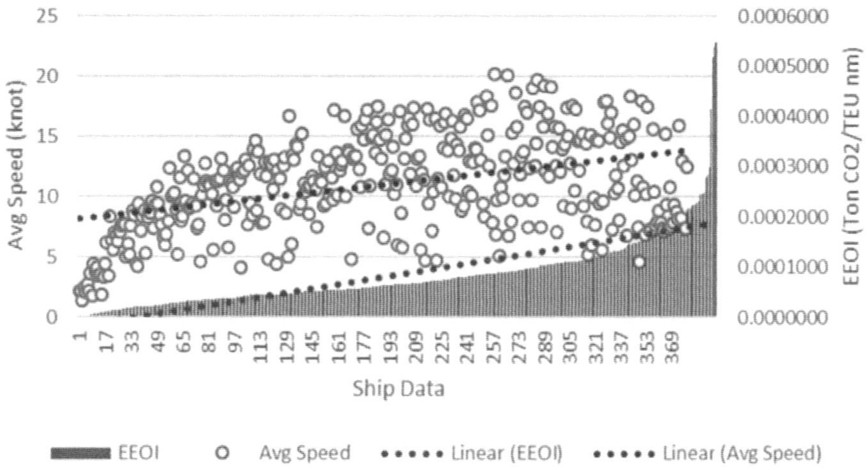

Figure 5. EEOI & Average Speed Last Trip

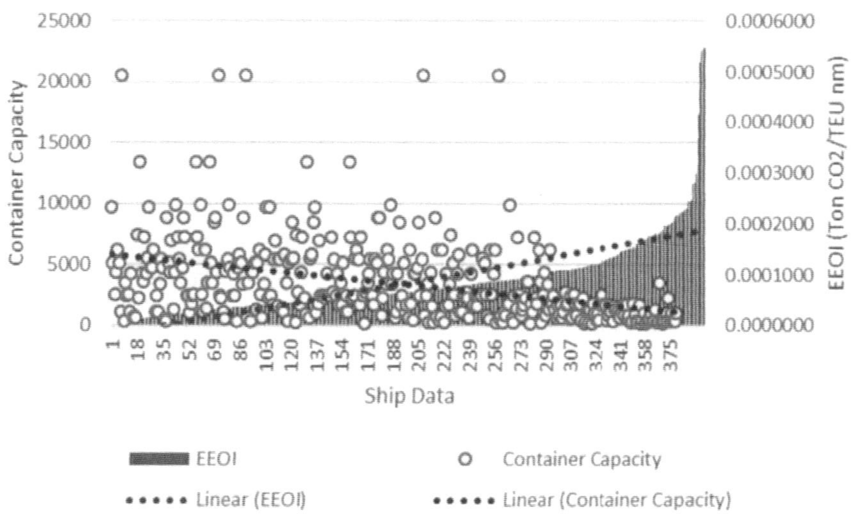

Figure 6. EEOI & Container Capacity

c) Figure 6. shows the correlation between EEOI and the amount of container capacity by the 100 ships. Analyzed from the graph, it shows the different relationships between these variables. The value of EEOI increases as the amount of cargo carried decreased significantly.

IOP Conf. Series: Earth and Environmental Science **1423** (2024) 012026 doi:10.1088/1755-1315/1423/1/012026

These relationships are shown by a negative gradient of the trend line. There is one ship that has the maximum value of container capacity with different EEOI value due to its sailing trip. Munich Maersk is the one that has the biggest container capacity, with 20,568 TEU capacity. The trip from Ningbo Zhoushan to Yangshan generate 0.0000045 Ton CO_2/TEU nm of EEOI, from Tianjin to Busan generate 0.0000340 Ton CO_2/TEU nm of EEOI, from Busan to Ningbo Zhoushan generate 0.0000384 Ton CO_2/TEU nm of EEOI, from Yangshan to Yantian generate 0.0000717 Ton CO_2/TEU nm of EEOI, and from Yantian to Tanjung Pelepas generate 0.0000919 Ton CO_2/TEU nm of EEOI. The highest EEOI value of a ship is done by Meratus Gorontalo, with 0.0005485 Ton CO_2/TEU nm, and the container capacity is 1005 TEU.

d) Figure 7. shows the correlation between EEOI and voyage distance by the 100 ships. Analyzed from the graph, it shows the different relationships between these variables. The graph shows that the voyage distance of the ship does have a slight increase correlated with the value of EEOI that increases dramatically. The longest trip done by a ship is from Busan to Newark, with 10,290.87 nm of distance, the ship is Clementine Maersk, and the EEOI value is 0.0000994 Ton CO_2/TEU nm. The highest EEOI value of a ship is done by Meratus Gorontalo, with 0.0005485 Ton CO_2/TEU nm, and the trip is from Port of Makassar to Semarang with 593.41 nm of distance.

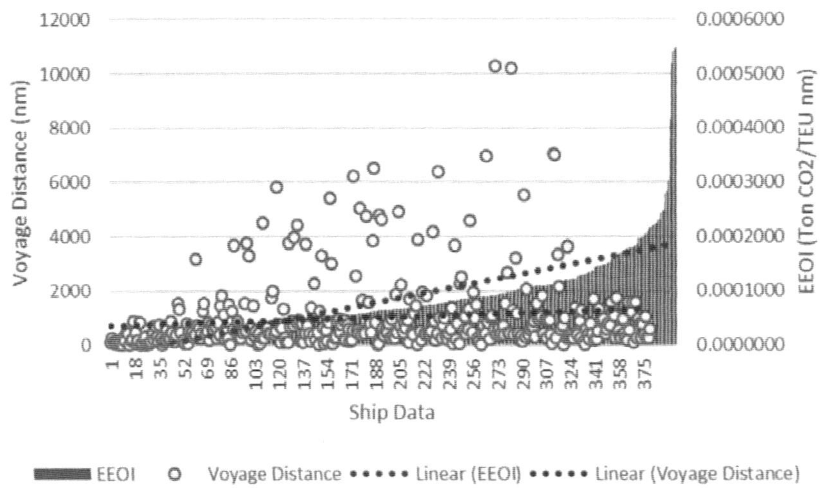

Figure 7. EEOI & Voyage Distance

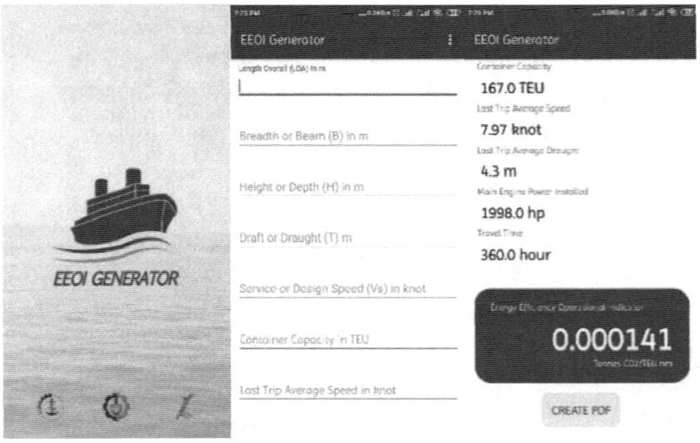

Figure 8. Development of EEOI Measurement Tool, Android Based Application

3.4. Development of EEOI Measurement Tool

The development of the EEOI measurement tool is generating EEOI measurement tools in the form of an android application. This apps will represent the EEOI measurement tools and could be used for determining the EEOI value from the ship based on the last trip/voyage of the ship. The design of this application must be simplified and comfortable for usage because it will determine how the user will interact with the application. The input consists of the variables that have been chosen, such as LOA (Length Overall) in meter, B (Breadth or Beam) in meter, H (Height or Depth) in meter, T (Draught) in meter, Vs (Design Speed or Service Speed) in knot, Container Capacity in TEU, Last Trip Average Speed in knot, Last Trip Average Draught in meter, Main Engine Power Installed in horsepower, and Travel Time in hours. The output design is the same as the input design but added the EEOI resulted value based on the MLR equation. The output could be resulted in negative terms due to the non-limitation to the input. The output also provided in the PDF form so that the EEOI resulted in value is easier to be saved and shared to the other device than android phones.

4. Conclusion

Estimation of the fuel oil consumption using the STEAM method resulted in fuel estimation per voyage/trip. The resulted value is compared with the actual fuel oil consumption based on the noon report of MV Meratus Benoa. Energy Efficiency Operational Indicator (EEOI) calculated using two methods, the STEAM method and the Multiple Linear Regression (MLR) equation. The STEAM method is based on the fuel oil consumption estimation and the MLR equation is generated from the STEAM EEOI value then generated with some chosen variables.

Improvement of cargo management and the sailing speed could be made based on the resulted EEOI value. It shows that the cargo is significantly affected the EEOI value. Ships that have less average container carried are higher at the EEOI value compared to the ships that carried more container. In terms of ship operational conditions, ship operators of the ships should monitor and evaluate which operational mode suits the best against the sea condition at that specific time. The development of the EEOI measurement tool resulted in generating android based application for estimating the EEOI value using MLR equation. The equation, as stated before based on the chosen variables, such as length overall, breadth, draught, depth, service speed, container capacity, last trip average speed, last trip average draught, main engine power installed, and travel time.

5. Acknowledgments

The authors thank to Department of Marine Engineering, and Institut Teknologi Sepuluh Nopember for supporting facilities this research, also to FleetMon for providing the data of this research.

References

[1] M. Sin and F. X. M. D. Oses, "Improvement of the Energy Efficiency of Vessels as a Measure for the Reduction of Greenhouse Gases Emission From Shipping", ResearchGate, 2014.

[2] A. T. Tien, "A Research on the Energy Efficiency Operational Indicator EEOI Calculation Tool on M/V NSU JUSTICE of VINIC Transportation Company," *Journal of Ocean Engineering and Science,* p. 13, 2016.

[3] H. Islam and C. G. Soares, "Effect of trim on container ship resistance at different ship speeds and drafts," *Ocean Engineering,* p. 106–115, 2019.

[4] J.-P. Jalkanen, A. Brink, J. Kalli, H. Pettersson, J. Kukkonen and T. Stipa, "A Modelling System for the Exhaust Emissions of Marine Traffic and its Application in the Baltic Sea Area," *Atmospheric Chemistry and Physics,* pp. 9209-9223, 2009.

[5] International Maritime Organization, "Guidelines for Voluntary Use of The Ship Energy Efficiency Operational Indicator (EEOI)," in *Marine Environment Protection Committee (MEPC) Conference*, London, 2009.

[6] The International Maritime Organization, "2016 Guidelines For the Development of a Ship Energy Efficiency Management System (SEEMP)," in *Marine Environment Protection Committee*, 2016.

[7] International Telecommunication Union, ITU-R Technical Characteristics for an Automatic Identification System Using Time Division Multiple Access in the VHF Maritime Mobile Frequency Band, Geneva: Electronic Publication, 2014.

[8] M. Chaal, "Ship Operational Performance Modelling for Voyage Optimization Thorugh Fuel Consumption Minimization," World Maritime University, Malmo, 2018.

[9] A. Coraddu, L. Oneto, F. Baldi and D. Anguita, "Vessels Fuel Consumption Forecast and Trim Optimisation: A Data Analytics Perspective," *Ocean Engineering,* pp. 351-370, 2017.

[10] L. M. G. and C. P. J., "Influence of NOx Emissions from Ships on Tropospheric Photochemistry and Climate," *Letters to Nature,* pp. 167-170, 1999.

[11] J.-P. Jalkanen, L. Johansson, J. Kukkonen, A. Brink, J. Kalli and T. Stipa, "Extension of an Assessment Model of Ship Traffic Exhaust Emission for Particulate Matter and Carbon Monoxide," *Atmospheric Chemistry and Physics,* pp. 2641-2659, 2011.

[12] K. Mersin, G. Alkan and T. Mısırlıoğlu, "A New Method for Calculating Fuel Consumption and Displacement of a Ship in Maritime Transport," *Cogent Engineering,* 2017.

[13] L. P. and B. Mo, "Emission Control Based Energy Efficiency Measures in Ship Operations," *Applied Ocean Research,* pp. 29-46, 2016.

[14] K. Wang, X. Yan, Y. Yuan and F. Li, "Real-Time Optimization of Ship Energy Efficiency Based on Prediction Technology of Working Condition," *Transportation Research Part D,* pp. 81-93, 2016.

[15] S. Zhang, Y. Li, H. Yuan and D. Sun, "An Alternative Benchmarking Tool for Operational Energy Efficiency of Ships and Its Policy Implications," *Journal of Cleaner Production,* 2019.

IOP Conf. Series: Earth and Environmental Science **1423** (2024) 012027 doi:10.1088/1755-1315/1423/1/012027

Assessing the Proposed Designation of Nusa Penida and Gili Matra in the Lombok Strait as a Particularly Sensitive Sea Area (PSSA)

Emmy Pratiwi[1], Fadilla I. Prastyasari[1], Danuja Wijayanto[1,2], AAB Dinariyana[1], and Dian Saptarini[3]

[1]Departement of Marine Engineering, Institut Teknologi Sepuluh Nopember, Surabaya, Indonesia
[2] Center for Excellence of Maritime Safety and Marine Installation (PUI KEKAL), Institut Teknologi Sepuluh Nopember, Surabaya, Indonesia
[3] Department of Biology, Institut Teknologi Sepuluh Nopember, Surabaya, Indonesia

*E-mail: emmy.p@its.ac.id

Abstract. The Lombok Strait, located in the Indonesian Archipelagic Sea Lane (IASL) II, serves as an international shipping route and is also crossed by ferries between Bali and Lombok Islands. The potential increase in international shipping traffic could pose a threat of marine pollution due to shipping activities, such as ship accidents and operational discharges from ships. There are two areas in the Lombok Strait that have been designated as national marine conservation areas by the Indonesia government, namely Nusa Penida Island and Gili Matra Island. The rich marine biodiversity of the Lombok Strait, especially in Nusa Penida and the Gili Islands, should be protected from the threats posed by international shipping activities. This study aims to assess the necessity of designating the Lombok Strait as a Particularly Sensitive Sea Area (PSSA). The newly established routing systems, Traffic Separation Schemes (TSS) in the Lombok Strait, can be adopted as Associated Protective Measures (APMs) within the designation of a PSSA. Risk assessment was implemented to demonstrate the exposed risks of international shipping activities that could threaten the marine ecosystem and evaluate the effectiveness of the TSS as an associated protective measure. Based on reviews to meet at least one of the criteria for designation as a PSSA, Nusa Penida and Gili Matra have its the significance of its ecological, socio-economic & culture, and marine traffic characteristics. The total annual frequency of ship collision is 3.32E-03. In conclusion, the implementation of TSS as the existing APM has proven effective in reducing the risk of ship collision in Lombok Straits by 60.3% compared to frequency ship collision before TSS.

1. Introduction

Indonesia has established three archipelagic sea lanes that are free to pass by international ships (freedom to passing). Indonesia Archipelagic Sea Lane (IASL) II located in Lombok Straits, connecting to the Indian Ocean. Lombok Strait serves as an international shipping route and is also crossed by ferries between Bali and Lombok Islands. The Lombok Strait is more suitable for large ships because its depth is deeper than the Malacca Strait, which reaches up to 250 meters

[1]. The international shipping vessels has the potential to be exposed to the risk of marine pollution due to shipping activities, such as ship accidents and operational discharges from ships [2]. Marine pollution has a major impact on the destruction of ecosystems and marine biota.

Moreover, at the southern opening of the Lombok Strait lies Nusa Penida Island, an island surrounded by high biological diversity, including coral reefs that contributing to the global coral triangle. At the northern opening of the Lombok Strait, coral reef ecosystems surround the Gili Islands. Nusa Penida Island and Gili Matra Island in the Lombok Strait that have been designated as national marine conservation areas by the Indonesia government. Nusa Penida Islands has been designated as a Marine Protected Area (MPA) by Minister of Marine Affairs and Fisheries Decree number 24/KEPMEN-KP/2014 then updated to 90/KEPMEN-KP/2018. Nusa Penida Island located in the southern opening of the Lombok Strait is an island surrounded by high biological diversity, including coral reefs and various coral species, contributing to the global coral triangle. Gili Matra has been designated as National Marine Conservation due to its coral reef ecosystems in 2009 by The Minister of Marine and Fisheries Number KEP.67/MEN/2009 then updated to number 34 in 2022.

However, designation as MPA which covers waters, flora and fauna as well as history and culture is only protected by national law and only applies nationally. Meanwhile, the risk that could cause disruption to marine ecosystems could come from threats due to international shipping vessels activities. Therefore, there is a need for international regulations to protect Nusa Penida and Gili Matra because their location is close to international shipping that passes through the Lombok Strait (Figure 1). International Maritime Organization (IMO) introduced the concept of PSSA as a global level of protection for an area that has the impact of vulnerability due to international shipping activities.

As stated in concept of PSSA from IMO Resolution A.982(24), a PSSA is defined as an area that needs special protection through action through IMO because of its significance for recognised

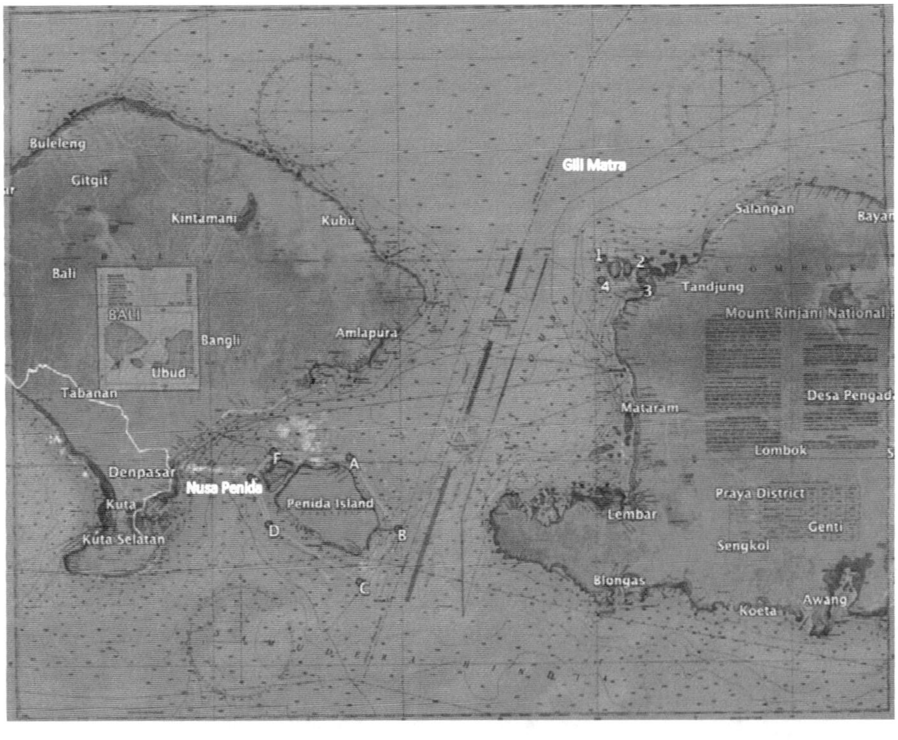

Figure 1. Location of Nusa Penida and Gili Matra in Lombok Strait as Proposed PSSA

ecological, socio-economic, or scientific attributes where such attributes may be vulnerable to damage by international shipping activities [3]. In proposal of PSSA to IMO, a state is also required to adopt the APMs to prevent the risks, or the threats posed by international shipping activities within the area to be designed as PSSA.

Currently, one of ship's routeing measure in IMO instruments, namely Traffic Separation Schemes (TSS), has been established in the Lombok Strait (NCSR 6/3/4). TSS in Lombok Strait can be adopted as APMs within the designation of a PSSA in Nusa Penida and Gili Matra. The designation of area as must meet at least one of the criteria as mentioned in paragraph 4 of Resolution A.982(24). These criteria consist of three categories: ecological, socio-economic, and scientific and educational criteria.

Based on these backgrounds, this study assesses the need of designation of PSSA in Lombok Strait by demonstrating the exposed risks of international shipping activities that could threaten the marine ecosystem. The first section presents reviews the ecological, socio-economic, and scientific significance of Nusa Penida and Gili Matra to meet the criteria for designation as a PSSA. The risk assessment in the term of frequency of ship collision will be modelled by the IALA Waterway Risk Assessment Program (IWRAP), a modelling tool recommended by IMO. In the next section, the evaluation of the effectiveness of the TSS as an APM are presented based on marine traffic risk assessment.

2. Reviews on ecological, socio-economic, and scientific and educational criteria in Nusa Penida and Gili Matra

Nuda Penida and Gili Matra Islands have been designated as MPA due to its high biodiversity of coral reef, mangroves, sea grass meadows and ich marine fauna. The Lombok Strait is included in the coral triangle area which contributes to the total area of coral reefs in Indonesia as the largest coral reef area in Southeast Asia [4]. Some of rare and endangered species also found in Lombok Straits, such as mola-mola (ocean sunfish), manta rays and whale sharks [5]. Marine tourism, fishing and seaweed cultivation are the main sectors that contribute to the local economy on Nusa Penida and Gili Matra Islands. In the designation of PSSA, Nusa Penida and Gili Matra Islands must at least meet one of criteria stated in Resolution A.982(24). The following are the general reviews of the analysis to fulfil the criteria for designating PSSA:

2. 1. Ecological criteria

Nusa Penida district including Nusa Ceningan and Nusa Lembongan has 1,419 hectares of coral reefs, 230 hectares of mangrove forests, 108 hectares of seagrass beds and 576 types of fish were found in Nusa Penida waters [6]. The uniqueness of Nusa Penida area is some of rare species are found living in Nusa Penida. One of the most unique and rare fish is Ocean sunfish, also known as Mola Mola as shown in Figure 2. Ocean sunfish or Mola Mola is the one of world's largest bony fish which size of at least 3 metres and weigh more than 900 kg [7]. Sunfish live in the deep sea, typically inhabit depths averaging between 400 and 600 meters. Mola alexandrini as one of Mola species is commonly saw in Nusa Penida. Mola alexandrini also called as bump-head sunfish because it has a large swelling on its forehead [8]. Mola mola is classified as "Vulnerable" in the IUCN Red List of Threatened Species (iucnredlist.org).

Figure 2. Ocean sunfish (Mola mola) [8]

The Nusa Penida MPA serves as critical habitat for the large local manta ray population, an also home for sea turtles and sharks [9]. The location of Lombok Straits in the main exit route for the Indonesian Throughflow (ITF) which connects Indonesian waters with Indian ocean cause the upwelling occurs. Upwelling from cooler sea water moving from the sea floor deeper to the surface influence the composition of coral and fish assemblages due to low temperature of water. Coral reef and other coastal habitats are also influenced by these upwelling. This phenomenon of upwelling driven by wind patterns and the steep bathymetry of the region where deep-sea habitats occur [10]. Therefore, Nusa Penida is a location for migrating sunfish, and the movement of cold water during the upwelling process also attracts other large pelagic species such as sharks and rays.

In conclusion regarding the ecological criteria, Nusa Penida has its significance in the sub criteria of uniqueness of sunfish as the largest bony fish, critical habitat in Nusa Penida as residential and migratory cetacean species, diversity in the term of coral species, fish, mangrove and seagrass in Gili Matra, representative of its coral triangle.

2. 2. Socio, Cultural, and Economic criteria

Economic activities for the income of local community in Nusa Penida and Gili Matra mainly depend on marine tourism, fishing and seaweed cultivation. Nusa Penida is very popular for marine tourism activities, especially for surfing, diving, snorkelling, sailing and also water sports with more than 839.790 tourists visiting Nusa Penida in 2019. While Gili matra is also globally marine tourism destination which attracts about 500,000 tourists yearly. Revenue from Gili Matra tourism contributes up to 70% of economy in the North Lombok. Fishing in Nusa Penida contributes 45.2% of Klungkung Regency's fish production [8], while seaweed cultivation contributes 99.34% of the seaweed production in the Bali Region [11].

Bali and Lombok community has a strong religious principal and local wisdom or philosophy of life. Society adheres to a principle called Tri Hita Karana (three reasons for prosperity) which is based on efforts to maintain harmony with God, with fellow humans and with nature or the environment. This religious principle is a strong reason for people, especially in Bali, to try to protect the environment. This could be a unique justification in terms of socio-cultural compared to other established PSSAs.

2. 3. Scientific and Educational criteria

Lombok Straits that located in ITF cause the upwelling can produce strong currents when combined with tides. It makes Lombok Straits especially Nusa Penida as the location for marine spatial planning to study the relationship between oceanographic characteristics and fish shape [12]. While the Gili Matra is location for satellite-based monitoring of turtle movements. Apart from that, the Wallace Line stretches from the South Philippines to Lombok Straits, which is the biogeographic boundary between the fauna of western Indonesia and eastern Indonesia [1], [13]. Therefore, research related fauna was also carried out in Nusa Penida.

3. Methodology of Ship Collision Risk Assessment

3.1. Theoritical Framework of Collision Modelling in IWRAP

In establishing a PSSA, a state must prove the vulnerably of its proposed area from the dangers posed by international shipping activities. An analysis of the vulnerability of Nusa Penida and Gili Matra to be designated as PSSA due to international shipping activities will be carried out by assessing the risk of ship collisions. One of the parameters in risk analysis is frequency assessment. Calculation of the frequency of collisions and grounding can be conducted based on the model developed by Fujii [14]. The model consists of two main parameters, namely the geometric number of candidate ship collisions or groundings, N_G, and the causal factor, P_C. The two parameters are multiplied so that the collision frequency, γ_{Col}, refers following formula:

$$\gamma_{Col} = N_G \cdot P_C$$

Ship collision is divided into two types of collision, collision along the route segment (*overtaking* and *head-on collision*) and collision when two routes cross, merge or intersect each other (crossing, merging or intersecting). The probability that ships will encounter a head-on collision depends on the distribution of lateral positions where the ships are sailing (Figure 3). The greater the value of μ, the smaller the probability of ship collision. Meanwhile in Figure 4, crossing collisions are not influenced by the distribution of ship traffic.

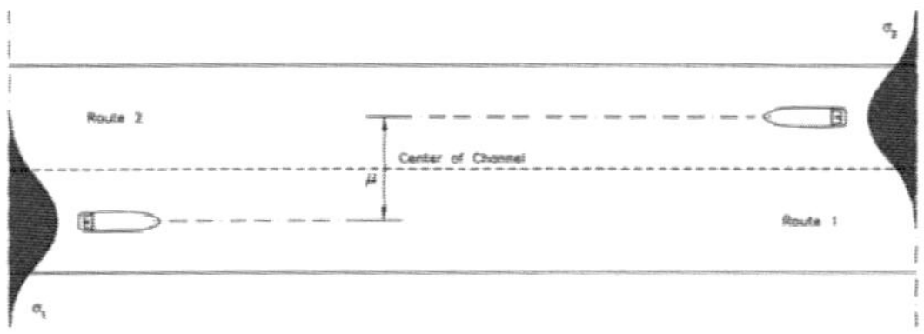

Figure 3. Head on Collision

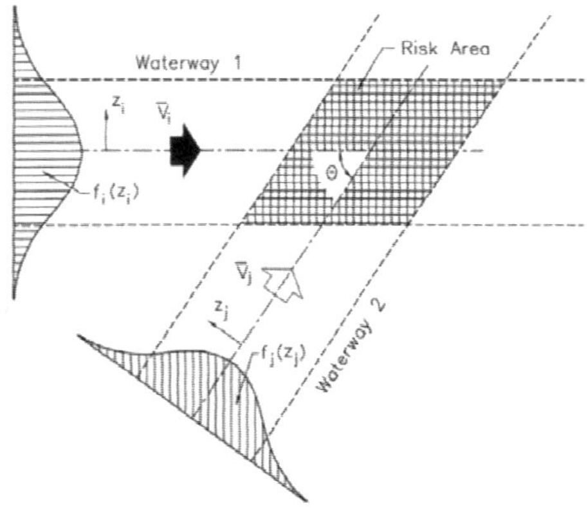

Figure 4. *Crossing Collision*

For head-on collisions, the number of geometric ship collision candidates for ships sailing along the route segment in directions (1) and (2) can be expressed as:

$$N_G^{head-on} = L_W \sum_{i.j} P_{Gi.j}^{head-on} \frac{V_{ij}}{V_i^{(1)} V_j^{(2)}} \left(Q_i^{(1)} Q_j^{(2)}\right)$$

Where $V_{ij} = V_i^{(1)} + V_j^{(2)}$ is the relative speed between ships and P_G is the probability that the two ships will collide with each other in a head on situation. P_G is expressed as:

$$P_{Gi.j}^{head-on} = \Phi\left(\frac{B_{ij}-\mu_{ij}}{\sigma_{ij}}\right) - \Phi\left(-\frac{B_{ij}-\mu_{ij}}{\sigma_{ij}}\right)$$

Where:

P_G = probability of collision

Φ = normal probability distribution function

$\mu_{ij} = \mu_i + \mu_j$ is the sailing distance between two ships

$\sigma_{ij} = \sqrt{\sigma_i^2 + \sigma_j^2}$ is the standard deviation of the joint distribution

$B_{ij} = \frac{B_i + B_j}{2}$ is the average width of the ship

For an overtaking collision, the number of geometric candidate ships that have the collision in direction (1) is expressed in the following formula with $V_{ij} > 0$:

$$P_{Gi.j}^{overtaking} = P\left[y_i^{(1)} - y_j^{(1)} < \frac{B_i^{(1)} - B_j^{(1)}}{2}\right] - P\left[y_i^{(1)} - y_j^{(1)} < -\frac{B_i^{(1)} - B_j^{(1)}}{2}\right]$$

For normally distributed variables the mean value in the equation is replaced by $\mu_{ij} = \mu_i^{(1)} + \mu_j^{(1)}$ to accommodate overtaking situations.

The frequency of crossing collisions depends on the angle between the two lanes. The geometric number of crossing collision candidates can be expressed with the following equation:

$$N_G^{crossing} = \sum_{i.j} \frac{Q_i^{(1)} Q_j^{(2)}}{V_i^{(1)} V_j^{(2)}} D_{ij} V_{ij} \frac{1}{sinsin\,\theta} \text{ for } 10° < |\theta| < 170°$$

Where $V_{ij} = \sqrt{\left(V_i^{(1)}\right)^2 + \left(V_j^{(2)}\right)^2 - 2V_i^{(1)}V_j^{(2)} \cos\cos\theta}$ is the relative speed between the two ships and D_{ij} is the apparent diameter of the collision (Figure 5).

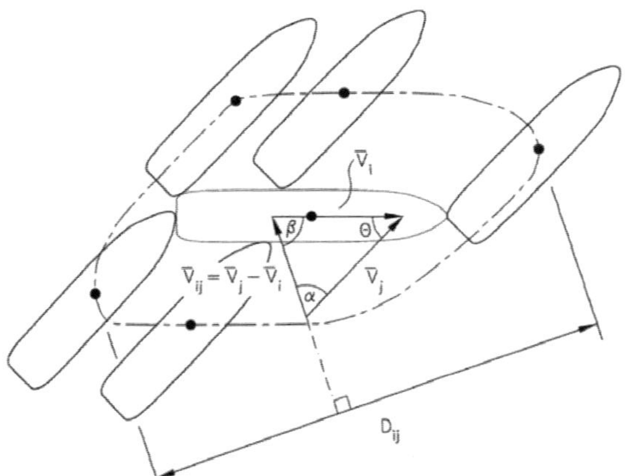

Figure 5. Definition and Calculation of Geometrical Collision Diameter D_{ij}

Analysis of the frequency of ship collisions was carried out using tools recommended by IMO in conducting risk assessment studies due to collisions, IWRAP (IALA Waterway Risk Assessment Program). In IWRAP, collision frequency analysis is modelled based on ship traffic volume, ship traffic composition, route geometry, and bathymetry.

IOP Conf. Series: Earth and Environmental Science **1423** (2024) 012027 doi:10.1088/1755-1315/1423/1/012027

3.2. Vessel Traffic Characteristic

Based on AIS data collected from Vessel Traffic Service (VTS) Benoa in 2023, the number of ships passing through the Lombok Strait was 4,885 ships with a total of 77,147 trips. Figure 6 presents a detailed breakdown of vessel passing through the Lombok Strait during 2023. Along the various vessel categories, passenger ships emerge as the predominant category, reflecting a significant volume of 25013 trips or 31% of total vessels. This statistic indicates the robust passenger ferry and maritime tourism sectors within the Bali-Lombok region. Pleasure craft also exhibit a notable presence with 1,793 trips recorded. General cargo ships contribute 12,964 trips and underlining the Lombok Strait's strategic role in freight transportations.

Tankers including Oil product, Chemical, Crude oil and Gas tankers follow closely behind, contributing a substantial 8837 trips to the overall traffic or 11.5% of total proportion by vessel type. This should be a concern, carrying large volumes of dangerous liquid cargo could possibly cause oil spill due to grounding or collision of those tankers. The number of traffic and the proportion of the traffic in the Lombok Strait in 2023 as shown in Figure 6.

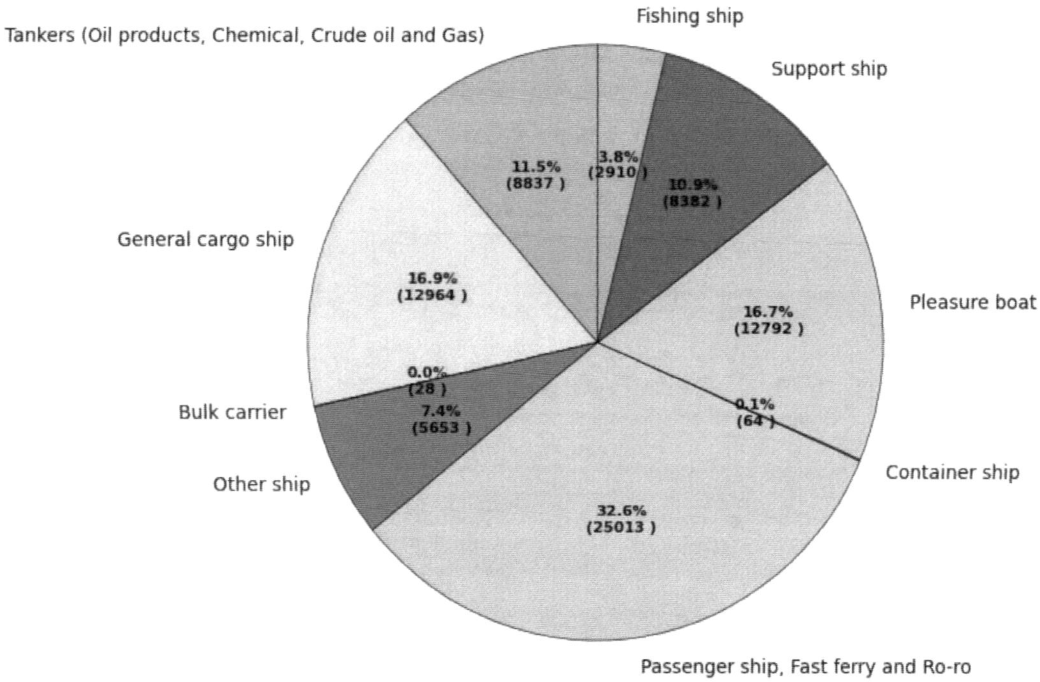

Figure 6 Number of each ship type and the proportion of the traffic in the Lombok Strait by vessel types

Figure 7 shows natural traffic in the Lombok Strait. It can be seen from the traffic pattern that there are ships passing north bound and southbound traffic lane, and there are ships with a high density also crossing Bali-Lombok, with the Padang Bai-Lembar and Padang Bai-Gili Matra routes.

IOP Conf. Series: Earth and Environmental Science **1423** (2024) 012027 doi:10.1088/1755-1315/1423/1/012027

Figure 7. Traffic pattern in Lombok Strait extracted from AIS data

4. Results of ship collision risk assessment due to international shipping activities in Lombok Strait

Based on AIS data about ship passing Lombok Straits, the collision frequency was modelled by IWRAP. In IWRAP modelling, an important initial stage is designing the *leg* to simulate ship traffic that represent the actual ship routes.

In the Lombok Straits has been implemented one of ship's routeing measures which is Traffic Separation Scheme (TSS). TSS in Lombok Straits was approved at the first session of the Sub-Committee on Navigation, Communications and Search and Rescue (refer NCSR 6/3/4), subsequently adopted by MSC 101 and entered into force on 1 July 2020. Therefore, in modelling the risk of collisions in IWRAP by determining the legs according to traffic flow in TSS and vessel crossing according to actual conditions as shown in Figure 8. Traffic data in 2023 from AIS represents the vessel traffic after TSS has been established in 2020.

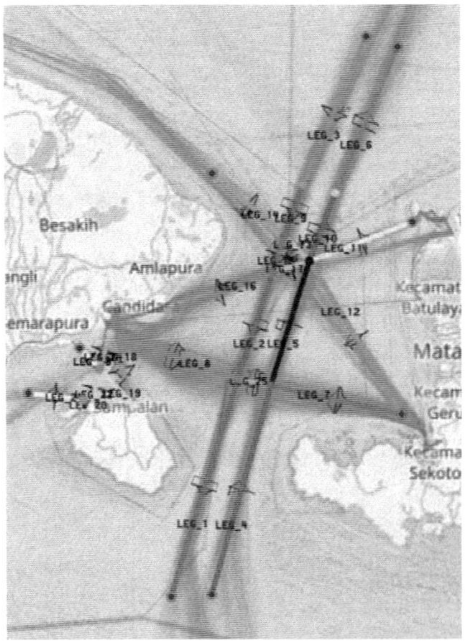

Figure 8. Determining Legs of Traffic Modelling in Lombok Straits with IWRAP

After creating the Leg in IWRAP, data is then input for each Leg including the number of ships passing in each traffic lane that has been modelled. The recapitulation results of the IWRAP simulation for the annual frequency of ship collisions due to ship traffic in TSS Lombok Strait. Table 1 shows a recapitulation of the total annual frequency of collisions for three types of collisions: Crossing, Head-on and Overtaking collisions. Based on studies conducted on existing conditions for all ships, the total frequency of collisions per year for head-on, overtaking and crossing collisions respectively is 1.90E-04, 1.14E-03 and 1.99E-03. Therefore, total annual collision frequency for all collision type is 3.32E-03.

The annual frequency then categorized under the acceptance criteria based on Frequency Index (FI) recommended by IMO regarding Guidelines for Formal Safety Assessment (FSA). The number of annual frequency ranging from 10^{-3} to 10^{-5} is categorized as Remote (FI=3) [15].

Table 1. Frequency of ship collisions in the TSS Lombok Strait

	After TSS: Traffic 2023			Total in Lombok Strait
Type of Collision	South bound TSS	North bound TSS	All TSS	
Head-on	2.68E-07	1.30E-07	3.98E-07	1.90E-04
Overtaking	9.09E-06	7.14E-06	1.62E-05	1.14E-03
Crossing	9.54E-05	1.51E-04	2.46E-04	1.99E-03
Total	1.05E-04	1.58E-04	2.63E-04	3.32E-03

5. Evaluation to prove the effectiveness of Traffic Separation Schemes (TSS) as APM within the designation of a PSSA

To protect the area within PSSA by preventing the pollutant from ships, a state is required to adopt APMs under the existing IMO instruments of general accepted international rules and standards of UNCLOS. However, the newly routeing systems, TSS in Lombok Strait has been established. TSS as one notable routing measures is proposed as the APM.

The risk assessment of ship collision was conducted to evaluate the effectives of TSS as APM within the designation of PSSA in Nusa Penida and Gili Matra. Based on ship collision frequency modelled in IWRAP, TSS Lombok Strait reduced the total frequency of ship collisions by 60.3%. AIS data from 2017 is used as the benchmark for calculating the collision frequency before the implementation of TSS and it is found to be 0.461 accident per year [16]. While the AIS data from 2023 is used to compare the calculation after the TSS implementation and the frequency becomes 0.0002634 accident per year.

6. Conclusion and Recommendation

In conclusion, Nusa Penida and Gili Matra must be considered to become a PSSA of the significance of its ecological, socio-economic & culture, marine traffic characteristics, and also an established TSS as APM. Along with submission of designation of Nusa Penida and Gili Matra as PSSA, the existing of TSS in Lombok Straits can be served as APMs in the proposal. This is because the implementation of TSS has proven effective in reducing the risk of ship collision in Lombok Straits which has the potential to cause damage to marine ecosystems with high biodiversity significance in Nusa Penida and Gili Matra.

The existence of TSS as an existing APM should make it easier to submit a PSSA proposal because the main purpose of the designation of an area as PSSA is to protect vulnerable areas, so APM is required, while the existing APM has been proven to reduce the risk due to international shipping to the areas. As analogous that Indonesia has achieved the environmental protection faster by applying TSS in Lombok Strait [17]. As long as at least one of criteria mentioned in the IMO Resolution A.982(24) as a PSSA are met, then the proposal to submit a PSSA in the Lombok Strait to establish Nusa Penida and Gili Matra has strong justifications to be accepted.

The vulnerability of the ecosystem and marine biota in the area proposed to become a PSSA is not only exposed to risks due to international shipping vessel activities in the form of ship accidents, but also due to ship operational discharge of gray and black water, sewage, ballast, etc. However, operational discharge that could cause environmental damage have not been modeled in this research. Therefore, it is recommended to estimate and model ship operations for further study in future research.

References

[1] A. J. Desmonda, 'The Prospect of Particularly Sensitive Sea Area (PSSA) Designation in Lombok Strait', vol. 4, no. 1, 2020.

[2] O. Bayazit and M. Kaptan, 'Evaluation of the risk of pollution caused by ship operations through bow-tie-based fuzzy Bayesian network', *Journal of Cleaner Production*, vol. 382, p. 135386, Jan. 2023, doi: 10.1016/j.jclepro.2022.135386.

[3] IMO, 'IMO Resolution A.982(24), Revised Guidelines for the Identification and Designation of Particularly Sensitive Sea Areas'. 2006.

[4] Asian Development Bank, 'State of the Coral Triangle: Timor-Leste', Asian Development Bank, Mandaluyong City, Philippines, 2014.

[5] Nyoman Darma, Riyanto Basuki, and Marthen Welly, 'PROFIL Kawasan Konservasi Perairan (KKP) Nusa Penida', 2010, doi: 10.13140/RG.2.1.1315.6966.

[6] ICCTF, 'Marine Protected Area (MPA) Nusa Penida', Indonesia Climate Change Trust Fund (ICCTF), Ministry of National Development Planning of the Republic of Indonesia.

[7] Britannica, 'The Editors of Encyclopaedia. "ribbonfish"'. Encyclopedia Britannica, Oct. 08, 2008. Accessed: Jun. 14, 2024. [Online]. Available: https://www.britannica.com/animal/ribbonfish

[8] Coral Triangle Center, 'Nusa Penida Is Now Home To The World's Largest Bony Fish – The Mola Alexandrini (Previously Known As Mola Ramsayi)'. [Online]. Available: https://www.coraltrianglecenter.org/2018/03/05/nusa-penida-is-now-home-to-the-worlds-largest-bony-fish-the-mola-alexandrini-previously-known-as-mola-ramsayi/

[9] E. S. Germanov *et al.*, 'Contrasting Habitat Use and Population Dynamics of Reef Manta Rays Within the Nusa Penida Marine Protected Area, Indonesia', *Front. Mar. Sci.*, vol. 6, p. 215, May 2019, doi: 10.3389/fmars.2019.00215.

[10] J. Wilson, A. Darmawan, J. Subijanto, A. Green, and S. Sheppard, 'Scientific design of a resilient network of marine protected areas. Lesser Sunda Ecoregion, Coral Triangle', Asia Pacific Marine Program, Report 2/11. 96 pp., Mar. 2011.

[11] Departement of Marine Affairs and Fisheries Denpasar, 'Management and Zoning Plan of Nusa Penida Marine Protected Area (MPA), Klungkung Regency, Bali Province', 2017.

[12] G. Sartori, M. L. Taylor, P. Sebastian, and R. Prasetyo, 'Coral reef carnivorous fish biomass relates to oceanographic features depending on habitat and prey preference', *Marine Environmental Research*, vol. 172, p. 105504, Dec. 2021, doi: 10.1016/j.marenvres.2021.105504.

[13] P. C. Van Welzen, J. A. N. Parnell, and J. W. F. Slik, 'Wallace's Line and plant distributions: two or three phytogeographical areas and where to group Java?: WALLACE'S LINE AND PLANT DISTRIBUTIONS', *Biological Journal of the Linnean Society*, vol. 103, no. 3, pp. 531–545, Jul. 2011, doi: 10.1111/j.1095-8312.2011.01647.x.

[14] Y. Fujii and K. Tanaka, 'Traffic capacity', *Journal of Navigation*, vol. 24, no. 04, pp. 543–552, 1971.

[15] I. IMO, 'Guidelines for Formal Safety Assessment (FSA) for use in the IMO rule-making process', *MSC/Circ 1023*, 2002.

[16] A. A. B. Dinariyana, M. A. Avesina, K. Sambodho, and F. Fahrirozan, 'Collision Risk Analysis and Oil Spill Dispersion Modelling in Lombok Strait', *IOP Conf. Ser.: Earth Environ. Sci.*, vol. 557, no. 1, p. 012002, Aug. 2020, doi: 10.1088/1755-1315/557/1/012002.

[17] S. Y. Kim, 'Problems and Processes of Restricting Navigation in Particularly Sensitive Sea Areas', *Int. J. Mar. Coast. Law*, vol. 36, no. 3, pp. 438–463, Jun. 2021, doi: 10.1163/15718085-bja10062.

Dual fuel retrofit LNG for Indonesia's Greenhouse Gas Strategy Case: Fulfilment of engine requirements on existing vessels in accordance with class requirements

M R F Hariadi[1], M A Kurniawan[1] and A A P Pratama[1]

[1] Biro Klasifikasi Indonesia, Jakarta, Indonesia

*E-mail: m.rizqi@bki.co.id

Abstract. The global maritime industry is transitioning towards sustainability to meet stringent environmental standards under IMO's MARPOL Annex VI and the Paris Agreement. LNG emerges as a promising marine fuel, significantly reducing sulfur oxides (SO_x), nitrogen oxides (NO_x), and particulate emissions compared to traditional fuels. Indonesia, a major LNG exporter, aims to utilize LNG domestically under its Long-term Strategy for Low Carbon & Climate Resilience 2050. Adopting LNG in Indonesia's maritime sector faces challenges, especially retrofitting existing diesel engines. Technical hurdles include ensuring engine compatibility, modifying fuel systems, and conducting comprehensive on-board testing due to limited infrastructure. Regulatory frameworks like Biro Klasifikasi Indonesia (BKI) Rules and the NO_x Technical Code 2008 amendments provide the guidance for emissions compliance and testing. This paper discusses methodologies to overcome these challenges, emphasizing robust testing protocols and sea trials to validate engine performance and emissions under real-world conditions. Collaboration among stakeholders is essential to facilitate Indonesia's transition to sustainable maritime practices. Future advancements in LNG technology and infrastructure will bolster Indonesia's energy security and contribute to global emissions reduction targets in maritime transport.

1. Introduction

LNG is a low-emission alternative fuel that complies with ECA (Emission Control Area) sulphur, NOX, and particulate standards in accordance with IMO MARPOL Annex VI standards. The technology for using LNG fuel in marine engines is very robust [1].

1.1. LNG demand in Indonesia

The Paris Agreement's goal is to limit the global temperature increase to well below 2°C and to pursue efforts to limit it to 1.5°C. This means that all sectors and countries will eventually have to reduce emissions to net-zero, with the main questions being when and how.

The revised IMO GHG Strategy includes an enhanced common ambition to reach net-zero GHG emissions from international shipping close to 2050, a commitment to ensure an uptake of alternative zero and near-zero GHG fuels by 2030 [2], as well as indicative check-points for 2030 and 2040. Each country has different strategies for reducing GHG emissions. In Indonesia, the

IOP Conf. Series: Earth and Environmental Science **1423** (2024) 012028 doi:10.1088/1755-1315/1423/1/012028

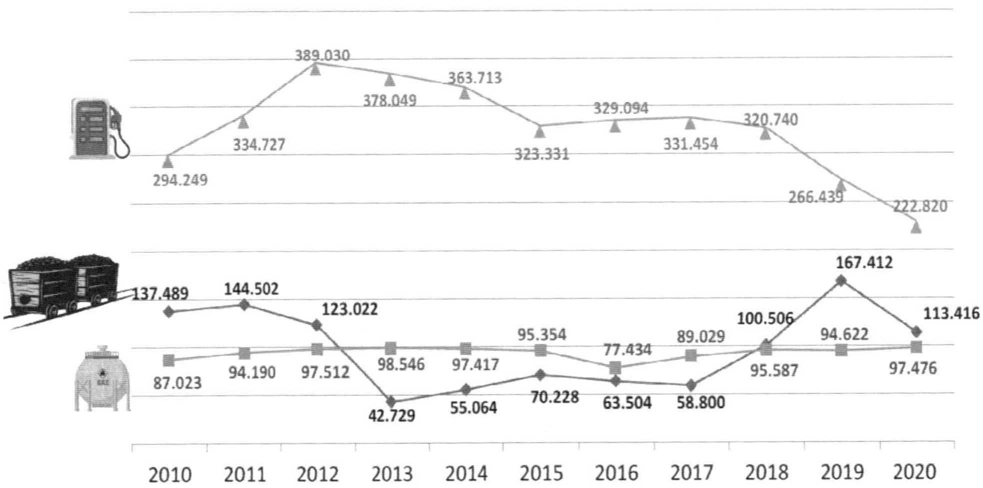

Figure 1. Energy demand in Indonesia based on Central Bureau of Statistics Indonesia (2021)

government plan to reach GHG target in 2060 or sooner. Based on 2021 data of Central Bureau of Statistics of Indonesia (Figure 1.) on energy demand it shows that from 2010 to 2020 energy from fuel oil sources is decreasing, but the demand for natural gas is stable [3]. Indonesia plans to use alternative energy fuels from LNG and methanol. The reasons for using LNG include:

a. Indonesia ranks 8th among global LNG exporters, with 14.6 billion m^3

b. The Indonesian government plans to halt new LNG export contracts in 2025-2026 to fulfill the domestic market. This is an opportunity to utilize LNG as an alternative fuel for ships in Indonesian waters.

c. Indonesia Long-term Strategy for Low Carbon & Climate Resilience 2050 (Indonesia LTS-LCCR) [4].

d. The Indonesian government is committed to reducing greenhouse gas emissions by increasing the use of renewable energy from 23% in 2025 to 29% in 2030 [4].

1.2. LNG Dual Fuel Mechanism

A dual-fuel engine can run on a combination of two different fuels, with the most common pairing being diesel and natural gas. These engines are designed to switch between diesel and natural gas depending on availability and operational needs. Dual-fuel engine retrofits seek to minimize changes to the internal combustion system. A commonly used method for dual-fuel conversion involves injecting LNG gas into the engine's intake manifold. The diesel at the nozzle acts as a pilot fuel [5].

1.3. LNG Retrofitting Challenges

In Indonesia, shipowners are more interested in retrofitting existing engines, which is more economically feasible than buying new dual-fuel engines. Here are the technical challenges of retrofitting existing engines:

1.3.1 Engine compatibility

Existing diesel engines are not designed to use LNG as fuel, so it is necessary to modify the engine in terms of the LNG fuel injection system, the composition of the mixture between LNG fuel and diesel fuel, and ensuring that the engine power does not decrease after modification. There is currently no age limit on engines suitable for dual-fuel retrofits. According to several class partners and BKI [6], there is also no requirement for engine age limits, but a statement of fact from the engine manufacturer is required to confirm that the diesel engine can use LNG fuel.

This issue may be overcome if the retrofitting or modification is carried out by an authorized engine maker. However, in Indonesia, many workshops operate without authorized engine makers, so special attention is needed on how to modify the engine and test it after retrofitting to a dual-fuel system.

1.3.2 Engine Testing

The test bench serves as a platform for testing and analyzing engine performance and emissions under various conditions. It primarily consists of the control room, the bench, and performance test equipment, which includes a dynamometer, a gas consumption meter, and a fuel consumption meter [7]. While new diesel engines are typically tested on these benches, it is challenging to test existing or retrofitted engines due to the limited availability of engine test bench facilities in Indonesia.

These challenges and limitations have led the authors of this paper to discuss the potential for on-board testing of retrofitted engines and the relevant test parameters.

2. Methodologies

The methodologies in this paper use gap analysis based on several literature reviews related to diesel engine testing on ships after modification, along with safety parameters that need to be addressed. The literature used includes BKI regulations, academic papers, and other technical documents.

Table 1. Item to be recorded

No	Item
1	Ambient air temperature
2	Ambient air pressure
3	Atmospheric humidity
4	Power and speed
5	Fuel index (or equivalent reading)
6	Maximum combustion pressures (only when the cylinder heads installed are designed for such measurement).
7	Exhaust gas temperature before turbine and from each cylinder
8	Charge air temperature
9	Charge air pressure
10	Turbocharger speed

3. Results and Discussions

3.1 BKI Rules and Guidelines

3.1.1 Rules for machinery Installations (Pt.1, Vol.III)

Related diesel engine testing in the BKI [8], where section 2.E.5 work Factory Acceptance Test (FAT), the objectives of which are to verify performance, fire safety and functionality to establish a reference or baseline for operation. During the machine test, the following must be recorded:

The engine manufacturer is responsible for measuring and recording the operational values at all stages of engine testing. All results should be compiled into an acceptance protocol issued by the engine manufacturer. Calibration records for the instruments must be provided to the attending BKI Surveyor upon request. Measurements at various load points should be conducted under steady-state operating conditions. However, time should be allowed for the Surveyor to perform visual inspections at all load points. The readings for MCR (100% power at the corresponding rpm) should be taken at least twice, typically with a 30-minute interval. Alternative test methods may be agreed upon between the manufacturer and the Society if they are determined to be equivalent in overall scope. Random checks of components to be presented for inspection after works trials are left to the discretion by BKI.

3.1.2 Guidance for Sea Trial of Motor Vessel

Guidance for Sea Trial of Motor Vessel [9] is implemented for ships undergoing major modifications (such as changing the main engine, altering propulsion systems, changing main dimensions, or other modifications that affect the ship's characteristics) must be conducted sea trial. During the classification survey of all ships, sea trials must be conducted under full load conditions, in the calmest possible sea and weather conditions, and in deep, unrestricted water. If it is not feasible to conduct sea trials under full load conditions, they may be performed under an appropriately loaded condition.

3.2 International codes

Amendments to the NO_x Technical Code 2008 to improve the recertification of existing engines onboard ships, PPR11 This document is submitted to IMO subcommittee pollution prevention and response. This document contained the co-sponsors suggest amendments to the NO_x Technical Code 2008 to define "retrofitting of existing engines" and to modify procedures to enhance the recertification process for existing diesel engines on ships [10]. This is to enable the retrofitting of modern technologies for better energy efficiency while adhering to nitrogen oxide emission standards.

In many retrofit cases, conducting the new parent engine emission test on a test bed is not feasible because the engine is already installed on the ship, or identical engines are unavailable due to the age of the model. Constructing a new, identical engine for testing is not practical, especially for large engines. Therefore, a significant challenge is conducting the new (dual fuel) parent engine emission test directly on board the ship. Retrofitting existing engines to reduce carbon intensity in international shipping is an emerging practice. Gaining more experience with retrofitting in the context of these amendments for re-certification of existing engines may provide valuable insights for future reviews. The example of reducing on load point and NO_x emission result on engine is shown on engine load test due retrofitting engine test on board:

With the revised weighting factors from Table 2, a cycle average NOx value of 13.41 g/kWh was obtained. Comparing this to the test value, the revised weighting factor results in a higher NOx value, indicating a worse-case scenario or a higher emissions target. Even with an additional 10% allowance, it still reflects a worse NO_x case compared to the cycle value of 12.02 g/kWh. The specific allowance amount needs to be determined by the administration or flag state.

3.3 Academic papers

3.3.1 Concept of a LNG gas handling system for a dual fuel engine

The key concern examined in this study revolves around the switch in engine operation mode from diesel to dual fuel within the engine laboratory housed at the department of maritime studies in Warnemünde. The specific diesel engine model under investigation is the MAN 6L23/30A [11]. In this investigation, the determination of fuel oil consumption and vaporizer is carried out based on the actual data obtained from engine operation. This necessitates running the engine to acquire specific fuel oil consumption (SFOC), engine water cooling temperature, and exhaust gas temperature readings.

Following the engine retrofit, a decline in engine power is evident from the findings presented in Table 3. Specifically, the power per cylinder volume calculation indicates a decrease in power output to 877 Kw for the MAN 6L23/30 A engine.

Based on the data in Table 3, the power per cylinder reduction of MAN 35/44 DF is greater compared to MAN 51/60 DF. This suggests that as power decreases, cylinder volume decreases but RPM increases, leading to an increase in power per cylinder reduction.

3.3.2 Experimental study on Diesel/LNG Dual Fuel Engine

The primary means of determining whether natural gas can serve as a feasible alternative fuel lies in the performance evaluation of a diesel/LNG dual-fuel engine [7]. The fundamental goal of adjusting the calibration of a diesel/LNG dual-fuel engine is to optimize the functionality of the engine's natural gas system through extensive testing, identifying various ECU parameters, and enhancing its efficiency and performance while ensuring compliance with emission regulations.

The engine bench calibration test findings reveal that the dual-fuel blending test prototype demonstrates excellent dynamics, economy, and emission performance. The maximum torque and specific fuel consumption are essentially identical to those of the original machine. Natural gas substitution can achieve a rate of 60%, and up to 78% in common working conditions, indicating that the calibration results satisfy the stipulated requirements.

Table 2. Combination load point

Power	100%	75%	50%	25%
Existing Nominal Weighting Factor	0.2	0.5	0.15	0,15
Engine Power	12400	9300	6200	3100
Nox Existing (gr/Kwh)	8.63	12.28	15.99	19.57
New Factor proposal	-	0.625	0.1875	0.1875
Nox New (gr/Kwh)	-	12.28	15.99	19.57

Table 3. Power Engine after Dual Fuel Retrofit

No	Engine Retrofit	Power reduction (%)
1	MAN 48/60 B to MAN 51/60 DF	19,056
2	MAN 32/44 CR to MAN 35/44 DF	28,87
3	Wartsila 32 to Wartsila 34 DF	23,53

4. Discussion

The global maritime industry is increasingly transitioning towards low-emission fuels to comply with stringent environmental regulations set forth by the IMO's MARPOL Annex VI and the Paris Agreement's ambitious climate goals. LNG emerges as a viable alternative fuel, offering significant reductions in Sulphur Oxides (SO_x), Nitrogen Oxides (NO_x), and particulate matter emissions compared to traditional marine fuels. This transition is particularly crucial for Indonesia, a major LNG exporter aiming to harness LNG domestically under its Long-term Strategy for Low Carbon & Climate Resilience in 2050.

The adoption of LNG as a marine fuel in Indonesia faces challenges, particularly in existing ships that requires retrofitting of their diesel engines to operate on LNG. Engine compatibility, LNG fuel system installations and safety system, and the availability of adequate testing facilities pose significant technical hurdles. The retrofitting process requires meticulous attention to ensure engine performance and emissions compliance, as outlined in regulatory frameworks such as the BKI Rules and Guidelines and international codes like the NO_x Technical Code 2008 amendments.

Testing of dual fuel retrofit engines can be carried out on board, utilizing test items similar to those on the engine test bed. Nevertheless, there are additional parameters apart from those specified in the BKI Rules for Machinery Installations (Pt.1, Vol.III) that should be recorded, including:

- Calculation of engine power per cylinder before retrofitting (refer to 3.3.1)
- Inlet and outlet water cooling temperatures before and after the retrofit
- SFOC of the engine before and after the dual fuel retrofit
- Exhaust gas temperature before and after the retrofit.

Looking forward, continued research and development in LNG technologies and infrastructure are essential to support Indonesia's energy diversification goals and contribute to global efforts in achieving net-zero greenhouse gas emissions in maritime transport by 2050. Addressing these challenges will not only bolster Indonesia's energy security but also position the country as a leader in sustainable maritime solutions regionally and globally.

References

[1] Marine Pollution Annex VI, Regulation 13, International Maritime Organizations (IMO), 2011
[2] Marine environmental Protection Committee (MEPC) 80, IMO, 2023
[3] Energy demand in Indonesia, Central Bureau of Statistics of Indonesia, 2021
[4] Indonesia Long-term Strategy for Low Carbon & Climate Resilience 2050, UNFCCC, 2021
[5] Rizqi Fitra, et al, "Conversion of Dual Fuel Opportunities in Indonesia Waterways", Springer International Publishing AG 2017
[6] Guidelines for the Use of Gas or other Low-Flashpoint Fuels for Ships (Pt.1, Vol.1), BKI, 2024
[7] Wang Zhiqiang, "Experimental Study on Diesel/LNG Dual Fuel Engine", 2nd International Conference on Electronics Technology, 2019
[8] Rules for Machinery Installations (Pt.1, Vol.III), BKI, 2024
[9] Guidance for Sea Trial of Motor Vessel (Pt.1, Vol.B), BKI, 2019
[10] Amendments to the NOx Technical Code 2008 to improve the recertification of existing engines onboard ships, PPR11, IMO, 2024
[11] Michael Rachow, et al, "Concept of a LNG gas handling system for a dual fuel engine", International Journal of Marine Engineering Innovation and Research, Vol. 1(4), Sept. 2017

IOP Conf. Series: Earth and Environmental Science **1423** (2024) 012029 doi:10.1088/1755-1315/1423/1/012029

Design of portable floating battery platform utilizing solar and wind energy sources

A.J. Kamal[1], E.S. Koenhardono[1*] and A. Kurniawan[1]

[1] Department of Marine Engineering, Institut Teknologi Sepuluh Nopember, Surabaya, Indonesia

[*] E-mail: eddy_koen@its.ac.id

Abstract. Overfishing is caused by excessive fishing, resulting in a significant decrease in fish populations. This has a detrimental impact on marine ecosystems, fish availability, and fishermen's income. Since 2008, the fishing area in the Madura Strait has experienced overfishing. In relation to the many fishing areas in Indonesia that have experienced overfishing, Government Regulation No. 11 of 2023 was issued on controlled and proportional fishing, carried out in measured fishing zones, based on fishing quotas to preserve fish resources and their environment, and equalize national economic growth. From the perspective of a Marine Engineer, one method to implement this government policy is by changing from fuel to batteries on fishing boats, as well as building energy stations throughout the waters of the Madura Strait. Therefore, this research aims to design a floating energy station based on solar panels and wind turbines to cover the needs of four 10 GT fishing boats with a battery capacity of 90 kWh. Based on numerical methods and trial and error according to weather data, the floating energy station uses 135 solar panels of 550 Wp and 16 wind turbines of 2 kW used to charge 300 batteries of 24 Volts, 100Ah. All batteries can be recharged in 8.5 hours.

1. Introduction

1.1 Backgrounds

The Madura Strait is a fishing area that is the workplace for approximately 92,480 fishermen with more than 9,000 fishing vessels. The water area of the Madura Strait, which is measured based on the concept of marine cadastre by utilizing base maps from the Geospatial Information Agency, is approximately 10,962 km^2[1]. Actually, the sustainable potential in the Madura Strait reaches 214,097 tons, but the production has reached 227,427 tons in 2008 [2]. This condition can be categorized as the Madura Strait has experienced over-fishing, even indications of over-fishing have been detected since 1997.

In addition, in 2013 there were conflicts between fishermen, in the form of violations of fishing areas due to, violation of catch zones and destruction of FADs / nets of other fishermen. The disparity in the use of fishing gear, the draining of FADs or fishing nets, social jealousy, and the existence of andon fishermen are potential causes of conflict in the Madura Strait [3]. Overfishing in the Madura Strait can also trigger conflicts between fishermen. The most recent conflict occurred with fishermen in Pamekasan Regency and Sampang Regency, East Java in 2020 [4]. These conflicts include resource management conflicts. Overfishing requires a comprehensive solution involving various parties. Some efforts that can be made are explained in below paragraphs.

The government should implement a measured fishing policy that ensures fishing does not exceed reproductive capacity. Restrictions on fishing tools and techniques should be enforced to avoid destructive fishing practices and prevent damage to marine ecosystems. Catch zone restrictions can reduce conflicts between fishers and limit the amount of fishing. Supporting fishing readiness with energy stations in fishing grounds [5], developing aquaculture to assist in the recovery of fish stocks and provide an alternative source of income for fishermen, and providing education and training for fishermen on sustainable fishing practices are also crucial. Lastly, enforcing fishing rules is essential to prevent overfishing.

Government Regulation No. 11 of 2023 supports controlled and proportional fishing to preserve fish resources and the environment, as well as economic equity. The technical approach from the Marine Engineer's point of view involves the use of full electric propulsion systems (batteries) on fishing boats and the construction of energy stations throughout the Madura Strait water area. This system limits fishing areas and the use of energy-intensive fishing gear. The floating battery station design uses renewable energy such as solar panels and wind turbines, allowing charging at sea without returning to port. This research is expected to provide an effective solution to overcome overfishing and fisher conflicts in the Madura Strait and support government policies.

1.2 Objectives

The purpose of this study is to determine the number of batteries to be installed on a floating battery station that will be used for battery-powered fishing boats with a size of 10 GT in a state of new production and the batteries that will be used on the floating battery station have the same specifications as the fishing boat. In addition, determining the number of solar panels that have been installed with solar tracking systems and wind turbines by directly using the efficiency of wind turbines used to charge batteries on floating battery stations; designing from floating battery stations to support batteries, solar panels, and wind turbines with sea wave conditions assumed to be absent or no sea wave analysis is carried out, structural strength analysis is not carried out, structural stability analysis is not carried out which will be placed around the Madura strait and can be moved.

2. Methodology

2.1 Problem Statement

This stage is an early stage in the construction of the research. In this stage, questions and problemsare specifically prepared to determine the specific objectives of this research. The content of the research is to overcome the statement of the problems mentioned earlier and it will be done by collectingsome information about the problems. Therefore, the purpose of this research can be understood at this stage.

2.2 Data Collection

The necessary data will be collected, these data include weather data that will be obtained from the National Aeronautics and Space Administration (NASA) website for the area around the Madura Strait, and ship size data that will be obtained from discussions with Surabaya Ministry of Marine Affairs and Fisheries (KKP) officers related to the catch of fishermen in Kedungcowek in 2022.

2.3 Calculation of energy requirements for fishing boats
In this section, the method used is the Numerical method. In a way, the number of fishing boats that use battery power will be calculated to get the total energy requirements needed. The total energy is used to obtain the total battery cells that will be installed on the floating battery charging station. For operations, the ship will be targeted to be able to exchange batteries at the floating battery charging station twice.

2.4 *Analyse the availability of renewable marine energy (solar and wind) in the Madura Strait fishing grounds.*
In this section, using numerical methods and trial and error methods. By determining the surface area and sweep area planned for solar panels and wind turbines. From the determination of the area, using weather data that has been processed, the weather data is used to obtain the availability of electrical energy that will be used to charge the battery on the floating battery charging station.

2.5 Determination of Solar Panel and Wind Turbine Modules according to Analysis of Energy Availability and Ship Operations.
In this section, after obtaining the results of the analysis of the availability of renewable marine energy and the power requirements of ship operations in the Madura Strait. then the number of solar panel modules and wind turbines that will be used to charge the battery at the battery charging station that will be used on the ship will be determined. Using numerical methods.

2.6 Floating Charging Platform Design Based on the Number of Panel Modules and Wind Turbines.
From the number of solar panel modules and wind turbines determined in the previous analysis, the shape of the platform will be designed using the AutoCAD program, the design includes, the size of the platform, the positioning of solar panels, and wind turbines, as well as the placement of battery cells that will be exchanged with battery cells from battery-powered fishing boats.

3. Analysis And Discussion

3.1 Ship data, battery charging station location, and Weather Data

3.1.1 Ship data
This research uses ship data from previous research. Two types of boats are used. boat of 3 GT and boat of 10 GT. The boats specifications are given in Table 1.

Table 1. Boats data that is the object of thesis research

Vessels specifications		
Type of vessel	3 GT	10 GT
Waterline Length	7,5 Meter	11,83 Meter
Breadth	1,75 Meter	3 Meter
Height	1,3 Meter	1,3 Meter
Draft	0,5 Meter	0,9 Meter
Service Speed	5 - 6 Knot	6 knot

The number of boat objects used in this study is based on data from discussions with officers from Ministry of Marine Affairs and Fisheries (KKP) Surabaya related to the catch of fishermen in Kedungcowek in 2022 which amounted to nine boats, consisting of 5 of 3 GT boats and 4 of 10 GT boats.

IOP Conf. Series: Earth and Environmental Science **1423** (2024) 012029 doi:10.1088/1755-1315/1423/1/012029

3.1.2 Battery Charging Station Location

In the Madura Strait there is a shipping channel east of Surabaya which is a busy and important shipping channel, the shipping channel in the Madura Strait is used to determine the location point of the battery charging station. In this study, 8 battery charging station location will be planned by considering the existence of the eastern shipping channel of Surabaya in the Madura Strait by figure 1. Determination of the distance of each battery charging station also takes into account the factor of the mileage of each boat. In this study, the boat that will be used is a 10 GT boat that has the farthest distance. Figure 2. Show each charging station is installed close to the ports of major cities such as the cities of Sampang, Sumenep, Probolinggo, and Situbondo with the distance from each nearest port in the city to the charging station is 13.89 nuatical miles.

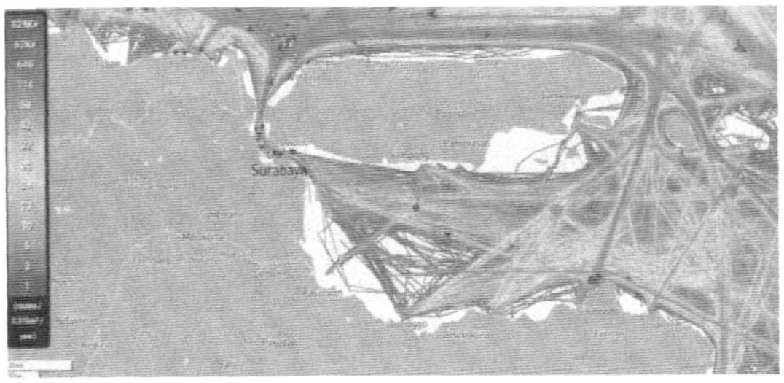

Figure 1. Shipping lanes in the Madura Strait

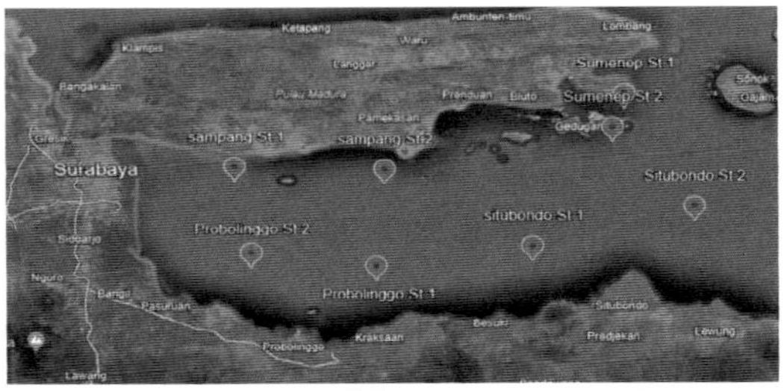

Figure 2. Location of charging stations in the Madura Sea

3.1.3 Weather Data

Weather data is obtained from the official National Aeronautics and Space Administration (NASA) website. The data used is irradiation data or solar radiation data and wind speed for 1 full day detailed by hour explained in table 2.

Table 2. Hourly average solar Direct Irradiation and wind speed data

Time	Aver DNI (kWh/m^2)	Aver WIND (m/s)
06:00	0,02	3,54
07:00	0,17	3,71
08:00	0,38	3,81
09:00	0,57	3,70
10:00	0,72	3,51
11:00	0,80	3,35
12:00	0,82	3,27
13:00	0,77	3,26
14:00	0,65	3,29
15:00	0,49	3,34
16:00	0,31	3,38
17:00	0,12	3,39
18:00	0,01	3,32
19:00	0,00	3,36
20:00	0,00	3,39
21:00	0,00	3,41
22:00	0,00	3,41

Time	Aver DNI (kWh/m2)	Aver WIND (m/s)
23:00	0,00	3,40
00:00	0,00	3,40
01:00	0,00	3,42
02:00	0,00	3,46
03:00	0,00	3,50
04:00	0,00	3,53
05:00	0,00	3,54

3.2 Analysis of Fish Boat Operational Power Requirements

Boats with a size of 10 GT have resistance data at each speed that has been studied in previous studies [6]. This data is used to determine the power required for the DC motor in the propulsion system. In determining the power required by DC motor propulsion at each speed can use the following formula:

$$P = \frac{\boldsymbol{Rt\ x\ Vs}}{\boldsymbol{Eff\ Propulsi}} \quad (1)$$

The propulsion efficiency is set at 60%, which consists of propeller efficiency, transmission efficiency, and electrical efficiency. Therefore, from formula (1), the power required to drive a 10 GT fishing boat is 4.73 kW. In addition, previous research [6] also explained the power requirements for supporting tools for fishing operations, navigation and communication tools, and lighting explained in table 3. so the overall power of the ship is 89,6 kW, for ships with a size of 10 GT using a battery cell with specifications of 3.2V, with a capacity of 100 Ah with the material is LiFePO4. then the number of battery cells is 280.

Table 3. Power requirements of each appliance

Equipment	Time	Unit power	Voltage	Overall	Usage time
Propulsion	10 Hours	5,5 kW	24V	55 kWh	day and night
Purse Seine	2,5 Hours	5 kW	24V	12,5 kWh	day and night
Anchor wind	0,1 Hours	1,50 kW	24V	0,15 kWh	day and night
Radio Navigation	11 Hours	0,025 kW	24V	0,28 kWh	day and night
Fish finder	9 Hours	0,006 kW	24V	0,05 kWh	day and night
Navigation Lamp	2 Hours	0,24 kW	24V	2,4 kWh	night
Emergency lamp	1 Hours	0,06 kW	24V	0,06 kWh	night

The station is designed to be able to serve 4 vessels with a size of 10 GT and each vessel can carry out fishing operations 2 times. so the number of battery cells to be installed is 2400 with details of 2240 battery cells to be exchanged with fishing boats and 160 batteries for reserves used for electricity needs at the station.

The first operation, the vessel departed from the home port to the charging station to make preparations for fishing. The ship went to the location with a service speed of 6 knots and traveled a distance of 14 nm. So it takes 2 hours 11 minutes the total power required by the ship in this operation is 10.32 kW. The next operation is the ship will do fishing around the charging station with a radius of 14 nm. This operation is carried out 2 times during the day and at night. In fishing operations, it does not use the full speed of the ship. The ship uses a speed of 5.5 knots. This is based on previous research [7]. The length of fishing operations is 8 hours based on previous research [6], so the total power required to drive the ship is 26.78 kW. The propulsion power of the ship during fishing plus the power of the equipment during the day is 39.76 kW and at night is 42.22 kW.

Table 4. Details of the power requirements of each fishing boat operation

Activities	Speed	Time	Power	Remaining power
trip to go, morning	6 knot	2,18 hours	10,32 kW	79,28 kW
Daytime Fishing	5,5 knot	8 hours	39,76 kW	39,52 kW
Night fishing	5,5 knot	8 hours	42,22 kW	52,88 kW
Return trip, evening	6 knot	2 ,18 hours	10,32 kW	42,56 kW

3.3 Availability of renewable marine energy (solar and wind) in the Madura Strait fishingrounds

In determining the power generated by solar energy. Using the previous weather data, the area of the panel module to be installed is based on 1 day can meet the power needs to charge the battery cells in the floating battery charging station so that the planned module area is 350 m^2. In determining the availability of power from solar energy, use the following formula:

$$P = Solar\ Irradiaton\ (kWh/m^2)x\ Plan\ Area\ (m^2) \quad (2)$$

The total availability of electrical power for 1 full day using solar energy sources is 2033.17 kW/day. This, already fulfills the power requirements for the battery cells provided at the floating battery charging station, with the total power at the charging station being 768 kW/day.

Table 5. Specifications of solar panels and sun tracking system

Merk	JA SOLAR	Merk	All earth solar tracker
Model	JAM72S30 525-550	Model	S24
Nominal power (Wp)	550	Length	7 meter
Efficiency (%)	0,213	Width	4,8 meter
Rated Curent (A)	13,11	Height	5,49 meter
Weight (kg)	28,6	number of Panel	12 panel
Area (m^2)	2,584386	Power	356 Watt DC

From Table 5 the total number of solar panels needed for the charging platform is calculated by dividing the planned solar panel area by the selected solar panel area, resulting in 135 panels with a total weight of 3873 kg. These panels have an efficiency of 21.3% in converting solar energy to electrical energy so the power that can be generated for 1 day based on the efficiency of the solar panel is 433,07 kW/day. The solar panels will be installed with a tracking system to maximize sunlight illumination. In addition, at least 12 solar tracking systems are required to accommodate 135 panels.

The wind turbine will be used to charge the battery used for electricity needs at the station, with a total of 160 battery cells and a total power of 51.2 kW. The availability of wind energy is determined by the wind speed in the area, and a formula is used to convert wind energy to electrical energy.

$$P_{wind} = 0{,}5\ x\ \rho\ x\ v^3 x\ A \qquad (3)$$

Where:
v = wind speed (m/s)
ρ = Air density, assumed to be 1225 kg/m^3
A = Wind turbine sweep area planned (m^2)

The planned sweep area of the wind turbine is determined by the battery used for the electricity requirement at the station. therefore, the planned sweep area is 150 m^2. Due to the low wind speed at the platform site, a large sweep area is required to ensure sufficient energy availability. By using formula (3) and the data in table 2 The results showed that the total wind energy available per day at the platform location was 88.31 kWh/day. The wind turbine used is a Vertical Axis Wind turbine, based on previous research [8] for wind turbine efficiency is 34.2%. then the power generated by the wind turbine based on efficiency is 29.89 kW/day.

Table 6. Specifications of wind turbine

Type	GV-2KW
Rated power	2000 W
Maks power	2500 W
Wind start speed	1 m/s
Rated Wind	10 m/s
wind speed work	1 - 25 m/s
Wind speed break	50m/s
Voltage option	24 - 380 V
weight	228 Kg
Area Sweep	9,62 m2

From table 6 the number of wind turbines needed on the charging platform is calculated by dividing the planned sweep area by the sweep area of each wind turbine. Based on this calculation, 16 wind turbines are required, with a total weight of 3555 kg.

The installation configuration of these wind turbines follows the guidelines from previous research [9].

where the distance of each wind turbine if given a distance with an angle of 45° then uses the formula:

$$\text{Relative distance} = 1{,}5D \qquad (4)$$

Where D is the diameter of the wind turbine rotor, so the distance of each wind turbine is 3.9 metres if arranged between wind turbines with a distance of 45°. For wind turbines if arranged parallel or 90°, use the following formula:

$$\text{Relative Distance} = 2{,}1D \qquad (5)$$

This means that the distance between wind turbines is 5.52 metres.

IOP Conf. Series: Earth and Environmental Science **1423** (2024) 012029 doi:10.1088/1755-1315/1423/1/012029

State of charge of the floating battery charging station. It can be seen from 06:00 to 18:00 that the power on the battery charging station is charged from the help of solar panels from 0% to 60% power. From 18:00 to 04:00, the power is slightly charged with the help of the wind turbine. The next 06.00 hours, the power at the charging station starts to rise again because the solar panel can work again. At 14.00 and 17.00 the power at the battery charging station began to drop due to the exchange of battery cells by battery-powered fishing boats according to the description table in table 4.16. at 22.00 until 01.00 it decreased in the state of charge due to the exchange of batteries at the battery charging station.

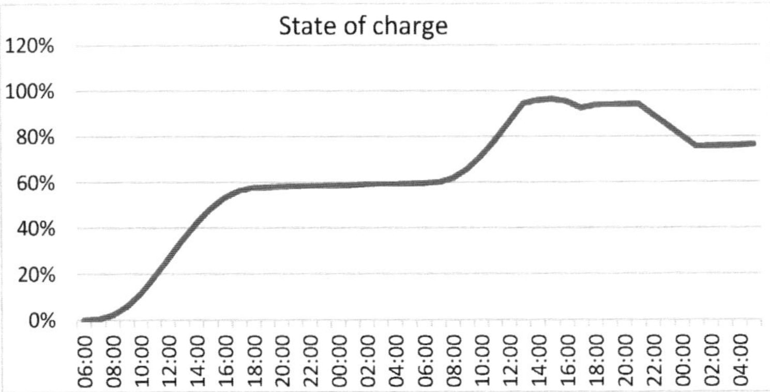

Figure 3. *State of charge*

3.4 Design of Floating Charging Platform according to Renewable Energy Availability Analysis

The design of the floating charging platform involves several components: batteries for boat operations, batteries for platform use, solar panels with a sun tracking system, and wind turbines. The platform is designed using AutoCAD software and its equipped with fish cages to accommodate live fishcatches or fish enlargement. the materials in the nets used in fish cages are the same as the HDPE fish cages. The results of the platform design from the AutoCAD software:

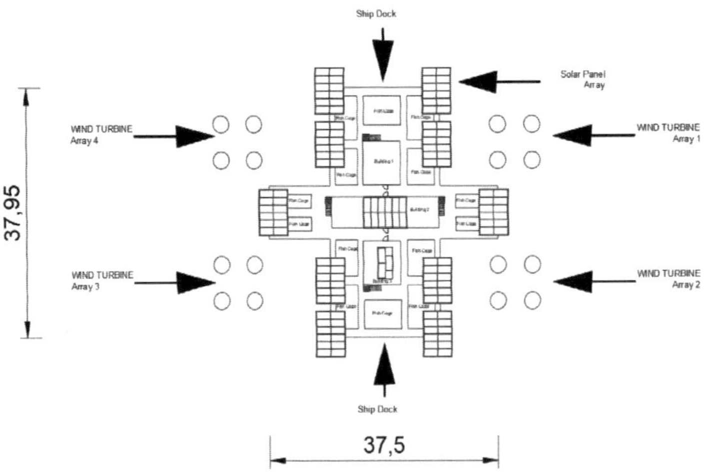

Figure 4. Floating charging platform model view from above.

The designed floating charging platform measures 37.95 meters by 37.5 meters and can accommodate a total of 14 fish cages for live catches and fish enlargement. The platform features threebuildings: one for the battery room, one for the empty room, warehouse, and battery room, and one forthe radio room, empty room, battery room, and a resting area for fishermen on the first and second floors.

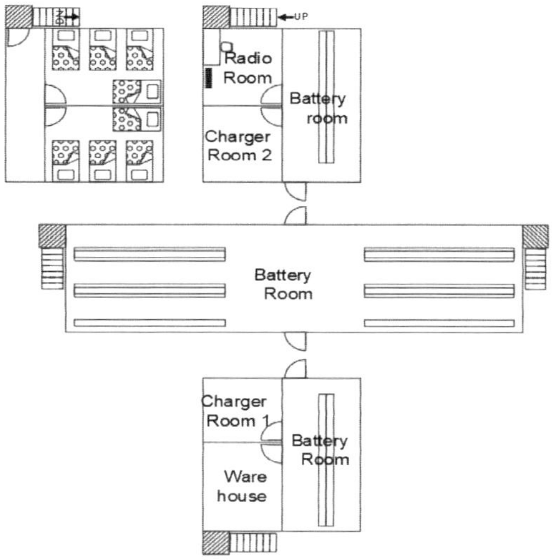

Figure 5. building layout model

For the determination of the battery charger through the following specifications of the solar charge controller and wind turbine.

Table 7. Specifications of charger controller

Model	M48200 – DNSOLAR	WGMC30K380 – wind turbine
Input Voltages	Maks 250V	Maks 550V
Input Power System	10 kW	30 kW
Input Current Maks	45A	Maks 90A
Output Current	200A	72A
Efficiency MPPT	99%	99%
Output Voltage	48V	0 – 550V

To set up the charging system using the charger controller, the solar panel modules are connected in series to increase the voltage and parallel to increase the current. The panels are connected in series 5 to get a voltage of 209.8V and parallel 3 for a current of 39.33A, so one charger controller requires 15 solar panel modules. For the M48200 charger, 9 charger controllers are required for 135 solar panel modules. The wind turbine produces a voltage of 48V according to the wind charger controller with a current of 41.67A. The wind turbine configuration is 8 in series for a voltage of 384V, then 2 in parallel for a current of 83.3A, requiring 1 charger controller. The next step is to design the wiring diagram for the energy transfer system from the solar panel and wind turbine to the battery. For the solar panel, a 2.5 mm^2 cable is used for a current of 13.11A with a 20A MCB. The wind turbine cable requires a cross section of 16 mm^2 for a current of 41.7A with a 60A MCB. The charger controller system consists of two parts: the first system uses two M48200 solar module charger controllers, each producing a current of 200A, using a 95 mm^2 cable with a

225A shield. These two chargers are connected to the distribution busbar, repeated 4 times for the other 8 charger controllers. Second system uses one M48200 solar module charger controller with a current of 200A (95 mm^2 cable, 225A MCB) and one WGMC30K380 wind turbine charger controller with a current of 72A (25 mm^2 cable, 100A MCB). Both are connected to the distribution busbar, from which the 68A current is distributed to the first system busbar using a 25 mm^2 cable with an 80A MCB. The distribution busbar and battery cell busbar use two 60 x 10 mm bars with conductive coating. The battery circuit consists of 15 battery cells in series to achieve the system voltage, paralleled by 40 for 48V input and 11.7A current. One battery pack takes 8.5 hours to charge. With 4 charger controller systems, a total of 2400 batteries can be charged in 8.5 hours.

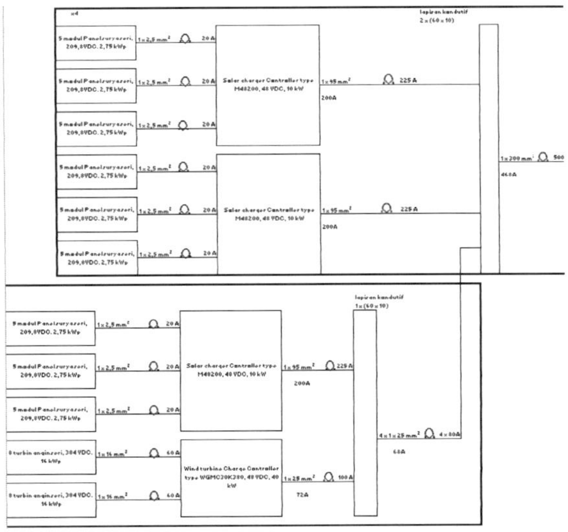

Figure 6. Explanation of charger controller system

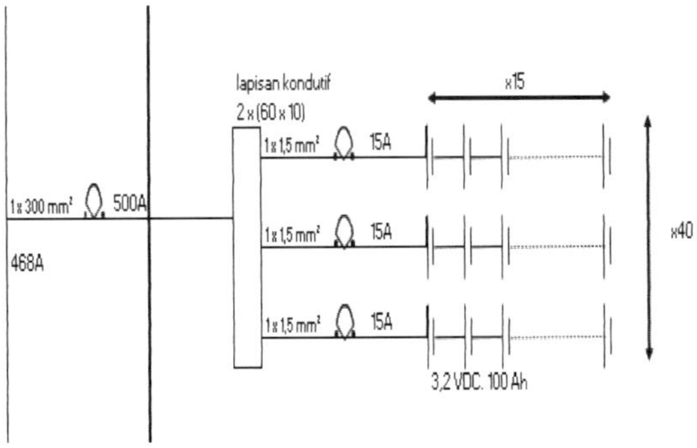

Figure 7. Wiring diagram of the charger controller busbar to the battery

4. Conclusion and Suggestion

4.1 Conclusion

The floating battery charging station is designed to serve battery-powered fishing boats that arrive twice a day, with each boat carrying 280 battery cells, making a total of 2400 battery cells in the station. The determination of the number of solar panels and wind turbines was done through numerical methods and trial and error, resulting in 135 solar panels with 550 Wp per panel and 16 wind turbines with 2 kW per turbine. The Homers Pro program shows the need for 328 solar panels in the first option and 255 solar panels in the second option, while wind turbines are not the main source due to low wind speed. The station has a size of 37.95 x 37.5 meters, equipped with 12 solar tracks for solar panels, and 3 buildings to protect the battery cells and charger controller. The power transfer from solar panels and wind turbines to battery cells uses 10 charger controllers, with 15 series and 40 parallel battery cells, and a charging time of 8.5 hours.

4.2 Suggestion

To improve efficiency, further research needs to be done on solar panels that have higher efficiency. In terms of wind turbines, it is recommended to use low rotor rotation specificationsfor low wind speeds. In addition, it is necessary to analyse other renewable energy sources tosupport a charging system that can meet the needs of the ship.

References

[1] Z. Hidayah, N. I. Nuzula, and D. B. Wiyanto, "Analisa Keberlanjutan Pengelolaan Sumber Daya Perikanan di Perairan Selat Madura Jawa Timur," *Jurnal Perikanan Universitas Gadjah Mada*, vol. 22, no. 2, p. 101, Dec. 2020

[2] Department of Marine and Fisheries of East Java Province. 2010. East Java Capture Fisheries Statistics.

[3] P. D. Suryanto and S. P. M. P. Moh. Bagus Gozhali Aniputro, "Model penyelesaian konflik nelayan di kawasan selat Madura berbasis pada kearifan lokal sebagai modal sosial," *lib.unair.ac.id*, 2013. https://repository.unair.ac.id/43093/

[4] Robertus Belarminus, "Wilayah Tangkapan Tak Jelas Picu Konflik Nelayan di Madura Halaman all - Kompas.com," *KOMPAS.com*, Nov. 02, 2020. https://regional.kompas.com/read/2020/11/02/17191731/wilayah-tangkapan-tak-jelas-picu-konflik-nelayan-di-madura?page=all

[5] K. Y. Yap, H. H. Chin, and J. J. Klemeš, "Solar Energy-Powered Battery Electric Vehicle charging stations: Current development and future prospect review," *Renewable and Sustainable Energy Reviews*, vol. 169, p. 112862, Nov. 2022

[6] Juniarko Prananda, Eddy Setyo Koenhardono, and Rivaldi Clarence Tjoa, "Design of an Optimum Battery Electric Fishing Vessel for Natuna Sea," *International Journal of Marine Engineering Innovation and Research*, vol. 4, no. 2, Sep. 2019

[7] Arham Rumpa, Najamuddin Najamuddin, and St. Aisjah Farhum, "PENGARUH DESAIN ALAT TANGKAP DAN KAPASITAS KAPAL PURSE SEINE TERHADAP PRODUKTIVITAS TANGKAPAN IKAN DI KABUPATEN BONE," *Jurnal IPTEKS Pemanfaatan Sumberdaya Perikanan*, vol. 4, no. 8, Sep. 2018

[8] Md. S. Mahmud, "Design and Analysis of a Vertical Axis Wind Turbine for Highway Application in Bangladesh," *Journal of Advanced Research in Dynamical and Control Systems*, vol. 12, no. SP7, pp. 805–817, Jul. 2020,

[9] Y. Chen *et al.*, "Investigation of pitch angles on the aerodynamics of twin-VAWT under staggered arrangement," *Ocean Engineering*, vol. 254, pp. 111385–111385, Jun. 2022

IOP Conf. Series: Earth and Environmental Science **1423** (2024) 012030 doi:10.1088/1755-1315/1423/1/012030

Development of Savonius Type Wind Turbine Model as A Source of Renewable Electrical Energy

Cahya Kusuma[1], Tri Sutrisno[2], Wawan Kusdiana[3], Aris Tri Ika R[4], and Ahmad Baidowi[5]

[1,2,3,4] Sekolah Tinggi Teknologi Angkatan Laut, Surabaya, Indonesia
[5] Institut Teknologi Sepuluh Nopember, Surabaya, Indonesia

kusumacahya99@gmail.com

Abstract. Wind turbines can be used as a renewable energy resource, especially in remote areas and outer islands where there is no electricity. Based on the calculations that have been carried out, it is found that the dimensions of the wind turbine model used in this research are a turbine height of 1 meter with a diameter of 0.5 meters. Meanwhile, the dimensions of the shaft used are 20mm. Model testing has been carried out in Juanda Surabaya, it is known that the generator power produced increases along with increasing wind speed which is also influenced by the rotation of the wind turbine. In the results of this research, the largest generator power was produced at a wind speed of 4.8 ms^{-1}, which produced a generator power of 12.6 watts. Meanwhile, the lowest power is 0 watts because a wind speed of 0.4 m/s is not able to rotate the wind turbine. From the data collection results, it was also found that the SCC could only charge the battery if the minimum voltage reached 13 volts. So the new SCC can charge the battery at a wind speed of 3.7 m/s with a voltage output of 13.1 volts and a current strength of 0.5 amperes. This research will use a Savonius wind turbine with 2 rotor blades with dimensions of 1 meter high and 0.5 meter diameter. The research results show that power is obtained at a generator voltage starting from 13.1 volts which produces a current strength of 0.7 amperes. It is at this voltage that the new Solar/Wind Charge Control can charge the battery.

1. Introduction

The availability of non-renewable natural resources such as gas, coal and petroleum is decreasing due to increasing population and increasing energy consumption.[1] Indonesia is the largest energy user with more than 36% of primary energy use in Southeast Asia.[2] Indonesia has set a target to achieve Net Zero Emissions (NZE) by 2060 or even sooner.[3][4] IESR modeling shows that Green House Gas (GHG) emissions from the energy sector are likely to increase from 743 MtCO2e in 2022 to 963 MtCO2e in 2030, with no signs of reaching its peak. Current energy policy is still far from sufficient to reduce the increase in emissions, only able to reduce 20% of projected emissions by 2030. The transition to a sustainable energy system requires a focus on four main pillars: renewable energy, electrification, reduction of fossil fuels, and fuel low carbon alternative. However, renewable energy growth remains slow, with a contribution of less than 10% to the primary energy mix in 2022, mainly from biodiesel for transportation.

The use of renewable energy in the electricity sector is also lagging behind, posing a risk in achieving the 2030 target. On the other hand, the use of fossil fuels is still dominant.[5][6] Coal contributes almost half of the primary energy mix.[7] [8]The electricity sector still relies on fossil fuel-based power plants, causing projected GHG emissions to increase. Despite efforts to increase the use of alternative fuels, such as bioenergy and low-carbon hydrogen, challenges remain. The energy transition requires large investments, but currently it is still far from the projected needs of the NZE.[9] The Indonesian government is targeting 31% of its energy supply renewable energy by 2050.[10][11]

One source of renewable energy is wind energy which can produce up to 16.7% of power.[12] One of the transitions being carried out is increasing the use of wind energy. Wind energy on a large scale is needed for areas that do not yet have electricity from the State Electricity Company. However, wind energy is also needed on a small scale, for example for outer, underdeveloped and remote islands and for security posts at the border. Without the support of electricity in operating the post, the post would not be able to work properly. Apart from lighting, electricity is also needed as support for communication tools, because in its operation the post reports activities and situations that occur in the environment where the post is located. So for these security posts, communication tools are very important. Apart from there being several areas that have not been reached by PLN, some areas that use diesel are still having problems in distributing the fuel.[13][14]

2. Methodology

A wind turbine is a system of objects that rotates around an axis due to the push of the wind. Windmills or wind turbines can also be used to convert wind energy from motion energy into mechanical and electrical energy.[15] Wind turbines themselves have different characteristics based on shape, type, diameter, number of blades and shape of the blades. Types of wind turbines can be grouped into two types according to the axis used, namely, vertical axis wind turbines or called Vertical Axis Wind Turbines (VAWT) and horizontal axis wind turbines or also called Horizontal Axis Wind Turbines (HAWT).[16] Wind Resources Potential are classified into 3, namely Marginal, Fair and Good. [17] This research was carried out in low rated wind speed at the Juanda area, Surabaya with an average wind speed of 3.0 ~ 4.0 ms^{-1}.[18]

2.1. Factors that influence wind turbines

a. Turbin Dimension

The turbine design used is based on wind speed in the surrounding conditions. The dimensions of a wind turbine can be found by assuming that the power produced is at the speed of the wind around us. With the power formula (P) for a wind turbine as follows:

$$P = Cpr\ \frac{1}{2}\rho A V^2 \tag{1}$$

where :

- P : Power (watt)
- Cpr : Rotor Power Coeficient (Cpr)
- ρ : Air Density (ρ average: 1,2 kg/m3)
- A : Rotor Area (m^2)
- v : Wind velocity (ms^{-1})

b. Rotor Power Coeficient (Cpr)

Rotor Power Coefficient (Cpr), the power coefficient will be calculated using strip theory for a certain rotor speed ratio. It gives the rotor power coefficient for different wind speeds at a fixed rotor speed or for different rotor speeds at one wind speed.

$$Cpr = \lambda C_q \tag{2}$$

where :

- Cpr : Rotor Power Coeficient

IOP Conf. Series: Earth and Environmental Science **1423** (2024) 012030 doi:10.1088/1755-1315/1423/1/012030

λ : Tip Speed Ratio
Cq : Rotor Torque Coeficient

c. Tip Speed Ratio (*TSR*)
Tip Speed Ratio is the ratio of rotor tip speed to free wind speed. For a certain nominal wind speed, the tip speed ratio will affect the rotor rotational speed.

$$\lambda = \frac{\pi D n}{60v} \tag{3}$$

where :
λ : Tip Speed Ratio
D : rotor diameter (m)
n : rotor rotation (rpm)
v : wind velocity (ms^{-1})

A graph of the relationship between tip speed ratio (TSR) and power coefficient (Cp) for various types of wind turbines can be seen in Figure 1.

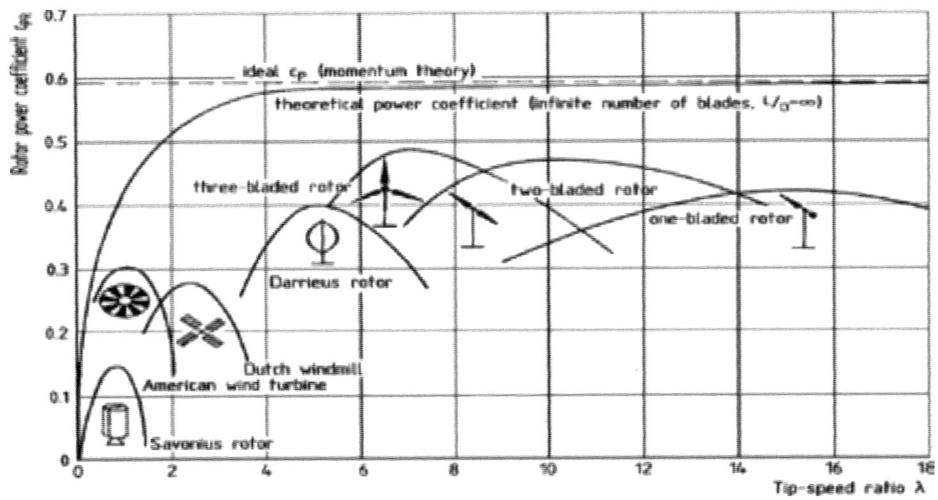

Figure 1. Relationship Between Cpr and TSR[19]

d. Rotor Torque Coeficient (Cq)
Rotor Torque Coefficient (Cq) is the torque produced by the turbine rotor which is used to calculate the Rotor Power Coefficient (Cpr). Rotor Torque Coefficient (Cq) can be found via the graph in Figure 2.

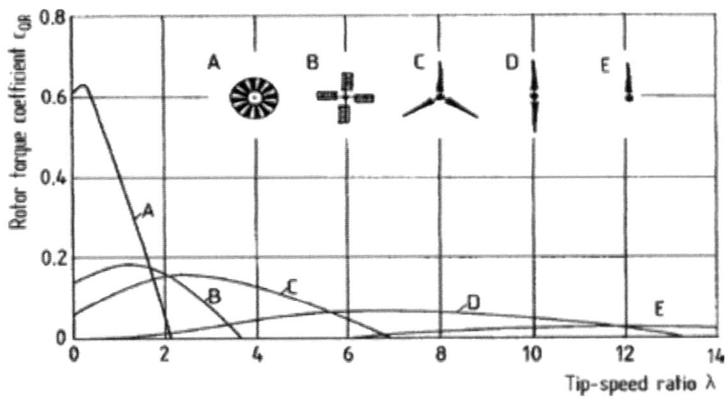

Figure 2. Rotor Coefficients of Some Wind Turbines

e. Turbine shaft strenght

Turbine shaft is a very important part, apart from the shaft functioning as a position for the blades, the shaft also functions as the main connecting device for energy changes, from kinetic energy to electrical energy which previously went through a generator. Based on this type of turbine, the shaft is installed vertically so that it receives a greater torsional load. Shaft diameter calculation:

$$Pd = fcP \quad (4)$$

$$T = 9.74 \,.\, 10^5 \frac{Pd}{n} \quad (5)$$

$$\tau_a = \frac{\sigma_b}{S_{f1} \,.\, S_{f2}} \quad (6)$$

$$ds = \left(\frac{5.1}{5} KtCbT\right) . \frac{1}{3} \quad (7)$$

f. Air power

The energy captured from the wind is its kinetic energy. Kinetic energy is energy that arises from moving objects. The basic formula for calculating wind energy is formula 8:

$$E = \frac{1}{2} mv^2 \quad (8)$$

where :

m : Moving air masses (kg)
v : Wind velocity (ms^{-1})

2.2 Wind Turbine Savonius

The Savonius wind turbine is a vertical axis wind turbine (VAWT) which can be used in low speed winds. This wind turbine was discovered by Sigurd J. Savonius in 1922. The construction of this turbine is very simple, the turbine is composed of two half-cylindrical blades. In a Savonius type wind turbine, the wind that blows on one of the rotor blades is expected to flow more to the other rotor blades. through the gap around the shaft, providing additional thrust on the rotor blades which causes the rotor to spin faster. The shape of the Savonius rotor is semi-circular (Savonius U), so that the air flow on both sides of the blade is the same, while in the second design (Savonius L) the air flow on the straight side of the blade is greater than on the side of the curved, quarter-circular blade. Figure 3 shows that the pushing force or drag force on the blade side which is in the same direction as the fluid flow is greater than on the blade side which is in the opposite direction of the flow, so it will produce a positive torque which will make the turbine spin. In theory it can be written as follows:

$$dF_x = (pdA)Cos\theta + (\tau_w dA)sin\theta \quad (9)$$

$$dF_y = -(pdA)Cos\theta + (\tau_w dA)cos\theta \quad (10)$$

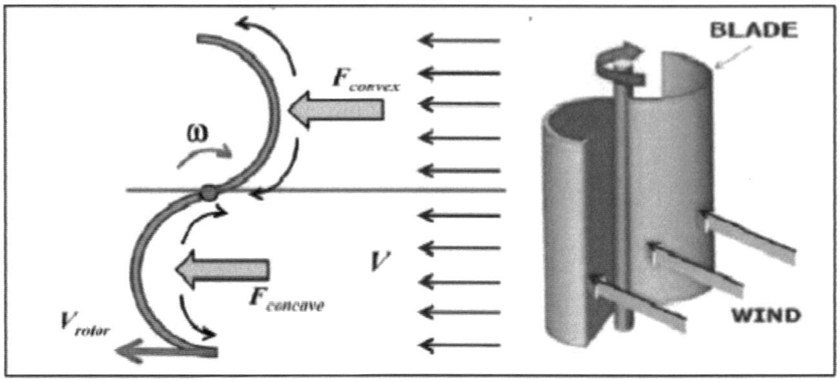

Figure 3. Drag Force in the Savonius Turbine[20]

3. Result and Discussion

Currently, police posts, border guards, lighthouses on the islands on the border are in great need of electricity support. In addition to lighting equipment, it is also a support for communication equipment, because in its operation, these posts report activities and situations that occur in the environment where the post is located. So for these posts, communication equipment is very important. Based on the importance of this, supporting equipment is needed to ensure that communication equipment is not constrained in its operation. One of them is to create a tool that can be used so that communication equipment is always ready. Therefore, electricity support is needed at these posts to charge their communication equipment. However, as we know, electricity support on the border is still very limited. In addition to several areas that have not been reached by electricity, some areas that use diesel are still constrained in the distribution of their fuel. The trial was carried out in Juanda considering that it is close to the coast such as border posts.

The working system of the turbine depends on the strength of the wind used. The movement of air flowing past and pushing the rotor blades will produce power received by the generator. The rotation generated from the wind turbine is received by the generator and is directly adjusted to the specifications of the generator so that it can produce the desired power. This power will later be transmitted to the previous battery via the Hybrid Charge Controller. The Hybrid Charge Controller functions as a power regulator that will be stored in the battery and also as a regulator of the direction of current from the generator to the battery. This power is stored in the battery so that the output voltage is stable or does not fluctuate. The power stored in the battery goes to the inverter. In the inverter, the electric current which was previously direct current (DC) is converted into alternating current (AC). This research uses a Savonius type wind turbine with 2 rotor blades.[21] In this way, the data that will be used is obtained as follows.

P = 15 watt (Model prototype requirements)
V = 4 m/s (data from measurements in the field)
n = 50 rpm (assumption)

Based on formula number 1, the P value is obtained

a. Tip Speed Ratio (TSR)
D = 50cm = 0,5m
n = 50 rpm
v = 4 ms^{-1}
Based on formula number 3, the value is obtained $\lambda = 0.32$

b. Rotor Torque Coeficient (Cqr)

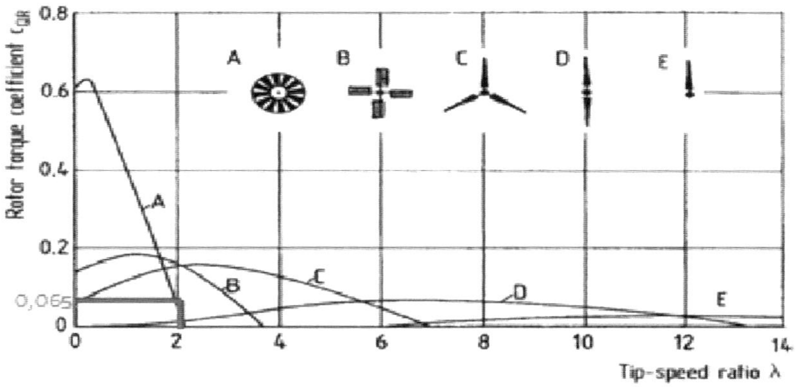

Figure 4. Rotor Coefficients of Some Wind Turbines

Based on Figure 4, the rotor coefficient relationship curve for several wind turbines is obtained for the Savonius type wind turbine rotor in area A:

$\lambda_1 = 2$ then $Cq_1 = 0{,}065$ for $\lambda_2 = 0{,}32$ then value $Cq_2 = \frac{\lambda 1}{\lambda 2} x\ Cq_1$

$Cq_2 = \frac{2}{0{,}32} x\ 0{,}065 = 0{,}41$

c. Rotor Power Coeficient (Cpr)

$\lambda = 0{,}32,\ Cq = 0{,}41$

$Cpr = \lambda\ x\ Cqr = 0{,}13$

Based on formula number 1, the area of the turbine blade A:

$$A = \frac{2P}{Cpr\ \rho\ v^3}$$

$$A = \frac{30}{8{,}32} = 3.6m^2$$

Based on the calculation results of the blade cross-sectional area (A) of $3.6m^2$, using 2 blades with a height of 1 meter, the blade dimensions are obtained:

$$d = \frac{2A}{2\pi L} = 0.5 \text{ meters}$$

The dimensions of the type of turbine used are the Savonius wind turbine with dimensions D x L, namely 0.5 x 1 meter

d. Perhitungan Shaft Strength Calculation

Calculation of a shaft with a power of 50 watts, turbine shaft rotation of 206 rpm, with a correction factor of 2.0. The design material used is steel bar ST 60.

P = 10 watt = 0.01 kW

n = 50 rpm

$f_c = 2.0$

ST 60 = 60 $kg.mm^{-2}$

Calculation of shaft diameter:

$Pd = f_c\ P = 0.02$

$T = 9{,}74\ x\ 10^5 \frac{Pd}{n}$

$T = 1948\ kgmm$

Material St 60 $\sigma_b = 60\ kg.mm^{-2}$, $S_{f1} = 6.0$; $S_{f2} : 2.0$

IOP Conf. Series: Earth and Environmental Science **1423** (2024) 012030 doi:10.1088/1755-1315/1423/1/012030

$$\tau_a = \frac{\sigma_b}{\text{Sfl } x \text{ Sf2}} = 5 \text{ kg/mm}^2$$
Cb = 2,0 ; Kt = 1,5
$$ds = (\frac{5{,}1}{5} Kt\ Cb\ T)^{1/3} = 18{,}13\text{mm}$$

The diameter of the shaft used in this wind turbine is 18.13 mm.

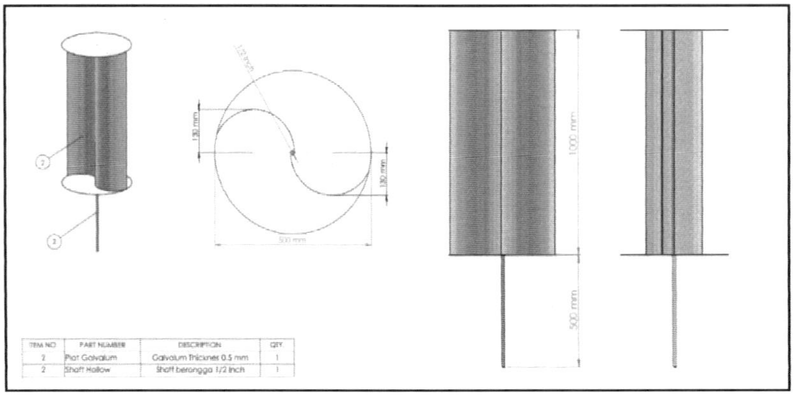

Figure 5. Turbine Blade Design

The wind turbine design drawing can be seen in Figure 5. The Savonius turbine model can be seen in Figure 6.

Figure 6. Savonius turbine model

The data taken are wind speed and temperature values measured using an anemometer. The rotation value on the wind turbine shaft and generator is measured using a tachometer. Current strength data is taken using a digital multimeter and also using a Solar Charge Controller which is integrated with data collection from the solar panel. Data collection was carried out on the heli deck of the Juanda simulator which was open and had no wind obstructions. Data collection starts at 07.00AM ~ 04.00 PM by looking at weather conditions and wind variations taken every 1 hour. Current measurement results in experiment no 1 shown at table 1.

Table 1. Current measurement results in experiment no 1

Time	RPM Turbine	RPM Gen	Wind Velocity (m/s)	Wind Power (watt)	Voltage (Volt)	Current (Ampere)	Tip Speed Ratio (TSR)	Power Coeff	Power (Watt)	Temp (°C)	Hum (%RH)	Weather
08.00	24	144	0.8	0.39	0	0	0	0	0	30.7	77	Overcast
09.00	15	90	0.5	0.15	0	0	0	0	0	28.0	78	Rain
10.00	15	90	0.5	0.15	0	0	0	0	0	28.1	94	Overcast
11.00	114	706	3.7	8.39	13.1	0.5	0.806	0.23	6.55	27.6	95	Overcast
12.00	36	216	1.2	0.88	0	0	0	0	0	27.5	95	drizzling
01.00	159	858	4.1	10.3	13.4	0.8	1.015	0.28	10.72	27.3	95	drizzling
02.00	206	957	4.8	14.11	14	0.9	1.123	0.21	12.6	27.5	99	Rain
03.00	0	0	0.4	0.1	0	0	0	0	0	27.5	99	Rain
04.00	15	90	0.5	0.15	0	0	0	0	0	27.7	99	drizzling
05.00	0	0	0.4	0.1	0	0	0	0	0	27.5	99	drizzling

Current measurement results in experiment no 2 shown in table 2.

Table 2. Current measurement results in experiment no 2

Time	RPM Turbine	RPM Gen	Wind Velocity (m/s)	Wind Power (watt)	Voltage (Volt)	Current (Ampere)	Tip Speed Ratio (TSR)	Power Coeff	Power (Watt)	Temp (oC)	Hum (%RH)	Weather
08.00	21	126	0.7	0.3	0	0	0	0	0	30.7	77	Sunny
09.00	27	162	0.9	0.5	0	0	0	0	0	30.9	78	Sunny
10.00	72	432	2.4	3.53	0	0	0	0	0	28.1	94	Overcast
11.00	63	378	2.1	2.7	0	0	0	0	0	27.6	95	Overcast
12.00	109	687	3.5	7.5	12.3	0	0.815	0	0	27.5	95	Gerimis
01.00	69	414	2.3	3.24	0	0	0	0	0	26.3	95	Rain
02.00	57	342	1.9	2.21	0	0	0	0	0	26.5	99	Rain
03.00	15	90	0.5	0.15	0	0	0	0	0	27.5	99	drizzling
04.00	21	126	0.7	0.3	0	0	0	0	0	27.7	99	drizzling
05.00	27	162	0.9	0.5	0	0	0	0	0	27.5	99	drizzling

Based on the experimental results in tables 1 and 2, it was found that the wind speed varied from 0 ms^{-1} to 4.8 ms-1 producing a generator power of 12.6 watts. Meanwhile, the lowest power is 0 watts because a wind speed of 0.4 m/s is not able to rotate the wind turbine. The Solar Charge Controller (SCC) can only charge the battery if the minimum voltage reaches 13 volts. The Solar Charge Controller can charge batteries at a wind speed of 3.7 m/s with a voltage output of 13.1 volts and a current strength of 0.5 amperes.

4. Conclusion

Based on the calculations that have been carried out, the dimensions of the wind turbine model used in this research are 1 meter high with a diameter of 0.5 meters. Meanwhile, the dimensions of the shaft used are 20mm. The generator power produced increases as the wind speed increases, which is also influenced by the rotation of the wind turbine. In the results of this research, the largest generator power was produced at a wind speed of 4.8 ms^{-1}, which produced a generator power of 12.6 watts. Meanwhile, the lowest power is 0 watt because a wind speed of 0.4 ms^{-1} is not able to rotate the wind turbine. The Solar Charge Controller can charge batteries at a wind speed of 3.7 m/s with a voltage output of 13.1 volts and a current strength of 0.5 amperes. In further research, it is expected to increase the number of trials and replace the turbine blade material with a lighter density material such as fiber composite. Trials are also carried out at the outermost island posts on the border according to actual conditions.

5. References

[1] N. Khurshid, C. E. Egbe, and N. Akram, "Integrating non-renewable energy consumption , geopolitical risks , economic development with the ecological intensity of wellbeing : evidence from quantile regression analysis," no. June, pp. 1–14, 2024, doi: 10.3389/fenrg.2024.1391953.

[2] International Energy Agency, "World Energy Outlook iea.org/weo," *Int. Energy Agency*, p. 763, 2017.

[3] "An Energy Sector Roadmap to Net Zero Emissions in Indonesia," *An Energy Sect. Roadmap to Net Zero Emiss. Indones.*, 2022, doi: 10.1787/4a9e9439-en.

[4] Kementrian ESDM, A. Wahyu Kencono, M. Dwinugroho, E. Satra Baruna, and N. Ajiwihanto, "Handbook Of Energy & Economic Statistics Of Indonesia 2023," p. 73, 2023.

[5] ASEAN Center for Energy, "The 5th ASEAN Energy Outlook," *ASEAN Cent. Energy*, pp. 1–137, 2017.

[6] E. Wahyuni, "Challenges Facing Indonesia's Energy Commitment," *Indones. Post-Pandemic Outlook Strateg. Towar. Net-Zero Emiss. by 2060 from Renewables Carbon-Neutral Energy Perspect.*, no. 2022, pp. 23–40, 2022, doi: 10.55981/brin.562.c3.

[7] Institute for Essential Services Reform (IESR), "Indonesia Energy Transition Outlook 2024 IESR Institute for Essential Services Reform," *Indones. Energy Transit. Outlook 2024*, vol. 4:2024, pp. 26–26, 2023, [Online]. Available: www.iesr.or.id

[8] G. Ganti, "Coal Phase Out," *Clim. Anal.*, 2020, [Online]. Available: https://climateanalytics.org/briefings/coal-phase-out/

[9] IEA, "Net Zero by 2050. A Roadmap for the Global Energy Sector. https://www.iea.org/reports/net-zero-by-2050," 2021.

[10] E. Erdiwansyah, M. Mahidin, and H. Husin, "Investigation of availability , demand , targets , and development of renewable energy in 2017 - 2050 : a case study in Indonesia," *Int. J. Coal Sci. Technol.*, vol. 8, no. 4, pp. 483–499, 2021, doi: 10.1007/s40789-020-00391-4.

[11] N. Reyseliani and W. W. Purwanto, "Pathway towards 100% renewable energy in Indonesia power system by 2050," *Renew. Energy*, vol. 176, pp. 305–321, 2021, doi: 10.1016/j.renene.2021.05.118.

[12] D. L. Pristiandaru and N. A. Pambudi, "Wind Energy in Indonesia: Current Status, Potential, Challenge, Opportunities, and Future Policy," *Indones. J. Energy*, vol. 2, no. 2, pp. 65–73, 2019.

[13] PLN, "Laporan Keberlanjtan PLN 2021 'Optimizing Strategy in Realizing Sustainable Energy Transition,'" p. 20, 2021.

[14] I. Journal *et al.*, "Implementation of the ASEAN Power Grid Interconnection West Kalimantan and Sarawak to Fulfill Electrical Energy Needs in Khatulistiwa 1,2,3,4)," vol. 3, no. 3, pp. 1587–1596, 2023.

[15] D. Chandra *et al.*, "Rancang Bangun Turbin Angin Savonius-Darrieus Tipe Cebong," *Pros. Semin. Nas. NCIET*, vol. 1, no. 1, pp. 518–527, 2020, doi: 10.32497/nciet.v1i1.169.

[16] F. Wenehenubun, A. Saputra, and H. Sutanto, "An experimental study on the performance of Savonius wind turbines related with the number of blades," *Energy Procedia*, vol. 68, pp. 297–304, 2015, doi: 10.1016/j.egypro.2015.03.259.

[17] S. Martosaputro and N. Murti, "Blowing the wind energy in Indonesia," *Energy Procedia*, vol. 47, pp. 273–282, 2014, doi: 10.1016/j.egypro.2014.01.225.

[18] A. Musyafa, I. Made Yulistya Negara, and I. Robandi, "A Wind Turbine for Low Rated Wind Speed Region in East Java," *Int. J. Acad. Res.*, vol. 3, no. March 2015, 2011, [Online]. Available: www.ijar.lit.az

[19] Erich Hau, *Wind Turbines Fundamentals, Technologies,Aplication, Economics 2nd edition*, vol. 2, no. 1. 2005.

[20] M. Hadi Ali, "Experimental Comparison Study for Savonius Wind Turbine of Two & Three Blades At Low Wind Speed," *Int. J. Mod. Eng. Res. www.ijmer.com*, vol. 3, no. 5, pp. 2978–2986, 2013.

[21] D. Rudianto, "Rancang Bangun Turbin Angin Savonius 200 Watt," *Conf. Senat. STT Adisutjipto Yogyakarta*, vol. 2, p. 71, 2016, doi: 10.28989/senatik.v2i0.35.

IOP Conf. Series: Earth and Environmental Science **1423** (2024) 012031 doi:10.1088/1755-1315/1423/1/012031

Analysis of camber ratio variation on B-Series propeller performance by using combined panel-vortex method and blade-element theory

M Indiaryanto[1], I K Suastika[2*], and T A Setyanto[3]

[1]Department of Ocean Engineering, Institut Teknologi Sepuluh Nopember (ITS), Surabaya, Indonesia
[2]Department of Naval Architecture, Institut Teknologi Sepuluh Nopember (ITS), Surabaya, Indonesia
[3] Research Center for Hydrodynamics Technology, National Research and Innovation Agency (BRIN), Surabaya, Indonesia

*E-mail: suastika@its.ac.id

Abstract. The camber ratio of a propeller significantly affects its hydrodynamic performance. The camber ratio particularly affects the lift force generated by the propeller blades, which ultimately affects the propeller thrust and torque. In this study, effects of camber ratio variations are investigated by using a combined method of panel vortex method and blade element theory. The camber ratios considered in this study are as follows: 0% (0.00, original foil), 1.6% (0.016), 2.2% (0.022), and 2.8% (0.028). A larger camber ratio results in a more convex propeller blade surface. Calculation results show that the larger the camber ratio, the higher the thrust and torque coefficients. The camber ratio of 2.2% (0.022) gives the highest propeller efficiency of 58.5% at the advance coefficient J = 0.7.

1. Introduction

An important component of the main conventional ship propulsion system is the propeller. To obtain an optimum ship propulsion performance, propeller selection is usually based on the B-series polynomial method and engine-propeller matching approach [1,2]. Further, propeller efficiency plays an important role in determining the ship propulsion performance, which ultimately affects the ship's fuel consumption. To improve the propeller hydrodynamic performance, the basic principles of propeller's mechanisms must be considered.

The propeller momentum theory (actuator-disc theory) describes the propeller as an ideal disc that induces fluid flow without considering the complexity of blade details [3]. The blade element theory models the propeller blade to consist of many elements, analyses each element, and integrates the elemental contributions to obtain the propeller's hydrodynamic performance parameters [4]. The lifting line theory describes the propeller blade as a lifting line, which generates vortices along the line [5]. The lifting surface theory considers the three-dimensional effects of the fluid flow around the blade, so that it allows for a more detailed analysis of the distribution of lift, drag, and induced flow across the blade surface [6]. The panel vortex method is a numerical method to simulate flow around the propeller blade, which will be elaborated below.

In this paper, the panel vortex method is combined with the blade element theory to analyze the propeller performance. Utilizing the present method, the effects of camber ratio on the propeller hydrodynamic performance are investigated.

2. Methodology

A method, which combines the panel vortex method with the blade element theory, is utilized in this study to investigate effects of the camber ratio on the thrust coefficient (K_T), torque coefficient (K_Q) and efficiency (η) of the propeller. Further, a B-series propeller is considered without considering cavitation effects.

2.1 Panel Vortex Method

The panel vortex method is a numerical method to simulate the fluid flow around the foil. In this method, the foil surface is divided into a series of elements, called "panels". The foil boundary is treated as flow streamlines, and the flow around the foil is modeled as a potential flow [7]. The foil, as shown in figure 1, is represented by many straight panels laying on the foil boundary, where a series of points are defined around the foil [8]. The method uses panels, called vortices, to accurately model the fluid flow around the foil, particularly to model the complex flow around the trailing edge. These vortices are placed on the foil surface and certain conditions must be met for the flow around the foil [9].

The end point of each panel is named 1, 2, 3... in the clockwise direction, starting from the trailing edge of the foil. To approximate the condition that the foil boundary is a streamline, zero normal velocity component is applied at the control point, which is the center point of the panel [8].

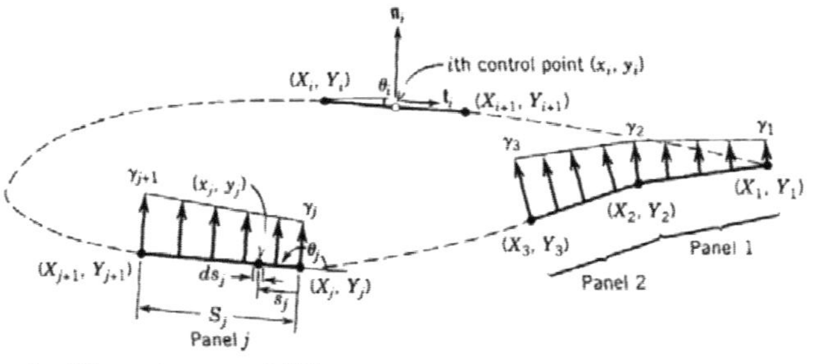

Figure 1. Foil replacement with vortex panel [8]

The i-th control point (x_i, y_i) velocity potential is given as,

$$\phi(x_i, y_i) = V_\infty(x_i\cos\alpha + y_i\sin\alpha) - \sum_{j=1}^{m} \int_{j}^{j+1} \frac{\gamma(S_j)}{2\pi} \tan^{-1}\frac{(y_i - y_j)}{(x_i - x_j)} ds_j \quad (1)$$

where

$$\gamma(S_j) = \gamma_j + (\gamma_{j+1} - \gamma_j)\frac{s_j}{S_j} \quad (2)$$

$\gamma(S_j)$ is the vortex strength distribution that varies linearly along the length of the panel, V_∞ is the velocity of uniform flow, α is the angle of attack, (x_i, y_i) is the arbitrary point on the j-th panel, m is the number of panel, S_j is the length of j-th panel, and s_j is the distance measured from the initial point to the panel control point.

The normal velocity component at the control point of each panel must be zero. Specifically, for the i-th panel, the normal velocity at the i-th control point must be zero, which indicates that the fluid flow is perfectly parallel to the surface at that point. So that,

$$\frac{\partial}{\partial_n}\phi(x_i, y_i) = 0 \quad i = 1,2, \ldots, m \tag{3}$$

After the differentiation and integration process has been carried out, the total potential velocity can be obtained as follows:

$$\sum_{i=1}^{m}\left(C_{n1_{ij}}\gamma'_j + C_{n2_{ij}}\gamma'_{j+1}\right) = \sin(\theta_i - \alpha) \quad i = 1,2, \ldots, m \tag{4}$$

in which, $\gamma' = \gamma/2\pi V_\infty$ is a dimensionless circulation density, θ_i is the angle between the i-th panel and the x-axis. The coefficients can be written as follows:

$$C_{n1_{ij}} = 0.5\, DF + CG - C_{n2_{ij}} \tag{5}$$

$$C_{n2_{ij}} = D + 0.5\frac{QF}{S_j} - (AC + DE)G/S_j \tag{6}$$

where

$$A = -(x_i - X_j)\cos\theta_i - (y_i - Y_j)\sin\theta_j \tag{7}$$

$$B = (x_i - X_j)^2 - (y_i - Y_j)^2 \tag{8}$$

$$C = \sin(\theta_i - \theta_j) \tag{9}$$

$$D = \cos(\theta_i - \theta_j) \tag{10}$$

$$E = (x_i - X_j)\sin\theta_i - (y_i - Y_j)\cos\theta_j \tag{11}$$

$$F = \ln\left(1 + \frac{S_j^2 + 2AS_j}{B}\right)0 \tag{12}$$

$$G = \tan^{-1}\left(1 + \frac{ES_j}{B + AS_j}\right)0 \tag{13}$$

$$P = (x_i - X_j)\sin(\theta_i - 2\theta_j) + (y_i - Y_j)\cos(\theta_i - 2\theta_j) \tag{14}$$

$$Q = (x_i - X_j)\cos(\theta_i - 2\theta_j) + (y_i - Y_j)\sin(\theta_i - 2\theta_j) \tag{15}$$

The Kutta conditions are applied to ensure smooth flow at the trailing edge of the foil:

$$\gamma_i + \gamma_{m+1} = 0 \tag{16}$$

To solve for the unknown (m + 1), the simultaneous system with equations in a simpler form can be written as follows,

$$\sum_{j=1}^{m+1} A_{n_{ij}}\gamma_j = (RHS)_i \quad i = 1,2, \ldots, m \tag{17}$$

The (m+1) simultaneous algebraic equations can be written in a matrix form as follows:

$$\begin{bmatrix} A_{n_{1,1}} & A_{n_{1,2}} & \cdot & \cdot & A_{n_{1,m}} & A_{n_{1,m+1}} \\ A_{n_{2,1}} & A_{n_{2,2}} & \cdot & \cdot & A_{n_{2,m}} & A_{n_{2,m+2}} \\ \cdot & \cdot & \cdot & \cdot & \cdot & \cdot \\ A_{n_{m-1,1}} & A_{n_{m-1,2}} & \cdot & \cdot & A_{n_{m-1,m}} & A_{n_{m-1,m+m+1}} \\ A_{n_{m,1}} & A_{n_{m,2}} & \cdot & \cdot & A_{n_{m,m}} & A_{n_{m,m+1}} \\ 1 & 0 & 0 & 0 & 0 & 1 \end{bmatrix} \begin{bmatrix} \gamma_1 \\ \gamma_2 \\ \vdots \\ \gamma_{m-1} \\ \gamma_m \\ \gamma_{m+1} \end{bmatrix} = \begin{bmatrix} RHS_1 \\ RHS_2 \\ \vdots \\ RHS_{m-1} \\ RHS_m \\ 0 \end{bmatrix} \tag{18}$$

where

$$RHS_i = \sin(\theta_i - \alpha) \qquad i = 1,2,\dots,m+1$$

for i < m+1;
$if\ j = 1 \qquad A_{n_{i1}} = C_{n1_{i1}}$
$if\ j = 2,3,\dots\dots m \qquad A_{n_{ij}} = C_{n1_{ij}} + C_{n2_{ij-1}}$
$if\ j = m+1 \qquad A_{ni_{m+1}} = C_{n2_{im}}$
$RHS_i = \sin(\theta_i - \alpha)$

for i = m+1;
$if\ j = 1\ and\ j = m+1\ A_{ni_1} = A_{ni_{m+1}} = 1$
$if\ j = 2,3,\dots\dots\dots\dots m \qquad A_{n_{ij}} = 0$ (19)
$RHS_i = 0$

With the calculation of the known circulation density, the speed at each control point can be calculated. Since there is only tangential velocity at these points, the dimensionless velocity, defined as $(\frac{\partial\Phi}{\partial t_i})/V_\infty$, can be written as

$$V_i = \cos(\theta_i - \alpha) + \sum_{j=1}^{m+1} A_{n_{ij}}\gamma_j \qquad i = 1,2\dots\dots m \tag{20}$$

Pressure coefficient at the i-th control point, defined by the equation,

$$C_p = 1 - \frac{V_i^2}{V_\infty^2} \tag{21}$$

The lift and drag coefficients can be obtained as follows:

$$\begin{aligned} C_D &= C_{F_x}\cos\alpha + C_{F_y}\sin\alpha \\ C_L &= C_{F_y}\cos\alpha - C_{F_x}\sin\alpha \end{aligned} \tag{22}$$

where

$$\begin{aligned} C_{F_x} &= C_p d_y \\ C_{F_y} &= -C_p d_x \\ d_y &= y\left(\frac{2}{C}\right) - y\left(\frac{1}{C-1}\right) \\ d_x &= x\left(\frac{2}{C}\right) - x\left(\frac{1}{C-1}\right) \end{aligned} \tag{23}$$

x is the length of the foil face, y is the length of the foil back, and C is the chord length of the foil.

2.2. Combined Panel-Vortex Method and Blade-Element Theory

The blade element theory was developed to calculate the thrust and torque generated by a propeller. The propeller's thrust and torque depend on the propeller's shape. In the development of the blade element theory the propeller blades are divided into several elements, each gives an elemental contribution to the total thrust and torque. The total thrust and torque can be found by integrating (summing) all the elemental contributions. In the following, the elemental contributions will be examined and also the integration of these to obtain the total thrust and torque [4].

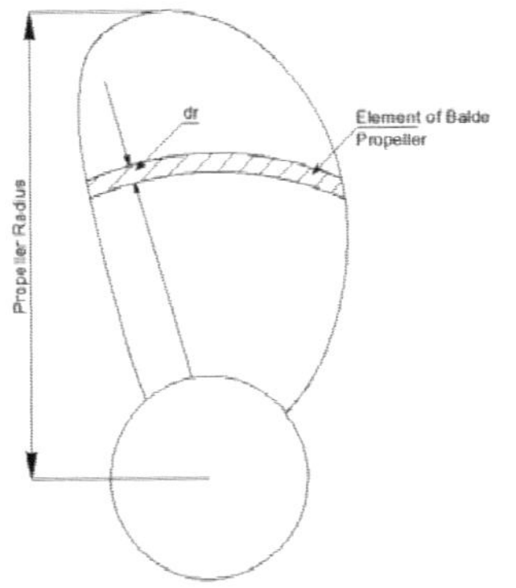

Figure 2. Illustration of a foil element on a propeller blade

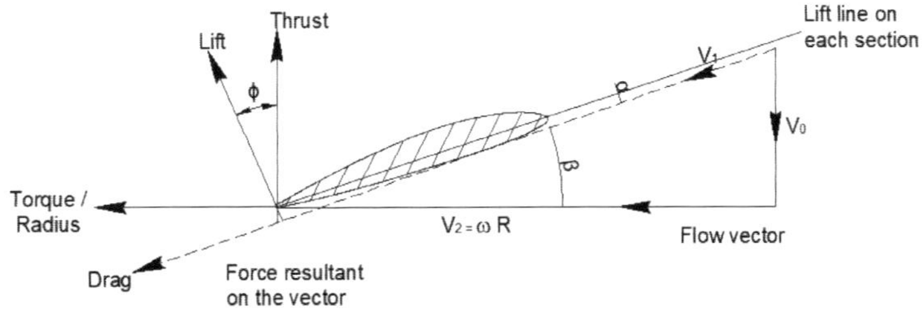

Figure 3. Illustration of force directions and velocity vector on a propeller foil

To calculate the thrust and torque of an element (see figure 2), the lift and drag coefficients of the foil are required. In this study, these values of lift and drag coefficients are calculated by using the panel vortex method. Integration of the elemental thrust and torque results in the total thrust and torque generated by the entire blade. If n is the number of blade elements and Z is the total number of propeller blades, the total thrust can be obtained by multiplying the total thrust of a blade with the number of propeller blades [10].

The flow around the propeller blade can be analyzed by considering the flow around the blade section. For the section as shown in figure 3, with section width dr and radius r, the flow around the blade section (foil) consists of V_0 (axial velocity parallel to the propeller shaft), V_2 (tangential velocity), V_1 (the resultant vector of V_0 and V_2). Further, θ is the pitch angle of the foil, and ϕ is the angular difference between the thrust direction and the propeller lift force, where $\phi = \theta - \alpha$ [11].

The elemental thrust and torque of the propeller blade section can be written as follows:

$$\begin{aligned} dT &= dL\cos(\phi) - dD\sin(\phi) \\ \frac{dQ}{r} &= dD\cos(\phi) + dL\sin(\phi) \end{aligned} \tag{24}$$

where

$$\begin{aligned} dL &= C_L \frac{1}{2}\rho V_1^2 c.\,dr \\ dD &= C_D \frac{1}{2}\rho V_1^2 c.\,dr \end{aligned} \tag{25}$$

In equation (26) ρ is the fluid density, c is the chord length of the element, and $c.dr$ is the lift-generating area of the blade element. Equations (25) and (26) can be rewritten as follows:

$$dT = \frac{1}{2}\rho V_1^2 c(C_L\cos(\phi) - C_D\sin(\phi)).Z.dr \tag{26}$$

$$\begin{aligned} \frac{dQ}{r} &= \frac{1}{2}\rho V_1^2 c(C_L\sin(\phi) + C_D\cos(\phi)).Z.dr \\ dQ &= \frac{1}{2}\rho V_1^2 c(C_L\sin(\phi) + C_D\cos(\phi)).Z.r.dr \end{aligned} \tag{27}$$

In the present method the thrust and torque are calculated by using a combination of panel vortex method and blade element theory. In the panel vortex method, the calculations focus on the circulation strength, which is converted into pressure coefficient, and lift and drag coefficients. In the blade element theory, the thrust and torque are calculated based on the lift and drag coefficients derived from the panel vortex method. As the flow is assumed to be inviscid, turbulence and viscosity effects are ignored in these calculations.

2.3 Verification of Calculation Results

To verify the results obtained from the present method (a combined method of panel vortex and blade element theory), results of calculations of a B-series propeller with a camber ratio of 0% (original foil) are compared with those results obtained from the polynomial method of B-series propeller. This selection is based on the extensive literature available for this type of propeller.

The propeller data are expressed in non-dimensional variables, namely, K_T (thrust coefficient), K_Q (torque coefficient), and η (efficiency) as a function of J (advance coefficient). The values of K_T, K_Q and η obtained from the present method are compared with those values obtained from the B-series propeller polynomial method. The expressions for J, K_T, K_Q and η are given as follows:

$$J = \frac{V_a}{nD} \qquad K_T = \frac{T}{\rho\, n^2 D^4} \qquad K_Q = \frac{Q}{\rho n^2 D^5} \qquad \eta = \frac{JK_T}{2\pi K_Q} = \frac{V_a\, T}{2\,\pi\, Q\, n} \tag{28}$$

where V_a is the advance velocity (m/s), n is the propeller rotation (rps), D is the propeller diameter (m), T is the propeller thrust (N), Q is the propeller torque (Nm), and ρ is the fluid density (kg/m^3).

Table 1 summarizes the main parameters of the B-series propeller considered in this study. In the calculations, the full-scale propeller dimensions are utilized. The J values are calculated from the propeller parameters listed in Table 1 by using Equation (35). Open water condition is assumed so that the wake fraction value (w) is zero, resulting in a J value of 0.7. This J value of 0.7 is referred to as the designed J.

Table 1. Main dimension of propeller B-series

Type	Unit	Propeller data
Dimension (D)	mm	1500
Blade number (Z)	-	4
Ratio of expanded area (A_e/A_o)	-	0.4
Pitch of ratio (P/D)	-	1.144
Rotation Propeller	Rpm	260
Max Thickness	mm	54.83

Referring to the ITTC, the Reynolds number must be first determined before carrying out the B-series polynomial calculations [14]. The Reynolds number is calculated at the radius r/R = 0.7 as follows:

$$Re_{0.7} = \frac{C_{0.7}\sqrt{V_a^2 + (0.7\pi nD)}}{v} \tag{29}$$

If Re < 2 x 10^6, the values of K_T and K_Q are given as follows:

$$K_T = \sum_{n=1}^{39} C_n^T\ (J)^{s_n}.(P/D)^{t_n}.(A_E/A_O)^{u_n}.(Z)^{v_n} \tag{30}$$

$$K_Q = \sum_{n=1}^{47} C_n^Q\ (J)^{s_n}.(P/D)^{t_n}.(A_E/A_O)^{u_n}.(Z)^{v_n} \tag{31}$$

In Equations (29-31) C is the regression coefficient of K_T and K_Q, V_a is the advance velocity, n is the rotation of propeller, D is the diameter of propeller, J is the advance coefficient, P/D is the ratio of pitch to the diameter, A_E/A_O is the ratio of blade area, and Z is the number of propeller blades. The value of the regression coefficients C, s, t, u, and v are summarized in Tables 2 and 3 [15].

Table 2. Regression coefficients and exponents for the calculation of K_T

n	C	s	t	u	v	n	C	s	t	u	v
		J	P/D	AE/AO	Z			J	P/D	AE/AO	Z
1	0.000880496	0	0	0	0	21	0.010465	1	6	2	0
2	-0.204554	1	0	0	0	22	-0.00648272	2	6	2	0
3	0.166351	0	1	0	0	23	-0.00841728	0	3	0	1
4	0.158114	0	2	0	0	24	0.0168424	1	3	0	1
5	-0.147581	2	0	1	0	25	-0.00102296	3	3	0	1
6	-0.481497	1	1	1	0	26	-0.0317791	0	3	1	1
7	0.415437	0	2	1	0	27	0.018604	1	0	2	1
8	0.0144043	0	0	0	1	28	-0.00410798	0	2	2	1
9	-0.0530054	2	0	0	1	29	-0.000606848	0	0	0	2
10	0.0143841	0	1	0	1	30	-0.0049819	1	0	0	2
11	0.0606826	1	1	0	1	31	0.0025983	2	0	0	2
12	-0.0125894	0	0	1	1	32	-0.000560528	3	0	0	2
13	0.0109689	1	0	1	1	33	-0.00163652	1	2	0	2
14	-0.133698	0	3	0	0	34	-0.000328787	1	6	0	2
15	0.00638407	0	6	0	0	35	0.000116502	2	6	0	2
16	-0.00132718	2	6	0	0	36	0.000690904	0	0	1	2
17	0.168496	3	0	1	0	37	0.00421746	0	3	1	2
18	-0.0507214	0	0	2	0	38	0.0000565229	3	6	1	2
19	0.0854559	2	0	2	0	39	-0.00146564	0	3	2	2
20	-0.0504475	3	0	2	0						

Table 3. Regression coefficient and exponent for the calculation of K_Q

n	C	s	t	u	v	n	C	s	t	u	v
		J	P/D	AE/AO	Z			J	P/D	AE/AO	Z
1	0.00379368	0	0	0	0	25	-0.0397722	0	3	2	0
2	0.00886523	2	0	0	0	26	-0.00350024	0	6	2	0
3	-0.032241	1	1	0	0	27	-0.0106854	3	0	0	1
4	0.00344778	0	2	0	0	28	0.00110903	3	3	0	1
5	-0.0408811	0	1	1	0	29	-0.000313912	0	6	0	1
6	-0.108009	1	1	1	0	30	0.0035985	3	0	1	1
7	-0.0885381	2	1	1	0	31	-0.00142121	0	6	1	1
8	0.188561	0	2	1	0	32	-0.00383637	1	0	2	1
9	-0.00370871	1	0	0	1	33	0.0126803	0	2	2	1
10	0.00513696	0	1	0	1	34	-0.00318278	2	3	2	1
11	0.0209449	1	1	0	1	35	0.00334268	0	6	2	1
12	0.00474319	2	1	0	1	36	-0.00183491	1	1	0	2
13	-0.00723408	2	0	1	1	37	0.000112451	3	2	0	2
14	0.00438388	1	1	1	1	38	-0.000029722	3	6	0	2
15	-0.0269403	0	2	1	1	39	0.000269551	1	0	1	2
16	0.0558082	3	0	1	0	40	0.00083265	2	0	1	2
17	0.0161886	0	3	1	0	41	0.00155334	0	2	1	2
18	0.00318086	1	3	1	0	42	0.000302683	0	6	1	2
19	0.015896	0	0	2	0	43	-0.0001843	0	0	2	2
20	0.0471729	1	0	2	0	44	-0.000425399	0	3	2	2
21	0.0196283	3	0	2	0	45	0.0000869243	3	3	2	3
22	-0.0502782	0	1	2	0	46	-0.0004659	0	6	2	2
23	-0.030055	3	1	2	0	47	0.0000554194	1	6	2	2

If the values $2 \times 10^6 < Re < 2 \times 10^9$ the K_T and K_Q values should be corrected using the following formulas [15]:

$$\begin{aligned}\Delta K_T = {} & 0.000353485 \\ & -0.00333758\,(A_E/A_O)J^2 \\ & -0.00478125(A_E/A_O)\,(P/D)J \\ & +0.000257792(\log R_n - 0.301)^2\,(A_E/A_O)J^2 \\ & +0.0000643192(\log R_n - 0.301)\,(P/D)^6J^2 \\ & -0.0000110636(\log R_n - 0.301)^2\,(P/D)^6J^2 \\ & -0.0000276305(\log R_n - 0.301)Z\,(A_E/A_O)J^2 \\ & +0.0000954(\log R_n - 0.301)Z\,(A_E/A_O)\,(P/D)J \\ & +0.0000032049(\log R_n - 0.301)Z^2\,(A_E/A_O)\,(P/D)^3J\end{aligned} \tag{32}$$

$$\begin{aligned}\Delta K_Q = {} & -0.000591412 \\ & +0.00696898(P/D) \\ & -0.0000666654\,Z\,(P/D)^6 \\ & +0.0160818(A_E/A_O)^2 \\ & -0.000938091(\log R_n - 0.301)\,(P/D) \\ & -0.00059593(\log R_n - 0.301)\,(P/D)^2 \\ & +0.0000782099(\log R_n - 0.301)^2\,(P/D)^2 \\ & +0.0000052199(\log R_n - 0.301)\,Z\,(A_E/A_O)^2 \\ & -0.00000088528(\log R_n - 0.301)^2Z\,(A_E/A_O)\,(P/D)J \\ & +0.0000230171(\log R_n - 0.301)Z\,(P/D)^6 \\ & -0.00000184341(\log R_n - 0.301)^2Z\,(P/D)^6 \\ & -0.00400252(\log R_n - 0.301)\,(A_E/A_O)^2 \\ & -0.000220915(\log R_n - 0.301)^2(A_E/A_O)^2\end{aligned} \tag{33}$$

$$\begin{Bmatrix}K_T \\ K_Q\end{Bmatrix} = \begin{Bmatrix}K_T(Re = 2x10^6) \\ K_Q(Re = 2x10^6)\end{Bmatrix} + \begin{Bmatrix}\Delta K_T \\ \Delta K_Q\end{Bmatrix} \tag{34}$$

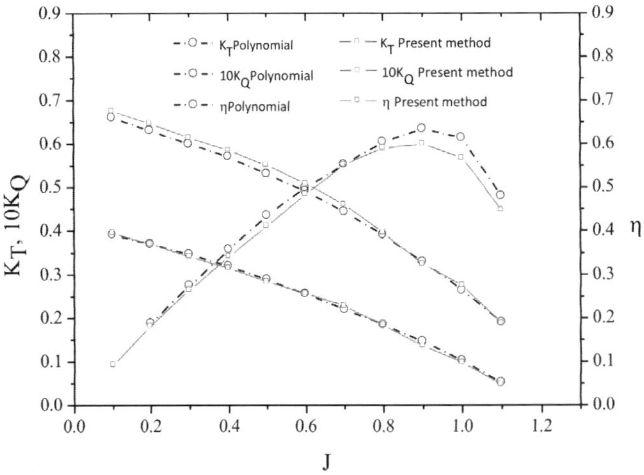

Figure 4. K_T, K_Q and η as function of J obtained from the present method compared with the results obtained from the polynomial method of B-series propeller for camber ratio of 0%

Figure 4 shows a comparison of K_T, K_Q and η diagrams obtained from the present method and the B-series polynomial method. The average percentage differences between the results obtained from the present method and the B-series polynomial method are respectively 2.63% for K_T, 2.72% for K_Q, and 3.92% for η. These percentage differences are less than 5%, showing that the present method gives sufficiently accurate results. In the following section, results of calculations utilizing the present method are presented for the B-series propeller with varying camber ratio.

3. Results and Discussion

To investigate effects of the camber ratio on the propeller hydrodynamic performance, the camber ratio of the B-series propeller was varied, and the values of J, K_T, K_Q and η are calculated using the present method. Variations in camber ratio are obtained by adjusting the centerline of the foil. A change in the camber ratio will particularly affect the lift generated by foil, which ultimately affects the foil's lift and drag coefficients [12].

Figure 5 shows a foil of B-series propeller with camber. A foil is said as having a positive camber if the top surface of the foil as shown in figure 5 is more convex than the bottom surface. Camber ratio is the ratio of the camber (f) to the chord length (c) at the point of maximum thickness. For NACA airfoils, typical camber ratios range from 0% to 6% [13].

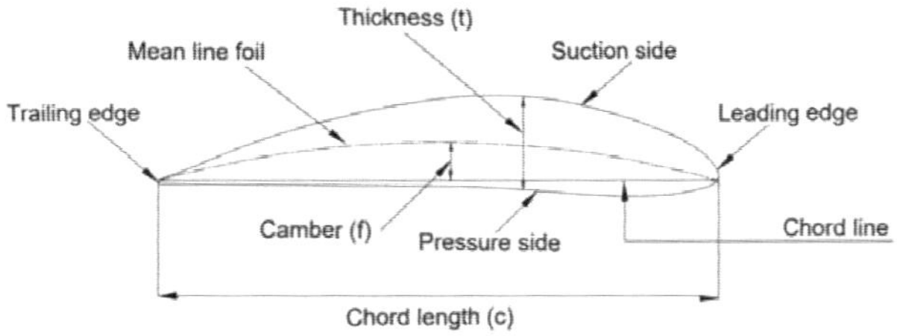

Figure 5. Foil of B-series propeller with camber

Changes in the camber ratio of the foil only affect the curvature of the propeller blade surface. The main dimensions of the propeller, such as diameter, blade element ratio, and pitch ratio, remain unchanged. The maximum thickness at each r/R of the propeller does not change either. The camber ratios considered in this study are 0% (0.00, original foil), 1.6% (0.016), 2.2% (0.022), and 2.8% (0.028). These camber ratios are summarized in Table 4. The increase in camber ratio results in a more convex propeller blade surface.

Table 4. Camber ratio foil of B-Series propeller

Camber Ratio (f/c)	0.00 (original foil)	0.016	0.022	0.028

Figure 6 a, b and c show K_T, K_Q and η diagrams obtained in this study for the four different camber ratios: 0%, 1.6%, 2.2%, and 2.8%. Figure 6 shows that K_T and K_Q decrease monotonically with increasing J, while η first increases, reaches a maximum value, and then decreases with increasing J. Figure 6 a and b show that the larger the camber ratio, the higher the thrust (K_T) and torque (K_Q) coefficients.

The value of η at J = 0.7 (designed J) is most important, for which the camber ratio of 2.2% gives the highest value of 0.585. For completeness, the values of K_T, K_Q and η at J = 0.7 are summarized in Table 5. Table 5 shows that the camber ratio of 2.2% (0.022) gives the highest efficiency of 0.585. The efficiency for the camber ratio of 1.6% (0.016) is 0.580 while the efficiency for the camber ratio of 2.8% (0.028) is 0.580.

IOP Conf. Series: Earth and Environmental Science **1423** (2024) 012031 doi:10.1088/1755-1315/1423/1/012031

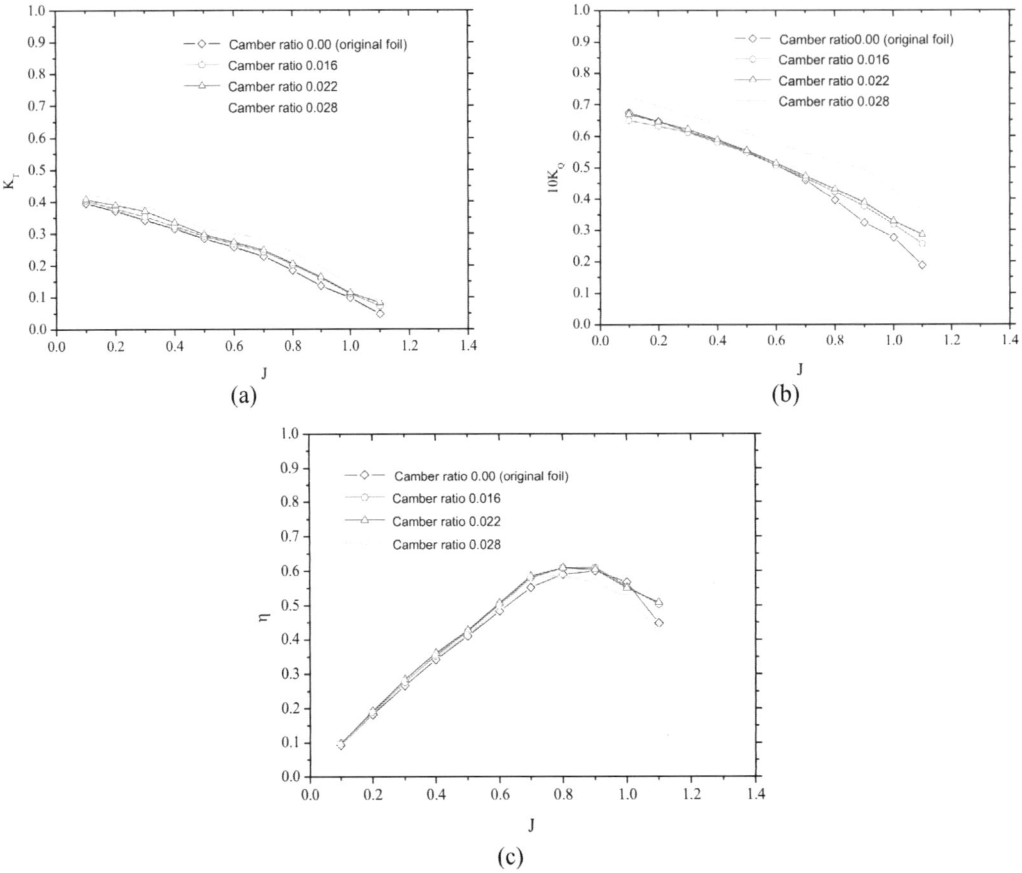

Figure 6. Comparison resulted of propeller thrust coefficient, K_T (a), torque coefficient, K_Q (b), and efficiency, η (c) from present method with varying camber ratio.

Table 5. Values of K_T, K_Q, and η for four different camber ratios at J = 0.7

Camber ratio (f/c)	K_T	$10K_Q$	η
0%	0.243	0.467	0.580
1.6%	0.243	0.467	0.580
2.2%	0.248	0.473	0.585
2.8%	0.284	0.546	0.580

4. Conclusions

This study investigates effects of the camber ratio on the B-series propeller hydrodynamic performance. A novel method was applied, which is a combination of the panel vortex method with the blade element theory. Cavitation phenomenon is not considered. Four camber ratios are considered: 0%, 1.6%, 2.2% and 2.8%. The increase in camber ratio results in a more convex propeller blade surface. The larger the camber ratio, the higher the thrust (K_T) and torque (K_Q) coefficients. Calculation results show that the camber ratio of 2.2% gives the highest propeller efficiency of 5.85%.

Acknowledgment

The authors acknowledge the financial support of the Indonesia Ministry of Finance (LPDP) for this study under the contract No. KEP-227/LPDP/LPDP.3/2022.

References

[1] Suratno, I M Ariana, and B Cahyono 2020 Optimization of Propeller Design Through Polynomial Approach to Optimize The Ship Energy Efficiency International Conference on Maritime Safety International Conference (MASTIC)
[2] I E Sandjaja, I M Ariana, Erwandi and M Indiaryanto 2022 Propeller Selection Optimization on Mini LNG Ship Based on Engine Propeller Matching Approach The 3rd International Conference Maritime Safety International Conference (MASTIC)
[3] R. Bontempo and M. Manna 2016 Analysis and Evaluation of the Momentum Theory Errors as Applied to Propellers AIAA Journal Vol. 54, No. 12 pp 3840-3848
[4] E. Benini 2004 Significance of Blade Element Theory in Performance Prediction of Marine Propellers Ocean Engineering Vol 31 pp 957–974
[5] A. Rahmana, R. Ullahb, and M. Karimb 2017 Marine Propeller Design Method based on Lifting Line Theory and Lifting Surface Correction Factors 10th International Conference on Marine Technology, MARTEC Procedia Engineering pp 174 - 181
[6] K.Koyama 2013 Relation Between The Lifting Surface Theory and The Lifting Line Theory in The Design of An Optimum Screw Propeller Journal of Marine Science and Technology Vol 18 pp145–165
[7] H. Liu 2018 Linear Strength Vortex Panel Method for NACA 4412 Airfoil International Conference Materials Science and Engineering
[8] A.M. Kuethe and C.Y. Chow 1998 Foundations of Aerodynamics Bases of Aerodynamic Design Fifth Edition, John Wiley & Sons Inc., Canada
[9] K.M. Sultan, A. M. Elshabli, M.A. Kashbour, S.A. Ben.Ateiga 2019 Performance Assessment of The Vortex Panel Method Libyan Journal of Science & Technology Vol 9 pp 122–128
[10] F. Mahmuddin 2017 Rotor Blade Performance Analysis with Blade Element Momentum Theory The 8th International Conference on Applied Energy Energy Procedia 105 pp 1123 - 1129
[11] R. MacNeill 2017 Blade Element Momentum Theory Extended to Model Low Reynolds Number Propeller Performance The Aeronautical Journal pp 1- 23
[12] I.H.Abbott and A.E.Von Doenhoff 1958 Theory of Wing Section Dover Publication Inc 180 Varick Street, New York
[13] N. M. Nouri, S. Mohammadi 2018 Numerical Investigation of The Effects of Camber Ratio on The Hydrodynamic Performance of a Marine Propeller Journal Ocean Engineering 148 pp 632–636
[14] ITTC Recommended Procedures and Guidelines. 2014. Open Water Test, Revision 3, page 3 of 11
[15] J. Carlton 2007 Marine Propellers and Propulsion 2nd ed. Oxford, UK (Butterworth- Heinemann

Development Study of Direct Damage Stability Algorithm Based on 3D Geometry Data Processing for Onboard Loading Instrument

Fajar Andinuari[1], Sjarief Widjaja[1] and Setyo Nugroho[2]

[1] Department of Naval Architecture, Faculty of Marine Technology, Institut Teknologi Sepuluh Nopember, Indonesia
[2] Department of Marine Transportation Engineering, Faculty of Marine Technology, Institut Teknologi Sepuluh Nopember, Indonesia

E-mail: fjar.nuar@gmail.com, sjarief@na.its.ac.id, snugroho@its.ac.id

Abstract. The maritime world experienced a very significant development such as the process of cargo planning is the existence of a system called loading instrument which is regulated by several regulations from both Classification and IACS. Damage stability calculation is also very important for loading instruments. Currently in Indonesia, there is no known algorithm to calculate direct damage stability from 3D models. And also there is still no loading instrument in Indonesia that applies the algorithm. Therefore, this research aims to develop a direct damage stability calculation algorithm for the development of loading instruments in Indonesia. Direct damage stability uses the lost buoyancy method which is calculated from 3D geometry model directly. The calculation process starts by dividing the ship geometry and compartment with the desired station. In general, the numerical integration used in the calculation process is the Simpson method. Then Newton's iteration is used to find an equilibrium condition by predetermined conditions. Finally, the calculation results will be compared with the cases of LR Rules and IACS Rules UR L5 as a form of validation. The calculation algorithm can later be applied to the onboard loading software computer needed by the ship for the loading process with damage applications.

1. Introduction

The maritime world experienced a very significant development when entering the 21st century. Various innovations are present in order to increase productivity in the maritime world, ranging from navigation, outfitting, machinery, safety equipment, etc. One of the things that is very concerned in this development process is safety. There are so many new regulations governing the safety of ships that have led to the emergence of many new tools/systems in the world of shipping and shipping. However, despite these improvements, ship accidents continue to occur. A total of about 7391 incidents of serious ship accidents occurred between 2000-2012 [1]. Most of the accidents were caused by grounding, collision, etc. Then the hull is damaged resulting in leakage on the ship. Automatically, the ship will experience periodic changes in draft, trim and heel and affect the stability of the ship. The worst thing is that the ship can capsize / capsize. Leaks in the ship's compartments will increase the influence on the weight, floatation point, and stability of the ship. It can also affect the strength of the ship, both lengthwise and transversely [2].

The calculation of ship stability damage has been done for many years. For example, research conducted by [3] on the appropriate methodology for damage safety assessment by considering environmental conditions. Then [4] described alternative methods, including conventional methods, time-domain simulation of leakage by assuming the leaking compartment loses buoyancy, and simulation of the model as an open compartment to get a realistic assessment when leaking at intermediate stages. [5]used numerical software to calculate leakage of intermediate stages during the design process. It was done for numerical investigation of marine accidents.

Today, digital twin technology continues to evolve. New rules have accommodated this technology, one of which is the calculation of direct damage stability from a 3D ship model. The emergence of loading instruments or loading computer software is a form of development of digital twin technology, where virtual data replaces the physical form of the ship itself. According to the International Association of Classification Societies (IACS) UR L5, there are several types of loading software, ranging from type 1, type 2, type 3, and type 4. Especially for types 3 and 4, direct damage stability calculation is required as *Table 1*.

Table 1. Types of loading instrument

Type	Definition
Type 1	Software calculating intact stability only (for vessels not required to meet a damage stability criterion)
Type 2	Software calculating intact stability and checking damage stability based on a limit curve (e.g., for vessels applicable to SOLAS Part B-1 damage stability calculations, etc.) or checking all the stability requirements (intact and damage stability) based on a limit curve.
Type 3	Software calculating intact stability and damage stability by the direct application of preprogrammed damage cases based on the relevant conventions or codes for each loading condition (for some tankers, etc.).
Type 4	Software calculating damage stability associated with an actual loading condition and actual flooding case, using direct application of user defined damage, for the purpose of providing operational information for safe return to port (SRtP).

Currently in Indonesia, there is no known algorithm to calculate damage stability directly from the 3D model. And also there is still no loading instrument in Indonesia that applies the algorithm. Therefore, this research is needed to develop a loading software computer that is able to apply the calculation of direct damage stability following the provisions of software type 3.

2. Methodology

2.1. Modeling and Importing 3D Model

The calculation process starts with 3D modeling. This is very important because it will be the database of the entire calculation. The accuracy of the calculation results will depend on this model. The 3D modeling is designed with the error limits required by IACS UR L5 regulations. For example, the validation of displacement value should not be more than 2%, for vertical center of buoyancy value maximum 1% / 5 cm, longitudinal center of buoyancy maximum 1% / 50 cm, etc. The model to be made is a half cylinder with a size of 200 x 30 x 15 m along with its tank and compartment following Lloyd Register Test Ship Data For Approval of Stability Calculation Program.

IOP Conf. Series: Earth and Environmental Science **1423** (2024) 012032 doi:10.1088/1755-1315/1423/1/012032

Then importing the model is done using the STL (stereolithography) file format with ASCII format. The STL format is popularly used for 3D printing or Computer-Aided Design (CAD). In theory, the result of STL modeling is a triangular surface called a facet. The shape of the half cylinder modeling can be seen in the following Figure 1.

Figure 1. STL Model of Half Cylinder 200 x 30 x 15 m and the tank view

2.2. Generating Hydrostatic and Tank Calibration Tables for All Trim

The overall calculation, both intact stability and damage stability, is taken from the output in the form of an offset table. Taking the station on the offset table is made with a tight interval so that it can produce good accuracy of the results. Then the results of the offset table will be used as a database as the basis for hydrostatic calculations and tank calibration. In this process, input values must be given. The input values that must be filled in are the mid-draft range and trim range. Then the hydrostatic table is calculated according to the draft and trim intervals that have been set. As for calculating the tank calibration table, input is required in the form of sounding range and trim range. After that, the tank calibration table can be calculated. The hydrostatic and tank calibration tables will be used as the first reference data in calculating the equilibrium conditions.

In general, the calculation methods commonly used in calculating hydrostatic tables and calibration tanks that are area-based or volumetric are the trapezoidal numerical integration method, Simpson 1 (1/3), or Simpson 2 (3/8). From the research results [6], it is proven that the numerical integration method that has the best approach is Simpson 1 (1/3). So the method used in the calculations in this paper is Simpson 1 (1/3). The formula for numerical integration of Simpson 1 (1/3) is as follows,

$$I_{S1} = \frac{h}{3}\left(f_0 + 4f_1 + 2f_2 + 4f_3 + 2f_4 + 4f_5 + \ldots + 4f_{N-3} + 2f_{N-2} + 4f_{N-1} + f_N\right) \quad (1)$$

Where the constant coefficients are 1, 4, 2, 4, 2, . . . , 4, 2, 1 are commonly referred to as Simpson factors [7].

2.3. Damage Stability Method

In many kinds of literature, it is explained that in damage stability analysis there are two methods, the lost buoyancy method and the added weight/added mass method. In the lost buoyancy method, the flooding compartment is considered not part of the ship, which means that floodwater can flow freely when the ship is moving. This method assumes that the time required to equalize the water level in the leaking tank/compartment with that of fresh water is infinite. Lost buoyancy is also known as the constant displacement method because in this analysis the displacement and weight point of the ship do not change.

Then in the other method, added weight, water that enters due to flood is considered part of the ship. In this case, it is considered as a liquid entering the tank/compartment, as opposed to the lost buoyancy method. For more details, the two methods can be seen in the following Figure 2.

Figure 2. Lost buoyancy (left) dan added weight (right) visualization
(Source:[4])

The damage stability calculation method used in this study is the lost buoyancy method following Lloyd Register Approval of Strength and Stability Calculation Program Ch. 5.4.8.12 which is also required by some statutes [8].

2.4. *Flooding Condition Optimization with Newton-Raphson Method*

To calculate the flooding condition of a tank/compartment, this paper uses the Newton-Raphson iteration approach. As is known, the most difficult part of calculating damage stability is when determining the waterplane area, weight points, and moments of inertia. The iteration process when determining these components will take a long time, so the Newton-Raphson method is used to streamline the calculation time. In general, the formula of the Newton-Raphson method is as follows,

$$h_{n+1} = h_n - \frac{f(h_n)}{f'(h_n)} \tag{2}$$

Where,

h = height of tank fluid (sounding) in intact condition [m]

$f(h_n)$ = V – V_{flood} = 0

V_{flood} = Volume after flooding

V = Calculated tank volume

The part that needs to be calculated using the Newton-Raphson method is the change in liquid height (sounding) due to flooding. Then the iteration value of the liquid height has the following equation:

$$h_{n+1} = h_n + \frac{V_{flood} - f(h_n)}{A_h} \tag{3}$$

Where,

A_h = projection of liquid area onto baseline [m^2]

The concept of the liquid entering the flooding compartment is considered to be replaced by seawater as a whole. So the density becomes the density of seawater. And if the liquid in the tank is higher than the surface of the sea water, the liquid in the tank flows out and down to the surface of the sea water.

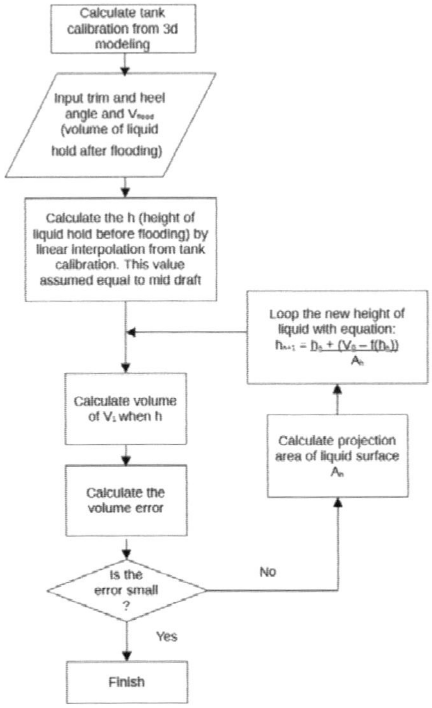

Figure 3. Algorithm to calculate flooding tank/compartment

The equilibrium condition is iterated using the Newton-Raphson method until a condition is reached,

F(Tmid, tan ϴ) = Δ - ρV = 0, and
F(Tmid, tan ϴ) = (LCB-LCG) – (TCB-TCG)tanϴ sinα cosα + (KB-KG) tanϴ $\cos^2\alpha$ = 0

Where,
ϴ = trim angle
α = heel angle

3. Result and Discussion

3.1. Volumetric Validation

From the results of the calculated offset table, we will then validate the volumetric values against the existing ship design software. First, what needs to be compared is the volume of the half cylinder hull at each waterline. Secondly, comparing the volume of the tank (FPT) at each sounding. The comparison table can be seen in the following table

Table 2. Comparison Table of Volume

Draft (m)	Volume (m^3)		Error (%)
	3D Ship Design Software	Proposed Method	
0	0	0	0.00%
1	1445.811	1444.93	0.06%
2	4047.422	4043.63	0.09%

IOP Conf. Series: Earth and Environmental Science **1423** (2024) 012032 doi:10.1088/1755-1315/1423/1/012032

3	7357.389	7349.71	0.10%
4	11205.42	11193.4	0.11%
5	15487.18	15470.6	0.11%
6	20128.02	20108.7	0.10%
7	25069.68	25050.1	0.08%
8	30264.09	30244.5	0.06%
9	35669.94	35650.2	0.06%
10	41250.64	41230.8	0.05%
11	46972.96	46952.9	0.04%
12	52806.1	52786.1	0.04%
13	58720.99	58700.9	0.03%
14	64689.75	64669.6	0.03%
15	70685.27	70315.7	0.52%

Table 3. Comparison Table of Tank Calibration (FPT)

Sounding (m)	**Volume (m3)**		**Error (%)**
	3D Ship Design Software	**Proposed Method**	
15	3527.562	3504.42	0.66%
14.753	3453.483	3455.74	0.07%
14.25	3302.936	3308.96	0.18%
13.5	3079.014	3084.62	0.18%
12.75	2855.979	2861.42	0.19%
12	2634.898	2639.92	0.19%
11.25	2415.948	2420.71	0.20%
10.5	2199.970	2204.42	0.20%
9.75	1987.592	1991.66	0.20%
9	1779.258	1783.12	0.22%
8.25	1576.041	1579.48	0.22%
7.5	1378.281	1381.53	0.24%
6.75	1187.241	1190.07	0.24%
6	1003.420	1006.04	0.26%
5.25	828.216	830.475	0.27%
4.5	662.556	664.596	0.31%
3.75	507.930	509.755	0.36%
3	366.194	367.718	0.42%
2.25	239.522	240.753	0.51%
1.5	131.118	132.076	0.73%
0.75	46.457	47.0413	1.26%
0	0.000	0	0.00%

3.2. Intact Stability Validation

In addition to comparing the hull and tank volumetric values, it is also necessary to validate the intact stability values at equilibrium conditions to ensure that the method used is concretely correct. The loading condition is made on half cylinder with displacement 25000 tons, LCG 105.974 m, VCG 4 m, and TCG 3.6 m, trim -1.824 m, heel 18.1 deg, and mid draft 6.436 m. The loadcase is included in the LR Test Ship (Strength and Stability Calculation Program Approval). The comparison of these conditions can be seen in the table below with error limits in accordance with IACS UR L5 regulations.

Table 4. Hydrostatic comparison when intact condition

No	Item	3D Ship Design Software	Proposed Method	Diff	Persen
1	Volume	24388.46	24342.74	45.72	0.2%
2	Displacement	24998.00	24951.30	46.70	0.2%
3	LCB	105.97	105.92	0.05	0.0%
4	KB	4.63	4.63	0.01	0.2%
5	TCB	3.40	3.39	0.01	0.2%
6	WPA	5033.74	5030.87	2.87	0.1%
7	LCF	101.49	101.48	0.01	0.0%
8	TCF	2.53	2.52	0.01	0.2%
9	TPC	51.60	51.57	0.03	0.1%
10	IT	266246.59	265770.16	476.44	0.2%
11	IL	16754039.72	16744806.72	9233.00	0.1%
12	BML	686.97	687.88	0.91	0.1%
13	BMt	10.92	10.92	0.00	0.0%
14	LKM	657.49	658.53	1.04	0.2%
15	TKM	15.00	15.07	0.07	0.4%
16	GML	687.62	688.60	0.98	0.1%
17	GMt	11.57	11.64	0.07	0.6%
18	MTC	859.47	859.08	0.39	0.0%

3.3. Further Development for Stability Algorithm and Validation

The stability calculation process, both intact and damaged, must be assessed or validated for accuracy. The validation process refers to 3D ship design software, LR rules, or manual stability booklet. The damage calculation stage in this discussion is still under development, both from the hydrostatic and the GZ after-flooding calculation. However, the proposed process and method are the basis of reference in conducting this research. From the previous volumetric validation and intact stability validation processes, results were obtained that met the validation comparison. This can be a foundation in the input-output data that will be used in the next damage stability calculation.

The damage validation comparison also comes from LR Test Ship (Strength and Stability Calculation Program Approval) just like the validation for intact stability. Damage Case used for validation material is with flooding in FPT as shown in the following figure below.

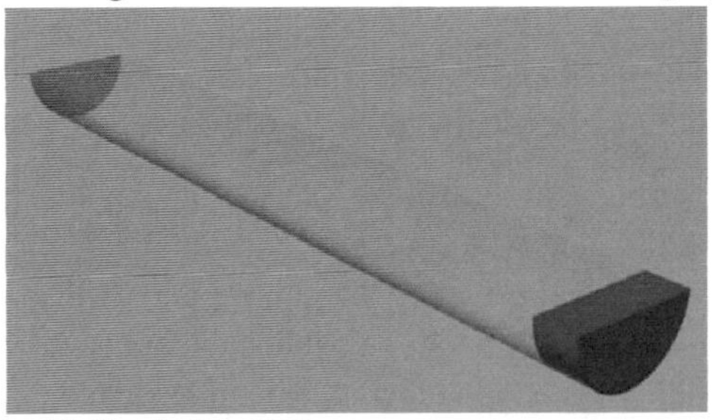

Figure 4. Half-cylinder model with flooded in FPT

In this calculation stage, the validation that can be compared is only in the hydrostatic section, where the Booklet (LR Test Ship) only shows the trim and draft components. From the results of the Damage Case that has been tested on the FPT, the trim and draft results are obtained in the following table,

Table 5. Result of trim and draft

No	Item	Unit	Booklet	Prop. Method 1	Diff.	Diff. (%)	Status
			[1]	[2]	|[1]-[2]|		
1	Trim	m	-1.48	-1.54			
2	Draft FP	m	9.40	9.44	0.04	0.4%	OK
3	Draft AP	m	7.92	7.90	0.02	0.3%	OK
4	Draft Midship	m	8.66	8.67	0.01	0.1%	OK

The stability calculation process can be seen in the following flowchart Figure 5,

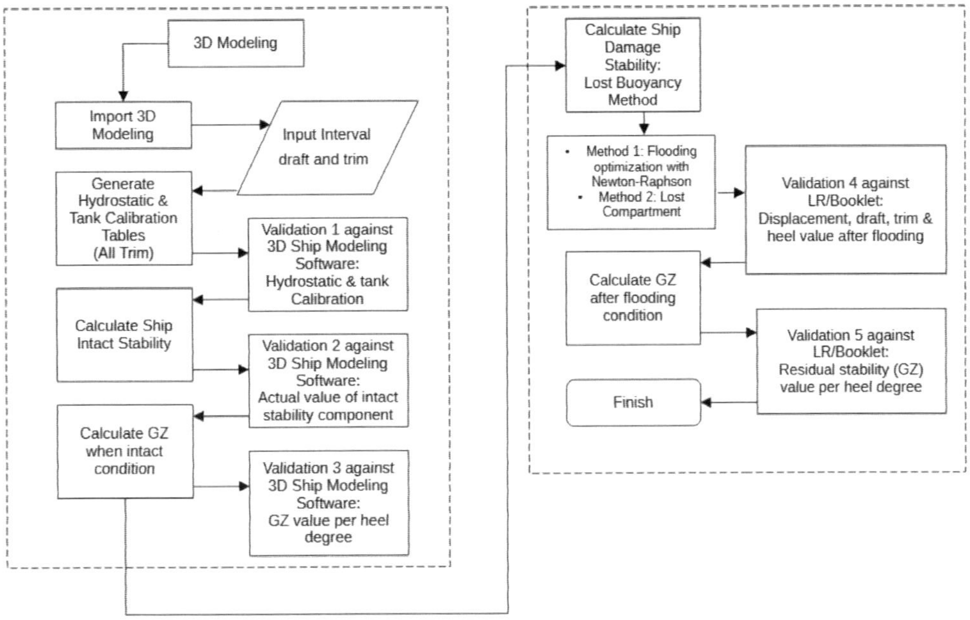

Figure 5. Stability calculation and validation flowchart

4. Conclusion

Based on the presented paper, we can conclude that the algorithm in calculating volumetric and intact stability is concrete. It is proved by validation comparison against 3D ship design software and LR approval which meets the requirements of IACS UR L5. As for the calculation of damage stability, it needs to be further developed related to the calculation of the GZ curve, although in some damage conditions some hydrostatic components are met such as trim and draft. And this developed stability calculation algorithm is expected to be used in the future to analyze stability applications.

Acknowledgement
This research is supported by PT. Pranala Digital Transmaritim. And the author would like to thank to Prof. Ir. Sjarief Widjaja, Ph. D. and Dr.-Ing. Setyo Nugroho as project advisors.

References

[1] X. Sun, Y. Ni, C. Liu, Z. Wang, and Y. Yin, "A practical method for stability assessment of a damaged ship," *Ocean Engineering*, vol. 222, Feb. 2021, doi: 10.1016/j.oceaneng.2021.108594.
[2] D. W. Pan, C. X. Lin, Z. J. Liu, and D. P. Sun, "Calculation on the uprighting process of a capsized ship," *Brodogradnja*, vol. 67, no. 3, pp. 115–132, 2016, doi: 10.21278/brod67307.
[3] J. Choi, D. Lee, H. Jin Kang, S.-Y. Kim, and S.-C. Shin, "Damage scenarios and an onboard support system for damaged ships," *International Journal of Naval Architecture and Ocean Engineering*, vol. 6, pp. 236–244, 2014, doi: 10.2478/IJNAO.
[4] P. Ruponen, T. Manderbacka, and D. Lindroth, "On the calculation of the righting lever curve for a damaged ship," *Ocean Engineering*, vol. 149, pp. 313–324, Feb. 2018, doi: 10.1016/j.oceaneng.2017.12.036.
[5] S. Krüger and T. U. Hamburg, "Numerical flooding simulations - a useful tool for marine casualty investigations," 2016.
[6] Md. N. Dhali, M. Farhad Bulbul, and U. Sadiya, "Comparison on Trapezoidal and Simpson's Rule for Unequal Data Space," *International Journal of Mathematical Sciences and Computing*, vol. 5, no. 4, pp. 33–43, Nov. 2019, doi: 10.5815/ijmsc.2019.04.04.
[7] P. A. Wilson, *Basic Naval Architecture*. Cham: Springer International Publishing, 2018. doi: 10.1007/978-3-319-72805-6.
[8] Lloyd Register, "Approval of Strength and Stability Calculation Programs," 2018.

IOP Conf. Series: Earth and Environmental Science **1423** (2024) 012033 doi:10.1088/1755-1315/1423/1/012033

The influence of time steps and turbulence model in numerical predicting total resistance on Japan Bulk Carrier with and without pre-duct

A. Munazid[1*], I M. Ariana[1] and I K. A. P. Utama[2]

[1] Department of Marine Engineering, Sepuluh Nopember Institute of Technology (ITS), Surabaya 60111, Indonesia
[2] Department of Naval Architecture, Sepuluh Nopember Institute of Technology (ITS), Surabaya 60111, Indonesia

*E-mail: alimunazid@gmail.com

Abstract. As a numerical fluid dynamic simulation method, Computational Fluid Dynamic (CFD) must guarantee the accuracy of the simulation results carried out, where the accuracy of the simulation results depends on: numerical method, grid generation, mesh independence, grid type, turbulence model, etc. This paper explains the influence of turbulence model and time step on the resistance prediction results of the Japan Bulk Carrier (JBC) ships, with CFD simulations using the ANSYS CFX software package. The model is simulated with several turbulence models including: k-Epsilon (k-ε), k-Omega (k-ω) and Shear Stress Transport k-ω (SST k-ω), apart from that the turbulence model is simulated in several time steps. From the results of the CFD modelling carried out, the influence of the turbulence model and time step on the accuracy of predicting ship resistance from the JBC ship was analysed, by comparing the CFD modelling results with experimental results from National Maritime Research Institute (NMRI). In general, the turbulence model and time step influence modelling and mesh selection, the correct time step provides accurate and convergent modelling results.

1. Introduction

Computational Fluid Dynamic (CFD) is a numerical simulation method of fluid dynamics [1], which in recent years has eperienced rapid improvement and has become one of the important techniques which plays an important role in ship design industry by successfully replacing experiments at the initial ship design stage [2]. In the case of ships hydrodinammics such us: resistance, seakeeping, propellers and wave load analysis, the application of potential flow codes, besides that CFD helps express dynamic fluid phenomena around the ship's hull, including interference and interaction of ship resistance components Ultra Large Container Ship [3], Container ship model [4], Catamaran [5], ducted propeller [6], Self-propulsion of bulk carrier [7], pre duct on traditional fishing vessel [8], pre-duct shape [9]. This is because the ability of the CFD program to analyze fluid flow and the characteristics resulting from moving objects in fluid is very good. Where the basic equations used in CFD are based on fluid dynamics equations, namely: continuity, momentum and energy equations. These equations are mathematical statments of three basic priciples of physics: conservation of mass, Newton's second law and conservation of energy.

With the high increase in CFD capabilities and supported by the rapid development of computer hardware and commercial CFD software that is widely used [10], such as ANSYS CFX, Fluent, NUMECA, etc. The current outlook and review of the CFD science community regarding the prediction results of simulations using CFD is the reliability and accurate the prediction results. This is something that is of great concern in fluid dynamic simulations. Several questions that the community often raises about CFD prediction include: How do the CFD simulation results reflect actual conditions or at least the same as the experiment? What are the accuracy and validity values of the CFD simulation results, are they close to actual conditions? What is the level of confidence in the prediction result from CFD simulations that have been widely carried out? These questions are often asked due to the level of uncertainty associated with the result of CFD prediction [11].

Sources of uncertainty in CFD simulation results include: numerical scheme [12], grid generation [5], mesh independency [13][14], grid type and turbulence model [15][16], turbulence model [17], set up: geometry, boundary condition, numeric and turbulence model [18], time step [19]. Thus, the source of accuracy CFD simulation results in predicting dynamic fluid simulations in other than the mesh and geometry models, the prediction of simulation results is influenced by the turbulence model and time step used. In CFD simulations with steady-state analysis, only the final state is obtained without any change in state over time, the state before the calculation reaches the final state does not have any physical meaning and only the final state has meaning. Whereas in transient analysis, the analysis is separated by a short time interval, namely situations where from one moment to the next moment is calculated and the procedure is repeated. The state of each calculation step are determined, and represent the state at the particular moment in time. Changes in state can be calculated until the phenomenon reaches the final state.

Thus, in this paper a study was carried out on how the turbulence model and time steps influence the accuracy of the simulation in predicting the value of ship resistance on ships with and without using pre-duct technology. In this case, it is Japan Bulk Carrier (JBC), to obtain an explanation of the influence of the turbulence model and time steps on the accuracy of predicting ship resistance with and without pre-duct. The JBC ship model was simulated with several turbulence model and time steps, then the accuracy of the CFD model was analysed by comparing it with the results of the JBC ship resistance experiment with and without pre-duct from National Maritime Research Institute (NMRI), from these results the accuracy of the turbulence model and time steps in predicting ship resistance was obtained

2. Japan Bulk Carrier (JBC)

Japan Bulk Carrier (JBC) is a large bulk cargo ship, designed and development by the National Maritime Research Institute (NMRI) in collaboration with Yokohama National University, Ship Building Research Center of Japan (SRC) and supported by Class-NK. With a full block coefficient, JBC has wake value that shows separation vortexes at stern, therefore the development of Energy Saving Devices (ESD) pre-duct to increase efficiency is possible [20]. The hull and main dimensions are provided by NMRI (2015)[21] as shown in **Table 1** and **Figure 1**(a). Appendages pre-ducts on the ship's hull, provided by NMRI (2015) in [21], consisting of ducts and struts, NACA-4420 foil profiles with a constraction angle of 20°, outlet diameter of 0.55 diameter of propeller and cord length of foil is 0.3 diameter of propeller as shown in **Figure 1**(b).

Table 1. Main dimensions of JAPAN Bulk Carrier

Particulars of Hull		**Full Scale**	**Model**	**Unit**
Length between perpendiculars	*Lpp*	280.00	7.000	m
Maximum beam of waterline	B_{wl}	45.00	1.250	m
Depth	*D*	25.00	0.694	m
Draft	*T*	16.50	0.412	m
Wetted surface area	*S*	19 556.00	12.220	m^2
Displacement volume	∇	178 369.90	2.787	m^3
Ship Speed	*V*	14.5 Knot	1,179 m/s	-
Block coefficient	C_B	0.858	0.858	-
Midship section coefficient	C_M	0.998	0.998	-
Slenderness	$L_{wl}/\nabla^{1/3}$	5.063	5.063	
Model Scale	λ		40	

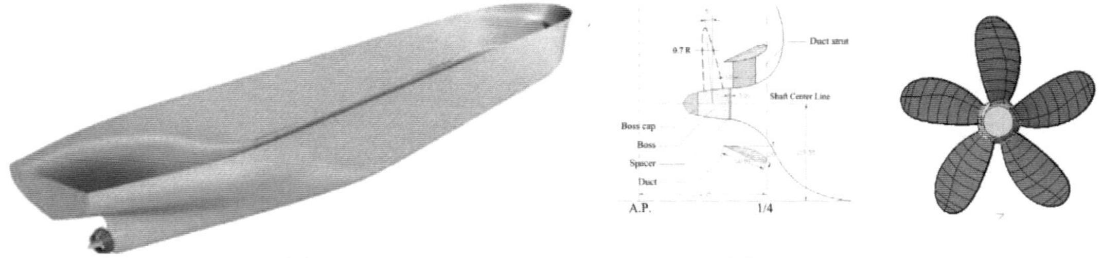

(a) (b) (c)

Figure 1. (a) The JBC reference hull, (b) Detail of pre-duct, (c) Propeller MP687

3. Computational Simulation

3.1 Mathematical Model

The Computational Fluid Dynamics (CFD) is numerical simulation model developed based on the Reynolds-Averaged Navier-Stokes (RANS). According F. Joel H. and P. Milovan (2002) in [22], the averaged continuity and momentum equation for incompressible flows may be given as in the following equations of mass and momentum. Turbulence models based on RANS equations are known as statistical turbulence models because they use statistical averaging procedures [23]. The Reynolds averaged equation as follow in Equations (1) and (2), the Reynolds averaged energy equation as follow in Equations (3).

$$\frac{\partial \rho}{\partial t} + \frac{\partial}{\partial x_j}(\rho U_j) = 0 \tag{1}$$

$$\frac{\partial \rho U_i}{\partial t} + \frac{\partial}{\partial x_j}(\rho U_i U_j) = \frac{\partial p}{\partial x_i} + \frac{\partial}{\partial x_j}\left(\tau_{ij} - \rho \overline{u_i u_j}\right) + S_M \tag{2}$$

$$\frac{\partial \rho h_{tot}}{\partial t} - \frac{\partial p}{\partial t} + \frac{\partial}{\partial x_j}(\rho U_j h_{tot}) = \frac{\partial}{\partial x_j}\left(\lambda \frac{\partial T}{\partial x_j} - \rho \overline{u_j h}\right) + \frac{\partial}{\partial x_j}\left[U_i\left(\tau_{ij} - \rho \overline{u_i u_j}\right)\right] + S_E \tag{3}$$

where τ is the molecular stress tensor, S_M is momentum source, h_{tot} is specific total enthalpy, S_E is energy source.

3.2 Turbulence Model

In fluid flow dynamics, there are two flow models: laminar flow and turbulent flow, when inertia and body forces are greater than viscous forces, then these two forces will enlarging the random disturbances that exist in each flow which then become unstable flows, interact with each other and combine into irregular random movements, then the flow becomes turbulent.

3.2.1 The k-Epsilon (k-ε)

The k-Epsilon (k-ε) model is a semi-empirical turbulence model with two equations developed by Launder and Spalding. The k-ε model is widely used in fluid flow simulations with various turbulent flows because it has adequate stability and accuracy and is more computationally economical. The turbulence model is based on the following two transport equations: turbulence kinetic energy (k) in equation (4) and turbulence dissipation rate (ε) in equation (5).

$$\frac{\partial}{\partial t}(\rho k) + \frac{\partial}{\partial x_j}(\rho U_j k) = \frac{\partial}{\partial x_j}\left[\left(\mu + \frac{\mu_t}{\sigma_k}\right)\frac{\partial k}{\partial x_j}\right] + P_k - \rho\varepsilon + P_{kb} \tag{4}$$

$$\frac{\partial}{\partial t}(\rho \varepsilon) + \frac{\partial}{\partial x_j}(\rho U_j \varepsilon) = \frac{\partial}{\partial x_j}\left[\left(\mu + \frac{\mu_t}{\sigma_\varepsilon}\right)\frac{\partial \varepsilon}{\partial x_j}\right] + \frac{\varepsilon}{k}(C_{\varepsilon 1}P_k - C_{\varepsilon 2}\rho\varepsilon + C_{\varepsilon 1}P_{\varepsilon b}) \tag{5}$$

where $C_{\varepsilon 1}$, $C_{\varepsilon 2}$, σ_ε are k-ε turbulence model constant, σ_k is turbulence model constant for the k equation, P_k is shear production of turbulence, P_{kb}, $P_{\varepsilon b}$ represent the influence of the buoyancy forces.

3.2.2 The k-omega (k-ω)

The k-omega (k-ω) is a turbulence model with first two equations, developed by Wilcox which involves the solution of the transport equations of turbulence kinetic energy (k) in equation (6) and turbulence frequency (ω) in equation (7).

$$\frac{\partial}{\partial t}(\rho k) + \frac{\partial}{\partial x_j}(\rho U_j k) = \frac{\partial}{\partial x_j}\left[\left(\mu + \frac{\mu_t}{\sigma_k}\right)\frac{\partial k}{\partial x_j}\right] + P_k - \beta'\rho k\omega + P_{kb} \tag{6}$$

$$\frac{\partial}{\partial t}(\rho \omega) + \frac{\partial}{\partial x_j}(\rho U_j \omega) = \frac{\partial}{\partial x_j}\left[\left(\mu + \frac{\mu_t}{\sigma_\omega}\right)\frac{\partial \omega}{\partial x_j}\right] + \alpha\frac{\omega}{k}P_k - \beta\rho\omega^2 + P_{\omega b} \tag{7}$$

where β', σ_ω are k-ω turbulence model constant, σ_k is turbulence model constant for the k equation, P_k is shear production of turbulence, P_{kb}, $P_{\omega b}$ represent the influence of the buoyancy forces.

3.2.3 Shear Stress Transport (SST)

The k-ω based SST model accounts for the transport of the turbulent shear stress and gives highly accurate predictions of the onset and amount of flow separation under adverse pressure gradient [24], The BSL model combines the advantages of the Wilcox (k-ω) and the k-ε model, but still fails to properly predict the onset and amount of flow separation from smooth surfaces. The reasons for this deficiency are given in detail in Metter [25] The main reason is that both models do not account for the transport of the turbulent shear stress. This results in an over prediction of the eddy-viscosity. Transport Equation for SST k-ω model as follow in Equations (8) and (9).

$$\frac{\partial k}{\partial t} + \frac{\partial k}{\partial x_j}(U_j) = \frac{\partial k}{\partial x_j}\left[(\upsilon + \sigma_\omega \upsilon_T)\frac{\partial k}{\partial x_j}\right] + P_k - \beta' k\omega \tag{8}$$

$$\frac{\partial \omega}{\partial t} + \frac{\partial \omega}{\partial x_j}(U_j) = \frac{\partial}{\partial x_j}\left[(\upsilon + \sigma_\omega \upsilon_T)\frac{\partial \omega}{\partial x_j}\right] + \alpha S^2 - \beta\omega^2 + 2(1 - F_1)\sigma_{\omega^2}\frac{1}{\omega}\frac{\partial k}{\partial x_i}\frac{\partial \omega}{\partial x_i} \tag{9}$$

where F_1 is blending function, υ_T is the formulation of the eddy-viscosity, S is an variant measure of the strain rate, .

3.3 Computational Domain and Boundary Condition

CFD modeling uses a transient model by applying a free surface, namely ship modeling using water and air fluid in the simulation process. Whereas for modeling the domain is defined as the scope or location for the ship which can be interpreted as the fluid through or around the ship. Domain size: upstream length 1.0Lpp, downstream length 3.0Lpp, top height from the waterline 0.5Lpp, bottom height from the waterline 1.5Lpp as in **Figure 2**. The modelling conditions: inlet as an incoming flow field with normal velocity, that outlet as the exit fluid field applies with normal velocity, side wall as domain boundary walls with symmetrical conditions, on the hull and pre-duct as walls which have roughness and influence the flow velocity (no-slip). The flow velocity of the model is the same as the model speed of 1.179 m/s.

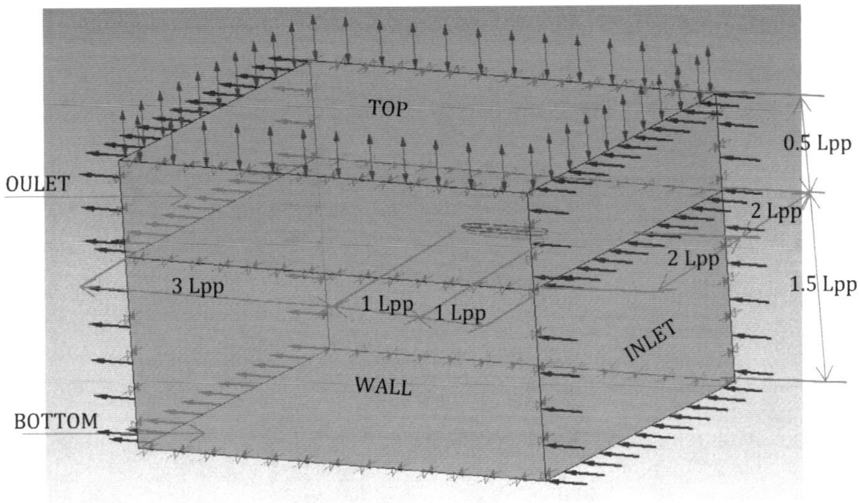

Figure 2. Computational domain of hull JBC

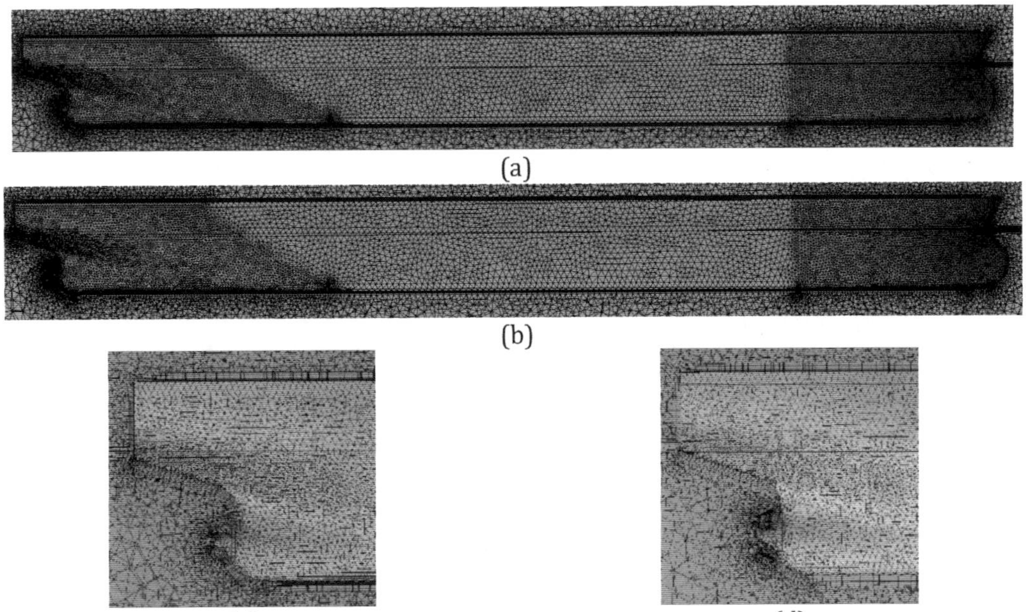

(a)

(b)

(c) (d)

Figure 3. Grid generation of model (a) hull without pre-duct, (b) hull with pre-duct, (c) stern without pre-duct, (d) stern with pre-duct

3.4 Grid Generation

Discretization of the model domain with a hybrid mesh, namely a combination of structural and non-structural triangular surface shapes and boundary layers reinforced with prisms surrounding the model with structural tetrahedral inflation, while those far from the model surface use non-structural. Grid generation model as in Figure 3. Because the function of the wall is to process the close to the wall, the thickness of the first layer near the wall surface. According to ITTC (2012) for boundary conditions near wall 30 < y+<100 [26]. Taner Çoşgun (2021) for areas below the water line (submerged) 30 < y+<100 [27], Ping-Chen Wu et. al. (2022) Average of y+ in the range 38-88 on JBC vessels [28]. The shape of the y+ contour from the hull and propeller modelling in **Figure 4**, and the y+ value of hull JBC in **Table 2**. From the modelling results the average y+ value in the range 62-64, meaning that the y+ value from the modelling carried out is generally still within the required limits.

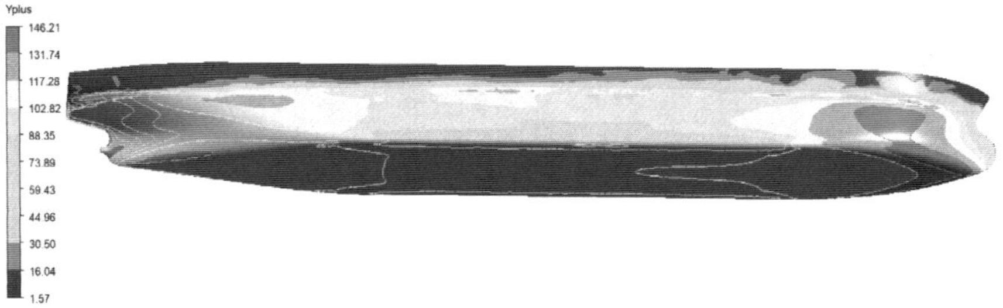

Figure 4. The y+ contour of hull JBC.

Table 2. The y+ value of JAPAN Bulk Carrier

Permodelan		Time Steps	Mesh			Average of y⁺		
			M-3	M-2	M-1	M-3	M-2	M-1
Bare Hull	BH-0	0.075	1,236,970	2,161,613	3,819,985	63.287	62.999	62.831
		0.050	1,236,970	2,161,613	3,819,985	63.814	63.534	63.375
		0.025	1,236,970	2,161,613	3,819,985	63.891	63.628	63.411
	BH-1	0.075	1,334,918	2,250,461	3,960,355	63.369	63.024	62.806
		0.050	1,334,918	2,250,461	3,960,355	63.886	63.548	63.322
		0.025	1,334,918	2,250,461	3,960,355	63.445	63.125	63.881

4. Result and Discussion

4.1 Mesh sensitivity

Mesh is a collection of elements forming a model, where the number and size of the mesh determine the precision and accuracy of the modelling results. In CFD simulation, models with a large number of meshes require a long simulation time and require a large memory capacity, and it is understood that a small mesh size increases the number of meshes. It is understood that the size of the mesh determines the degree of roughness of the model. Thus, mesh sensitivity is process of the testing the level of sensitivity of the number of mesh to simulation results. In general, in CFD modelling, the sensitivity test aims to see that mesh changes are not significant to changes in simulation results, which is known as CFD model verification.

CFD simulation verification is estimating the error/uncertainty of the modeling [29], verification is an effort to ensure the CFD model is correct and appropriate [14] [13], sources of error and uncertainty in CFD model: mesh discretization, iterative convergence and computer iteration[30]. According to IKAP Utama, (2014) [31] and K. Suastika, et al. (2022) [32] performance differences in mesh increments of less than 2% modeling are considered stable/convergent. Roache (1994) General Richardson Extrapolation (GRE) method based on the convergence and smoothness of the model [13], Grid convergence recommended by ITTC CD model[33], model convergence ratio (R_i) in Equations (10), modelling errors or uncertainties ($\delta^*_{RE_{i,1}}$) General Richardson Extrapolation (GRE) method in Equations (11) and modelling accuracy (p_i) in Equations (12). There are four possible CFD model convergence conditions: Monotonic divergence (R_i>1), Monotonic convergence (1>R_i>0), Oscillatory convergence (0>R_i>1) and Oscillatory divergence (R_i<-1).

$$R_i = \frac{\varepsilon_{i.21}}{\varepsilon_{i.32}} \tag{10}$$

$$\delta^*_{RE_{i,1}} = \frac{\varepsilon_{i.21}}{r_i^{p_i} - 1} \tag{11}$$

$$p_i = \frac{\ln\left(\varepsilon_{i.32}/\varepsilon_{i.21}\right)}{\ln(r_i)} \tag{12}$$

where: $\varepsilon_{i.21}$ is differences in performance or simulation results from medium to fine mesh, $\varepsilon_{i.32}$ is difference in performance or simulation results from coarse to medium mesh

Grid Convergence Index (GCI) is standard for assessing mesh/grid convergence and is a descriptive uncertainty method for hydrodynamic models developed y ITTC with including safety factor (FS) values which aim to focus model on increasing safety factors [34][35][36] in Equations (13).

$$U_i = GCI_i = F_S\left|\delta^*_{RE_{i,1}}\right| \tag{13}$$

Bare hull modeling verification is an error evaluation from the CFD simulation results o the total resistance of a ship with and without pre-duct which is not influenced by the presence of the propeller at the stern of the ship (ship model without propeller), expressed in the total ship resistance coefficient ($C_{T(BH)}$), which determined according to Equations(14).

$$C_{T(BH)} = \frac{R_{T(BH)}}{\frac{1}{2}\rho S V^2} \tag{14}$$

where: $R_{T(BH)}$ is total resistance of ship without a propeller with and without pre-duct, ρ is fluid density, S is the surface area of the model that is submerged in fluid, V is fluid low velocity.

Table 3. Grid independent of JAPAN Bulk Carrier

Modeling		Time Steps	$C_{T(BH)}$			Difference (%)		Convergence
			M-3	M-2	M-1	$D_{i.32}$	$D_{i.21}$	
Bare Hull	BH-0	0.075	4.785	4.333	4.268	9,45	1,479	Yes
		0.050	4.778	4.325	4.261	9.46	1.480	Yes
		0.025	4.773	4.321	4.257	9.47	1.481	Yes
	BH-1	0.075	4.635	4.291	4.242	7,42	1,141	Yes
		0.050	4.627	4.284	4.235	7.42	1,142	Yes
		0.025	4.622	4.279	4.230	7.42	1,142	Yes

Bare hull modelling consists of three grid densities: fine grid in notation (M-1), medium grid in notation (M-2) and coarse grid in notation (M-3). In **Table 3** explains the independent grid of the total resistance coefficient ($C_{T(BH)}$), the difference between coarse and medium mesh results in notation $(D_{i.32})$ is less than 9.47%**,** while the mesh is medium and fine in notation $(D_{i.21})$ is less than 1.48%, this explains that the model is generally convergent, as stated by IKAP Utama et al. (2018) [31] and K. Suastika, et al. (2022) [32] that the difference in results between meshes is less than 2% in considered convergent (more stable). In **Figure 5**, by comparing the results of the CFD and EFD simulation, the percentage difference ($\Delta 10^{-3}C_T$(%) = (EFD-CFD)/EFD*100) between M-2 and M-1 is less than 2%, this explains that the CFD modeling results are convergent and in accordance with the EFD.

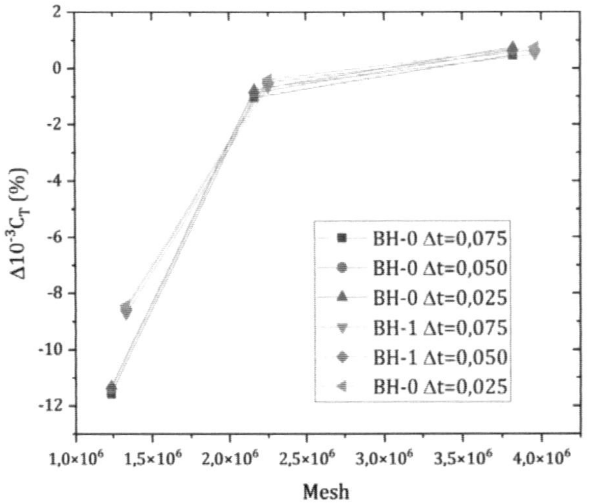

Figure 5. Differences between CFD and EFD Simulations Results

Bare Hull model verification parameters in grid convergence studies in accordance with ITTC recommendations in **Table 4**. Bare hull model convergence ratio(R_i) 0.1418 to 0.1427, this means that the bare hull model is monotonic convergence, that is, the difference in results for fine and medium mesh is smaller than for medium and coarse mesh. Meanwhile, the modeling accuracy value (p_i) 3.427 to 3.445 according to Gokce, et al. (2019) [7] because the modeling accuracy value (p_i) is greater than 1, the CFD modeling accuracy is considered to be quite high. From these results it can be concluded that the CFD simulation of the JBC bare hull model with and without pre-duct is convergent.

Table 4. Bare hull model verification parameters

Parameters	Model	Time Steps	$\varepsilon_{i.32}$	$\varepsilon_{i.21}$	p_i	R_i	$\delta^*_{RE_{i,21}}$	GCI_i
10^{-3} $C_{T(BH)}$	BH-0	0.075	0.4526	0,0643	3,427	0.1420	3.028%	0.263%
		0.050	0.4523	0,0642	3,429	0.1419	3.023%	0.263%
		0.025	0.4521	0,0641	3,430	0.1418	3.020%	0.263%
	BH-1	0.075	0.3438	0,0490	3,445	0.1427	2.340%	0.204%
		0.050	0.3433	0,0489	3,445	0.1427	2.336%	0.204%
		0.025	0.3429	0,0489	3,445	0.1427	2.334%	0.204%

4.2 Time steps sensitivity

In CFD simulation, the transient method is to simulate the model in real time, starting at t=0 and using additional time / time step (Δt), which is the time step obtained for each calculation or simulation process. Similar to mesh, time step is one of several components that affects the accuracy of CFD simulations. CFD is a discretization method, it is understood that the mesh is a discretization of the geometry while the time step is a discretization of the solution steps, the smaller the time step, the more process iterations and it takes longer.

ITTC recommends the time step value *(Δt) = 0.005 ~ 0.01 L/v*, where *L* is the length of model, *v* is the speed of model, based on the results of the time step calculation for the JBC CFD simulation is 0.029 to 0.059. In the convergent mesh condition, the medium mesh (M-2) is simulated at time step (Δt) 0.025 to 0.075, in the **Figure 6** at time step less than 0.05 (following ITTC recommendation) has a very small difference in CFD and EFD (convergent), while time step more than 0.05 (more than ITTC recommendation) has a value that is quite large difference. Thus the time step value that is greater than the recommendation has a result that is quite large difference to the results and it is possible that it does not converge.

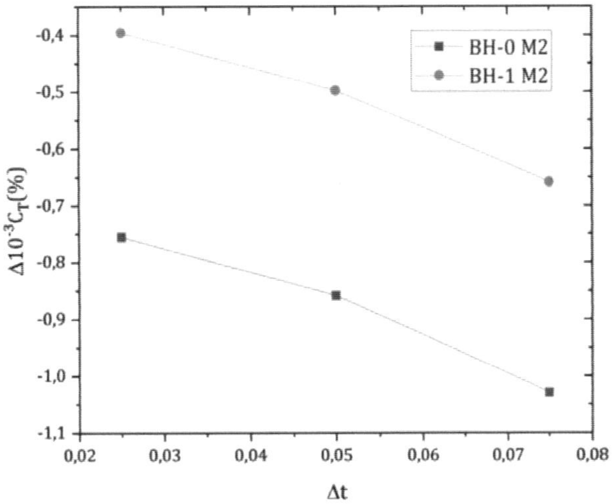

Figure 6. Time steps sensitivity.

4.3 Turbulence model sensitivity

Turbulence models are a source of CFD uncertainty and complex phenomena [18], components that influence model accuracy [14], determine model accuracy and reliability [17]. Turbulence models based on the RANS equation are known as statistical turbulence models because they use statistical averaging procedures[23]. In the time step 0.05, simulated in several turbulent model: k-Epsilon (k-ε), k-omega (k-ω) and SST, as for the modeling results in the **Figure 7**, based on the difference in CFD and EFD results for the three turbulence models, the SST Turbulence model has better results, with a difference o less than 1% for JBC modeling with and without pre-duct.

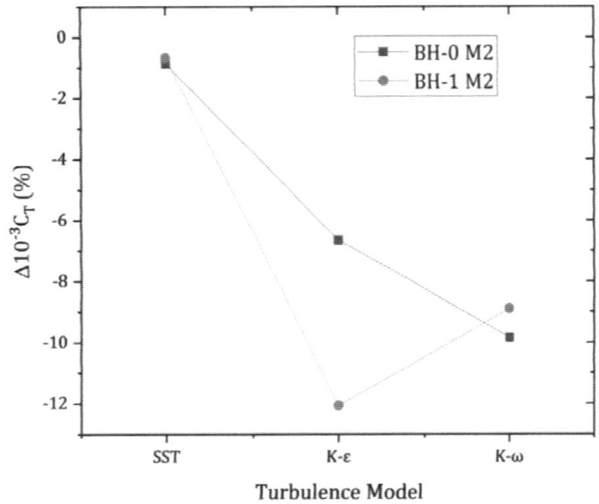

Figure 7. Turbulence model sensitivity.

The SST turbulence model is closer to the modeling results, because according to Metter [83] the SST model is a model that is widely used in hydrodynamics and hybrid models, the k-ε model in free stream is ideal for flow far from the wall and the near-wall model for flow in viscous sublayers. The SST model calculates turbulent shear stress transport and predicts very accurately the onset and amount of flow separation under adverse pressure gradients [84], because it does not calculate turbulent shear stress transport, the k-ε and k-ω models cause predictions of excess eddy viscosity.

4.4 Discussion

CFD is a simulation method that focuses on discretization, where it needs to be understood that the discretization carried out in CFD modelling includes geometry which is known as meshing, while the simulation operation steps are known as time steps. In general, the more rigid, in this case, the smaller the time step, the more convergent and stable the results will be, the same goes for geometric discretization, where the larger the mesh size, the more convergent and stable the modelling will be. The more rigid the discretization used, the better it will be, but there are consequences that must be accepted, namely the simulation time is very long and requires a computer with good performance. The turbulence model used does not have a big influence, and the speed of the model is not fast, this means that the turbulence model does not have a very big influence.

5. Conclusion

From the modelling results that have been carried out on JBC ships with and without pre-duct, because CFD simulation is an analysis method with a discretization approach, in addition to the size of the mesh affecting the modelling results, time step (discretization of the simulation process steps) and the turbulence model used greatly affect the results. Very good modelling is modelling with a time step set up that follows the ITTC recommendations and the SST Model, this is because with a time step of 0.05 it simulates rigidly in the operating steps and the SST model also models near and far fluid flow better and SST takes into account turbulent shear stress transport, this is what causes the SST model and the right time step to produce better results.

References

[1] V. Bertram, Practical Ship Hydrodynamics Practical Ship Hydrodynamics, Second Edi. Butterworth-Heinemann is an imprint of Elsevier, 2012.

[2] G. H. Kim and S. Park, "Development of a numerical simulation tool for efficient and robust prediction of ship resistance," Int. J. Nav. Archit. Ocean Eng., vol. 9, no. 5, pp. 537–551, 2017, doi: 10.1016/j.ijnaoe.2017.01.003.

[3] M. Kim and D.-W. Park, "A Study on the Green Ship Design for Ultra Large Container Ship," J. Korean Soc. Mar. Environ. Saf., vol. 21, no. 5, pp. 558–570, 2015, doi: 10.7837/kosomes.2015.21.5.558.

[4] A. G. Elkafas, M. M. Elgohary, and A. E. Zeid, "Numerical study on the hydrodynamic drag force of a container ship model," Alexandria Eng. J., vol. 58, no. 3, pp. 849–859, 2019, doi: 10.1016/j.aej.2019.07.004.

[5] R. Deng, D. bo Huang, J. Li, X. kai Cheng, and L. Yu, "Discussion of grid generation for catamaran resistance calculation," J. Mar. Sci. Appl., vol. 9, no. 2, pp. 187–191, 2010, doi: 10.1007/s11804-010-9080-2.

[6] Y. Ding, B. Song, and P. Wang, "Numerical investigation on the performance of ducted propeller," Int. J. Nav. Archit. Ocean Eng., vol. 7, pp. 795–804, 2015, doi: 10.1051/matecconf/201713807002.

[7] M. K. Gokce, O. K. Kinaci, and A. D. Alkan, "Self-propulsion estimations for a bulk carrier," Ships Offshore Struct., vol. 14, no. 7, pp. 656–663, 2018, doi: 10.1080/17445302.2018.1544108.

[8] A. Munazid, I. M. Ariana, I. K. Aria, and P. Utama, "Study of Pre-Swirl Duct on Traditional Fishing Vessels as An Energy Saving Device (ESD)," p. 2.

[9] A. Munazid, M. Ariana, and I. K. A. P. Utama, "CFD Analysis on the Development of Pre-Duct Shape to Improve Propeller Performance," CFD Lett., vol. 16, no. 2, pp. 118–132, 2024, doi: 10.37934/cfdl.16.2.118132.

[10] H. L. Liu, M. M. Liu, L. Dong, Y. Ren, and H. Du, "Effects of computational grids and turbulence models on numerical simulation of centrifugal pump with CFD," IOP Conf. Ser. Earth Environ. Sci., vol. 15, no. PART 6, pp. 0–7, 2012, doi: 10.1088/1755-1315/15/6/062005.

[11] B. R. (Rabi) Baliga and I. Y. Lokhmanets, "Generalized Richardson extrapolation procedures for estimating grid-independent numerical solutions," Int. J. Numer. Methods Heat Fluid Flow, vol. 26, no. 3/4, pp. 1121–1144, Jan. 2016, doi: 10.1108/HFF-10-2015-0445.

[12] Y. Bazilevs, M. Hsu, J. Kiendl, R. Wüchner, and K. Bletzinger, "3D Simulation of Wind Turbine Rotors at Full Scale. Part II: Fluid – Structure Interaction Modeling with Composite Blades," Int. J. Numer. Methods Fluids, vol. 65, no. October 2010, pp. 236–253, 2011, doi: 10.1002/fld.

[13] K. M. Almohammadi, D. B. Ingham, L. Ma, and M. Pourkashan, "Computational fluid dynamics (CFD) mesh independency techniques for a straight blade vertical axis wind turbine," Energy, vol. 58, pp. 483–493, 2013, doi: 10.1016/j.energy.2013.06.012.

[14] A. Seeni, P. Rajendran, and H. Mamat, "A CFD mesh independent solution technique for low reynolds number propeller," CFD Lett., vol. 11, no. 10, pp. 15–30, 2019.

[15] M. Morgut and E. Nobile, "Influence of grid type and turbulence model on the numerical prediction of the flow around marine propellers working in uniform inflow," Ocean Eng., vol. 42, pp. 26–34, 2012, doi: 10.1016/j.oceaneng.2012.01.012.

[16] A. Sikirica, Z. Carija, L. Kranjcevic, and I. Lucin, "Grid type and turbulence model influence on propeller characteristics prediction," J. Mar. Sci. Eng., vol. 7, no. 10, 2019, doi: 10.3390/jmse7100374.

[17] S. Boughou, A. A. Omar, O. A. Elsyed, and M. Aldheeb, "Numerical Investigations of Aerodynamic Characteristics Prediction of High-lift Low Reynolds Number Airfoil," CFD Lett., vol. 14, no. 2, pp. 111–121, 2022, doi: 10.37934/cfdl.14.2.111121.

[18] F. R. Menter and R. Lechner, "Best Practice : RANS Turbulence Modeling in Ansys CFD." Ansys Germany GmbH, 2021, [Online]. Available: https://www.ansys.com/resource-center/technical-paper/best-practice-rans-turbulence-modeling-in-ansys-cfd.

[19] S. Dendy, K. A. P. U. I, and Mukhtasor, "The influence of time step setting on the CFD simulation result of vertical axis tidal current," vol. 19, no. 2, pp. 3399–3409, 2022.

[20] F. Furcas, G. Vernengo, D. Villa, and S. Gaggero, "Design of Wake Equalizing Ducts using RANSE-based SBDO," Appl. Ocean Res., vol. 97, no. February, p. 102087, 2020, doi: 10.1016/j.apor.2020.102087.

[21] NMRI, "Tokyo 2015 A Workshop on CFD in Ship Hydrodynamics," 2015. https://t2015.nmri.go.jp/.

[22] F. Joel H. and P. Milovan, Computational Methods for Fluid Dynamics, Third, Rev. New York, 2002.

[23] J. John D. Anderson, Computational Fluid Dynamics: The Basics with Applications, vol. 332, no. 7555. New York, United States of America, 1995.

[24] T. D. Canonsburg, ANSYS CFX-Solver Modeling Guide, vol. 15317, no. November. United States of America, 2011.

[25] F. R. Menter, "Two-Equation Eddy-Viscosity Turbulence Models for Engineering Applications," AIAA J., vol. 32, no. 8, 1994, doi: 10.2514/3.12149.

[26] ITTC, "Practical Guidelines for Ship CFD Applications," ITTC – Recomm. Proced. Guidel. ITTC, pp. 1–8, 2011.

[27] T. Çoşgun, "Numerical Self-Propulsion Assessment of a Generic Submarine Model at Various Forward Speeds," J. Eta Marit. Sci., vol. 9, no. 3, pp. 192–199, 2021, doi: 10.4274/jems.2021.24654.

[28] P. C. Wu, C. W. Chang, and Y. C. Huang, "Design of Energy-Saving Duct for JBC to Reduce Ship Resistance by CFD Method," Energies, vol. 15, no. 17, 2022, doi: 10.3390/en15176484.

[29] F. Stern, R. V Wilson, H. W. Coleman, and E. G. Paterson, "Verification and validation of CFD simulations," Iowa City, 1999.

[30] L. Larsson, M. Visonneau, and F. Stern, Eds., Numerical ship hydrodynamics. An assessment of the Gothenburg 2010 Workshop, 1st ed. Springer Dordrecht Heidelberg, 2014.

[31] I. K. A. P. Utama and B. L. Richard, Kapal Berlambung Banyak: Tinjauan Hambatan, Kebutuhan Tenaga Penggerak, dan Efisiensi Energi Kapal. Brillian Internasional Surabaya, 2023.

[32] K. Suastika, G. E. Nadapdap, M. H. N. Aliffrananda, Y. A. Hermawan, I. K. A. P. Utama, and W. D. Aryawan, "Resistance Analysis of a Hydrofoil Supported Watercraft (Hysuwac): A Case Study," CFD Lett., vol. 14, no. 1, pp. 87–98, 2022, doi: 10.37934/cfdl.14.1.8798.

[33] ITTC, "Uncertainty Analysis in CFD Verification and Validation Methodology and Procedures," ITTC – Recomm. Proced. Guidel., p. 12, 2008, [Online]. Available: http://ittc.info/media/4184/75-03-01-01.pdf.

[34] P. J. Boache, "Perspective: A method for uniform reporting of grid refinement studies," J. Fluids Eng. Trans. ASME, vol. 116, no. 3, pp. 405–413, 1994, doi: 10.1115/1.2910291.

[35] P. J. Roache, "Quantification of uncertainty in computational fluid dynamics," Annu. Rev. Fluid Mech., vol. 29, pp. 123–160, 1997, doi: 10.1146/annurev.fluid.29.1.123.

[36] C. Ismail B., "Procedure for Estimation and Reporting of Discretization Error in CFD Applications," J. Fluids Eng., no. 130, pp. 1–70, 2008.

IOP Conf. Series: Earth and Environmental Science **1423** (2024) 012034 doi:10.1088/1755-1315/1423/1/012034

Influence of faceplate thickness reduction on the strength of sandwich structure under static compression loading

H Siswanti[1*], M Musta'in[1], A M Mulananda[1], A Nasrudin[1] and D R Aldara[1]

[1] Department of Maritime Technology, Politeknik Negeri Madura, Sampang, Indonesia

*E-mail: henisiswanti31@gmail.com

Abstract. Sandwich structures are increasingly used to replace conventional stiffened plate ship structures. The sandwich plate thickness design is determined based on the strength index (R) value. The sandwich plate thickness design is flexible because the core and faceplate thicknesses can be changed according to the desired safety factor. This faceplate and core thickness difference has influenced structural strength, deformation, weight, and other structural characteristics. This study analysed the effect of different faceplate thicknesses on the performance of sandwich structures on barge decks subject to static compression loading. The performance of the sandwich structure on the deck carrying bulk cargo under several loading conditions was evaluated using the finite element method to obtain a general consideration in determining the faceplate thickness for the sandwich structure. Reducing faceplate thickness significantly reduces weight, but it is also essential to prioritize strength to ensure the sandwich structures are safe, efficient, and compliant with all necessary standards and regulations.

1. Introduction

Classification rules regulate a ship's material specifications, component configurations, and structural strength to ensure the safety of ships and cargoes. Classification societies have developed special rules for sandwich materials to ensure that the strength of sandwich structures is equivalent to that of conventional stiffened plate structures. Rules and standards for applying sandwich materials to ship structures include mechanical characteristics, material certification, minimum requirements, test standards, minimum thickness, and strength criteria [1]. Sandwich materials are developed as lightweight materials for engineering applications. Faceplates separated by thick core material will increase the moment of inertia of the cross-section so that the plate's bending stiffness and flexural strength increase significantly without a significant increase in the structure's weight [2]. In marine applications, sandwich plates were developed for ship materials because they reduce the structure's weight by 4-9% compared to conventional stiffened plate structures [3, 4, 5, 6]. Reducing the structural has implication for increases payload [3]. Applying sandwich structures also increases the stiffness-to-weight ratio [7, 8]. Previous studies show stress reduction in sandwich applications in several parts of the ship, such as the main deck structure [4, 9], car deck [5, 10], and ramp door [11]. Previously, the use of sandwich plates of double-hull tankers [12], dredging ships [13], and tank construction [14] also reduced stress and structural deformation.

The sandwich plate for ship structures is a hybrid metal-concrete metal sandwich type consisting of a steel faceplate and an elastomeric polyurethane matrix core. This type of sandwich structure replaces conventional stiffened plate structures. Sandwich structures' weight reduction effect is due to low-density cores and the tendency to use thinner faceplates [15]. The sandwich plate has three layers: upper faceplate, bottom plate, and core. The thickness of each layer of the sandwich plate is t1, t2, and tc. Based on the Lloyd's Register rules, the faceplate and core thickness are determined based on the strength index (R). The strength index (R) for new construction is less than or equal to 1 (R ≤ 1) [16]. The thickness allowance variables t_{aR}, t_{a1}, and t_{a2} vary depending on the structure's location. Meanwhile, the minimum thickness of each faceplate is shown in Table 1. With the minimum faceplate thickness rule for sandwich structures, designers flexibly choose the faceplate thickness as long as the result R≤1. Reduced faceplate thickness resulted in a more significant structure weight reduction. However, it is necessary to consider the structural strength due to the reduction of faceplate thickness to optimize the structural design [17, 2].

Table 1. Minimum thickness of faceplates [16]

Item	Minimum thickness (mm)	
	New construction	Overlay construction
t_1 min	3,0	3,0
t_2 min	3,0	50% of the as-built thickness

Determining the design loads to satisfy the Rule's requirements is a fundamental part of ship structure design. In addition to structural loads, the ship receives external loads, such as wave-induced and cargo loads. Most of the external loads are compression loads [18]. Several types of ships, such as container ships, log carriers, car carriers, and barges, carry cargo on their decks. The cargo load must be considered in the deck structure design. Therefore, the cargo load must also be appropriately considered when applying a sandwich structure to replace conventional stiffened plate structures. This study aims to evaluate the effect of selecting faceplate thickness on the sandwich structure applied to a barge's deck carrying bulk cargo. The finite element method analyzed several alternative sandwich structure designs with different strength index (R) values due to reduced faceplate thickness and different loading conditions. This evaluation ensures that the sandwich structure exposed to the cargo load meets the strength criteria. So, the optimum faceplate thickness can be selected to produce a lightweight structure without sacrificing strength.

2. Method

2.1 Data collection

In this study, a hybrid sandwich type consisting of a steel faceplate and an elastomeric polyurethane matrix core was applied to the barge's deck structures. Table 2 shows the main dimensions of the barges, and Table 3 shows the material sandwich properties.

Table 2. Main dimension of the barges

Length Overall (Loa) (m)	82.29	m
Length between perpendicular (LBP) (m)	79	m
Breadth (B) (m)	24.38	m
Height (H) (m)	4.87	m

Table 3. Sandwich Material Properties [19]

Part	Faceplate	Core
Material	Steel	Polyurethane Elastomer
Density (Kg/m³)	7850	1119
Young's Modulus (Gpa)	206	0,901
Poisson's Ratio	0,3	0,36

The model for the finite element analysis is a 15.8 m long cargo hold on an amidship. The conventional deck structure uses a 15 mm steel plate thickness. The deck structure is changed to a sandwich structure by replacing the deck plate with a sandwich plate and removing deck longitudinal/stiffeners. Meanwhile, the configuration of other components, such as girder, transverse, and bulkhead, is maintained the same as the original configuration. The structural components of the deck structure are shown in Table 4.

Table 4. Deck structure components

Component	Stiffened plate	Sandwich	Units (pcs)
Deck plate	15 mm	N/A	1
Faceplate	N/A	4,5,6 mm	2
Core	N/A	15 mm	1
Center Girder	(T) 210x10/150x10	(T) 210x10/150x10	1
Side Girder	(T) 210x10/150x10	(T) 210x10/150x10	6
Deck Longitudinal	(L) 120 x 80 x 10	N/A	30
Deck Transverse	(T) 210x10/100x10	(T) 210x10/100x10	5

The cargo hold is divided into three longitudinal compartments separated by a longitudinal bulkhead. The conventional stiffened plate deck structure model is shown in Figure 1, and the sandwich structure model is shown in Figure 2.

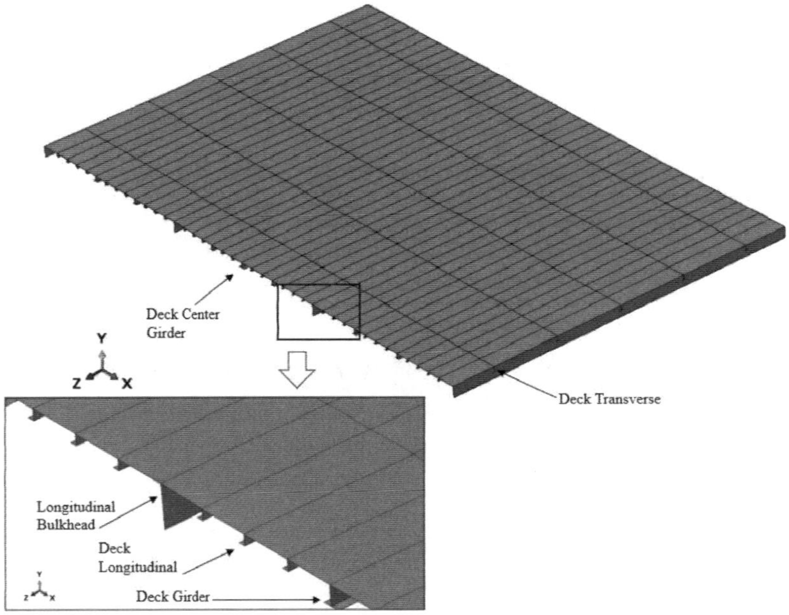

Figure 1. Stiffened plate deck structure

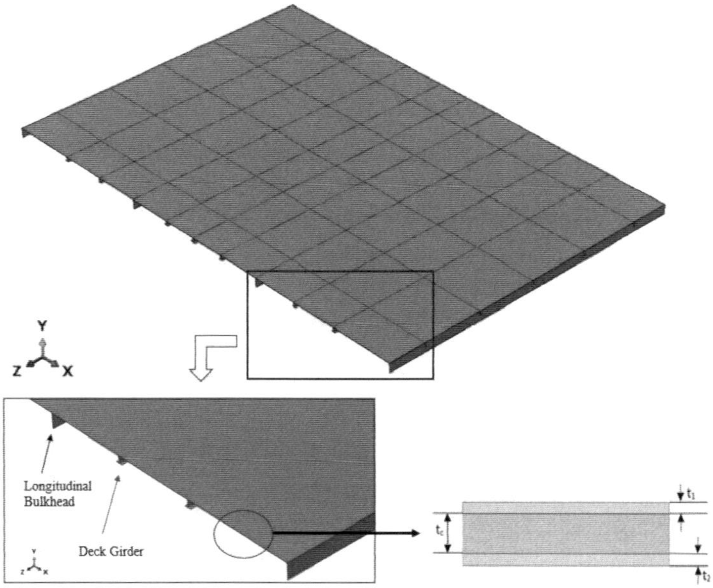

Figure 2. Sandwich deck structure

Sandwich plate thickness calculation refers to the rules in Lloyd's Register, which regulate the application of sandwich plates to ship structures. The thickness of the sandwich plate is based on the Strength Index (R) value, which must be equal to or less than 1 (R≤1) [16]. The strength index can be determined using equation (1), with each thickness allowance value being t_{aR} 2.5, t_{a1} 2.5, and t_{a2} 0.

$$R = 0.01\,AR\left[0.1\frac{b^2}{d(t_1+t_2)} + 11.7\left(\frac{bt}{d^2}\right)^{1.3}\right]kP_{eq}, R \tag{1}$$

It is assumed that the core material uses a minimum thickness of 15 mm to produce the equivalent modulus and R≤1; the possible faceplate thickness is 6, 5, and 4 mm, and the strength index (R) is 0.620, 0.764, and 0.980, respectively. The stiffened plate and sandwich structure are subjected to the static compressing load assumed to be the external cargo load. The strength assessment was conducted at 50%, 75%, and 100% loading conditions. The equation for static analysis can be seen in equation (2). Equations (3), (4), and (5) are variables associated with static analysis.

$$\{F\} = [K]\{U\} \tag{2}$$

$$\sigma = \frac{F}{A} \tag{3}$$

$$\tau = \frac{F}{A} \tag{4}$$

$$\varepsilon = \frac{\delta}{L} \tag{5}$$

where: $\{F\}$ is force (N); $[K]$ is stiffness (N/m), $\{U\}$ is displacements (m); σ is the normal stress (Pa); τ is the shear stress (Pa); A is the surface area (m^2); ε is the strain; δ is the deformation (m); L is the length (m).

2.2 Method

Structural strength assessment was conducted using the finite element method. An eight-node quadrilateral shell element (SC8R) is used to model the faceplates, and an eight-node hexahedral solid element (3D8I) is used for core material. The faceplates and core were contacted using tie constraints, which depict no relative movement between the two types of elements. The boundary condition of the four sides of the structural model is fully clamped. Contact between elements on the faceplate and core uses tie constraints, assuming there is no relative movement between the two types of elements. The boundary condition of the four sides of the structural model is fully clamped, representing the deck structure tied to the transverse bulkhead and the side shell. Three sandwich structure models with different faceplate thicknesses were analyzed and compared with the initial stiffened plate model.

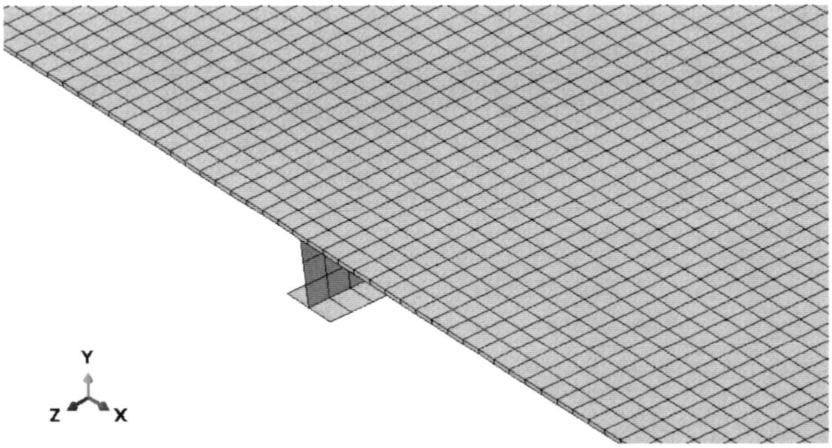

Figure 3. Meshing on the structure model

In finite element analysis, the structure is partitioned into a finite number of small elements. In this study, the type of mesh used is structural mesh as shown at Figure 3. Each element is analysed to obtain the global results. The number and size of elements affect the analysis results. The analysis must be done using the appropriate size and number of elements to get accurate analysis results. Therefore, it is necessary to conduct a convergence analysis to get the optimal number and size of elements. The convergence study was conducted on stress and deformation variables. A valid value is obtained when convergence is less than 5%, as per Wang's study, in a similar case study [20]. The graph of the convergence study is illustrated in Figure 4.

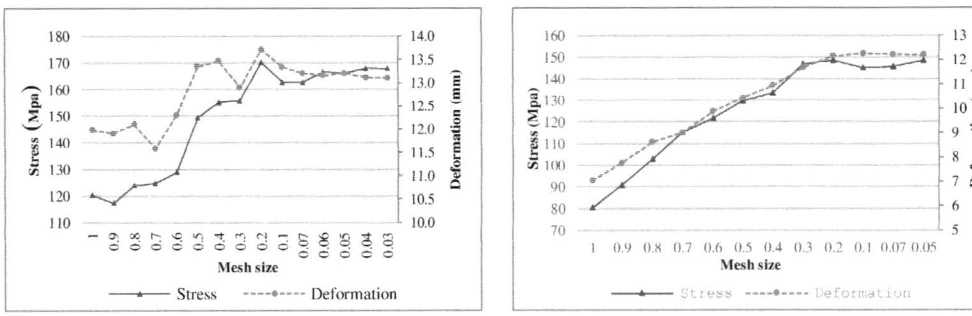

(a) **(b)**

Figure 4. Convergence analysis of (a) stiffened plate structure model; (b) sandwich structure model

3. Result and Discussion

Finite element analysis investigates structural strength by comparing the von Misses stress between each structure model. The stress indicated by von Misses is a scalar quantity obtained from the stress acting on a structure. Table 4 shows the von Misses stress of each model at several loading conditions. The stress in all sandwich models is lower than the stiffened plate structure model. It shows that the application of sandwich structure reduces the stress, this occurs in all loading conditions. Several previous studies noticed that the sandwich plate application on the deck structure increased the strength compared to the stiffened plate structure [4, 9]. The constitutive relationship between reducing the faceplate thickness of sandwich-A, sandwich-B, and sandwich-C against a change in Stress value was also analyzed. Reducing the thickness of the faceplate in the sandwich structure model does increase the stress by 10-12%. However, the stress is still smaller than the stiffened plate structure.

Structural strength is determined by comparing structural stress with material yield strength. For safety factors, the classification recommends the permissible stress as the design criteria [21]. The yield strength of this structure is assumed to refer to the steel yield strength, which has a nominal upper yield point of 235 N/mm^2. For steel structures with yield properties less than 235 N/mm^2 must be regarding the material factor (k) [21]. Since the sandwich structure's yield is estimated below the steel yield strength, its permissible stress is assumed to be 175 N/mm^2. This value is maximum stress limit for the designed structure. Table 5 also shows that the Stress of the stiffened plate structure is close to the permissible Stress value, whereas it is below that value in all sandwich models. This analysis indicates promising results in terms of structural strength in deck structures.

Table 5. Comparison of von Misses stress

Structure	R	Perm Stress	Stress at loading capacity					
			100%	Diff	75%	Diff	50%	Diff
		Mpa	Mpa	%	Mpa	%	Mpa	%
Stiffened plate		175	166.20		124.70		83.1	
Sandwich-A	0.620	175	145.70	-12.33	109.30	-12.35	72.86	-12.32
Sandwich -B	0.764	175	147.10	-11.49	110.30	-11.55	73.55	-11.49
Sandwich -C	0.980	175	148.80	-10.47	111.60	-10.51	74.39	-10.48

One of the advantages of structural strength analysis using numerical simulation with FEM is that it can estimate the position of the structure's stress concentrations. The different colours of the ribbon diagram show different stress values throughout the structure. Figure 5 shows the stress contours of the stiffened plate structure and typical sandwich structure analysed in this study. Figure 5 shows that the stress concentration in the stiffened plate and sandwich structures is similar at the intersection between the two main stiffeners. Even though the location of stress concentration is relatively the same, the sandwich structure's strength is better because its stress value is smaller. Reducing the faceplate thickness and different loading conditions in the sandwich structure also does not change the stress concentration's location but only affects the stress value. This result differs from previous studies, which stated that sandwich application could change or eliminate stress concentration [6, 9]. Since structural design depends on ship type, stiffener configuration, location, and loading condition, this difference can be caused by different stiffener configurations and loading positions in each structural design.

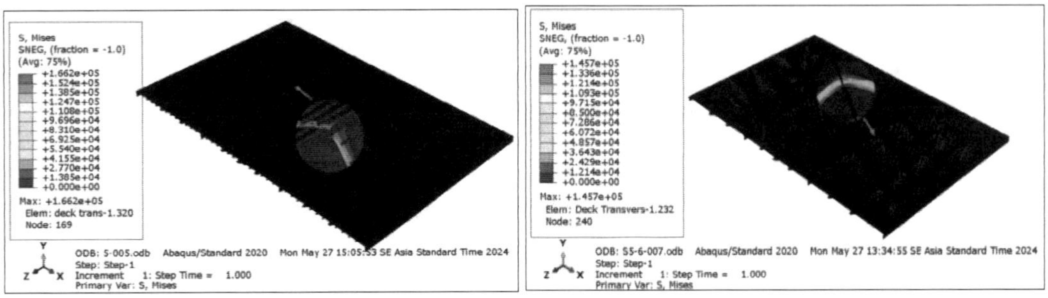

(a) **(b)**

Figure 5. Stress contour of (a) stiffened plate structure; (b) typical sandwich structure.

Structural strength parameters must also describe the structure's resistance to changes in shape due to its working load. Therefore, besides stress value, structural strength also represents the structure's resistance to deformation under a load. Figure 6 compares each structure's deformation at any loading condition. Similar studies have shown that the application of sandwich structures results in a decrease in deformation [4, 6, 9, 12]. The outstanding reduction of deformation due to the application of sandwich structure is affected by the dissimilarity of thickness between solid and sandwich plates [2]. Applying a sandwich plate increases the moment of inertia. Since the deformation of the structure is associated with the structural stiffness, the stiffness is directly proportional to the moment of inertia of the plate cross-section. So, the sandwich plate's deformation should decrease compared to the solid plate. Even though this deformation reduction is relatively small, sandwich plate application is promising to improve structural strength. However, reducing faceplate thickness can increase deformation, as shown in the Sandwich-C model. A thin faceplate is ineffective in improving structural strength and increases the deformation.

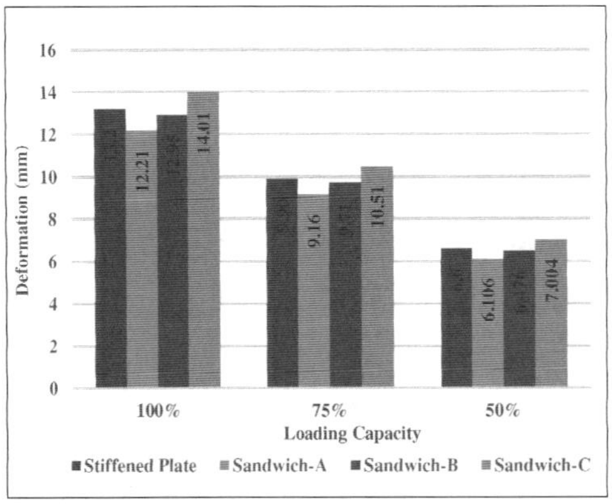

Figure 6. Comparison of deformation

The sandwich and stiffened plate structure show different deformation contours. Figure 7 shows that the sandwich structure has a smaller deformed area than the stiffened plate structure. Two faceplates separated by a thick core material expand the cross-section modulus, improving the plate's bending stiffness. Besides the decrease in deformation value, the deformed area decreases instead. Applying this sandwich plate eliminates deformation concentrations, as the results of Tuswan's [10] and Ismail's [6, 9] previous studies. Reducing faceplate thickness resulted in a reliable increase of deformation. The sandwich-C model with the thinnest faceplate produces more significant deformation, even more significant than the stiffened plate structure. Even though the strength index meets the criteria (R<1), using a faceplate that is too thin is ineffective in correcting structural deformation. Therefore, in the design of sandwich structures, it is necessary to analyse deformation changes due to faceplate thickness reduction. Reducing the faceplate thickness too much may result in a less-than-optimal design and decrease its safety factor.

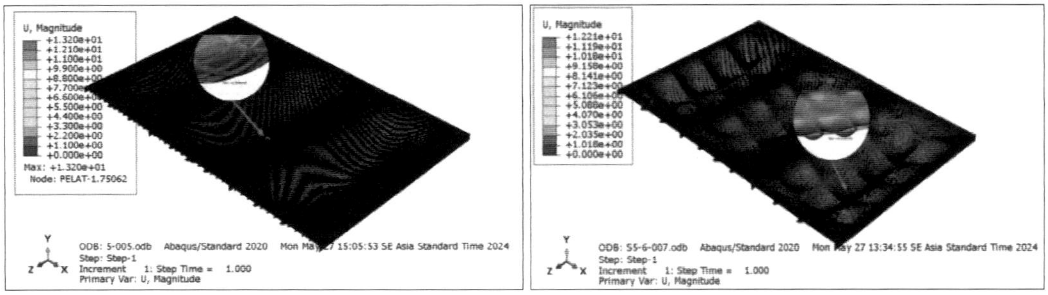

(a) **(b)**

Figure 7. Deformation contour of **(a)** stiffened plate structure; **(b)** typical sandwich structure

The sandwich structure application aimed to produce a lightweight structure. Furthermore, structural weight estimation was conducted to ensure that the sandwich application in this study reduced structural weight.

Table 6. Comparison of structural weight

Model	**Structural Weight**	**Difference weight**
	ton	**%**
Stiffened Plate	58.37	
Sandwich-A	48.51	16.91
Sandwich-B	42.46	27.27
Sandwich-C	36.41	37.63

Structural weight estimation is performed to see changes in the structure's weight for each model. The weight estimation is carried out only for the deck structure. Table 6 compares the structural weight of the stiffened plate and sandwich structure. The structural weight reduction from replacing solid plates with sandwich plates and omitting deck longitudinal stiffener in sandwich-A is 16.91%. The faceplate thickness in this sandwich-A is 6 mm. Reducing the faceplate thickness by 1 mm, respectively, in sandwich-B and sandwich-C reduces more weight. Compared to stiffened plate structures, it is about 27.27 and 37.63% reduction weight. The results of this study are close to the results of the previous study, which state weight reduction by 16-17% [22]. This weight reduction is significant compared to previous studies, which showed only 4.2-8.8% [6]. This significant weight reduction is primarily due to eliminating the longitudinal deck stiffener and reducing faceplate thickness. However, although reducing faceplate thickness significantly reduces weight, selecting the thinnest faceplate is sometimes impossible. In the previous analysis, the thinnest faceplate on the C-sandwich produced more significant deformation, so this design was considered less than optimal. Although it provides more weight reduction, reducing faceplate thickness does not always make an optimum design.

Furthermore, ship designers must consider the ship's specific operational requirements and loading conditions, conduct thorough analyses and adhere to regulatory standards to ensure that any reduction in faceplate thickness does not compromise the vessel's safety and performance. The practical approach to choosing the sandwich structure's appropriate faceplate thicknesses is defining the operational requirements and ensuring compliance with the standards. Do the finite element analysis to simulate the realistic load conditions, including static, dynamic, and impact loads, to mimic actual operational conditions. Examine the distribution of stresses, strains and deflection, then adjust the thickness of faceplates and core based on FEA results to meet performance criteria. Suppose the sandwich structure is exposed to extreme loading; making a scale model and testing it to realistic load conditions may be necessary to validate the design assumption. By following these steps, designers can ensure that the structures are safe, efficient, and compliant with standards.

The finite element analysis in this study still needs to be improved because it only analyzes static loads. To optimize the design, further analysis should be carried out using dynamic loading and fatigue conditions to mimic realistic operational loading. However, this study initially illustrates selecting the appropriate faceplate thickness for a sandwich structure.

4. Conclusion

The primary observation regarding the effect of different faceplate thicknesses on the performance of sandwich structures on barge decks subject to static compression loading is summarized. The strength of three sandwich structures with varying faceplate thicknesses at different loading conditions was analysed and compared to the conventional stiffened plate structure. A comparison of von Misses stress and deformation parameters shows that the sandwich structure is quite promising for application on barge decks with cargo loading. However, the thickness of the faceplate needs to be appropriately analysed to produce an optimum design. It is also essential to prioritize strength rather than structural weight reduction. Although it provides more weight reduction, reducing faceplate thickness does not always make an optimum design. Reducing the faceplate thickness too much may decrease its safety factor. The designer must choose the faceplate thickness by defining requirements, performing FEA, optimizing the design, and validating through testing to ensure the sandwich structures are safe, efficient, and compliant with all necessary standards and regulations.

References

[1] G. Palomba, G. Epasto and V. Crupi, "Lightweight sandwich structures for marine applications: a review," *Mech of Adv Mat and Struct,* 2021.

[2] SANDCORe, "Best Practice Guide for Sandwich Structures in Marine Applications," University of Newcastle upon Tyne, New Castle, 2013.

[3] N. Momcilovic and M. Motok, "Estimation of ship lightweight reduction by means of application of sandwich plate system," *FME Trans vol. 37, no. 3,* p. 123–128, 2009.

[4] S. Sujiatanti, A. Zubaydi and A. Budipriyanto, "Finite element analysis of ship deck sandwich panel," *Applied Mechanics and Materials 874,* pp. 134-139, 2018.

[5] J. Andric, S. Kitarovic, V. Radolovic and P. Prebeg, "Structural analysis and design of a car carrier with composite sandwich deck panels," *Ships and Offshore Structures,* 2019.

[6] A. Ismail, A. Zubaydi, B. Piscesa and Tuswan, "Study of Sandwich Panel Application On Side Hull Of Crude Oil Tanker," *Journal of Applied Engineering Science,* 2021.

[7] C. Borsellino, L. Calabrese and A. Valenza, "Experimental and numerical evaluation of sandwich composite structures," in *Composite Science Technology*, 2004.

[8] J. Yang, L. Maa, R. Schmidt and G. Qi, "Hybrid lightweight composite pyramidal truss sandwich panels with high damping and stiffness efficiency," *Composite Structures,* vol. 148, p. 85–96, 2016.

[9] A. Ismail, A. Zubaydi, B. Piscesa and E. Panangian, "A strenght analysis of conventional and sandwich plate deck using Finite Element Method," in *IOP Conference Series: Materials Science and Engineering, Vol 1034*, 2020.

[10] Tuswan, A. Zubaydi, A. Budipriyanto and S. Sujiatanti, "Comparative study on ferry ro-ro's car deck structural strength by means of application of sandwich materials," in *The 3rd International Conference on Marine Technology vol. 1*, 2018.

[11] T. Tuswan, A. Zubaydi, B. Piscesa and A. Ismail, "Influence of Apllication of Sandwich Panel on Static and Dynamic Behaviour of Ferry Ro-Ro Rampdoor," *JOURNAL OF APPLIED ENGINEERING SCIENCE,* vol. 11, no. 1, pp. 208-216, 2021.

[12] M. Brooking and S. Kennedy, "The performance, safety and production benefits of SPS structures for double hull tankers," in *Proceedings of the RINA Conference on Double Hull Tankers*, London, UK, 2004.

[13] J. Kortenoeven, B. Boon and A. De Bruijn, "Application of sandwich panels in design and building of dredging ships," *J. Sh. Prod., vol. 24, no. 3,,* p. 125–134, 2008 .

[14] K. Niklas and J. Kozak, "Experimental investigation of Steel–Concrete–Polymer composite barrier for the ship internal tank construction," *Ocean Engineering., vol. 111,* p. 449–460, 2016.

[15] T. J. Grafton and J. R. Weitzenbo€ck, "Steel-concrete-steel sandwich structures in ship and offshore engineering," in *Advance Marine Structure Proceedings of the 3rd International Conference on Marine Structures*, 2011.

[16] Lloyd's-Register, Rules for the Application of Sandwich Panel Construction to Ship Structure, Lloyd's Register, 2020.

[17] G. Palomba, G. Epasto and V. Crupi, "Lightweight sandwich structures for marine applications: a review," *Mechanics of Advanced Materials and Structures,* 2021.

[18] S. E. Hirdaris, W. Bai, D. Dessi, A. rgin, X. Gu, O. A. Hermundstad, R. Huijsmans, K. Iijima, U. D. Nielsen, J. Parynov, N. Foncesa, A. Papanikalaou, K. Argyriadis and A. Incecik, "Loads for use in the design of shipsand offshore structures," *Ocean Engineering,* vol. 78, pp. 131-174, 2014.

[19] A. Ismail, A. Zubaydi, B. Piscesa and Tuswan, "A novel fiberglass-reinforced polyurethane elastomer as the core sandwich material of the ship–plate system," *Journal of the Mech Behavior of Mat,* vol. 32, 2023.

[20] I. T. Wang, "Numerical and experimental verification of finite element mesh convergence under explosion loading," *Journal of Vibroengineering,* vol. 16, p. 1786–1798, 2014.

[21] BKI, Rules for the Classification and Construction of Seagoing Steel Ship, Vol II (Rules for Hull), Biro Klasifikasi Indonesia, 2020.

[22] Tuswan, K. Abdullah, A. Zubaydi and A. Budipriyanto, "Finite-element analysis for structural strength assessment of marine sandwich material on ship side-shell structure," in *Materials Today: Proceedings, vol. 13, no. 1*, 2019.

Integration of Hand Motion Sensor, Artificial Intelligence, and QR Code for Real-Time Monitoring of Welder's Performance and Welding Quality: A Conceptual Framework

Adi Sasmito Aji [1*], Triwilaswandio Wuruk Pribadi [2$], and Imam Baihaqi [2#]

[1] Engineering Master Programme student, Department of Naval Architecture, Faculty of Marine Technology, Institut Teknologi Sepuluh Nopember, Surabaya, Indonesia
[2] Department of Naval Architecture, Faculty of Marine Technology, Institut Teknologi Sepuluh Nopember, Surabaya, Indonesia

e-mail: *adisasmitoaji@gmail.com, $triwilas@na.its.ac.id #imam.baihaqi@its.ac.id

Abstract. The welding quality, as the product of manual or semi-automatic welding process, is highly affected by the hand motion of the welders. In the shipbuilding industry, the welding inspection process is commonly performed after the welding process is finished by the welding inspectors. The research aims to develop a real-time monitoring system for welder's performance by recording their hand motion. The hand motion sensors and QR codes are integrated through the Internet of Things (IoT) to monitor the welder's performance and weldment quality. The real-time monitoring concept was developed by designing the monitoring concept utilising the QR code and Android mobile smartphone. The welder's hands are equipped with the Inertial Measurement Unit (IMU) sensor to record their hand motion. The motion data are transferred directly from the IMU sensor to the smartphone storage through a Bluetooth connection. The data then are uploaded to the cloud storage through the internet connection. The stored recorded data was then analysed and compared with the qualified welders' hand motions. After welding ends, the welder reports the visual appearance of the weldment and its location as identified by the QR code through the Android system. Finally, the functionality of the application prototype is tested, and the results show that the system can be used practically for real-time monitoring of welder performance and welding quality.

Keywords: hand motion sensor; artificial intelligence; welder's performance; real-time monitoring.

1. Introduction

The shipyard's performance still heavily depends on its personnel, as they are the most influential factor in its success [1]. Moreover, the welding process in shipbuilding significantly determines the quality and accuracy of ship hull construction. This process still heavily depends on the expertise of the welders. It has been acknowledged that the wrist-hand movement skills of welders are a crucial factor in achieving professional competency in the welding process.

However, the system used to assess welders' proficiency relies on traditional methods, primarily visual inspection of welding outcomes, supplemented by both destructive and non-destructive testing techniques. Currently, there are no measures in place to evaluate the quality of welders' output by monitoring their hand movements in real-time.

The existing study on the real monitoring system has been applied to recognised the physical activity of human [2–4]. However, there are still limited studies applying the recognition of physical movement to monitor productivity and quality output, such as in welders during the shipbuilding process. Pribadi and Shinoda [5] have developed a system for real-time monitoring by integrating Inertial Measurement Unit (IMU) motion sensors affixed to the wrist (as a wearable device), along with Internet of Things (IoT) infrastructure and Artificial Intelligence (AI) technology. This system leverages the Support Vector Machine (SVM) method and Deep Learning CNN-LSTM to assess the activities and performance of welders, thereby enabling the classification of each welder's proficiency level. However, its application in shipbuilding processes still faces limitations. It cannot identify the specific part of the construction or joint design where welding is performed, the Welding Procedure Specification (WPS) used, the layer being welded, or conduct subsequent visual inspections of the resulting weld.

Furthermore, QR-Code technology has been developed to track prefabricated assets through a digital system, which can be integrated with blockchain and smart contract technology [6]. In this context, QR-Code technology can also assist welders in recalling and accessing Welding Procedure Specification (WPS) information during the shipbuilding process. By utilizing an application system to access and retrieve welding parameters based on QR-Code technology, welders can be better supported in adhering to WPS guidelines. This approach is expected to aid in performing welding according to WPS standards and facilitate quality control in monitoring welding progress and results in compliance with WPS.

2. State of the Art

2.1 Deep Learning (DL) and Convolutional Neural Networks (CNN)

Deep Learning constitutes a segment of Machine Learning (ML) advocating for the recycling of attributes and, with increasingly intricate layer architectures, acquires more generalized features at advanced layers of representations. Convolutional Neural Networks (CNNs) stand as the predominant manifestation of deep learning, extensively deployed across various scientific and industrial domains in recent years, particularly in image scrutiny. While any neural network comprising two or more concealed layers qualifies as a deep neural network, CNNs have emerged as highly efficacious instruments in computer vision applications, such as object recognition and image categorization. Fundamentally, the convolutional layers, traditionally employing two-dimensional matrices as filters, possess adaptability to accommodate varied inputs by adjusting to alternative dimensions. The fundamental concept underlying the proposed network revolves around executing convolution across the spectral dimension, wherein each image pixel is encoded as a vector housing reflectance values across different wavelengths. Consequently, CNNs are adept at extracting comprehensive relational patterns from image spectra, leveraging their inherent potential to discern nuanced underlying associations.

A CNN differs from a traditional neural network by incorporating distinct layers between the input and output that are not entirely interconnected. Within CNNs, the concealed layers comprise numerous convolutional and pooling layers. Following these layers, one or multiple fully connected layers are appended and linked to the output layer. The integration of CNN with Long Short-Term Memory (LSTM) facilitates the classification or characterization of time series signals, such as hand motion. This amalgamation embodies a form of recurrent neural network (RNN) architecture adept at learning from experience to classify, process, and forecast time series data, particularly in scenarios with extended time lags of uncertain magnitude between significant events [7].

Such wearable IMU sensors and the AI applications commonly used in other fields have been reported in several studies identified and listed as follows. The SVM, the MLP and the CNN-LSTM indicate the commonly used machine learning techniques. Accelerometer sensors have been utilized for data collection from smartphones [2] as well as wearable devices [3]. These sensors can also track shifts in user position by integrating acceleration data twice [8]. Additionally, acceleration data can be leveraged to categorize user activities, commonly employing Support Vector Machine (SVM) and Multi-Layer Perceptron (MLP) models [2,4]. Deep learning methodologies, including Deep Convolutional Neural Networks (CNN) [9] and Long Short-Term Memory (LSTM) models [7], have also been employed for acceleration data classification, facilitated by cloud infrastructure [10]. Furthermore, classification processes have been extended to incorporate accelerometer and gyroscope measurements [9], alongside magnetic field data from magnetometers [8]. Electrocardiograms (ECGs) [11] and passive infrared sensors (PIR sensors) [12] have also been investigated for recognizing human activities in medical care settings, encompassing activities such as falling, moving, sleeping, and walking.

2.2 Welding Procedure Specification

A Welding Procedure Specification (WPS) delineates a written procedure detailing the welding process, validated to furnish reference parameters aligning with the standards referenced during its qualification. It serves to furnish welding parameters to welders, ensuring adherence to applicable standards. A comprehensive WPS entails the description of essential variables, non-essential variables, and supplementary essential variables (if necessary) for each welding parameter stipulated within the document.

2.3 QR-Code

Codes printed on a more compact medium. QR codes possess distinct characteristics when juxtaposed with traditional barcodes, notably their capability to store extensive volumes of data. A QR code can encode various types of data including numeric, characters, kanji, hiragana, symbols, and binary, and can present up to 7,089 characters in a single symbol. In contrast, a regular barcode can only store information up to 20 digits. QR codes have error correction capabilities, enable data restoration even in instances where the symbol is tarnished or impaired. They can be read in any direction (360 degrees). The ability of QR codes to be read quickly in various directions (omni-directional) is a result of the limited technological features of linear one-dimensional (1D) barcodes, also known as classic or conventional barcodes. QR codes can store large amounts of data independently and encode various types of data, including numeric, characters, kanji, hiragana, symbols, and binary, up to 7,089 characters in a single symbol, unlike regular barcodes that can only store information up to 20 digits.

Based on the existing literature, the application of motion recognition is largely confined to physical human recognition. Its implementation in the shipbuilding industry, particularly for real-time monitoring of welders' performance, is still very limited. Additionally, the integration of wearable devices with QR codes for real-time monitoring systems has not yet been developed in this field. This paper introduces an innovative real-time monitoring system for welders' performance. It includes performance recognition integrated with Welding Procedure

Specification (WPS) compliance, considering factors such as construction and joint design where welding is performed.

3. Framework Development

The integrated framework for real-time monitoring of the welding process using hand motion sensors and QR codes is illustrated in Figure 1. It begins with the recording of welding parameters through motion sensors placed on the welder's hand. This motion data is then uploaded to a cloud system, where it is analyzed and its characteristics recognized through a deep learning process based on the framework developed by Pribadi & Shinoda [5]. Detailed explanations of how the motion data is captured, stored in the cloud, and processed are thoroughly described in the referenced study.

The motion data stored in the cloud system is then analyzed and compared with the Welding Procedure Specification (WPS). This includes welding parameters such as welding speed, welding position, and the type of welding technique used (manual or semi-automatic). The deviation of welder performance can be examined through the given WPS parameters. For example, if the welder speed is over of under the standard speed ranges, then the welding performance can show the inappropriate performance.

The QR code placed on the hull block construction can identify the type of block or construction, its position, and the work progress. This data is used to determine which block is being welded by the welders. Additionally, the QR code can capture data such as images of the completed weld, which are taken by the welders using an Android mobile phone and then stored in the cloud system. Both the data from the welder's motion sensors and the QR code are processed through web servers and can be monitored on a desktop computer to control performance and track progress.

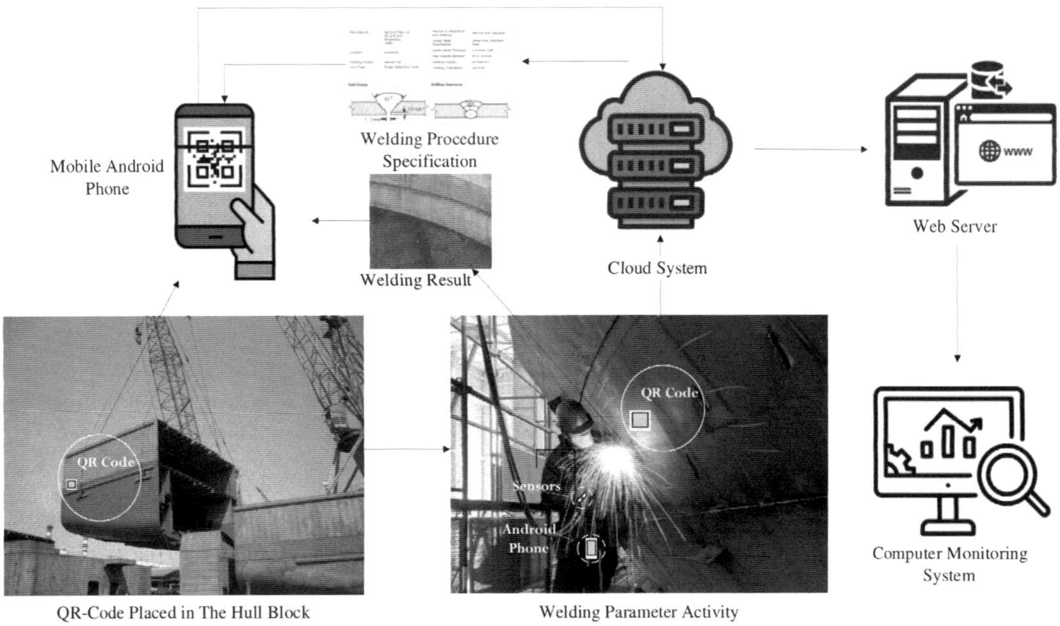

Figure 1. Integration Diagram of Hand Motion Sensor, Artificial Intelligence, and QR Code.

The QR code encoding is performed for each production process and location, utilizing a scheme consisting of 13 alphanumeric digits. The initial 11 digits define the production process stages and number the piece parts for welding joints in shipbuilding. The first 2 digits represent the dock location where the ship is being built. The following 2 digits are abbreviations for the ship production process: SA for sub-assembly, AS for assembly, and ER for erection. The next 2 digits code the section or panel of the block, such as the double bottom. Subsequent digits specify panel numbers, blocks, sub-blocks, and types of welding. The next step involves processing this information to be integrated into the QR code. This data includes the welding map, which encompasses 3D construction, significant welding information, the geo-position of the block (coordinate system on the block), and the Welding Procedure Specification (WPS). This data is intended to form a database that provides detailed welding information when the QR code is scanned. The flowchart diagram depicting the procedural workflow for using QR codes in the welding process is presented in Figure 2.

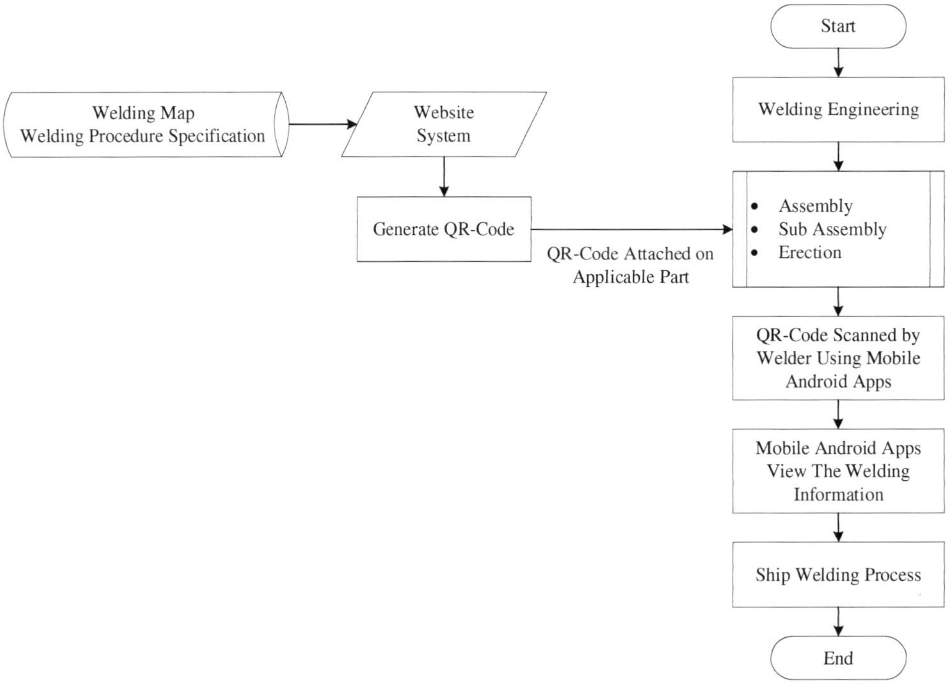

Figure 2. Workflow for Using QR Code in the Welding Process.

4. Case Study Progress Results and Analysis

In this article, the case study is an ongoing process. An example of one of the QR codes is presented in Figure 3, illustrating the detailed monitoring results of the welding process. This includes the WPS number, applicable part, the welding process, and other relevant information, culminating in the QR code generation. These data serve as parameters that welders must adhere to, ensuring that the welding process is executed correctly and in the appropriate location with the proper welding parameters. On the other hand, the analysis of the framework developed based on each component are discussed in the following sub-section.

Welding Detail

Search:

No. WPS	Applicable Part	Thickness	Welding				Throat	Geo-Positions	Certificate	Code	QR Code	
			Process	Type	Material	Joint						
001/WPS/SMAW/3F	Tank Top to Inner Bottom PS 1,PS 2,PS 3, PS 4	8 cm to 10cm	SMAW	Fillet	Ship Plate Grade A to Grade A	Continous Weld	3	PS.Inner Bottom.1 (X= - 0,6 m , Y= 0, Z = 1,2 m) PS.Inner Bottom.2 (X= - 1,2 m , Y= 0 m, Z= 1,2m) PS.Inner Bottom.3 (X= -1,8 m , Y = 0m , Z = 1,2 m) PS.Inner Bottom.4 (X = - 2,4 m , Y = 0 m , Z = 1,2 m)	3F	01.SA.DB.117.01.01	Generate	

Figure 3. Example of Welding Detail monitoring results.

4.1. Hand Motion Sensors, Artificial Intelligence and QR codes

Extensive research has been conducted on the use of wearable sensors for real-time monitoring of shipyard personnel. Preliminary investigations revealed a correlation between delays in ship construction activities and subsequent delays in new shipbuilding projects within Indonesian shipyards. As a result, there emerged a need for monitoring, particularly of fabrication workers, to enhance productivity. This research aimed to explore the effectiveness of integrating accelerometer and gyroscope sensors for fabrication workers. Initial findings indicated that employing this sensor combination, along with Arduino as the Microcontroller Unit (MCU), facilitated the recognition of hand movement gestures exhibited by welders [13].

Subsequently, the research outcomes were applied to the training of novice welders. Wearable sensors were employed to collect data on wrist movements from aspiring welders, and this data was analyzed to assess their proficiency levels in comparison to those of experienced welding instructors. The evaluation of welders' skill levels was conducted using the Support Vector Machine (SVM) method, renowned for its effectiveness in appraising welding trainees and aiding in skill enhancement efforts [5]. This research highlighted a significant weakness: the system could not identify the specific part of the construction and joint design where the welding was performed, the WPS used, the welding layer, or the visual result of the weld. Consequently, further research is required to address these shortcomings. There is ongoing research on utilizing QR code technology to assist welders in remembering and accessing WPS information during the shipbuilding process.

Enhancements from previous research have been made by improving the recognition accuracy of weld quality using the Support Vector Machine method with Deep Learning CNN-LSTM. In this study, an application was designed to integrate wearable sensors and Android smartphones into a cloud- and web-based monitoring system. This system can display activity logs of welders during the ship production process and allows for remote monitoring.

4.2 Integrated of Hand Motion-AI-QR Code

Figure 1 illustrates the concept of combining hand motion sensor technology, artificial intelligence, and QR code technology. QR codes are placed in designated welding areas on the ship, which, when scanned with the designed application, present WPS data and welding location information. Once scanned, the QR code notifies the cloud that welding has begun. During welding, the welder's activity is monitored using a wearable hand motion sensor attached to their wrist and an Android device, which sends real-time data to the cloud. After completing the welding activity, the welder is required to take a photo of the weld using the developed Android application. This photo is then uploaded to the cloud, allowing for a visual inspection of the completed weld.

The integration of hand motion sensor technology, artificial intelligence, and QR code technology presents a viable framework for the real-time monitoring of welders. The sensors attached to the welder's wrist continuously collect hand movement data as long as the sensor is powered. Additionally, the QR code on each part requiring welding, when scanned via a smartphone, indicates that welding activity has started with the specified WPS, including location and joint design information. As the welder performs the welding, the welding time is recorded until the welder takes a photo of the weld. This allows for visual inspection of the weld and analysis of the welder's productivity.

The use of QR codes in the WPS application system is centered around a web-based application where authorized administrators input welding map and WPS data. Initially, data input is performed on the computer system or website, related to welding maps and welding procedure specifications for each section that will undergo welding in the construction of new ships, before being converted into a QR code. The authorized administrator responsible for data input is the welding engineer. After data input, the administrator generates QR codes, which are then attached to each section to be welded.

4.3 Deep Learning Data Collection and Analysis

During welding operations, wrist sensors capture hand movement data, which is analysed using deep learning CNN-LSTM. Data collection involves recording welders' activities using a MetaMotion sensor manufactured by MbientLab Inc., facilitating real-time monitoring of motion and environmental conditions. The collected data is stored on a smartphone and subsequently transferred for processing on a computer. The output from the wearable sensor is structured in CSV format (comma-separated values). Upon data retrieval, the data is organized sequentially by accelerometer, gyroscope, and magnetometer readings, along with details regarding activity identity, session identity, and subject identity.

The analysis then examines the relationship between motion discontinuities and weldment discontinuities by reviewing welding and sensor data graphs. This involves identifying correlations between defects in the welding results and their causes. After completing the welding process, the welder must take a picture of the welding result using a smartphone and upload it to the system. This image serves as a comparison between the system's analysis and the actual field implementation.

5. Conclusions

This study presents the progress of developed framework of integrated hand motion sensors with the QR-code through artificial intelligent using deep learning. It's integration suggests that the framework can be effectively utilized for real-time monitoring of welder activities and weld quality. Incorporating QR codes with hand motion sensors and artificial intelligence advances prior research by enabling the system to not only monitor welder activities and weld quality but also detect compliance with Welding Procedure Specifications (WPS), construction and joint design, productivity, and visual weld quality. This includes identifying defects such as undercut, concavity, porosity, incomplete penetration, incomplete fusion, and over spatter.

The future plan for this research is to apply this framework to demonstrate its effectiveness and the anticipated outcomes. The next phase will involve capturing data from various parts of the construction, different welding parameters, and multiple welders to validate the framework's effectiveness.

Acknowledgements

This work was supported by research, innovation and entrepreneurship grant project Higher Education for Technology and Innovation (HETI) Asian Development Bank Loan No. 4110-INO 2024, product innovation scheme.

References

[1] Baihaqi I, Lazakis I and Kurt R E 2023 Development of a novel integrated value engineering and risk assessment (VENRA) framework for shipyard performance measurement : a case study for an Indonesian shipyard *Ships Offshore Struct.* **XX** 1–16

[2] Bayat A, Pomplun M and Tran D A 2014 A study on human activity recognition using accelerometer data from smartphones *Procedia Comput. Sci.* **34** 450–7

[3] Attal F, Mohammed S, Dedabrishvili M, Chamroukhi F, Oukhellou L and Amirat Y 2015 Physical human activity recognition using wearable sensors *Sensors* **15** 31314–38

[4] Gyllensten I C 2010 *Physical Activity Recognition in Daily Life Using a Traxial Accelerometer* (Skolan för datavetenskap och kommunikation, Kungliga Tekniska högskolan)

[5] Pribadi T W and Shinoda T 2022 Hand motion analysis for recognition of qualified and unqualified welders using 9-DOF IMU Sensors and Support Vector Machine (SVM) approach *Int. J. Technol.* **13** 38–47

[6] Van Groesen W and Pauwels P 2022 Tracking prefabricated assets and compliance using quick response (QR) codes, blockchain and smart contract technology *Autom. Constr.* **141** 104420

[7] Chen Y, Zhong K, Zhang J, Sun Q and Zhao X 2016 LSTM networks for mobile human activity recognition *2016 International conference on artificial intelligence: technologies and applications* (Atlantis Press) pp 50–3

[8] Yun X, Bachmann E R, Moore H and Calusdian J 2007 Self-contained position tracking of human movement using small inertial/magnetic sensor modules *Proceedings 2007 IEEE International Conference on Robotics and Automation* (IEEE) pp 2526–33

[9] Jiang W and Yin Z 2015 Human activity recognition using wearable sensors by deep convolutional neural networks *Proceedings of the 23rd ACM international conference on Multimedia* pp 1307–10

[10] Keshavarzian A, Sharifian S and Seyedin S 2019 Modified deep residual network architecture deployed on serverless framework of IoT platform based on human activity recognition application *Futur. Gener. Comput. Syst.* **101** 14–28

[11] Qiu H, Qiu M and Lu Z 2020 Selective encryption on ECG data in body sensor network based on supervised machine learning *Inf. Fusion* **55** 59–67

[12] Lee W, Lin K-Y, Seto E and Migliaccio G C 2017 Wearable sensors for monitoring on-duty and off-duty worker physiological status and activities in construction *Autom. Constr.* **83** 341–53

[13] Pribadi T W and Shinoda T 2020 Application of Wearable Device for Real Time Monitoring System of Shipyard's Fabrication Workers *In Proceedings of the 3rd International Conference on Marine Technology (SENTA 2018)* (SCITEPRESS–Science and Technology Publications, Lda.) pp 63–72

Deep Learning CNN-LSTM Approach for Identifying Welder's Hand Motion Gestures Using Wearable Sensors

Vialdo Muhammad Virmansyah [1], Triwilaswandio Wuruk Pribadi[2$], Imam Baihaqi[2#]

[1] Student of Enggineering Magister programme of Department of Naval Architecture Faculty of Marine Technology, Sepuluh Nopember Institute of Technology, Surabaya 60111, Indonesia
[2] Department of Naval Architecture, Faculty of Marine Technology, Sepuluh Nopember Institute of Technology, Surabaya 60111, Indonesia

Email: vialdomvirmansyah@gmail.com*, triwilas.its@gmail.com$, imam_b@na.its.ac.id#

Abstract. A welder plays a crucial role in the construction of a new ship. Monitoring the performance of welders involved in the construction of a ship is particularly important for controlling dimensions. he competency and consistency of welders play a vital role, with specific hand movement patterns contributing to welding quality. This study aims to identify individual welders' hand movement patterns using Deep Learning with CNN-LSTM configurations. Data was collected through IMU motion sensors attached to welders' wrists, capturing acceleration, angular speed, magnetic force, and electric current changes. The data was classified using CNN-LSTM, which showed higher accuracy than SVM methods. Experimental results indicated an overall classification accuracy of 99.73% for identifying individual welders and 97.07% for determining welding positions.

Keywords: Wearable Device, Deep Learning, Convolutional Neural Network-Long Short Term Memory

1. Introduction

In ship construction, welding is a crucial process. The welding process in ship construction must be carefully considered to ensure the strength and quality of the welded joints. This involves the use of welding procedures and techniques, the right materials, the right welding methods in accordance with the Welding Procedure Description (WPS), which has been qualified, close monitoring during the process, and inspection after the welding process. The welder's competency and consistency during the welding process must be maintained. A welder has a certain pattern of hand movements that forms a pattern that produces the same welding quality.

The welder's performance and skills were successfully identified using the Support Vector Machine (SVM) method, where the welder's hand movements were monitored using an IMU motion sensor on the wrist. Using this method, typical welder hand movements can be classified. The IMU installed on the wrist can monitor hand movements in terms of acceleration and angular speed of hand movement (in the 3 directions of the X,Y,Z axes), magnetic force, and changes in electric current due to the proximity of the welding metal electrode (in the 3 directions of the X,Y axes). ,Z). Meanwhile, the SVM can be used to classify the welder's hand movement patterns based on data provided by the IMU. The SVM will be trained using previously collected data containing the welder's hand movements that have been correctly classified manually. Currently, the SVM method can only recognize the quality of the welding activities of qualified and unqualified welders [1]. Meanwhile, the Deep Learning method with an input layer configuration, three hidden layers and one hidden layer with each layer consisting of 100 nodes and an activation function using sigmoid can recognize two levels of qualifications (qualified and unqualified) with higher accuracy than the SVM method [2]. Currently, human activity recognition such as walking, sitting, and standing can be identified with higher accuracy using the CNN-LSTM method. This model can automatically extract features and classify them into several parameters [3]. Pribadi and Shinoda [2] have developed a real-time monitoring system by integrating an Inertial Measurement Unit (IMU) motion sensor mounted on the wrist (as a wearable device) with Internet of Things (IoT) infrastructure and Artificial Intelligence (AI) technology. This system utilizes the Support Vector Machine (SVM) method and Deep Learning CNN-LSTM to assess the activities and performance of welders, enabling the classification of each welder's skill level. However, its application in the welding process has not yet been able to recognize individual welders through their hand movement patterns.

The aim of this research is to recognize the hand movement patterns of individual welders wearing motion sensors, enabling the identification of welders performing welding activities using the Deep Learning method CNN-LSTM.

2. Methods

There are several steps in the experimental data collection. The experimental welding data collection involves welding activity data for FCAW in positions 1G, 2G, and 3G as depicted in Figure 2. Among these steps are attaching sensors to the welder's body, data acquisition, preprocessing, and classification of the data.

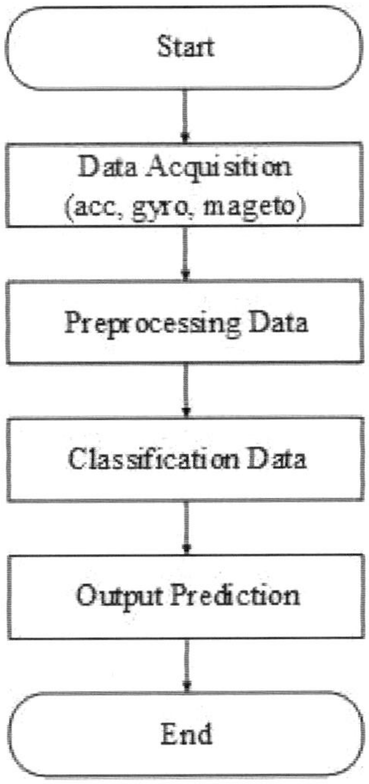

Figure 1. Methods of research

IOP Conf. Series: Earth and Environmental Science **1423** (2024) 012036 doi:10.1088/1755-1315/1423/1/012036

(a) (b) (c)

Figure 2. (a) 1G Weld Groove, (b) 2G Weld Groove, (c) 3G Weld Groove

2.1. Welding Activity

In the manufacturing industry, welding processes can be classified into two main categories. The first category includes processes in which the weld metal is melted by passing through an electric arc. This category encompasses shielded metal arc welding (SMAW), gas metal arc welding (GMAW), flux cored arc welding (FCAW), and submerged arc welding (SAW), all of which are commonly utilized in shipbuilding in developing countries. The second category involves welding techniques in wich the weld metal does not pass through an arc. In these methods, an electric arc is used to heat the filler material and base metal, with the welder using both hands. This category includes gas tungsten arc welding (GTAW), plasma arc welding, and plasma cutting.

With the progression of industrial technology, thicker plates are increasingly used in massive welded structural components. Owing to the cross-sectional area with V-shaped grooves, the conventional welding technique employed for welding plates encounters numerous challenges, such as diminished production efficiency, excessive utilization of welding materials, and subpar welding quality stability, etc [4]. However, for welding with grooves that possess narrow gaps, FCAW and GMAW welding apparatuses have emerged as options for welding gear with heightened production efficiency [5], Two characters signify the position and type of welding being executed: 1 symbolizes the downhand position, 2 symbolizes the horizontal position, 3 symbolizes the vertical position, and G represents the groove welding type.

2.2. Determine sensor position

Human movement refers to changes in position or posture relative to one's environment. This definition has evolved to acknowledge that movement is not uniform; human movement is highly diverse and necessitates extensive analysis and precise identification. Furthermore, the concept of human movement is not solely confined to the physical realm. Instead, it requires a comprehensive "context" or "form of life" encompassing sociological, environmental, psychological, mechanical, physiological, and anatomical factors. These elements must be integrated into the overall understanding of movement. The physical dimension of life, including health, physicality, energy, and vitality, significantly contributes to quality of life. [6].

The placement of sensors is a crucial factor in supporting the recording of user movement data. Therefore, when determining sensor placement, it's important to observe and record the user's body movements while they are performing their tasks. The research began with a simulation of welding movements performed by a welder, in accordance with the Scope of Work for Welders. The study results showed that the most frequently moving part of the body is the forearm.

2.3. Data acquisition

Data acquisition begins by preparing for the collecting of welding activity data conducted experimentally in the laboratory. Data acquisition comprises the welding process itself, followed by slag cleaning and grinding. For the experimental base metal specimens, carbon steel material (A36) was used. These specimens consist of two iron plates with a V groove shape, with a bottom size of 3 mm and a top size of approximately 32 mm. The dimensions of each specimen are 150x300x12mm. Several welders were assigned to weld three specimens each, corresponding to welding positions 1G, 2G, and 3G. The welding process was performed using a Weiro WM350F FCAW welding machine, and the electrodes used were FLUX ESAB Weld 71T-1 with a diameter of 1.2 mm. This experiment was conducted at the Ship Production Technology and Management Laboratory, Sepuluh Nopember Institute of Technology. The welding results were visually inspected by a certified welding inspector. Following this, the data are segmented into multiple data windows, each spanning 10 seconds, denoted as subsets. Each welder is then assigned a distinct name tag for identification.

2.4. Pre-procesing data

Second, the raw data are sorted into 3 sensors, each with 3 axes (x, y, and z). The data is identified based on the welding activity performed, the welding position, and the welder identity. Subsequently, these data will be imported into a database to enable the classification process with deep learning. Segmentation is performed when deep learning is executed, with a sensor output frequency of 25 Hz, meaning that every second the sensor will produce 25 data points.

The raw sensor output data that will undergo classification by deep learning will first be processed by assigning identifiers for welding activities, welder identities, and welding position identities. This is aimed at enabling deep learning to recognize the data. Activity identifiers are labeled as "actid," welder identifiers as "subid," and welding position identifiers as "posid."

Table 1. Pre-Processed Data

Elapsed (s)	Accel x (g)	Accel y (g)	Accel z (g)	Gyro x (deg/s)	Gyro y (deg/s)	Gyro z (deg/s)	Magnt x (T)	Magnt y (T)	Magnt z (T)	Actid	Subid	Posid
742	0,241	0,5	0,79	0,488	-0,122	-0,122	-5,84E-05	-2,66E-05	1,34E-04	Weld	Imam	1G
742,04	0,249	0,505	0,788	0,366	-0,122	-0,183	-6,39E-05	-3,10E-05	1,44E-04	Weld	Imam	1G
742,08	0,254	0,505	0,787	0,61	0,915	-0,061	-3,40E-05	-3,06E-05	1,32E-04	Weld	Imam	1G
742,12	0,237	0,495	0,794	-0,061	0,793	-0,244	-2,48E-05	-3,94E-05	1,25E-04	Weld	Imam	1G
742,16	0,25	0,505	0,792	0,305	0,854	0	-4,29E-05	-5,09E-05	1,29E-04	Weld	Imam	1G
742,2	0,242	0,505	0,787	0,61	1,585	0,549	-7,16E-05	-6,49E-05	1,39E-04	Weld	Imam	1G
742,24	0,247	0,503	0,792	-0,061	1,524	0,366	-6,24E-05	-4,91E-05	1,32E-04	Weld	Imam	1G
742,28	0,242	0,5	0,79	0,122	2,134	0,549	-6,72E-05	-4,35E-05	1,18E-04	Weld	Imam	1G
742,32	0,239	0,499	0,789	0,305	1,524	0,305	-7,35E-05	-4,43E-05	1,33E-04	Weld	Imam	1G
742,36	0,24	0,501	0,784	-0,183	2,012	0,061	-7,20E-05	-3,73E-05	1,42E-04	Weld	Imam	1G

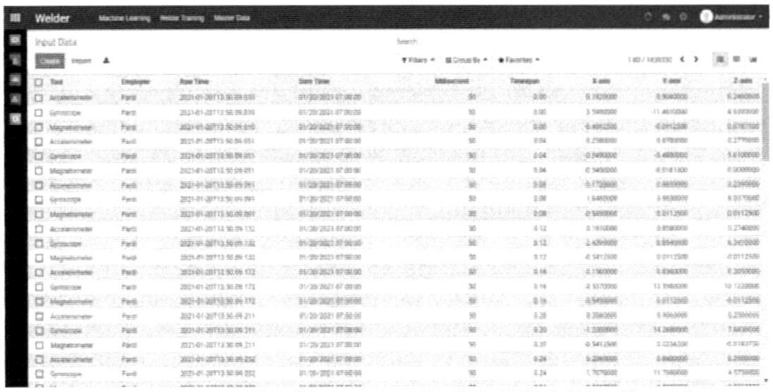

Figure 3. An example of Data Input for DL CNN-LSTM Algorithm

After labeling the data, it is used as a sample to recognize the hand movements of welders. The data will be imported into the web database, and once the data is in the database, a learning process using deep learning can be initiated.

2.5. Classification of Data Using Deep Learning

Deep Learning is a part of artificial intelligence and machine learning thats involves creating multi-layered neural networks to achieve high precision in tasks like object detection, speech recognition, and language translation. Unlike traditional machine learning techniques, deep learning automatically handles data like images, videos, or text without requiring pre-established code of feature extraction rules. [7].

Deep learning comprises a collection of algorithms within machine learning that grasp various levels of abstraction. It typically employs artificial neural networks, where the layers within the statistical model correspond to distinct levels of concepts. Advanced concepts stem from simpler ones, and these foundational concept aid in defining numerous higher level concepts [8].

Human activity recognition involves delineating, evaluating, and comprehending the actions and objectives of one or more individuals or agents [9]. IMU, human activity recognition has emerged as a method for monitoring both human and device activities, representing a significant area of inquiry [10]. The integration of deep learning into human activity recognition has streamlined this task, marking a positive trend in machine learning recently [11]. The prominence of CNN and RNN architectures has increased alongside the rise of deep learning, especially in leveraging deep learning models to train time-series data from inertial sensors, an area still under active investigation [12][13]. Data used in human activity recognition usually manifest as signal graphs based on time sequences, a method supported by previous research employing deep learning for signal processing [14][15].

The used Deep Learning model is CNN-LSTM. CNNs bear a resemblance to classical neural networks but possess varying layers between the input (sensor data) and output that are not entirely interconnected. The hidden layers within CNNs consist of diverse combinations of convolutional layers and pooling layers. Following the organization of these layers, the subsequent step involves linking one or more fully connected layers to generate the output layer.

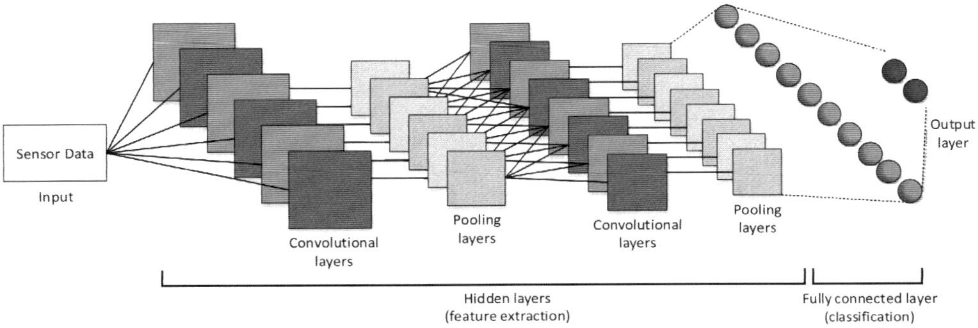

Figure 4. CNN with convolutional layers and pooling layers

The convolutional layer comprises a set of filters linked to local zones within the sensor data. These filters automatically extract features from the input (sensor data), and their size is reduced by the pooling layer. By incorporating more layers, the network can glean additional information about features, dissecting sensor data into smaller interconnected parts (features) to maximize information for improved classification levels. Through iterations and backpropagation, the filter structure undergoes modifications to optimize the network's performance. For the classification or characterization of time series signals like hand movements, CNNs have been paired with LSTM.

LSTM, a type of Recurrent Neural Network (RNN) architecture, is adept at learning from experience to classify, process, and forecast time series data, particularly when there are lengthy time intervals of unknown sizes between significant events. Utilization of accelerometer sensors entails leveraging data from smartphones obtained via wearable devices. Algorithms like Deep Convolutional Neural Networks can classify accelerometer data, and with the advancement of cloud infrastructure, alternative approaches to feature extraction processing employing LSTM models and RNN models can be adopted. The classification process may also involve integrating accelerometer data with gyroscope data and magnetic field data measured by a magnetometer.

Third, the data classification process uses a deep learning model consisting of CNN and LSTM, which focuses on a data-driven approach used to sequentially learn the characteristics of different sensor data [9]. By implementing the CNN-LSTM method for human activity recognition, accuracy can be enhanced. This is because CNN's ability to exploit network robustness in feature extraction is combined with LSTM's capability for forecasting and classification. When unified, the CNN-LSTM model exhibits both spatial and temporal depth, surpassing other deep learning algorithms that utilize signal data as input [10]. Deep Learning with a CNN-LSTM configuration is used to process signal data from welding activities. The experimental signal data uploaded to the database is executed with customized Python code to identify individual welders. To achieve better results, it's essential to adjust several parameters within deep learning, including the number of hidden layers used, number of nodes, epochs, activation functions, optimizers, and loss functions. Adjusting these parameters effectively can greatly impact the performance and accuracy of the model. After parameter adjustment, the classification learning process is initiated.

3. Result

3.1. The results of data analysis of the welder's hand movements

The researcher conducted data analysis using CNN-LSTM to differentiate welder personnel in FCAW welding. The data used in the deep learning classification process comprised welding activity data collected by three welder personnel. The parameters employed were as follows: four hidden layers, each with 100 hidden nodes, 500 epochs, a window data size of 250, a ReLU activation function, a adam optimizer, and mean squared error loss function.

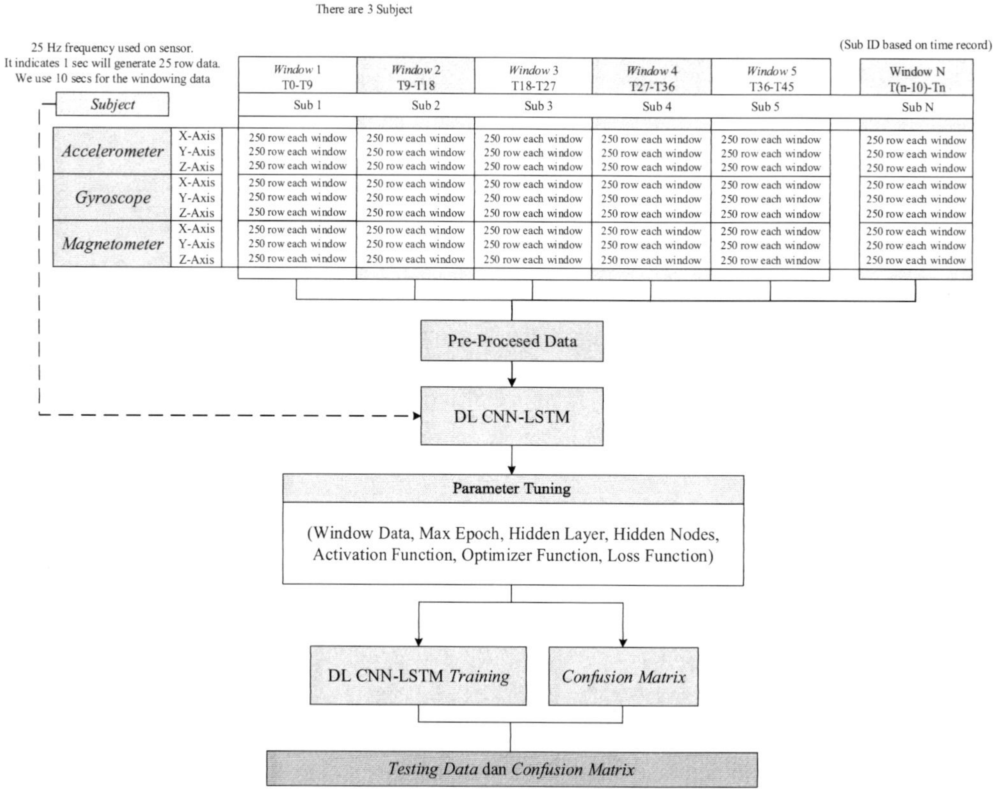

Figure 5. Flowchart of the Deep Learning Process

Figure 6. Configuring the Training Parameters

Table 2. Deep Learning Learning Results

Epoch	Data Length	Accuracy
500	373	99,73%

Table 3. Confusion Matrix (Include)

	9	10	11
9	25 (44,64%)	0 (0,00%)	0 (0,00%)
10	0 (0,00%)	13 (23,21%)	0 (0,00%)
11	0 (0,00%)	0 (0,00%)	18 (32,14%)

Data Length: 173
True Positive: 97,69%

Table 4. Confusion Matrix (Exclude)

	9	10	11
9	14 (21,21%)	4 (6,06%)	2 (3,03%)
10	7 (10,61%)	15 (22,73%)	2 (3,03%)
11	9 (13,64%)	3 (4,55%)	36 (15,15%)

Data Length: 66
True Positive: 59,09%

In the deep learning algorithm, a total of 373 data points were processed. In the table, the numbers highlighted in blue on the left represent the True Class, indicating the actual identities of the welders identified by the researcher during welding. Conversely, numbers highlighted in orange represent the predicted class, indicating the welder identities predicted by the deep learning model. The overall accuracy of recognizing welder hand movements from the total data above is 99.73%.

For the included test data, consisting of 173 data points, the total accuracy result is 97.69%. These results are derived from the combined percentage predictions of welder personnel. Code 9 represents the first welder personnel, with 25 data points identified, constituting 44.64% of the total 173 included test data. Code 10 corresponds to the second welder personnel, with 13 data points identified, comprising 23.21% of the total 173 included test data. Code 11 represents the third welder personnel, with 36 data points identified, accounting for 32.14% of the total 173 included test data.

For the excluded test data, which included 66 data points, the total accuracy result was 59.09%. These results are derived from the combined percentage predictions of the welder personnel. Code 9 corresponds to the 1st welder personnel, with 14 data points identified, constituting 21.21% of the total 66 excluded test data. Code 10 represents the second welder personnel, with 15 data points identified, comprising 22.73% of the total 66 excluded test data. Code 11 represents the third welder personnel, with 36 data points identified, accounting for 15.15% of the total 66 excluded test data.

The researcher also conducted data analysis on the welding position of each welder. The data codes are as follows: code 10 represents 1G welder 1 data, code 11 represents 1G welder 2 data, code 12 represents 1G welder 3 data, code 13 represents 2G welder 1 data, code 14 represents 2G welder 2 data, code 15 represents 2G welder 3 data, code 16 represents 3G welder 1 data, code 17 represents 3G welder 2 data, and code 18 represents 3G welder 3 data. The model results are presented in Table 5.

Table 5. Deep Learning Results

Epoch	Data Length	Accuracy
500	307	97,07%

The information for reading the table above is the same as that in the previous table. The results of the algorithm model as a whole obtained an accuracy of 97.07%, the results of the Include test were 93.55%, and the results of the exclude test were 27.27%

The results obtained in Tables 2, 3, and 4 are algorithm models that can be used to predict the welder, while in tables 5, 6, 7 are the algorithm models used to predict the welding position performed by each welder. This was done to prove whether the model could predict the welder and the welding position. the researchers used pre-existing welder data. Some prediction results for the welder are shown in the following summary Table 8.

Table 6. Confusion Matrix (Include)

	10	11	12	13	14	15	16	17	18
10	7	0	0	0	0	0	0	0	0
11	0	8	0	0	0	0	0	0	0
12	0	0	11	0	0	0	0	0	0
13	0	0	0	9	0	0	0	0	0
14	0	0	0	0	7	0	0	0	0
15	0	0	0	0	0	10	0	0	0
16	0	0	0	0	0	0	11	0	0
17	0	0	0	0	0	0	0	15	0
18	0	0	0	0	0	0	0	0	9

Data Length: 93
True Positive: 93,55%

Table 7. Confusion Matrix (Exclude)

	10	11	12	13	14	15	16	17	18
10	6	2	0	1	0	0	0	3	1
11	4	3	1	0	0	1	4	0	1
12	2	1	4	1	3	0	3	3	4
13	6	1	3	2	1	1	1	0	1
14	1	0	0	0	6	2	0	4	1
15	1	0	0	1	1	9	2	2	0
16	3	1	1	2	1	1	3	1	0
17	0	0	4	1	1	1	0	3	1
18	1	3	2	1	2	0	1	4	0

Data Length: 132
True Positive: 27,27%

Table 8. Prediction Test for each Welder

Prediction Test Using the LSTM Model 99,73%	
Test Subject	Percentage (%)
Welder 1	58,22
Welder 2	63,01
Welder 3	67,81

Table 9. Prediction Test on each Position of the Welder

Prediction Test Using the LSTM Model 97,07%		
Test Subject	Predicted Subject	Percentage (%)
1G, Welder 1	1G, Welder 1	85,42
1G, Welder 2	1G, Welder 2	47,92
1G, Welder 3	1G, Welder 3	37,50
2G, Welder 1	2G, Welder 1	37,50
2G, Welder 2	2G, Welder 2	43,75
2G, Welder 3	2G, Welder 3	77,08

3G, Welder 1	3G, Welder 1	64,58
3G, Welder 2	3G, Welder 2	52,08
3G, Welder 3	3G, Welder 3	31,25

4. Conclusion

Hand movements performed by welders are categorized into two groups: classification of each welder and classification of the welding positions of each welder. The classification model algorithm achieved an accuracy of 99.73% for welder classification and 97.07% for the classification of positions among welders. This model serves as a tool for assessing welder data and welding positions using training data. However, the prediction results for welders are relatively low, averaging at 63.01%. Particularly, the prediction results for welder positions in cases 1G-Welder 3, 2G-Welder 1, and 3G-Welder 3 exhibit very low percentages, indicating the model's inability to accurately predict positions among welders. Further research is warranted to explore variations in the number of window data, nodes, layers, activation functions, optimization functions, and loss functions to enhance prediction outcomes.

References

[1] Pribadi, T. W., & Shinoda, T. (2022). Analysis of Wrist Hand Motion for Monitoring of Basic Welder Training using Wearable Sensors. IOP Conference Series: Earth and Environmental Science, 972(1).

[2] Pribadi, Triwilaswandio Wuruk, & Shinoda, T. (2022b). Hand Motion Analysis for Recognition of Qualified and Unqualified Welders using 9-DOF IMU Sensors and Support Vector Machine (SVM) Approach. International Journal of Technology, 13(1), 38–47.

[3] K. Xia, J. Huang and H. Wang, "LSTM-CNN Architecture for Human Activity Recognition," in IEEE Access, vol. 8, pp. 56855-56866, 2020

[4] Ramakrishnan, M., & Muthupandi, V. (2013). Application of submerged arc welding technology with cold wire addition for drum shell long seam butt welds of pressure vessel components. International Journal of Advanced Manufacturing Technology, 65(5–8), 945–956.

[5] Wang, J., Sun, Q., Feng, J., Wang, S., & Zhao, H. (2017). Characteristics of welding and arc pressure in TIG narrow gap welding using novel magnetic arc oscillation. International Journal of Advanced Manufacturing Technology, 90(1–4), 413–420.

[6] Bartlett, R.M. (2007) Introduction to Sports Biomechanics: Analyzing Human Movement Patterns, 2nd edn, Abingdon, UK: Routledge.

[7] Pumsirirat. (2018). Credit Card Fraud Detection using Deep Learning based on Auto-Encoder and Restricted Boltzmann Machine. International Journal of Advanced ComputerScience and Applications (IJACSA), 9(1), 1825–1844.

[8] Deng. (2014). Deep Learning: Methods and Applications. Foundations and Trends in Signal Processing, 7(3), 197–387

[9] Shiranthika, C., Premakumara, N., Chiu, H. L., Samani, H., Shyalika, C., & Yang, C. Y. (2020). Human Activity Recognition Using CNN & LSTM. Proceedings of ICITR 2020 - 5th International Conference on Information Technology Research: Towards the New Digital Enlightenment.

[10] Mutegeki, R., & Han, D. S. (2020). A CNN-LSTM Approach to Human Activity Recognition. 2020 International Conference on Artificial Intelligence in Information and Communication, ICAIIC 2020, 362–366.

[11] Hammerla, N. Y., Halloran, S., & Ploetz, T. (2016). Deep, Convolutional, and Recurrent Models for Human Activity Recognition using Wearables. IJCAI International Joint Conference on Artificial Intelligence, 2016-Janua, 1533–1540.

[12] Kwapisz, J. R., Weiss, G. M., & Moore, S. A. (2011). Activity recognition using cell phone accelerometers. ACM SIGKDD Explorations Newsletter, 12(2), 74–82.

[13] Oniga, S., & Suto, J. (2014). Human activity recognition using neural networks. Proceedings of the 2014 15th International Carpathian Control Conference (ICCC), 403–406.

[14] Shevitski, B. S. (2020). Digital Signal Processing Using Deep Neural Networks. Unknown, 1–21.

[15] Yu, D., & Deng, L. (2011). Deep Learning and Its Applications to Signal and Information Processing [Exploratory DSP. IEEE Signal Processing Magazine, 28(1), 145–154.

IOP Conf. Series: Earth and Environmental Science **1423** (2024) 012037 doi:10.1088/1755-1315/1423/1/012037

Critically analysis for fuel oil storage facilities using FMECA method

B Wijanarto[1*], M B Zaman[1], B Cahyono[1]

[1]Department of Marine Engineering, Institut Teknologi Sepuluh Nopember (ITS), Surabaya, 60111, Indonesia

*Email : wijanarto.bagus@gmail.com

Abstract. Fuel oil storage facilities, such as the PT.X Fuel Terminal, have an important role in maintaining a stable and safe energy supply for the community. Smooth operations and reliability of equipment in the storage facility are crucial factors in ensuring an adequate and safe fuel supply. In carrying out its operations, the landfill facility is faced with various challenges. One of the main challenges is maintaining complex and diverse equipment. Equipment such as storage tanks, connecting pipes, pumps and other equipment must be maintained regularly to prevent failures that could cause operational disruptions and potential safety risks. However, traditional maintenance approaches based on routine schedules are often inefficient and less responsive to the actual condition of the equipment. In this research, criticality analysis at the fuel oil terminal was carried out. This aims to carry out condition monitoring. By analyzing criticality, it will be easier to make decisions on the next steps. In this research, the FMECA method was used. This method is very good for finding criticality values in the system.

1. Introduction

Industrial fuel storage tanks, also known as petroleum tanks, can store a variety of liquids. Generally used to store non-organic and organic liquids. They can also contain vapors as well as various flammable liquids. This tank needs to be maintained as well as possible. Maintenance strategies need to be taken so that they can be implemented well. In storage tanks, there are several critical components that must be watched out for. Condition monitoring needs to be done to make maintenance easier[1][2].

An oil storage tank is a reservoir or container that temporarily holds oil during various stages of processing into various types of oil products, or before it is consumed or used. The materials and structures of industrial oil storage tanks are based on their standards as well as other safety, environmental and legal requirements of the storage site.

Failure Mode Effect and Criticality Analysis (FMECA) is used as a methodology to identify and analyze all potential failure modes of various parts of the system, the effects of such failures on the system, how to avoid failure and/or reduce the impact of failure on the system[3]. FMECA was originally developed by the National Aeronautics and space Administration (NASA) which aims to improve and verify the reliability of Hardware programs MIL-STD-785 space entitled Reliability Program for Systems and Equipment Development and Production reviews procedures for performing FMECA on equipment or or system. MIL-STD-1629 is a military standard that specifies requirements and procedures for conducting FMECA, to evaluate and document impacts the potential of any functional or hardware failure on mission success, security personnel and systems, system maintenance and performance.

This study aims to conduct a critically analysis for fuel oil storage tank using the FMECA method. Traditional FMECA has proven to be very good which results in a level of risk by using RPN. The results of this study are expected to be used to provide input to stakeholders in the Pertamina Company to be give priority to the safety[4][5].

2. Research Area

The research location in this paper is at PT.X's fuel oil storage facilities. Data was taken in several areas in Indonesia. Several components of this tank facility need to be maintained properly. The critical point is the basis for carrying out the maintenance.

Figure 1. Fuel Oil Storage Facilities

Figure 1 shows the fuel oil storage facilities owned by PT.X, Indonesia. These storage facilities have an important role in storing fuel for distribution for industrial purposes, transportation, and public facilities. PT.X's Fuel Terminal is a fuel oil storage facility that has high operational complexity. To ensure the reliability and availability of fuel supply, Fuel Terminal needs to implement an effective and efficient maintenance strategy. However, the use of condition monitoring and predictive systems at the PTG.X Fuel Terminal is still limited. Therefore, this research aims to conduct a study at the Fuel Terminal by designing a condition monitoring and predictive system that can support the implementation of RBM and increase operational reliability.

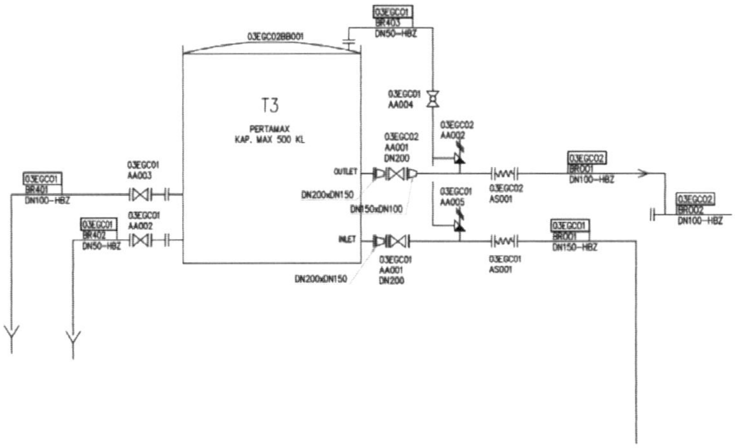

Figure 2. P&ID Fuel Oil Storage Facilities

Figure 2 shows the P&ID storage tank with a tank capacity of 500 kl which is used to accommodate Pertamax products. In the tank there is several equipment and connected pipe lines. The following is the equipment contained in the tank:

1. Slot Dipping Device (SDD)
2. Breather Valve
3. Automatic Tank Gauge (ATG)

The pipe lines connected to the tank are as follows:

1. Inline pipe with size 8"
2. Out pipe measuring 8”
3. 4" and 2" drain lines
4. Pressure Relief Valve with size 2”

Fuel Terminal is designed to handle various types of fuel oil, such as gasoline, diesel, kerosene and aviation fuel, as well as providing safe storage facilities that comply with safety standards. Fuel Terminal has complex infrastructure, including storage tanks, connecting pipes, pumps, measuring equipment and control systems. The condition and reliability of this equipment is very important to maintain smooth operations and prevent potential failures that could cause negative impacts, including stopping the fuel supply.

3. Method

FMECA (Failure Mode and Effects Analysis) is an engineering technique used to identify the potential failure modes, prioritize, and reduce problems from a system, design, or process. The failure mode is defined as the manner in which a component, system, subsystem, or process could potentially fail to meet the design intent[6][7].

FMECA method established the risk priorities of failure modes through the risk priority number (RPN), which is the product of the occurrence (O), severity (S), and detection (D) of a failure [11]. The formula of RPN is below:

$$RPN = O \times S \times D \tag{1}$$

Occurrence is define how often the possible causes of failure occur. This occurrence value is given for each cause of failure. Severity is an assessment related to how likely the occurrence of impacts arising from failures or defects that occur. Detection is an assessment of detection, how far the cause of failure can be detected. Consists of a rating of 1-10. The estimation by bexpert using a scale from 1 to 10 could be shown in Tables 2-4 [8][9][10].
The Failure Modes Effects and Criticality Analysis (FMECA) procedure can generally be done includes several systematic steps including:

- Defining the system, which includes identifying internal functions and interfaces, the performance of which expected in varying degrees of complexity, system limitations and definitions failure.
- Perform functional analysis, which illustrates interrelated operational activities, and functional entity dependencies.Identify all potential failure mode for the item.
- Identify failure modes and their impacts, all potential failure modes of items and interfaces are identified and their impact on direct functions, items and systems should be clearly defined..
- Determines the severity rating (S) of the failure mode, which refers to how serious it is impact or effect of failure mode.Determine the appropriate recomendation based on analysis of risk.
- Determine the occurrence rating (O) from the frequency of failure modes and analysis criticality of failure mode. Assuming that the system components tend to be failure in various ways, this information is used to illustrate the most critical aspect of system design.
- Determine the Detection rating (D) from the design control criteria for failure mode occurrence.
- Risk Priority Number (RPN) is the result of multiplying the weights of Severity, Occurance and Detection. These results will be able to determine critical components.

After going through all items for each failure, a rating of severity, occurance and detection are assigned. The failure modes with higher RPN are assumed to be more important and will be given higher priorities for correction[11][12].

Table 1. Occurrence of a failure

Rating	Probability of occurrence	Failure probability
9–10	Very High: failure is almost inevitable	> 1 in 2 1 in 3
7–8	High: repeated failures	1 in 8 1 in 20
4–6	Moderate: occasional failure	1 in 80 1 in 400 1 in 2000

2–3	Low: relatively few failures	1 in 15,000 1 in 150,000
1	Remote: failure is unlikely	< 1 in 1,500,000

Table 2. Severity of a failure

Rating	Effect	Severity of effect
9–10	Very High	very serious effect on system
7–8	High	serious effect on system
4–6	Moderate	moderately serious effect on system
2–3	Low	little serious effect on system
1	Remote	no serious effect on system

Table 3. Detection of a failure

Rating	Detection	Likelihood of detection
9–10	Remote	Remote of detect for potential cause failure mode
7–8	Low	Low of detect for potential failure mode
5–6	Moderate	Moderate of detect for potential failure mode
3–4	High	High of detect for potential failure mode
1–2	Very High	Very high of detect for potential failure mode

IOP Conf. Series: Earth and Environmental Science **1423** (2024) 012037 doi:10.1088/1755-1315/1423/1/012037

4. Result and Discussion

By using the FMECA method, the critically analysis for fuel oil storage facilities has been carried out. By determining the RPN value, the risk level was determined. Information about the level of risk was needed. It could be used to mitigate and provide input to stakeholders to improve the safety of fuel oil storage facilities. The RPN value is seen in Table 4.

Table 4. Risk level based on FMECA

Scenario	Process	Occurance	Severity	Detection	RPN	Rank
1	Pressure Relief Device	7	8	4	224	5
2	Distribution Piping System	9	8	5	360	3
3	Single Point Monitoring	5	6	6	180	7
4	Filling Shet System	4	6	5	120	8
5	Pump	9	9	8	648	1
6	Loading Unloading Piping System	8	9	7	504	2
7	Jetty	7	7	4	196	6
8	Storage Tank	8	7	5	280	4
9	Fire Fighting System	4	5	5	100	9
10	Proving and Metering System	3	4	5	60	10

Table 4 shows the results of the RPN. From the RPN calculation results, the criticality value for each component in the fuel oil storage terminal is obtained. There are 10 components taken, then the RPN value is calculated. From these results, it was found that the most critical value was the pump. Then the loading unloading piping system, then the Distribution Piping System is in 3rd place and the Proving and Metering System occupies the lowest rank with the lowest RPN value. With this RPN value, the component criticality value can be a basic reference for developing a good maintenance strategy. Maintenance strategies are the key to maintaining reliability, component life and maintaining system operational stability.

5. Conclusions

Critically analysis for fuel oil storage terminal has been carried out using the FMECA method. Failure mode and effects critically analysis (FMECA) is a technique used to improve a system that has been proven to improve safety. FMECA is a team-based, systematic and proactive technique that is used to prevent problems from processes or products before they occur. FMECA could provide an overview not only of what problems might occur but also of the severity of the consequences.

Based on results, pump has the highest RPN value. There are 10 components taken, then the RPN value is calculated. From these results, it was found that the most critical value was the pump. Then the loading unloading piping system, then the Distribution Piping System is in 3rd place and the Proving and Metering System occupies the lowest rank with the lowest RPN value. With this RPN value, the component criticality value can be a basic reference for developing a good maintenance strategy.

Acknowledgments

The authors thank to the team at marine engineering department Institut Teknologi Sepuluh Nopember (ITS) who have discussed and supported this research.

References

[1] Scheu, M.N.; Tremps, L.; Smolka, U.; Kolios, A.; Brennan, F. A systematic Failure Mode Effects and Criticality Analysis for offshore wind turbine systems towards integrated condition based maintenance strategies. Ocean. Eng. 2019, 176, 118–133.

[2] Sayareh J, Ahouei V R 2013 Failure Mode and Effects Analysis (FMEA) for reducing the delays of cargo handling operations in marine bulk terminal. J.Mar. Research.10 (2) 43-50

[3] Ookalkar A D, Joshi A G and Ookalkar D S 2009 Quality improvement in Haemodialysis Process Using FMEA. Int.J. Quality & Rel.Man 26 (8) 817-830

[4] Kai, M.T.; Lim, C.P. Fuzzy FMEA with a guided rules reduction system for prioritization of failures. Int. J. Qual. Reliab. Manag. 2006, 23, 1047–1066

[5] W Y Kwok, Alan H S and Duffy V G Using AHP for determining priority in a safety management system J. Emerald Group publishing Limited 14 430-45

[6] Liu, H.C.; You, J.X.; Duan, C.Y. An integrated approach for failure mode and effect analysis under interval-valued intuitionistic fuzzy environment. Int. J. Prod. Econ. 2017, 207, 163–172.

[7] Aswin, K.R.; Renjith, V.R.; Akshay, K.R. FMECA using fuzzy logic and grey theory: A comparitve case study applied to ammonia storage facility. Int. J. Syst. Assur. Eng. 2022, 13, 2084–2103

[8] Gupta, G.; Mishra, R.P. A Failure Mode Effect and Criticality Analysis of Conventional Milling Machine Using Fuzzy Logic: Case Study of RCM. Qual. Reliab. Eng. Int. 2017, 33, 347–356.

[9] Carpitella, S.; Certa, A.; Izquierdo, J.; La Fata, C.M. A combined multi-criteria approach to support FMECA analyses: A real-world case. Reliab. Eng. Syst. Safe 2018, 169, 394–402

[10] Certa, A.; Hopps, F.; Inghilleri, R.; La Fata, C.M. A Dempster-Shafer Theory-based approach to the Failure Mode, Effects and Criticality Analysis (FMECA) under epistemic uncertainty: Application to the propulsion system of a fishing vessel. Reliab. Eng. Syst. Safe 2017, 159, 69–79.

[11] Kerk, Y.W.; Tay, K.M.; Lim, C.P. An Analytical Interval Fuzzy Inference System for Risk Evaluation and Prioritization in Failure Mode and Effect Analysis. IEEE Syst. J. 2017, 11, 1589–1600.

[12] Siswantoro, N., Semin, S., Zaman, M., Criticality Assessment for Marine Diesel Engine Using Failure Mode and Effect Criticality Analysis (FMECA) Approach: Case Study on Lubricating Oil System, (2020) International Review of Mechanical Engineering (IREME), 14 (4), pp. 258-263.https://doi.org/10.15866/ireme.v14i4.18060

IOP Conf. Series: Earth and Environmental Science **1423** (2024) 012038 doi:10.1088/1755-1315/1423/1/012038

Redesign the air ventilation system in the engine room 48 meter patrol ship

C Kusuma[1], M A Widayat[2], W Kusdiana[3], and A T Ika R[4], A Baidowi[5]

[1,2,3,4] Mechanical Engineering Sekolah Tinggi Teknologi Angkatan Laut, Surabaya, Indonesia
[5]Marine Engineering Institut Teknologi Sepuluh Nopember, Surabaya, Indonesia

*kusumacahya99@gmail.com

Abstract. The patrol boat with a Length Overall of 48 meters has the main function as a coastal patrol boat. The air ventilation system installed on patrol ships currently reaches an average temperature of 320°K ~ 345°K with an average speed of 0 ~ 2.5 m/s and the airflow is not distributed throughout the engine room. The temperature in the engine room is above the ISO 8861:1998 standard where the maximum temperature is 318.15 K (45°C). Based on calculations according to ISO 8861, the total air capacity requirement is 10.806 m^3s^{-1} or 38.901 m^3/hour, while the currently installed air blower capacity only has an air capacity of 3.988 m^3s^{-1} or 14.358 m^3h^{-1}. The existing air ventilation system consists of 2 air blowers that suck in outside air. Computing Fluid Dynamics (CFD) is used to calculate the temperature distribution of the engine room between the existing ventilation system and the ventilation system based on ISO 8861 calculations. Based on the CFD simulation results, a blower modification with a capacity of more than 10.806 m^3s^{-1} is carried out so air can be distributed throughout the room at an average speed 2.5 ms^{-1} ~ 8 ms^{-1} with engine temperature at a height of 1-2 meters is 303°K ~ 315°K according to ISO 8861:1998 standards.

1. Introduction

The 48 meter patrol boat is a coastal patrol boat. As a patrol vessel, speed is a primary requirement.[1][2][3] But when operated, the engine room temperature reaches 46°C - 48°C. To overcome this heat, a ventilation system is needed to supply fresh air into the engine room and circulate hot air released by the engine and equipment to the engine room.[4] outside the engine room, so that the temperature in the engine room does not exceed 45ºC. Overheating conditions have a hot impact on the equipment in the engine room, especially the main engine control system (engine controller) and other auxiliary equipment.[5] This control system is very vulnerable to high temperatures so it can cause the equipment in the machine room not to work optimally and the machine cannot be operated in optimal conditions. [6][7][8].From the results of the analysis during the ship's operation, the temperature increase was caused by the air ventilation system in the engine room which was not following the needs of the supply fan and exhaust fan in the engine room which was not balanced. Based on ISO 8861:1998, it is stated that the normal temperature conditions permitted in the engine room are no more than 45°C.[9][10] This is because the threshold limit for air expansion is 45°C. Overheating in the main engine room also makes the crew uncomfortable in manning the engine, or checking equipment in the engine room.[4] Ventilation planning according to ISO 8861:1998 standards will be adjusted to room conditions and all equipment so that it can handle high temperatures. It is hoped that the ventilation redesign will be able to reduce high engine room temperatures, so that the equipment works optimally and the crew is comfortable manning the equipment.

2. Methodology

The methodology used in this research is CFD-based simulation. Computational Fluid Dynamics (CFD) analysis is a calculation method with dimension, area and volume control by utilizing computer computing to calculate the separating elements. The principle is that the space containing the fluid must be calculated and divided into several parts, this is called a cell and this process is called meshing. the divided part is the control calculation that will be carried out by the application or software. one of the CFD programs is numeca. Numeca has products for processing design, meshing and solvers for CFD processing software. Some of the software that will be used in making this research. One of the products from Numeca that has specifications for identifying fluid flow rates. Based on ISO 8861:1998 concerning "Shipbuilding, Engine-Room Ventilation In Diesel, Engined Ships, Design Requirements And Basis Of Calculations", the outside air environmental temperature can be taken for engine rooms to be around ±35°C with Relative Humidity (RH) 70% and pressure 101.3 kPa. The temperature rise of the inlet air into the engine room to the intake casing is ±12.5 K (the increase in air temperature in the engine room is the difference between the inlet and outlet temperatures measured at design conditions). When the machines and equipment in the engine room are turned on, the engine room temperature reaches 10~12° C higher than the ambient air temperature outside or around 45 ~ 47° C. According to ISO 8861, the total airflow capacity in the engine room must be greater than the two calculation formulas as follows:

$$Q = qc + qh \quad \text{or} \quad Q = 1{,}5\ x\ qc \tag{1}$$

where,
Q : Total air capacity (m^3s^{-1})
Qc : Air flow for the engine combustion process (ms^{-1})
Qh : Air flow for circulation of heat emissions (ms^{-1})

2.1. Air flow for the engine combustion process (m^3s^{-1})

Air flow for the combustion process is the addition of the main diesel engine (qdp), diesel generator (qdg), and air flow for the boiler (qb) in cubic meters per second (m^3s^{-1}).[10]

$$qc = qdp + qdg + qb \tag{2}$$

where,
qdp :Air flow for main diesel engine combustion (ms^{-1})
qdg :Air flow for diesel generator combustion (m^3s^{-1})
qh :Air flow for boiler (ms^{-1})

2.2. Air flow for main diesel engine combustion (qdp)

$$qdp = \frac{Pdp\ x\ mad}{\rho} \tag{3}$$

where:
Pdp : Standard service power of the main diesel engine at maximum continuous power output (kW).
m_{ad} : Combustion air requirements in diesel engines (kg $kW^{-1}s^{-1}$)
ρ : Air density (kgm^{-3})

2.3. Air flow for diesel generator combustion (qdg)

$$qdg = \frac{Pdg\ x\ mad}{\rho} \tag{4}$$

P_{dg} is the power of the diesel generator at maximum output (kW) is the combustion air requirement in the diesel engine (kg $kW^{-1}s^{-1}$).

The air flow for the boiler (qb) with the total steam capacity of the boiler known, the formula is used:

$$qb = \frac{ms\ x\ mfs\ x\ maf}{\rho} \tag{5}$$

The air flow for the boiler (qb) with the capacity of the boiler or thermal fluid heater is known in kilowatts, the formula is used:

$$qb = \frac{Q\ x\ mfs\ x\ maf}{\rho} \tag{6}$$

Where,

Q : *Max. continuous rating for boiler* (kW)
mfs : Total capacity *steam boiler* (kgs^{-1})
maf : Fuel consumption (fuel kg / kg uap)
ρ : Air density ($kg.m^{-3}$)

2.4. Air flow for circulation of heat emissions (ms^{-1})

The amount of air flow required to evacuate heat emissions in the engine room is calculated in cubic meters per second (ms^{-1}) using the following formula:

$$qh = \frac{\emptyset dp + \emptyset dg + \emptyset b + \emptyset p + \emptyset g + \emptyset el + \emptyset ep + \emptyset t + \emptyset o}{\rho\ x\ c\ x\ \Delta T} - 0{,}4\ (qdp + qdg) - qb \tag{7}$$

Where,

$\emptyset dp$: heat emissions from the main diesel propulsion engine (kW)
Ødg : heat emissions from diesel generator engines (kW)
Øb : heat emissions from boilers and thermal fluid heaters (kW)
Øp : heat emissions from steam and condensate pipes (kW)
Øg : heat emissions from electric air-cooled generators (kW)
Øel : heat emissions from electrical installations (kW)
Øep : heat emissions from exhaust pipes including flue gas fired boilers (kW)
Øt : emissions from hot tanks (kW)
Øo : emissions from other components
qdp : air flow for the main combustion engine diesel propulsion (m^3s^{-1})
qdg : air flow for combustion generating diesel engines (m^3s^{-1})
qb : air flow for combustion boilers (m^3s^{-1})
ρ :1.13 kg/m^3 (air density, in ± 35 ° C, 70% RH and 101.3 kPa)
c :1.01 kJ / (kg°K) (specific heat capacity of air)
ΔT :12.5 K (increase in air temperature in the engine room)

2.5. Heat emissions from the main diesel propulsion engine

$$\emptyset dp = Pdp\ x \frac{\Delta hd}{100} \tag{8}$$

Pdp : total power from basic diesel (kW)
Δhd : heat loss from basic diesel (%), if specification data is not available then it can be taken 2%

2.6. Heat emissions from diesel generator engines

$$\emptyset dg = Pdg\ x \frac{\Delta hd}{100} \tag{9}$$

Pdp : total power from basic diesel (kW)

IOP Conf. Series: Earth and Environmental Science **1423** (2024) 012038 doi:10.1088/1755-1315/1423/1/012038

Δhd : heat loss from basic diesel (%), if specification data is not available then 2% can be taken

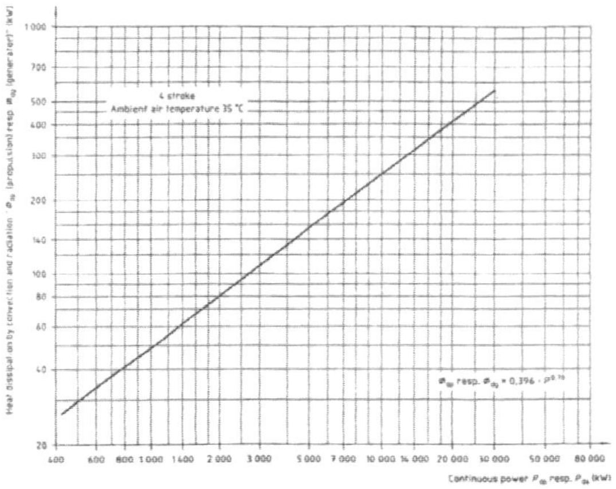

Figure 1. Heat emissions from Diesel engine 4 Stroke[11]

Specific percentages are not available in the engine information, so the determination of heat emissions from the main propulsion diesel engine uses Figure 4.2. From the figure, the heat emission value for a total engine power of 2684 kW is 100 kW.

2.7. Heat emissions from boilers and thermal fluid heaters

If the total steam capacity of the boiler is known, the following formula is used:

$$\emptyset b = ms \times mfs \times h \, x \frac{\Delta hd}{100} \times B1 \tag{10}$$

When the demand for heat is covered by a thermal fluid heater or the boiler capacity is known in kilowatts, the following formula is used:

$$\emptyset b = Q \times B1 \, x \frac{\Delta hb}{100} \tag{11}$$

where,

ms : total steam capacity (kW/s)

mfs : fuel consumption (kg fuel / kgof steam)

h : the lower calorific value of the fuel (kj/kg)

Δhb : the maximum Heat Loss of the boiler or thermal fluid heater (%).

B1 : a constant that applies to the location of boilers and other heat exchangers in the engine room is the maximum continuous rating of a thermal fluid heater or boiler (KW).

2.8. *Heat emissions from pipe steam and condensate*

$$\emptyset p = msc \, x \frac{\Delta hp}{100} \tag{12}$$

where,

msc : Total steam consumption (kW)

Δhp : Total heat loss from pipe steam and condensate (%)

2.9. Heat emissions from electricity generator

$$\emptyset g = Pg \times (1 - \frac{\eta}{100}) \tag{13}$$

where,

Pg : Generator power (kW)

η : Generator efficiency (%)

2.10. Heat emissions from electrical installations

Heat emissions from electrical installations are calculated in kilowatts and adjusted by one of the following two methods:

a. Complete details of the electrical installation are known, the heat emission must be taken as the sum of the simultaneous heat emissions
b. For conventional ships where details about the electrical installation are unknown, the heat emission is taken as 20% of the rated power of the electrical and lighting apparatus used at sea

2.11. *Heat emissions from exhaust pipes include flue gas fired boilers*

Heat emissions from exhaust pipes and gas-fired boiler exhaust can be determined from figure 2, in kilowatts per meter of pipe. If a specific Drawing is not available, AT = 250°K can be used for 2-stroke engines and AT = 320°K for 4-stroke engines.

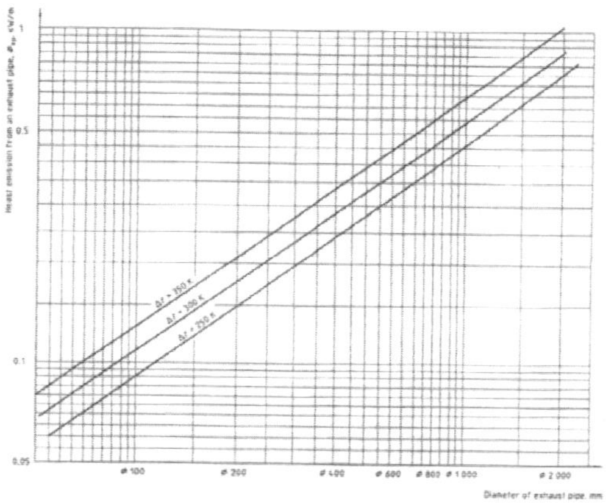

Figure 2. Diesel Engine Exhaust Pipe Heat Transfer

The main engine Cummins Model KTA 50-M2 has an exhaust gas pipe diameter of 550 mm with Δt = 329 K. Based on these data, the heat emissions released from the exhaust gas pipe are known by reading the graph in Figure 2. Based on the graph data in Figure 2, the value of Øep M/E for the main engine is 0.38 kW. The length of the exhaust gas pipe on the main engine is 6 m, so Øep = 2.28 kW

2.12. Heat emissions from hot tanks

Data collection on heat emissions from hot tanks is carried out by direct measurements on the tanks in the main engine room.

2.13. Heat emissions from other components

Heat emissions from other components must be added in calculating the air flow for evacuation of heat emissions. The results of this addition are determined in kilowatts. The components referred to here are the compressor, steam turbine, gear box, separator, heat exchanger, pipes and hydraulic system, etc. An electric motor which is the energy source of the equipment will release heat when working. The heat released by the electric motor can be calculated using the equation:

$$Q = 864 \times N \times \left(\frac{1-\eta}{\eta}\right) \tag{14}$$

where:
Q : generated heat load (kcal/h)
N : motor power (HP)
H : motor efficiency (%)

3. Result and Discussion

Based on the analysis results during the ship's operation, the temperature increase was caused by the air ventilation system in the engine room that did not match the needs of the supply fan and the engine room exhaust fan was not balanced (ISO 8861:1998). That the normal temperature conditions permitted in the engine room are no more than 45°C. This is because the air expansion threshold is 45°C, for the current conditions the temperature of the patro boat 48m engine room reaches 46°C - 48°C. The methodology used in this study is a simulation based on using software with appropriate specifications. Computational Fluid Dynamics (CFD) analysis is a calculation method with control of dimensions, area and volume by utilizing computer computing to calculate the separating elements.[13] [14][15][16].

3.1. Modeling

The model in the research was carried out in the engine room. Air speed is planned using the existing blower inlet and speed from ISO 8861 calculations. The engine room of the 48m patrol boat is located on the lower deck plan with a length of 636 cm, a width of 711 cm and a height of 240 cm from the double bottom tank plan. Engine room layout shown in figure 3.

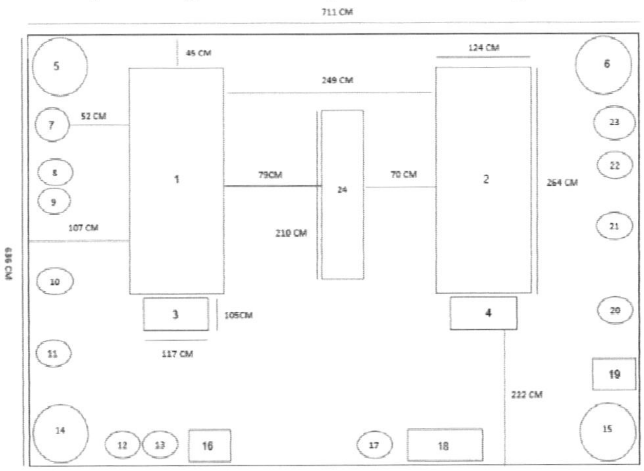

Figure 3. Engine room layout

Engine room layout on as shown at figure 3. Main Engine room equipment shown in table 1.

Table 1. Main Engine Room Equipment

No	Unit	Merk/Type
1&2	Main engine	1398KW/1950RPM
3&4	Gear Box I & II	ZF FRIEDRICHSHAFEN AG
5&6	Press Blower I & II	3 KW/380-400 V/60 Hz/ 3 PHASE
7&23	Separator I & II	MAB 103B-24
8&9	Sea water pump I & II	ALDF3-7 FSWSC
10&20	AC pump I & II	65ZX30-15 3 KW
11	Fire Pump	CA 50/7A 5.5 KW/7 HP
12&13	Fuel transfer pump I & II	CA 50/3A 2.2 KW
14&15	Suction Blower I & II	3 KW/380-400 V/60 Hz/ 3 PHASE
16&19	Oil Tank I & II	-

No	Unit	Merk/Type
17	OWS Pump	CDZF-0.5 220V – 750W
18	OWS Tank	-
21	Fresh Water Transfer Pump	65ZX30-15 3 KW
22	Bilge Pump	65ZX30-15 3 KW
24	Water Canoon	FAWDE 110 KW

Temperature in the engine room shown in table 2

Table 2. Temperature in the engine room.

Auxiliary engine	Temperature (°C)			Air Speed (ms^{-1})		
	800 RPM	1000 RPM	1580 RPM	800 RPM	1000 RPM	1580 RPM
Engine Air Inlet port side	46.86	47.25	44.90	4.43	6.53	16.86
Engine Air Inlet starboard	46.87	47.45	44.95	4.96	5.98	18.69
The blower presses the bow ps	46.88	47.35	41.25	5.92	5.81	7.00
The blower presses the bow stbd	46.89	47.35	40.85	6.96	6.49	6.26
Blower Suction aft ps	46.90	48.35	46.25	5.92	4.72	5.33
Blower Suction aft stbd	46.91	48.35	45.65	5.91	5.36	5.91

Air speed in the engine room(off mode) shown in table 3.

Table 3. Air speed in the engine room(off mode).

Auxiliary engine	Air Speed When The Main Engine Is Off (ms^{-1})
Engine Air Inlet port side	6.83
Engine Air Inlet stbd	7.47
The blower presses the bow ps	7.32
The blower presses the bow stbd	5.34

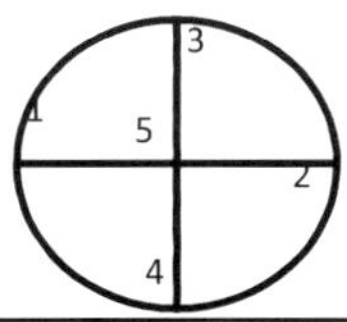

Calculation of airflow discharge shown in table 4.

Table 4. Calculation of airflow discharge

Formula	Rotation 800 Rpm (m^3s^{-1})	Rotation 1000 Rpm (m^3s^{-1})	Rotation 1580 Rpm (m^3s^{-1})
$Q_1 = A.V_{R1}$	1.845	1.810	2.181
$Q_2 = A.V_{R2}$	2.156	2.022	1.951
$Q_3 = A.V_{R3}$	1.673	1.334	1.506
$Q_4 = A.V_{R4}$	1.670	1.515	1.670

Calculation of the mass flow rate shown in table 5.

IOP Conf. Series: Earth and Environmental Science **1423** (2024) 012038 doi:10.1088/1755-1315/1423/1/012038

Table 5. Calculation of the mass flow rate(kgs-1)

Formula	Rotation 800 Rpm (m^3s-1)	Rotation 1000 Rpm (m^3s-1)	Rotation 1580 Rpm(m^3s-1)
$\dot{m}_1 = Q_1.p_1$	2.079	2.040	2.412
$\dot{m}_2 = Q_2.p_2$	2.430	2.279	2.157
$\dot{m}_3 = Q_3.p_3$	1.885	1.503	1.666
$\dot{m}_4 = Q_4.p_4$	1.882	1.707	1.847

3.2. Existing Engine room analysis results

Existing Engine room analysis results shown in figure 4,5, and 6.

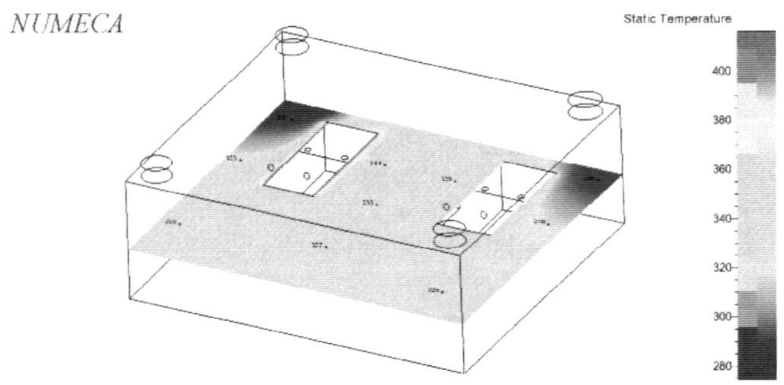

Figure 4. Engine room air temperature contour(altitude 1 meter)

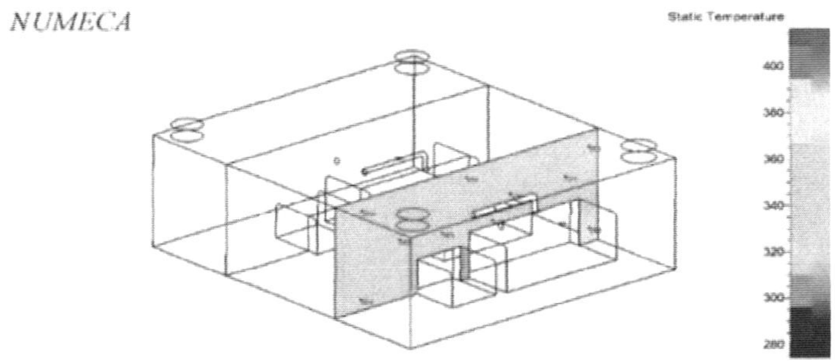

Figure 5. Engine room air temperature contour at Center Line

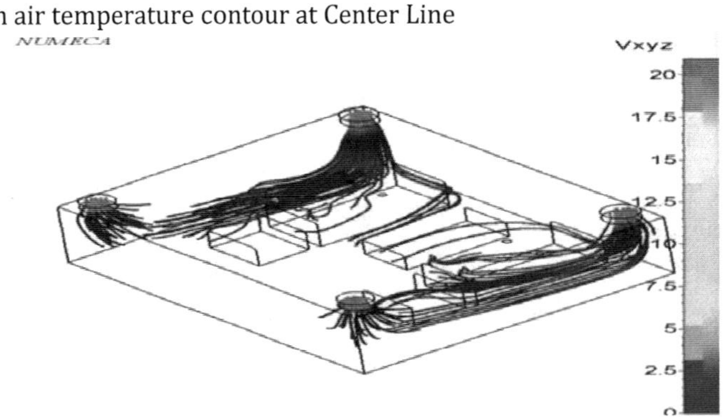

Figure 6. Velocity Vector Engine room

The average temperature at an altitude of 1 meter in the engine room is around 320 °K - 345 °K. Meanwhile the average temperature at an altitude of 1.5 - 2 meters in the engine room is around

IOP Conf. Series: Earth and Environmental Science **1423** (2024) 012038 doi:10.1088/1755-1315/1423/1/012038

320 °K - 345 °K. The average air velocity 0 – 2.5 m/s, where the air distribution is uneven at several points in the engine room. The air temperature in the engine room exceeds the ISO 8861 standard where the maximum temperature is 45°C or 318.15 °K.

3.3. *Calculation of air requirements According to ISO 8861*

According to ISO 8861, the total airflow capacity in the engine room must be greater than the two calculation formulas as follows:

$$Q = qc + qh \quad \text{or} \quad Q = 1{,}5\ x\ qc \tag{15}$$

Air flow for the engine combustion process (m^3s^{-1})

$$qc = qdp + qdg + qb \tag{16}$$

Air flow for main diesel engine combustion (qdp) is qdp = 4,750 m^3s^{-1}.The 48 meter Patrol Boat has a Diesel generator located outside the main engine room, dg = 0 m^3s^{-1}, qb = 0 m^3s^{-1}. The 48 meter Patrol Boat does not have a boiler. qc = 4,750 m^3s^{-1}

3.4. *Air flow for heat emission circulation* (m^3s^{-1})

$$qh = \frac{\emptyset dp + \emptyset dg + \emptyset b + \emptyset p + \emptyset g + \emptyset el + \emptyset ep + \emptyset t + \emptyset o}{\rho\ x\ c\ x\ \Delta T} - 0{,}4\,(qdp + qdg) - qb \tag{17}$$

If specific heat emission data is not available, Ødp to determine the heat emission from a diesel engine shown in Figure 7.

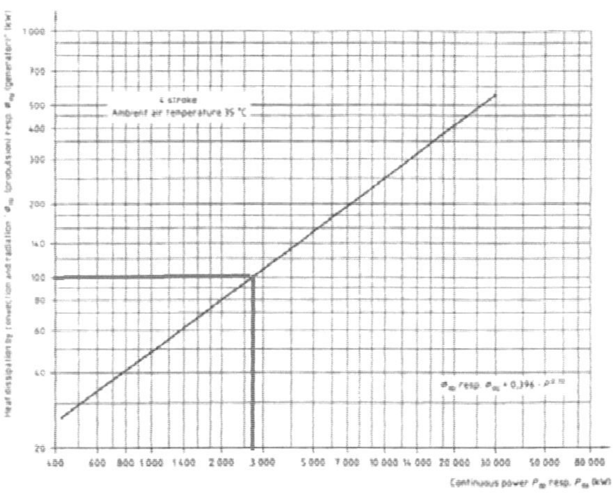

Figure 7. Main Engine Heat loss 4 Stroke

Base on Figure 7, the heat emission value in the total engine power is 100 kW. The diesel generator on the 48 meters Patrol Ship is outside the main engine room and the ship does not use a boiler so the values (Ødg), (Øb), (Øp), (Øg) = 0 kW. Because the electrical installation data is unknown, the second method is used, where heat emission is taken at 20% of the value of electric power and lighting when the ship is operating or sailing ($\emptyset el$) = 0 kW. Based on data, the main engine has an exhaust gas pipe diameter of 550 mm with Δt 329 so that the heat emissions released from the exhaust gas pipe can be seen using the graph shown in figure 8.

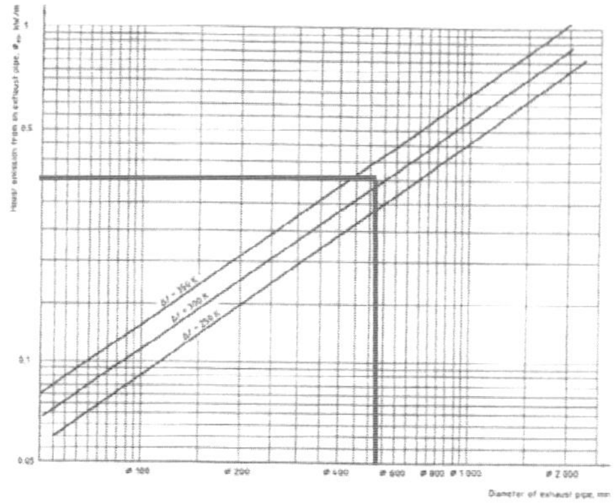

Figure 8. Heat Emissions From Engine Exhaust Pipes

Base on figure 8. the Øep M/E for the main engine is 0.38 kW. The length of the exhaust pipe in the main engine and main diesel generator is 6 m each, so the total gas is $Øep$ = 2.28 kW

a). From data collection on the ship, the total area of the overall tank in the engine room is 11.9 m^2 and with a temperature of 43ºC. according to the ISO 8861 table, for a tank with a temperature of 43ºC and without insulation it has a constant of 0.14 is Øt = 1.666 kW

b). Heat emission from other components auxiliary $Q =$ 713.367 W = 0.713 kW

From the calculation above, it can be seen the heat emission in other components with data on motor power and motor efficiency shown in table 6.

The airflow for circulating heat emission is $qh =$ 6.056 m^3s^{-1}

Base on the calculation details above, the air requirements in the ship's engine room according to ISO 8861 must be greater than Q calculations:

Q = 4.750 + 6.056 = 10.806 m^3s^{-1}

Table 6. Power motor and heat release of other components

Auxilary Engine	Qty	Power Motor (N)			Released Heat Load			
		kW	HP	η	kkal/h	kal/s	watt	KW
Separator Pump	2	0.75	1.005	0.70	1775296	493138	809095	0.809
AL pump	2	0.55	0.737	0.77	620701	172417	282886	0.283
Bilge Pump	1	3.00	4.023	0.85	858758	238544	391381	0.391
AC pump	2	3.00	4.023	0.85	3089710	858253	1408143	1408
BB Transfer Pump	2	2.20	2.950	0.85	1349476	374855	615027	0.615
AT Transfer Pump	1	3.00	4.023	0.85	248280	68967	113154	0.113

3.5. *Air analysis results based on ISO 8861*

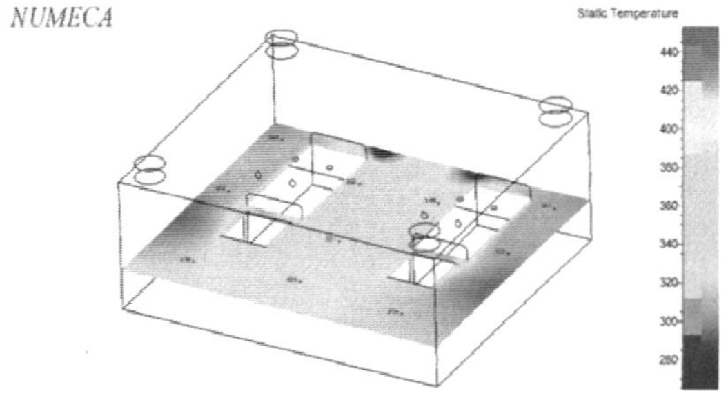

Figure 9. Engine room air temperature contour altitude 1 meter

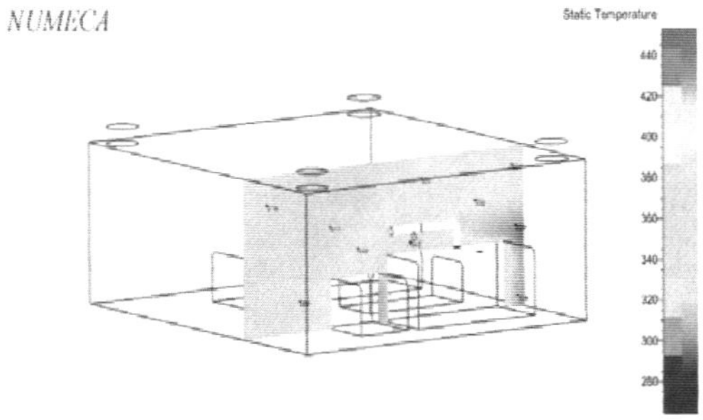

Figure 10. Engine room air temperature contour altitude at the center line

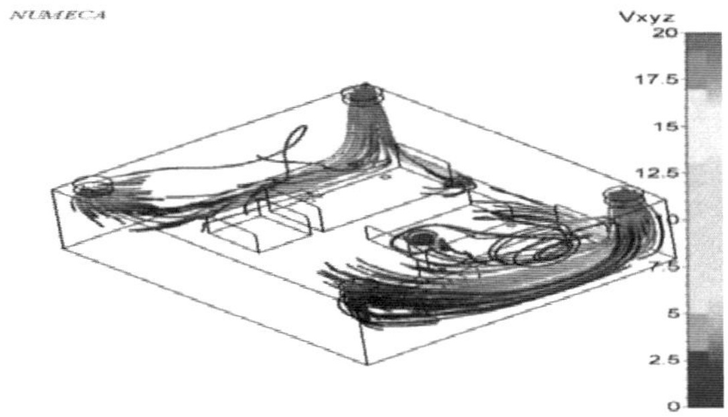

Figure 11. Velocity Vector Engine room

The average temperature in the engine room at an altitude of 1-2 meters is 303°K ~ 315°K. This indicates that the temperature in the engine room meets ISO 8861 standards, which has a maximum temperature of 45°C or 318.15°K. Air can be distributed in throughout the room with an average speed of 2.5 m/s ~ 8 m/s, where the maximum air distribution in the engine room.

4. Conclusion

Based on the results of the CFD analysis and calculations that have been carried out, the following conclusions can be drawn:

a. The air ventilation system installed on the 48 meter Patrol Ship produces an average air temperature of 319.3 °K – 346 °K in the engine room, the average air speed is 0 – 2.5 m/s, and the air flow cannot be distributed throughout engine room so it does not meet the criteria of the ISO 8861:1998 standard.
b. The capacity of the air blower that is still installed is 3.988 m^3s^{-1} or 14.358 m^3h^{-1} for air needs in the engine room and maintaining air temperature according to standards, it is unable to meet the air needs in the engine room and a new replacement is recommended. Based on ISO 8861:1998 standards, the total air capacity requirement in the engine room is 10.806 m^3s^{-1}or 38.901 m^3h^{-1}.
c. Based on analysis using CFD and ISO 8861:1998 standards, modifications to the blower were recommended and an air supply value of 10.806 $m^3s^{-1}$was achieved.

References

[1] C. Kusuma, I. M. Ariana, and W. H. Nugroho, "An Innovative Proposal for Increasing the Speed of a 60 Metres Fast Patrol Vessel," *Int. Rev. Mech. Eng.*, vol. 15, no. 9, pp. 486–496, 2021, doi: 10.15866/ireme.v15i9.21411.
[2] C. Kusuma, I. M. Ariana, W. H. Nugroho, E. B. Djatmiko, A. A. Masroeri, and Sutardi, "Design Propeller Of Fast

Missile Boat 60 M By Using Gawn Series," *IOP Conf. Ser. Mater. Sci. Eng.*, vol. 1052, no. 1, p. 012026, 2021, doi: 10.1088/1757-899x/1052/1/012026.

[3] C. Kusuma, I. M. Ariana, and B. Ali, "Redesign KCR 60m Bow with Axe Bow Type to Reduce Ship Resistance," *IOP Conf. Ser. Earth Environ. Sci.*, vol. 557, no. 1, 2020, doi: 10.1088/1755-1315/557/1/012033.

[4] C. Kusuma, J. N. Pramono, W. Kusdiana, and I. S. Arief, "Analysis of air ventilation system in the engine room patrol boat 60 meters after repowering," *AIP Conf. Proc.*, vol. 2592, no. March 2015, 2023, doi: 10.1063/5.0115311.

[5] W. K. Horng and H. J. Shaw, "A study on the integrated design of engine room ventilation," *J. Mar. Sci. Technol.*, vol. 25, no. 1, pp. 70–83, 2017, doi: 10.6119/JMST-016-0913-1.

[6] E. M. W. Hario Pramudito, Alam Bahermasyah, *Modifikasi Desain Sistem Ventilasi Kamar Mesin Tanjung Sole*, no. 0. 2016. doi: 10.12962/j23373539.v8i2.49592.

[7] A. P. Nugroho, M. M. E. Prayitno, and A. Gafur, "Kajian Numerik Tata Letak ventilator Ruang Kamar Mesin Di Kapal Coaster 2000 GT," *Vam*, 2018.

[8] P. Manik *et al.*, "Perancangan Kapal General Cargo 1500 Dwt Rute Pelayaran Jakarta-Surabaya," *Kapal*, vol. 10, no. 2, pp. 108-115–115, 2013.

[9] International Standard, "ISO 8861 Shipbuilding — Engine-room ventilation in diesel-engined ships — Design requirements and basis of calculations," vol. 1998, 1998.

[10] "Rules for the Classification and Construction Part 2. Inland Waterways Ship Volume Iii Rules for Machinery Installations 2015 Edition Biro Klasifikasi Indonesia," vol. III, 2015.

[11] M. Vollmer and K.-P. Möllmann, *Some Basic Concepts of Heat Transfer*. 2010. doi: 10.1002/9783527630868.ch4.

[12] J. P. (Jack P. Holman, *Heat transfer —10th ed.* 2010. doi: 10.1016/b978-034073162-8/50017-5.

[13] P. Puspitasari, "Analisa Supply Aliran Udara Terhadap Variabel Suhu, Tekanan Dan Kecepatan Udara Pada Kamar Mesin Kapal Tanker 6500 Dwt Menggunakan Computational Fluid Dynamics," *J. Tek. ITS*, pp. 1–12, 2010.

[14] S. B. Cem Taşdemir, "Cfd Analysis of Ventilation System," *1st Int. Sh. Mar. Technol. Congr. (SHIP-MAR 2016)*, vol. 41, no. December, pp. 337–348, 2017.

[15] N. S. Hendri, A. adi, O. S. Suharyo, and A. D. Susanto, "The Air Flow Analysis in Engine Rooms at Frigate Class Ship with CFD Approach (Computational Fluids Dynamics)," *Int. J. Recent Eng. Sci.*, vol. 5, no. 4, pp. 11–18, 2018, doi: 10.14445/23497157/ijres-v5i4p103.

[16] X. Jian, R. Ping, and D. Konovessis, "CFD analysis of natural gas dispersion in engine room space based on multi-factor coupling," *Ocean Eng.*, vol. 111, pp. 524–532, 2016, doi: 10.1016/j.oceaneng.2015.11.018.

IOP Conf. Series: Earth and Environmental Science **1423** (2024) 012039 doi:10.1088/1755-1315/1423/1/012039

Repowering MP Bandar Madani: "Transforming into a Fully Electric Pilot Boat to Support Pare-Pare Green Port"

Ibrahim Shaleh[1], E S Koenhardono[1*], and S Sarwito[1]

[1] Department of Marine Engineering, Institut Teknologi Sepuluh Nopember, Surabaya, Indonesia

*E-mail: koenhardono@gmail.com

Abstract. Parepare Port is within the working area of PT Pelabuhan Indonesia (Persero) which serves the export and import cargo. In 2021, 99% of visiting ships entering Parepare port will be more than 500 GT, so it is mandatory to use pilot services to enter the port. Currently ship guiding services are provided by MP. Masela and MP. Bandar Madani. However, M.P. Bandar Madani experienced a drowning incident in September 2021. Even though the engine had been repaired, the condition was not as before. PT Pelindo Marine Service as the pilot boat operator will repower the MP Bandar Madani into an electric pilot boat. This initiative aims to enhance pilotage services and contribute to the sustainability goals of Pare-pare Port, thereby transforming it into a Green Port. This paper discusses the technical analysis of repowering at MP Bandar Madani to improve pilot services and support the Pare-pare Port as a Green Port.

1. Introduction

PT Pelindo Marine Service has a fleet of pilot boats, namely MP. Bandar Madani, which has not been able to operate optimally, due to a sinking incident in September 2021 while docked at Pare-Pare Port. In 2022, the MP Bandar Madani outboard motor underwent a General Overhaul to restore its previous performance. Due to less-than-optimal performance, the role of MP Bandar Madani was replaced by MP Masela which operates at Pare-pare Port. Due to the increasing need for pilotage at Pare-Pare Port, an assessment was carried out again in 2023 on MP Bandar Madani with the results in Figure 1. Based on the assessment results, repowering of MP Bandar Madani needs to be carried out. In line with the commitment of PT Pelabuhan Indonesia (Persero) to develop a green port, this repowering must be able to reduce carbon emissions and greenhouse effects. This will support the creation of Pare-Pare Port as a green port that is environmentally friendly and sustainable. Therefore, this paper will analyze the repowering of MP Bandar Madani which has an electric propulsion system and battery energy source.

IOP Conf. Series: Earth and Environmental Science **1423** (2024) 012039 doi:10.1088/1755-1315/1423/1/012039

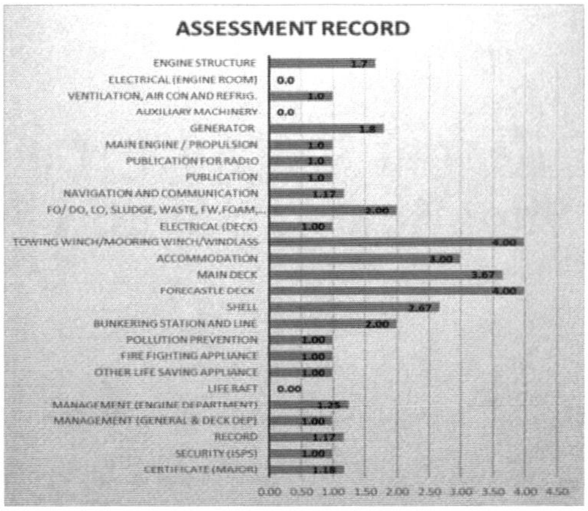

Figure 1. MP Assessment Results. Bandar Madani in 2023

2. Methods

Analysis of renewable energy-based pilot boats The planning in this research can be explained in in the form of a flow diagram of thought compiled sequentially in the following flow diagram:

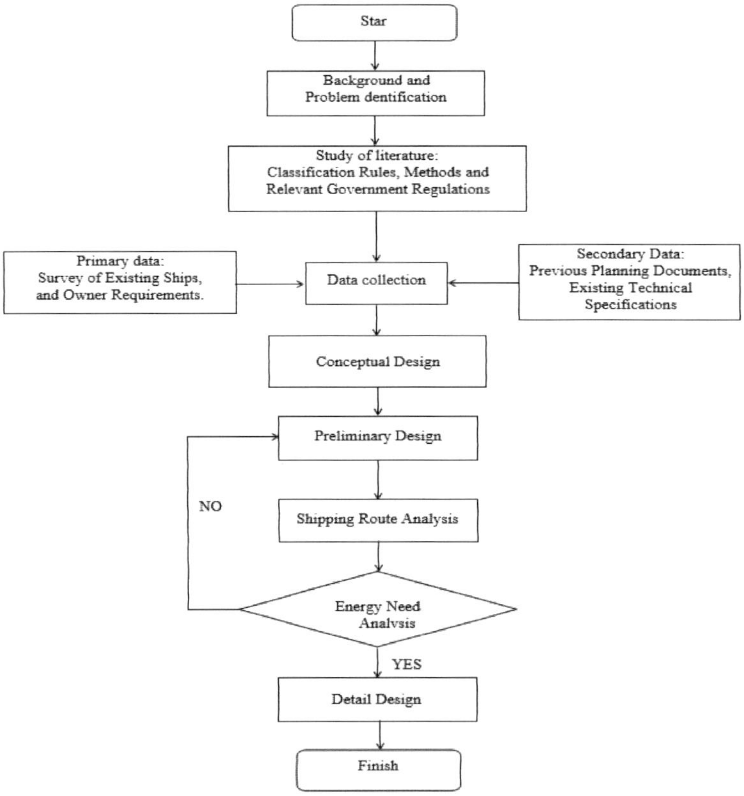

Figure 2. Methodology research.

IOP Conf. Series: Earth and Environmental Science **1423** (2024) 012039 doi:10.1088/1755-1315/1423/1/012039

3. Result and Discussion

Carry out repeated measurements of the ship to find out the ship's dimensions perfectly. The results of these measurements are then depicted using the Maxsurf application. As for The variables to be achieved in the Maxsurf application are:

Table 1. The results of the main dimensions of the ship are as follows

Particular	Dimension
LOA	7 m
LWL	6,6 m
LPP	6,2 m
B	2,1 m
H	1 m
T	0,45 m
Displacement	3,133 ton
Fn	0,6184
Cf	0,00205

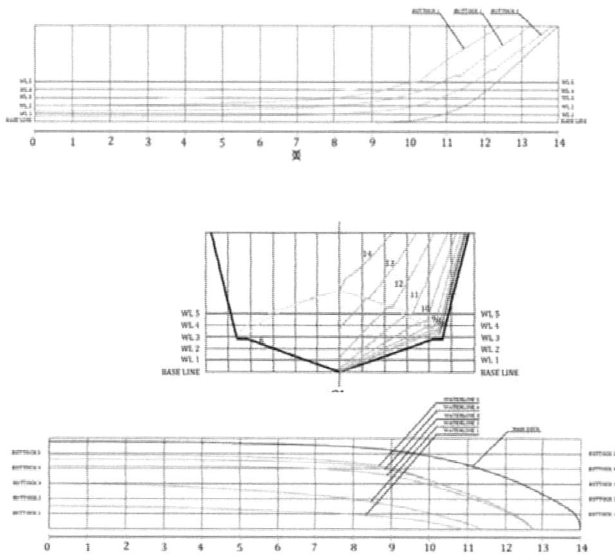

Figure 3. Lines Plan

Determination of ship resistance methods with Maxsurf

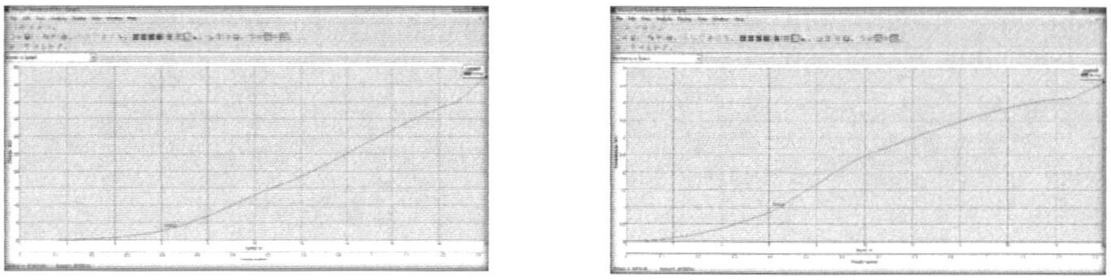

(a) (b)

Figure. 4 Power, Speed and Resistance with Method Holtrop (Existing Boat)

Power Calculations

EHP = RT Service x Vs
= 4.6 x 10.269 m/s
= 47,237 HP

THP = EHP / Hull Efficiency
= 47.237 / 1
= 47,237 HP

DHP = THP / PC
= 47.237 / 1.139
= 41,472 HP

SHP = DHP / $\eta_s\eta_b$
= 41.472 / 0.98
= 42,319 HP

From the power of 42,319 HP obtained from calculations using the Holtrop method, 15% was added for sea trial conditions to obtain the required power for the new main engine:

P = 42.319 x 1.15
= 48,666 HP

P = 48,666 HP / 95%
= 51,228 HP
= 38,20 kW

After obtaining the BHP value, the next step is to select the main engine as the main engine of the ship. The main engine used on this pilot boat is an outboard engine because of the ease of installation and engine size which is relatively small so it doesn't take up too much space. The selection of the main machine is carried out taking into account the weight of the machine, engine power and the price of the machine. From the catalogue (Torqeedo, 2024) several main engines and their specifications are obtained. The list of main machines is as follows seen in Table 1:

IOP Conf. Series: Earth and Environmental Science **1423** (2024) 012039 doi:10.1088/1755-1315/1423/1/012039

Table 2. Outboard Electric Engine Specification

Type Main Engine	Input Power (kW)	Ouput Power (kW)	Max Propeller Speed (rpm)	Voltage (volt)	Weight (kg)
Deep Blue 25 RL	27.5	25	2400	350	139
Deep Blue 50 RL	54.5	50	2400	350	139
Green Thruster	26	22	1800	80	94
Green Thruster	50	42	1800	144	98
Aries 50 RL	50	42	1750	48	120

Source : www.torqeedo.com

From Table 2, the Torqeedo Deep Blue 50 RL boat engine was selected considering its power was close the electric engine looks like Figure 5 as follow:

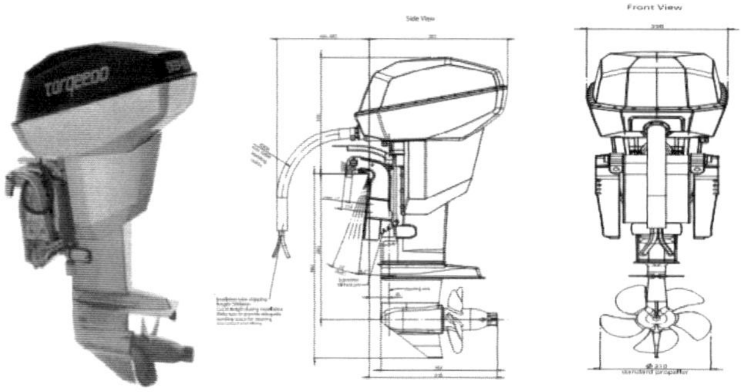

Figure 5. Electric Engine Torqeedo Deep Blue 50 RL

Table 3. Battery Specification

Type Main Engine	Battery	Protection Category	Voltage (volt)	Capacity (kWH)	Weight (kg)
Deep Blue Batteries 40	Litihium	IP 67	352	38	284
Deep Blue Batteries 80	Litihium	IP 67	348	77.6	562
AGM	Litihium	-	80	48	
AGM	Litihium	-	144	28.8	
Mastervolt MLI	LiFePO4	-	48	36	342

Source : www.torqeedo.com

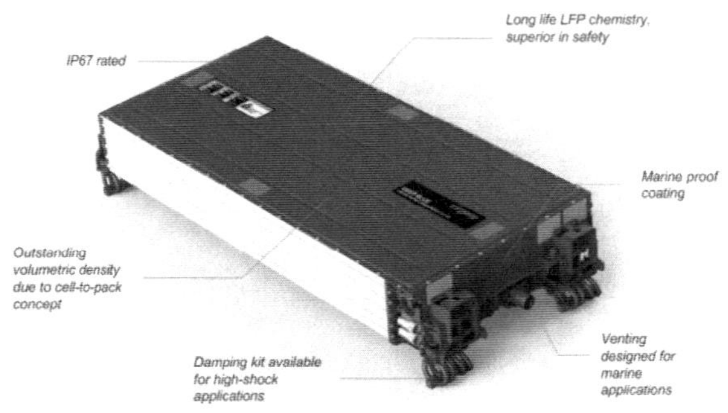

Figure 6. Deep Blue Battery 80

Table 4. Result Recalculation Displacement

Component	Wight (ton)
DWT	0.29
LWT	3.094
Ship Weight (DWT + LWT)	3.384
Displacement (LWL x B x T x Cb x ρ)	3.481
Weight Difference	0.096
MARGIN	**2.77%**

Stability Calculations

The ship that has been designed must be theoretically proven that the ship meets shipping stability standards using the International Code of Safety for High Speed Craft (2000) standards. Stability calculations are carried out with the help of Maxsurf Stability Enterprise Education Version software. The stability criteria used in software calculations are (HSC) Code 2000. Table 4 is a summary of the calculation results which have been compared with the limitations :

Table 5. Result Stability

Data	Loadcase I	Weight (kg)	Wight (ton)
E 0-30° (m.deg)	7.993	≥ 3.1513	Accepted
Hw (deg)	2.2	≤16	Accepted
Areal / Area2 shall not be less than	299.55	≥ 100%	Accepted

IOP Conf. Series: Earth and Environmental Science **1423** (2024) 012039 doi:10.1088/1755-1315/1423/1/012039

The ship's weight consists of the LWT (Light Weight Tonnage) component as shown in Table 5. From the calculations carried out, the ship weight margin was obtained at 2.77%. Meanwhile, the maximum allowable ship weight margin is 10%, so the ship weight calculation is accepted.

In the BLDC motor wiring diagram for electric guided ships using a 3 phase BLDC motor equipped with a hall sensor and controller and inverter system. The following is a wiring diagram image in Figure 7 below :

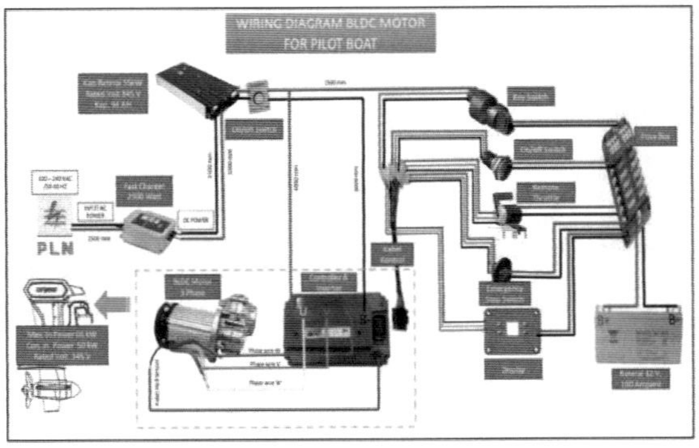

Figure 7. Wiring Diagram BLDC Motor

Power Management Systems (PMS) are an important part of the control equipment in a ship and usually distribute power to various control stations that can operate together and share information with each other or independently.

The ship's electrical power source is from a PLN source which is supplied to the battery as an energy source for the electric motor. Electrical energy usage management can be displayed in the following table 6 :

Table 6. Result Power Management System

Route	Engine Duration (minute)	Power Requirements (kW/minute)	Battery Charging Duration (hours)
Pasenger Ship Services (in)	30	14.23	5
Pasenger Ship Services (out)	30	14.23	5
Pertamina Ship Services (in)	68	21.74	8
Pertamina Ship Services (out)	58	21.69	8
Cargo Ship Services (in)	48	18.57	7
Cargo Ship Services (out)	48	18.57	7
PLTD Barge Services (in)	150	42.02	15
PLTD Barge Services (out)	148	42.01	15

In the process of 3D modeling of the hull using Maxsurf Modeler, the hull and main deck models were obtained. Then, to display the hull shape in a solid way, use the rendering menu on the available toolbar to get a shape like in Figure 8

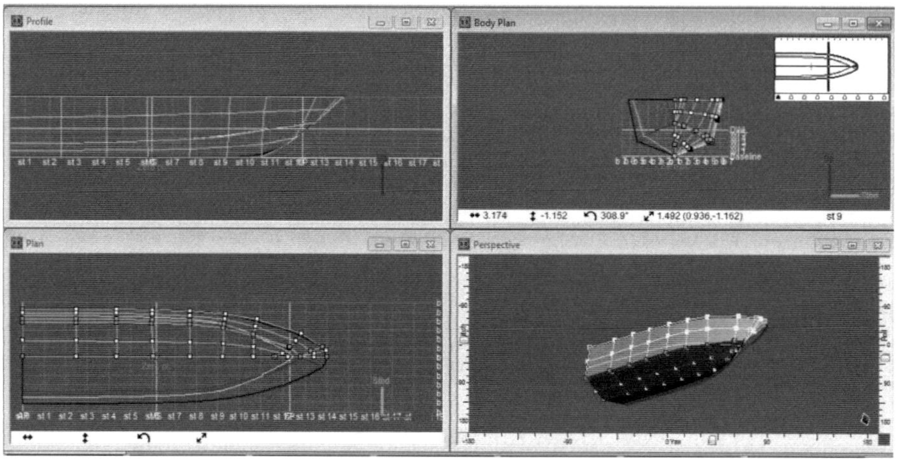

Figure 8. Modeling 3D Pilot Boat

Next, the detailed 3-dimensional design of the electric guided ship, files from Maxsurf, are exported to Sketchup 2021 software. The 3-dimensional design of the ship can be seen in Figure. 9

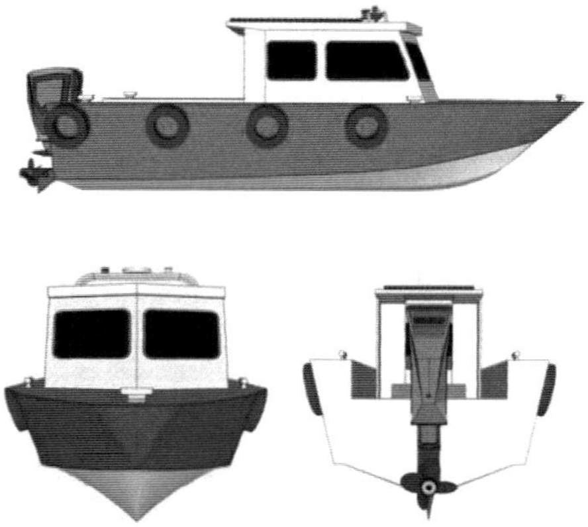

Figure 9. Modelling 3D Pilot Boat

The image above is the result of a calculation of the total demand multiplied by the factor for each emission. The orange graph represents fuel consumption before it is converted to electricity, while the blue colour is after electricity is applied.

Analyse the economic feasibility to find out whether using the battery will make a profit and return on investment in how many years. The data needed to calculate the economic feasibility of this ship is the investment costs of batteries along with electric motors and operations.

IOP Conf. Series: Earth and Environmental Science **1423** (2024) 012039 doi:10.1088/1755-1315/1423/1/012039

Table 7. Total Electricity Cost

Month	Pasengger Ship (in)	Passenger Ship (out)	Pertamina Ship (in)	Pertamina Ship (out)	TUKS Cargo Ship (in)	TUKS Cargo Ship (out)	PLTD Barge(in)	PLTD Barge (out)
Jan	203,904	71,282	277,022	245,449	53,117	17,246	60,991	17,246
Feb	234,489	61,099	234,403	163,633	70,823	34,492	-	-
March	295,660	101,831	340,950	278,175	123,939	51,738	-	-
April	285,465	81,465	255,712	179,996	106,234	68,985	-	-
May	173,318	20,366	63,928	65,453	-	-	-	-
June	214,099	91,648	234,403	196,359	70,823	51,738	-	-
TOTAL	1,366,155	366,593	1,555,584	1,194,518	389,524	224,200	304,953	86,231

Table 8. Electric Motor Investment Costs

Component	Qty	Unit Price	Amount
Electric Motor Type Deep Blue 50 RL	1	457.000.000	457.000.000
Deep Blue Batterie 80	1	1.038.100.000	1.038.100.000
Power Cable – negatice	1	17.800.000	17.800.000
Power Cable – positive	1	17.800.000	17.800.000
Battery suspection kit DB80	1	52.000.000	52.000.000
DB Battery Refrigeration Unit	1	238.000.000	238.000.000
Shore Power Distribution L 1-ph	1	28.800.000	28.800.000
Charger 3,3kW for DB HV Battery	2	35.800.000	71.600.000
DB Conbox S Std	2	144.000.000	288.000.000
Throttle Twin Top Mounting	1	50.000.000	50.000.000
DB Touch Display 7"	1	33.000.000	33.000.000
Router Deep Blue	1	48.000.000	48.000.000
12 V Distribution Box	1	12.800.000	12.800.000
TOTAL (without Tax)			2.352.900.000

Table 9. Total Fuel Cost

Month	Pasengger Ship (in)	Pasengger Ship (out)	Pertamina Ship (in)	Pertamina Ship (out)	TUKS Cargo Ship (in)	TUKS Cargo Ship (out)	PLTD Barge (in)	PLTD Barge (out)
Jan	3,676,000	1,195,600	5,392,400	3,843,000	878,400	268,400	915,000	268,400
Feb	4,227,400	1,024,800	4,562,800	2,562,000	1,171,200	536,800	-	-
March	5,330,200	1,708,000	6,636,800	4,355,400	2,049,600	805,200	-	-
April	5,146,400	1,366,400	4,977,600	2,818,200	1,756,800	1,073,600	-	-
May	3,124,600	341,600	1,244,400	1,024,800	-	-	-	-
June	3,859,800	1,537,200	4,562,800	3,074,400	1,171,200	805,200	-	-
TOTAL	25,364,400	7,173,600	27,376,800	17,677,800	7,027,200	3,489,200	915,000	268,400

Table 10. Total Call Services

Month	Pasengger Ship (in)	Pasengger Ship (out)	Pertamina Ship (in)	Pertamina Ship (out)	TUKS Cargo Ship (in)	TUKS Cargo Ship (out)	PLTD Barge(in)	PLTD Barge (out)
Jan	20	7	13	15	3	1	1	1
Feb	23	6	11	10	4	2	0	0
March	29	10	16	17	7	3	0	0
April	28	8	12	11	6	4	0	0
May	17	2	3	4	0	0	0	0
June	21	9	11	12	4	3	0	0
TOTAL	138	42	66	69	24	13	1	1

Table 11. Result Comparation Cost

Services	Equipment Engine Price (Rp)	Electric Engine Price (Rp)	Result Comparation Price (Rp)
Passenger ship (in)	25,364,400	1,406,936	23,957,464
Passenger ship (out)	7,173,600	427,692	6,745,908
Pertamina ship (in)	27,376,800	1,406,418	25,970,382
Pertamina ship (out)	17,677,800	1,129,065	16,548,735
TUKS Cargo ship (in)	7,027,200	424,935	6,602,265
TUKS Cargo ship (out)	3,489,200	224,200	3,265,000
PLTD ship (in)	915,000	60,991	854,009
PLTD ship (out)	268,400	17,246	251,154
TOTAL (without Tax)	89,292,400	5,097,482	84,194,918

4. Conclusion

From the results of ship optimization data, it was found that there was an increase in ship load of 434 kg after replacing the electric motor engine, installing batteries and other electrical equipment. A ship speed of 20 Knots can be achieved when the battery is fully charged 77.6 kW with a power load of 37.1 kW and can operate for 2 hours

There is a cost efficiency in converting petrol motorbikes to electric motorbikes amounting to IDR 84,194,918 for 6 months of operational activities from January, February, March, April, May and June 2023 By utilizing speed changes you can save energy consumption. The lowest speed, namely 5 knots, can load less energy with a range of 324.65 nm and can run for up to 65 hours

With a fully electric drive configuration, it can reduce exhaust emissions by around 100%. This is in line with the government's efforts to reduce emissions.

5. Acknowledgments

Expressions of thanks to the Regional General Manager 4 Parepare Port and the Supervisor of PT Pelindo Marine Service Parepare Branch who have contributed to the preparation of this research so that this paper can be published.

References

[1] Ajioka, Yushifumi & Kiyoshi Ohno. (2013). Hitachi Review (Vol. 2). Japan.
[2] Allan, Robert G. (1996). Application and Advantages of Catamarans for Coastal Patrol Vessels.
[3] ASVglobal. (2017). Data Sheet. Retrieved February 15, 2017 from ASV Global web site: https://www.asvglobal.com/
[4] Dubrovsky, Victor A. (2009). Multi-Hulls: Some New Options as the Result of Science Development. Rusia.

[5] Bertram, Volker. Unmanned Surface Vehicles – A Survey. Retrieved April 26, 2017 from web site:https://www.researchgate.net/publication/228393498_Unmanned_Surface_Vehic les-A_Survey
[6] Bassham, Bobby A. ; [2003]; "An Evaluation of Electric Motors for Ship Propulsion"; Naval Postgraduate School, Monterey, California.
[7] Catamaransite. (2017). Catamaran Design Formulas. Retrieved April 26, 2017 from web site: http://www.catamaransite.com/catamaran_hull_design_formulas.html
[8] HSC Annex 7 Stability of Multihull Craft. (2000). International Code of Safety for HighSpeed Craft. London.
[9] International Maritime Organization (IMO). (Consolidated Edition 2009). International Convention for the Safety of Life at Sea, 1974, as amended (SOLAS 1974). London: IMO Publishing.
[10] Kementerian Perhubungan. (2009). Standar Kapal Non-Konvensi Berbendera Indonesia. Jakarta.
[11] Molland, M., & Insel, A. F. 1992. An Investigation Into the Resistance Components of High Speed Displacement Catamarans. RINA.
[12] Nazarov, Albert. (2010). Power Catamarans: Design for Performance. Thailand.
[13] Rawson, K.J. and Tupper, E.C. (2001). Basic Ship Theory (5th ed., Vol. 1). Oxford: Butterworth-Heinemann.
[14] Sahoo, Prasanta K & Marcos Salas & Schwetz. (2007). Practical Evaluation of Resistance of High-Speed Catamaran Hull Forms-Part I. Australia.
[15] Satriawansyah, M. H. (2016). Tugas Akhir. Desain Kapal Penumpang Katamaran Untuk Rute Dermaga Boom Marina, Banyuwangi-Pelabuhan Benoa. Surabaya, Institut Teknologi Sepuluh Nopember.
[16] Taggart, R. (1980). Ship Design and Construction. New York: The Society of Naval Architects and Marine Engineers.
[17] Torqeedo. (2017). Propulsion System. Retrieved March 2, 2017 from Torqeedo web site: http://www.torqeedo.com/en
[18] UKessays. (2015). Advantages Of Aluminium As A Shipbuilding Material Engineering Essay. UK: UKessays.
[19] Watson, D. (1998). Practical Ship Design (Vol. 1). (R. Bhattacharyya, Ed.) Oxford: Elsevier
[20} Whitelegg, Ian, Bucknall, Richard [2013]; "Electrical Propulsion in the Low Carbon Econmy"; Low Carbon Shipping Conference, London.

Boil of Gas Formation Rate Analysis in LNG Storage Tank on LNG Fuelled Harbor Tug Vessel

I M Ariana [1*], A Z M Fathallah[1], H A Zaky[1] and T A Akbar[1]

[1] Marine Engineering Department, Institut Teknologi Sepuluh Nopember, Surabaya 60111, Indonesia

*E-mail: ariana@its.ac.id

Abstract. To accelerate the achievement of the net zero emission program which also one of the contributors of the climate resilience aspect in the Sustainable Development Goals (SDG), the use of natural gas as ship fuel is one solution during the energy source transition period. Storing natural gas in the liquid phase (LNG) can reduce tank volume requirements by up to 600 times compared to the gas phase. However, the formation of BOG (Boil of Gas) can increase the pressure in the tank. In this research, the rate of BOG formation in tanks designed for LNG storage on Harbor Tug vessels was analysed using the CFD (Computational Fluid Dynamics) method. The simulation results of the BOG formation rate when the ship is not operated and at full speed are 0.61 m3/minute, - 0.43 m3/minute, respectively. The storage tank is able to withstand gas pressure for up to 50 days when the ship's system is not operated.

1. Introduction

In the recent years the urge to shift the conventional fossils fuel with high emission footprint into more environmentally friendly fuels is inevitable. This is due to the fact that the amount of CO2 emissions over the past decade is increasing dramatically coming all the way from 21 Megatons in 1990 to 33 Megatons in 2019 [1]. Beside one of the objectives of Sustainable Development Goals (SDG) is to achieve the climate resiliency. One of the alternative substitutions of conventional fossil fuels is Natural Gas (NG) which offer lower price per energy units with higher thermal efficiency as well. Beside the facts that NG also produces 50-60% less CO2 emissions compared to conventional fossils fuels and produces less heat-trapping gas by 15-20% compared with conventional combustible fuels [2].

However, using NG as alternative fuels require the storage of the NG in Liquefied Natural Gas form or LNG. The simplification explanation of NG liquefaction process is achieved by cooling down the NG up to -160 °C. Considering the main component of NG is methane which under the cryogenic condition the methane change its phase to liquid state. Under the liquid state the methane decreases its volumetric properties and shrink up to 1/600 from the gaseous state and made it easier to store. However, due to its cryogenic properties LNG is vulnerable to thermal shifting and exposure in its storage, which made the LNG undergoes weathering or heat stratification. This is where the condition of the methane changes from liquid phase to gas phase, this is called BOG (Boil Off Gas) and happened due to the gradient of the temperature between inside and outside LNG storage [3].

During utilization of LNG, BOG production usually takes between 0.10% - 0.15% volumetric capacity of the LNG and caused by various factors such as thickness of the tank [4]. The gas produced from BOG shall be treated so that it will not cause any problem mainly to the tanks damage due to pressure build up, beside the excessive BOG production will be considered as loss in operation. The importance of BOG handling makes the utilization of LNG become challenge for most operation, especially when dealing with dynamic condition such as LNG as ship fuel. This paper will conduct an analysis on how the BOG formation and rate in LNG storage tank of a Harbour Tug with dual fuel engine with the power of 2 x 2,500 HP through numerical simulation of Computational Fluids Dynamic (CFD).

2. Theoretical Background

2.1 The LNG, LNG Storage Tank and Boil Off Gas

As explained in the introduction, the LNG mainly consist of methane (CH_4) which made >90% of the composition depend on the specific LNG properties. However, before being delivered to the user, LNG must go through several processes to comply the specific requirements and specification. The process of treating the NG usually involving purification of other substances (H_2S, CO, H_2O) and heavier hydrocarbon compound such as propane (C_3H_8) and butane (C_4H_{10}).

After the serial proses of gas treating and liquefaction, the LNG will proceed transportation and storing activity. The studied case on this paper with Harbour Tug as an object shall comply the certain regulation provided by International Code of Safety for Ship Using Gases or Other Low-flashpoint Fuels (IGF Code). In summation there are three recommended materials for the LNG fuel tanks, namely: 9% Nickel Steel, Austenitic Stainless-Steel type: 304, 304L, 316, 316L, 321 and 347, and Aluminium Alloy 5083.

For this research the material used is Austenitic Stainless-Steel type 304, for the thickness and size specifications of the tank's inner part is using 11 mm plate thickness, while the head side of the tank used is 17 mm, for the outer part of the tank is using carbon steel material with a wall thickness of 5 mm while the thickness of the tank head is 5 mm.

Due to temperature gradient between LNG and its ambient it is nearly impossible to suppress the heat loss of insulation material of the storage tank. This is not possible due to heat losses from the tank or its insulation. Estimating the boil-off gas is crucial when designing the tank, as it is linked to the tank's pressure build up. Therefore, the boil-off rate must be determined to ensure the tank's safety. Initially, nitrogen and methane, which are both highly volatile, will evaporate first compared to the other heavier hydrocarbon compound.

2.2 Calculation Model

This subchapter is intended to describe the principal equation in determining the BOG behaviour in the simulation. This research which emphasizes on CFD simulation rely on several basic equations models to conduct the numerical simulation. The equation used are namely Continuity Equation, Phase Transfer Equations, Turbulence Equations, and Surface Tension Effects. Each of the equation will be explained briefly as follows:

1. Continuity Equation

   From the law of conservation of mass, the value of the continuity equation is obtained as follows:

$$\frac{\partial p}{\partial t}+\frac{\partial(\rho V_x)}{\partial x}+\frac{\partial(\rho V_y)}{\partial y}+\frac{\partial(\rho V_z)}{\partial z}=0 \tag{1}$$

where:
V_X, V_Y, and V_Z = *Velocity components in vector*
ρ = *density*
x, y, and z = Global Cartesian Coordinates
t = Time

2. Turbulence Equation
The turbulence equation must be seen from the effect of viscosity, with this it can be seen whether the fluid flow experiences turbulence. The equation used in the case calculation of this study is Standard K-w [5]. The standard turbulence equation is as follows:

$$\mu_e=\mu+C_{\mu\rho}\frac{k^2}{\epsilon} \qquad K_e=K+\frac{\mu_t C_p}{\sigma_t} \tag{2}$$

where:
μ_e = *Viscosity effectiveness*
K_e = *Conductivity effectiveness*
σ_t = *Prandtl Turbulent Value*

3. Momentum Equation
In the Newtonian fluid formulation, the relation between stress and the value of the rate of deformation in the fluid is defined as follow:

$$\tau_{ij}=-P\delta_{ij}+\mu\left(\frac{\partial_{u_i}}{\partial_{x_j}}+\frac{\partial_{u_j}}{\partial_{x_i}}\right)+\delta_{ij}\,.\,\lambda\frac{\partial_{u_i}}{\partial_{x_i}} \tag{3}$$

where:
τ_{ij} = *Stress Tensor*
u_i = *Orthogonal Velocity* ($u_1=V_x$, $u_2=V_y$, $u_3=V_z$)
μ = *Dynamic Velocity*
λ = *Coefficient Velocity Value*

4. Energy Equation
The energy equation applied in CFD calculations is based on the conservation of energy law, specifically adapted for the boiling process using the Lee method. The equations used are:

$$\frac{\partial}{\partial t}(\rho H)+\nabla.(\rho\vec{v}T)=\nabla.(k\nabla T)+S \tag{4}$$

where:
H = *Enthalpy* ($H=h+\Delta H$)
ρ = *Density*
$\vec{v}$ = *Fluid Velocity*
S = *Source term*

5. Phase Change Equation
The flow in the pressurized tank is a two-phase flow, with two distinct and immiscible phases separated by a free surface. To model these different phases, the Volume of Fluid (VoF) method

is used. To model the boiling of gas in LNG-liquid in the CFD application tank, use the following equation.

$$\dot{m}_{lv} = \begin{cases} \gamma \alpha_l \rho_l \frac{T - T_{sat}(p)}{T_{sat}(p)}, & if\ T > T_{sat} \\ \gamma \alpha_v \rho_v \frac{T - T_{sat}(p)}{T_{sat}(p)}, & if\ T \leq T_{sat} \end{cases} \tag{5}$$

where:
γ = *vaporization rate* (s^{-1})
T_{sat} = *Saturation Temperature (k)*
ρ = *Density*

6. Surface tension effect
The Surface Tension Effect is one of the important equations for capturing bubble dynamics. Each CFD application provides guidance on the importance of this effect. The criteria in this equation depend on the Reynolds number (Re), the capillary number (Ca) and the Weber number (We). For the case where Re is very low i.e., Re<<1, the quantity considered is the capillary number. When Re>>1, the quantity to consider is the Weber number. The effect of surface tension is negligible if Ca or We are much larger than unity. The equation of the surface tension effect is explained as follows:

$$Re = \frac{\rho VL}{\mu}, Ca = \frac{\mu V}{\sigma}, We = \frac{\rho LV}{\sigma} \tag{6}$$

where:
Re = *Reynold Number*
Ca = *Capiler Number*
We = *Weber Number*
ρ = *Density*

Based on research conducted by [1] the Reynolds value will be greater or smaller than the Weber value depending on the heat flux parameter, or it can also be from the heat ingress whether it is larger or smaller.

7. Volume of Fluid
The phase change behaviour of LNG into gaseous phase requires some phases change volumetric calculations. To calculate the interphase volume of fluids the equation used is explained as follows:

$$\frac{1}{\rho_q}\left[\frac{\partial}{\partial t}(\alpha_q \rho_q) + \nabla.(\alpha_q \rho_q v_q) = \sum_{p=1}^{n}(\dot{m}_{pq} - \dot{m}_{qp})\right] \tag{7}$$

where:
ρ = *Density*
n = *Number of phases*
v = *Flow speed*

3. The Modelling Setups

3.1 LNG Storage Tank Material Properties

In order to conduct the simulation of LNG storage tank, there is some adjustment in the CFD model that need to be undertaken. The primary rules for the geometry In CFD application are including ensuring the model is rigid, no overlapping parts and ensuring the volumetric capacity of the tank match the parameter for each layer of the tank. The detailed geometry of the tank in the CFD simulation are taken from the previous research conducted by [6] which follow the dimensional principle explained as follows:

Volume	: 89 m^3
No. of Layers	: 2 layers (inner and outer layers)
Type	: Torispherical head type.
Material	: Inner side is Stainless Steel 304 with wall thickness of 11 mm and a head thickness of 17mm. Outer part it is a type of Carbon Steel with wall thickness of 5 mm and a head thickness of 5 mm.

One consideration as mentioned in the specification, the volume considered is 89,3 m^3 are obtained from the software generated value with a slight different value compared to the [6]. The illustration of 3D model of the tank can be seen in the Figure 1.

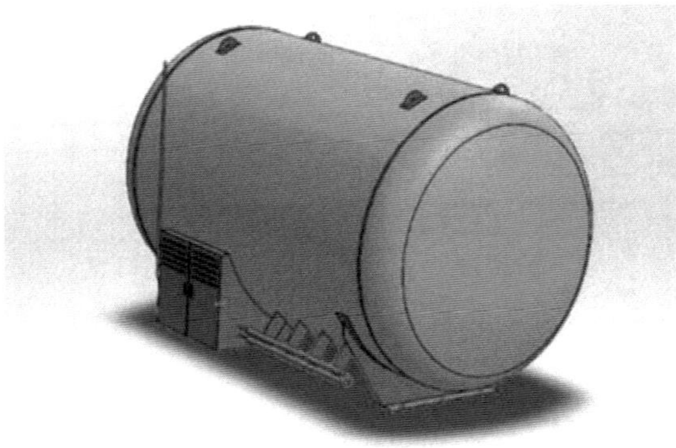

Figure 1. LNG Storage Tank for Harbour Tug 2 x 2500 HP

However, the tank must withstand the specified LNG characteristic/specification, which on this case will follow the specific properties of LNG is explained as follows:

- Density : 424.2 Kg/m^3
- Specific Heat : 3480 j/g-°C
- Thermal Conductivity : 0.1859 W/m-K
- Viscosity : 0.000117
- Molecular Weight : 16.043
- Temperature Ref. : -164.5°C

Meanwhile for the NG the specification is explained as follows:

- Density : 1.811 Kg/m3
- Specific Heat : 2217.4 j/g-oC
- Thermal Conductivity : 0.014 W/m-K
- Viscosity : 4.32E-06
- Molecular Weight : 16.043
- Temperature Ref. : -161 °C

IOP Conf. Series: Earth and Environmental Science **1423** (2024) 012040 doi:10.1088/1755-1315/1423/1/012040

3.2 Simulation Validation

The validation process of simulation on this research will be used to determine the possible deviation of the result. For this research, it uses residual validation which provided by the CFD software in which the value can be obtained when the model is running. The validation process will be obtained during two conditions, firstly at ship is not in operation, secondly, when the ship under operating condition. The result from validation process is shown in the Figure 2 for ship not in operation and Figure 3 for ship under operating condition.

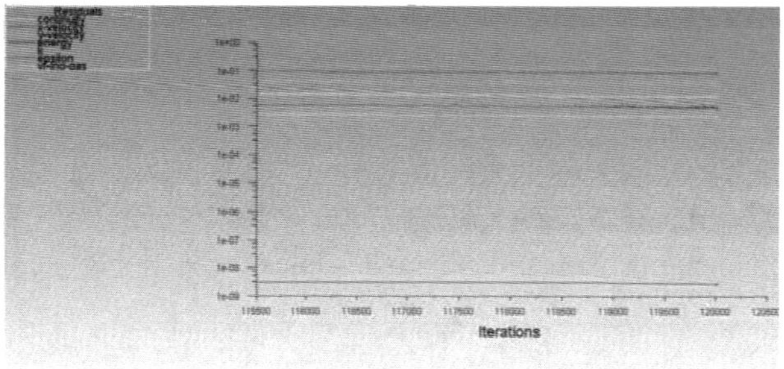

Figure 2. The graph of residual value in trials when the ship is not operating

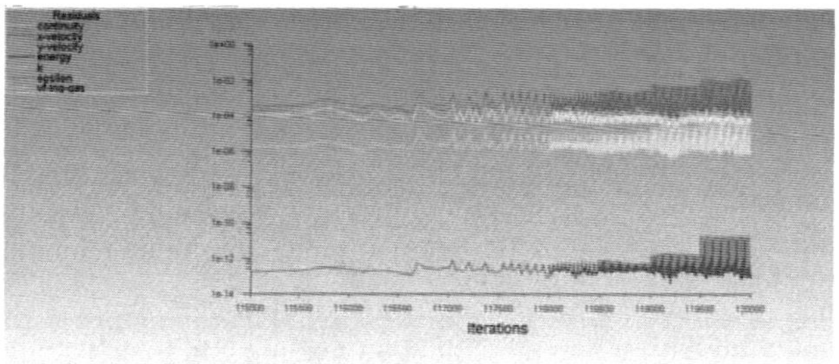

Figure 3. The graph of the residual value in the experiment when the ship under operating condition

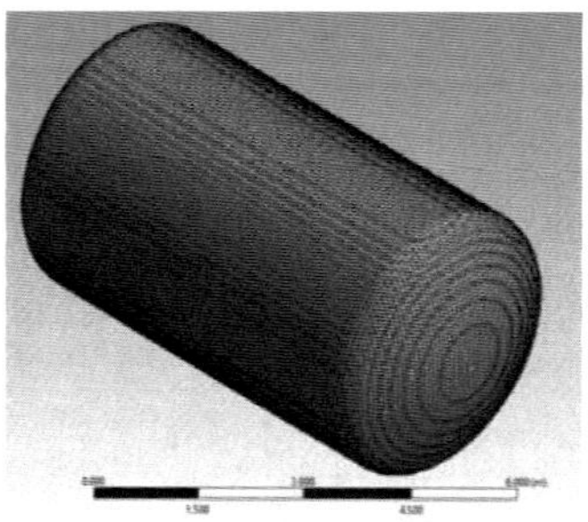

Figure 4. 3D Meshing on Pressure LNG storage tank

3.3 Tank Model Meshing

Meshing is an important step in CFD model since in the numerical method the point in the tank model geometry will become the pinpoint of the calculation. Each point also connected, and the type of connection will determine the result of meshing process. The number of final mesh also will determine the accuracy of simulation result. The meshing initially coming from the 3D model as seen in the Figure 4 and then proceeded into the 2D meshing for ship not in operation and under operation as seen in the Figure 5 and 6.

For meshing in 2D geometry, only the Face Sizing method is used, where the average skewness value is 3.3733e-02 and the maximum value is 0.56205, with the number of nodes 23,401 and elements 23,151. For meshing in the experimental ship operation, the average skewness value for meshing is 4.482e-02 and a maximum value of 0.55166 with a total of 23,457 nodes and a total of 23,205 elements.

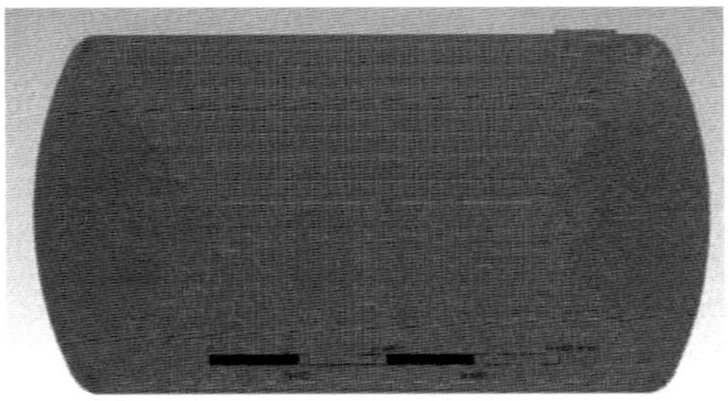

Figure 5. Meshing on Pressure LNG storage tank when the ship is not operating

Figure 6. Meshing on Pressure LNG storage tank when the ship is under operating

4. Simulation Result for Non-Operating Ship

This research emphasizing on the effect of BOG onboard LNG storage tank for Harbor Tug, in which the simulation will represented in the actual condition by varying the scenarios as similar as possible. The variations of Harbor Tug and the effect of the BOG will be mainly driven by the volumetric quantity. In this condition this research has four variables of volume of the tank contents in the simulations, which are: 85% volume, 75% volume, 50% volume, and 25% volume. Another consideration is when the ship is not under operating condition, the initial pressure difference (t=0) is 1 ATM. Beside the mentioned condition in there are also several simulation variables need to be considered, the detailed variables for non-operating ship condition are explained as follows:

Temperature Atmosphere	: 32 ^{o}C
Temperature LNG-liq	: -164,4 oC
Temperature NG-gas	: -160,435 oC
Gravity	: X (0) Y (-9.81) Z (0)
Energy	: On
VoF	: On
Viscosity	: Standard K-w
Surface Tension	: On (Constant = 0.014)
Heat Flux	: 0,025 W/m2K
Pressure	: 1 ATM

In order to explain the simulation result for various conditions will be explained in the following subchapter. Please note that the depiction of the simulation will be represented in the graphical chart of volume fraction of an LNG against time and pressure gradient inside the tank for certain conditions.

4.1 85% LNG Tank Volume Capacity

For 85% tank volume capacity the graphical result of volume fraction against time can be seen on the Figure 7 meanwhile illustration of CFD simulation for volume fraction and pressure gradient inside the tank can be seen in the Figure 8.

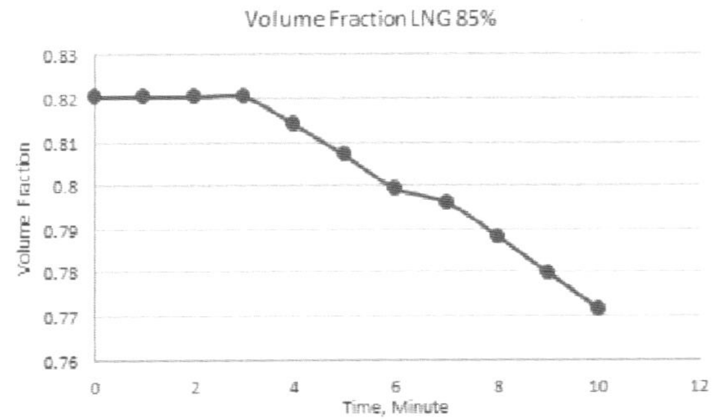

Figure 7. Graph of Changes in the Volume of Liquid LNG Fraction For 10 Minutes

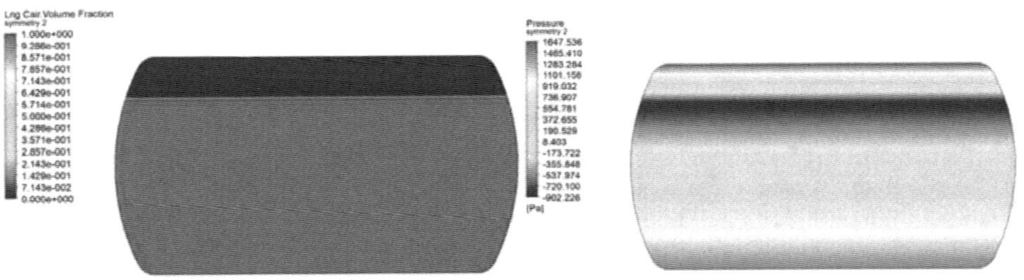

Figure 8. Volume fraction LNG storage tank (85%) and Pressure Gradient of LNG Storage Tank at 0s time

The results from the simulation indicate that atmospheric heat transfer into the liquid LNG induces the BOG. This thermal input causes the volume fraction of the LNG to decrease from 0.8204 to 0.7713, while the pressure increases from 276.8 Pa to 443.3 Pa. The observed boiling rate is quantified at 0.447 m^3/min, corresponding to a volumetric reduction rate of 0.492% per minute.

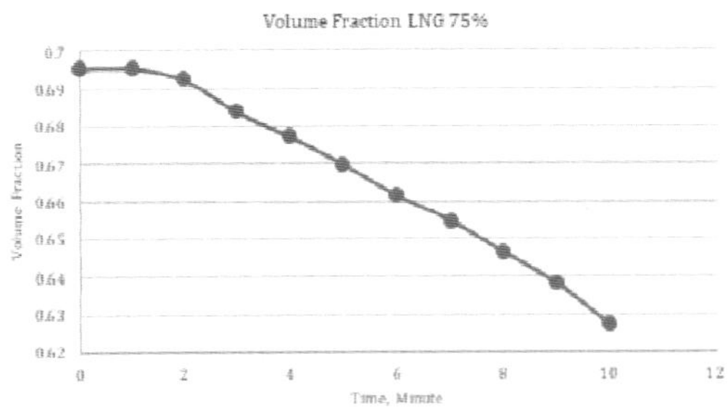

Figure 9. Graph of Changes in the Volume of Liquid LNG Fraction for 10 Minutes at 75% Tank Volume

IOP Conf. Series: Earth and Environmental Science **1423** (2024) 012040 doi:10.1088/1755-1315/1423/1/012040

4.2 75% LNG Tank Volume Capacity

For 75% tank volume capacity the graphical result of volume fraction against time can be seen on the Figure 9 meanwhile illustration of CFD simulation for volume fraction and pressure gradient inside the tank can be seen in the Figure 10.

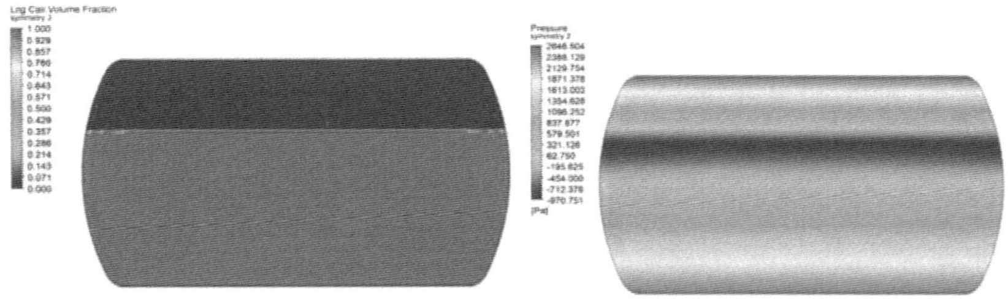

Figure 10. Volume fraction LNG storage tank (75%) and Pressure Gradient of LNG Storage Tank at 0s time

The results from the simulation conducted with a 75% volume indicates that atmospheric heat transfer to the LNG induces BOG. This thermal exchange results in a reduction of the volume fraction of the LNG from 0.695 to 0.627, accompanied by an increase in pressure from 761.2 Pa to 1145.9 Pa. The rate of boil-off is determined to be 0.6075 m^3/min, equivalent to a volumetric reduction rate of 0.685% per minute.

4.3 50% LNG Tank Volume Capacity

For 50% tank volume capacity the graphical result of volume fraction against time can be seen on the Figure 11 meanwhile illustration of CFD simulation for volume fraction and pressure gradient inside the tank can be seen in the Figure 12.

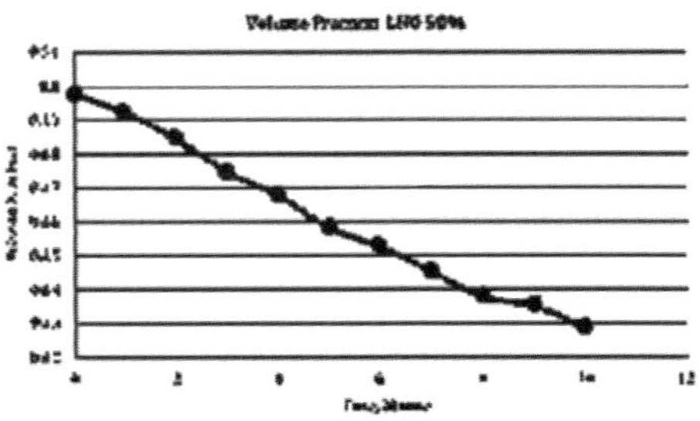

Figure 11. Graph of Changes in the Volume of Liquid LNG Fraction for 10 Minutes at 50% Tank Volume

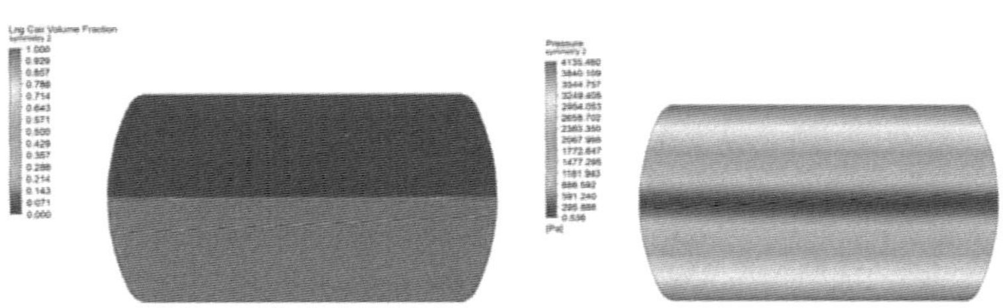

Figure 12. Volume fraction LNG storage tank (50%) and Pressure Gradient of LNG Storage Tank at 0s time

The results from the simulation of 50% volume LNG experiment indicate that atmospheric heat transfer to the LNG induces BOG. This process results in a decrease in the volume fraction of the liquid LNG from 0.498 to 0.429, while the pressure decreases from 1996.4 Pa to 1495.6 Pa. The boil-off rate is quantified at 0.612 m^3/min, corresponding to a volumetric reduction rate of 0.686% per minute.

4.4 25% LNG Tank Volume Capacity

For 25% tank volume capacity the graphical result of volume fraction against time can be seen on the Figure 13 meanwhile illustration of CFD simulation for volume fraction and pressure gradient inside the tank can be seen in the Figure 14.

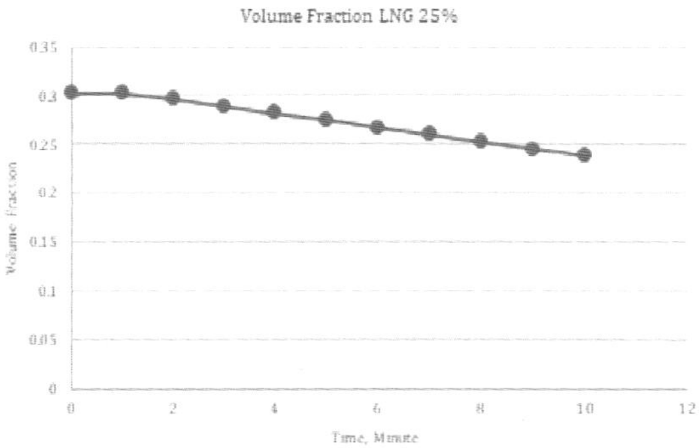

Figure 13. Graph of Changes in the Volume of Liquid LNG Fraction for 10 Minutes at 25% Tank Volume

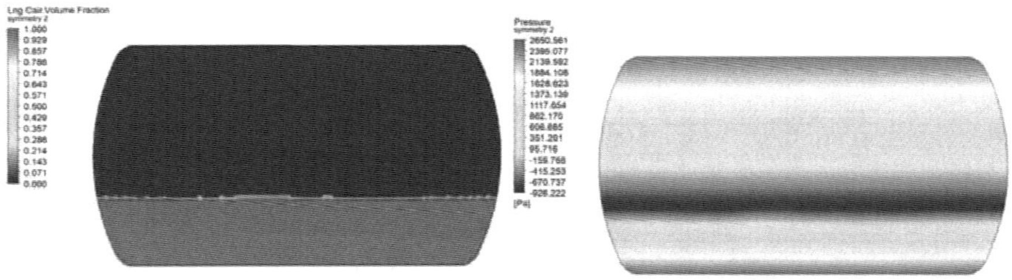

Figure 14. Volume fraction LNG storage tank (25%) and Pressure Gradient of LNG Storage Tank at 0s time

The results from the simulation of 25% volume LNG experiment indicate that atmospheric heat transfer to the LNG induces BOG. This process results in a decrease in the volume fraction of the liquid LNG from 0.303 to 0.238, while the pressure decreases from 758.6 Pa to 476.9 Pa The boil-off rate is quantified at 0.585 m^3/min, corresponding to a volumetric reduction rate of 0.645% per minute.

5. Simulation Result for Ship Under Operation

This chapter emphasize on the simulation as closely as possible to the conditions of ships under sailing condition with dynamic effect. In this condition, this research conducts 5 variables of flow rate value that adjusted according to the ship mode. The 5 modes namely: full speed mode,

IOP Conf. Series: Earth and Environmental Science **1423** (2024) 012040 doi:10.1088/1755-1315/1423/1/012040

economy speed mode, slow speed mode, manoeuvring mode, and ship towing. In operating conditions, the ship has a different initial pressure (t = 0) compared to the non-operating condition which is 7.5 Bar. The summarized variables setup for ship under operations scenario is explained as follows:

Temperature Atmosphere	: 32 ^{o}C
Temperature LNG-liq	: -164,4 ^{o}C
Temperature NG-gas	: -160,435 ^{o}C
Gravity	: X (0) Y (-9.81) Z (0)
Outlet to GVU	: ship operation mode adjustment
Energy	: On
VoF	: On (adjustment volume)
Viscous	: Standard K-w
Surface Tension	: On (Constant = 0.014)
Heat Flux	: 0,025 W/m2K
Pressure	: 7.5 Bar

5.1 Full Speed Mode

The graphical result of volumetric fraction shifting over time for full speed mode is shown in the Figure 15, meanwhile illustration of CFD simulation for volume fraction and pressure gradient inside the tank can be seen in the Figure 16.

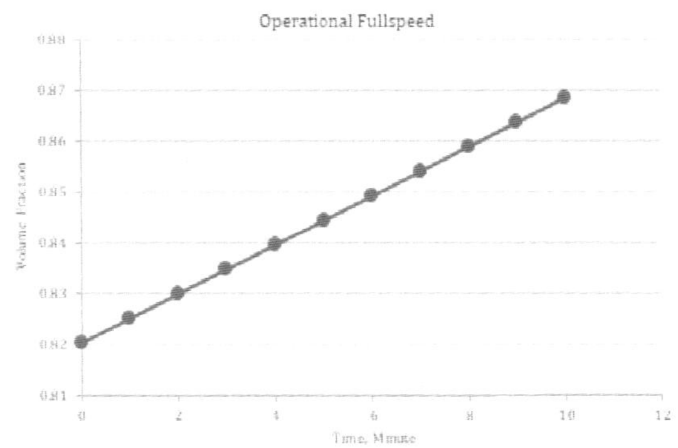

Figure 15. Changes in the Volume of LNG Fraction For 10 Minutes at Full Speed

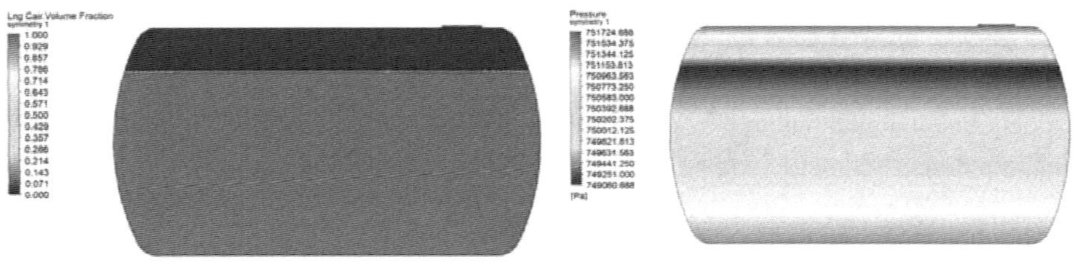

Figure 16. Volume fraction LNG storage tank (Full Speed) and Pressure Gradient of LNG Storage Tank at 0s time

The results of the simulation conducted under full-speed ship conditions reveal that condensation occurs in the LNG, attributable to a reduction in gaseous LNG at a rate of 1.66×10^{-5}

m^3/s. Consequently, the volume fraction of the liquid LNG increases from 0.8203 to 0.8682, while the pressure decreases from 750,265 Pa to 750,146 Pa. The observed boil-off rate is -0.428 m^3/min.

5.2 Economic Speed Mode

The graphical result of volumetric fraction shifting over time for economic speed mode is shown in the Figure 17, meanwhile illustration of CFD simulation for volume fraction and pressure gradient inside the tank can be seen in the Figure 18.

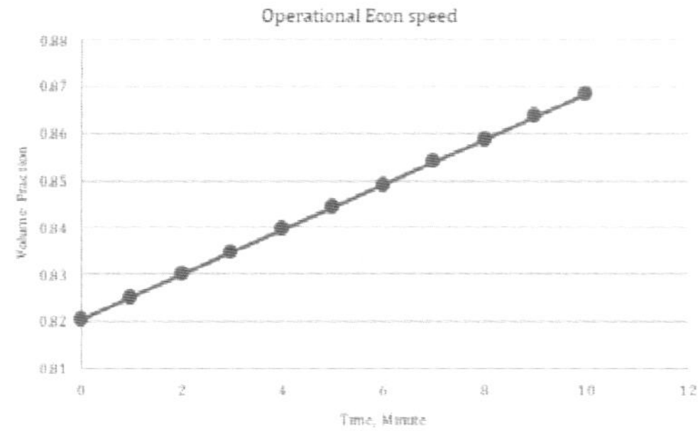

Figure 17. Changes in the Volume of LNG Fraction For 10 Minutes at Economic Speed

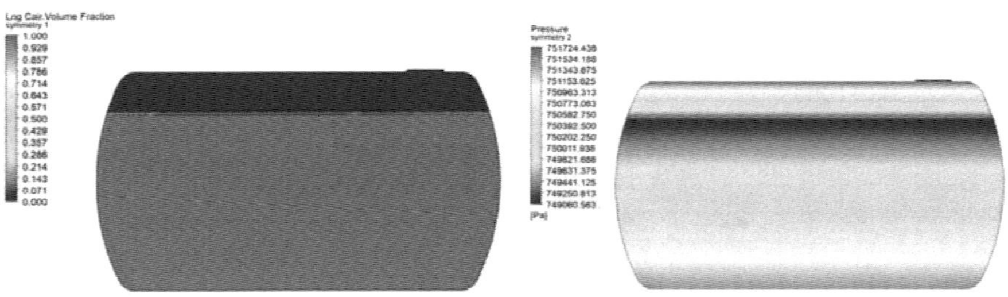

Figure 18. Volume fraction LNG storage tank (Economic Speed) and Pressure Gradient of LNG Storage Tank at 0s time

The results of the LNG experiment conducted under economic speed ship conditions reveal that condensation occurs in the LNG, attributable to a reduction in gaseous LNG at a rate of 2.67 $\times 10^{-5}$ m^3/s. Consequently, the volume fraction of the liquid LNG increases from 0.8203 to 0.8684, while the pressure decreases from 750,265 Pa to 750,145 Pa. The observed boil-off rate is -0.430 m^3/min.

5.3 Slow Speed Mode

The graphical result of volumetric fraction shifting over time for slow speed mode is shown in the Figure 19, meanwhile illustration of CFD simulation for volume fraction and pressure gradient inside the tank can be seen in the Figure 20.

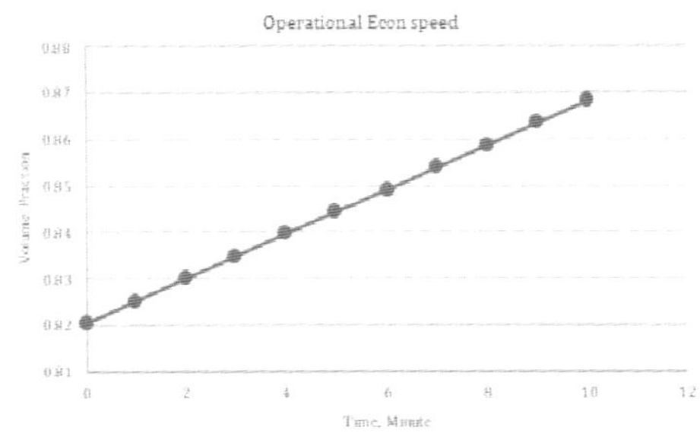

Figure 19. Changes in the Volume of LNG Fraction For 10 Minutes at Slow Speed

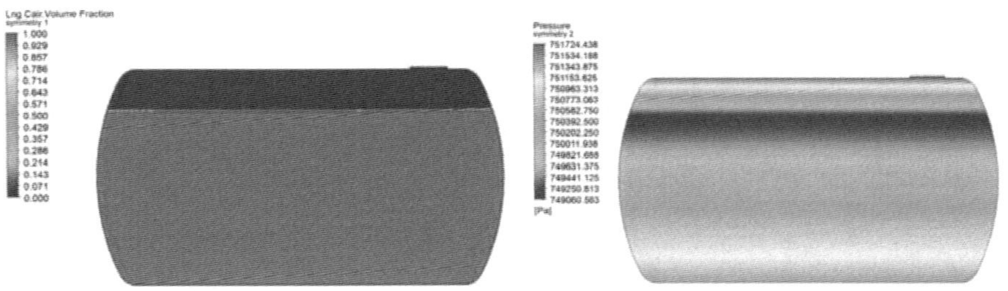

Figure 20. Volume fraction LNG storage tank (Slow Speed) and Pressure Gradient of LNG Storage Tank at 0s time

The results of the LNG experiment conducted under slow speed ship conditions reveal that condensation occurs in the LNG, attributable to a reduction in gaseous LNG at a rate of 5.50×10^{-5} m^3/s. Consequently, the volume fraction of the liquid LNG increases from 0.8203 to 0.8691, while the pressure decreases from 750,265 Pa to 750,144 Pa. The observed boil-off rate is -0.436 m^3/min.

5.4 Manoeuvre Mode

The graphical result of volumetric fraction shifting over time for manoeuvre mode is shown in the Figure 21, meanwhile illustration of CFD simulation for volume fraction and pressure gradient inside the tank can be seen in the Figure 22.

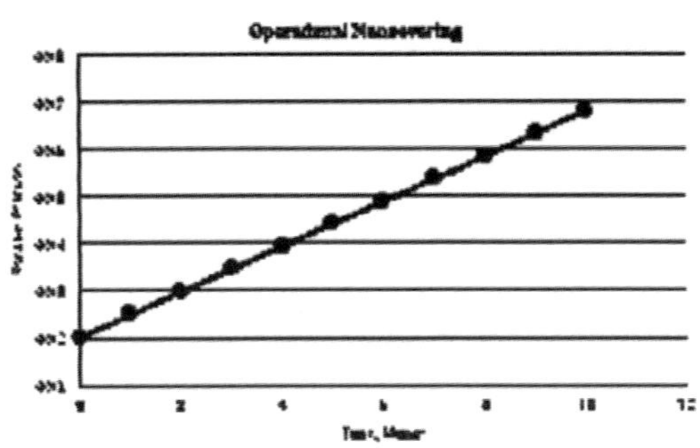

Figure 21. Changes in the Volume of LNG Fraction For 10 Minutes at Slow Speed

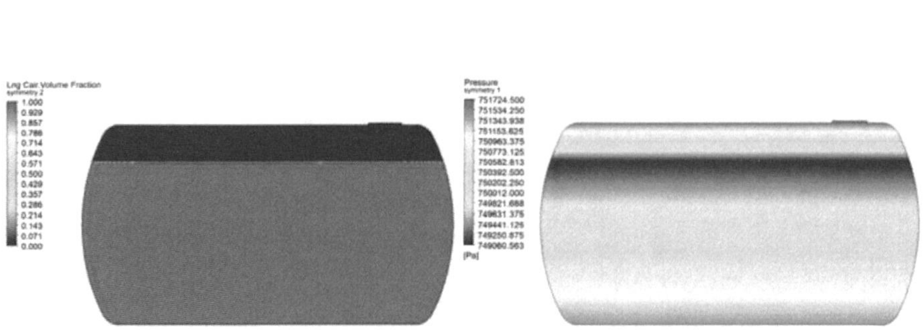

Figure 22. Volume fraction LNG storage tank (Manoeuvre Mode) and Pressure Gradient of LNG Storage Tank at 0s time

The results of the LNG experiment conducted under manoeuvre speed ship conditions reveal that condensation occurs in the LNG, attributable to a reduction in gaseous LNG at a rate of 7.82 × 10^{-5} m^3/s. Consequently, the volume fraction of the liquid LNG increases from 0.8203 to 0.8680, while the pressure decreases from 750,265 Pa to 750,146 Pa. The observed boil-off rate is -0.426 m^3/min.

5.5 Towing Mode

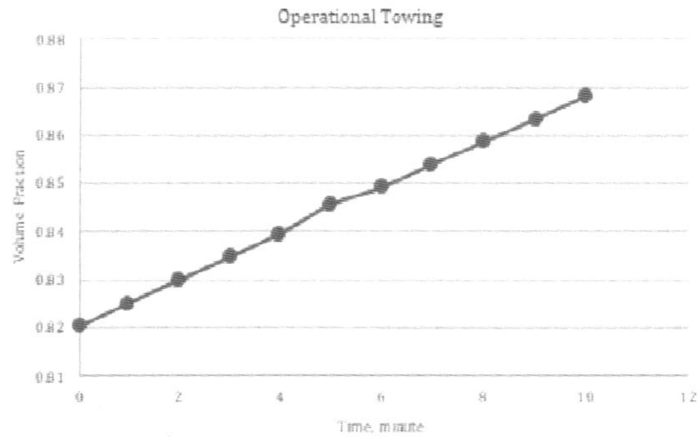

Figure 23. Changes in the Volume of LNG Fraction For 10 Minutes at Slow Speed

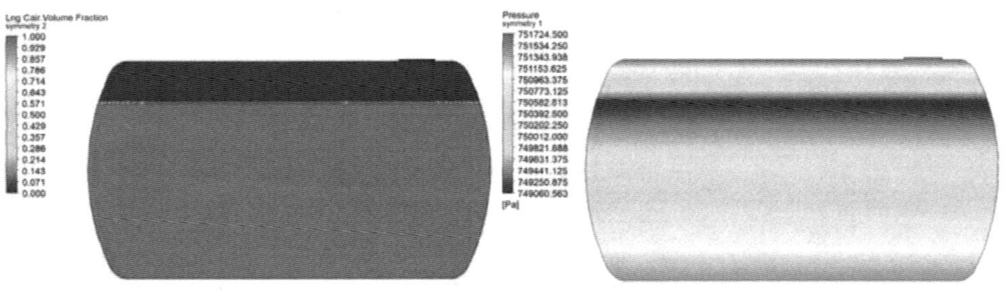

Figure 24. Volume fraction LNG storage tank (Manoeuvre Mode) and Pressure Gradient of LNG Storage Tank at 0s time

The graphical result of volumetric fraction shifting over time for towing mode is shown in the Figure 23, meanwhile illustration of CFD simulation for volume fraction and pressure gradient inside the tank can be seen in the Figure 24.

The results of the LNG experiment conducted under manoeuvre speed ship conditions reveal that condensation occurs in the LNG, attributable to a reduction in gaseous LNG at a rate of 7.82 × 10^{-5} m^3/s. Consequently, the volume fraction of the liquid LNG increases from 0.8203 to 0.8680, while the pressure decreases from 750,265 Pa to 750,146 Pa. The observed boil-off rate is -0.426 m^3/min.

6. Simulation Result Analysis

6.1 Analysis on Ship non-operating

Simulations on non-operating condition reveal that BOG rates increase as the LNG volume in the tank decreases, demonstrating an inverse relationship between BOG rates and tank volume. The experimental BOG rates observed were 0.447 m^3/min for a full tank condition, 0.6075 m^3/min for 75% volume, 0.612 m^3/min for 50% volume, and 0.585 m^3/min for 25% volume. Notably, the BOG rate for the 25% volume condition is lower than for the 50% condition, likely due to lower pressure in the one-quarter full scenario. Rate dynamics significantly influence BOG rates; increased gas concentration within the tank leads to higher pressure, which in turn affects the rate of boil-off gas generation.

6.2 Analysis on Ship Under Operation

During ship operations, an unexpected anomaly was observed: an increase in the concentration of liquid LNG, as evidenced by a negative BOG rate, indicating a reduction in the gas phase and the occurrence of condensation. The experimental BOG rates were recorded as -0.428 m^3/min at full speed, -0.430 m^3/min at economy speed, -0.436 m^3/min at slow speed, -0.426 m^3/min at manoeuvring speed, and 0.427 m^3/min while towing.

This inverse relationship between LNG fuel demand and BOG rates suggests that higher fuel consumption results in increased condensation within the LNG tank. This can be attributed to the direct proportionality between BOG rates and tank pressure: higher LNG gas demand in different operational modes leads to a faster pressure decrease in the tank, which subsequently affects the BOG rate.

Additionally, the heat radiated by LNG gas significantly influences the BOG rate. Experimental results show that higher LNG gas outflow enhances the energy within the liquid LNG, promoting condensation of the gas phase. This dynamic can vary over time as atmospheric heat input into the liquid LNG can intermittently increase the energy of the gas phase concentration.

7. Conclusion

The conclusions derived from the Boil-off Gas (BoG) experiments conducted on the LNG tank of the Harbor Tug 2 x 2500HP vessel are as follows:

When the Harbor Tug 2 x 2500HP vessel is in a non-operational, steady-state condition, the BOG values vary with different LNG tank volumes. Specifically, the BOG rates are 0.447 m^3/min for a full tank (85%), 0.6075 m^3/min for a three-quarters full tank (75%), 0.612 m^3/min for a half-full tank (50%), and 0.585 m^3/min for a quarter-full tank (25%). These variations occur due to the combined effects of heat ingress and tank pressure. Notably, the highest BOG rate is observed at the half-full condition (50%), where the pressure is maximized due to both the vapor pressure of LNG and the hydrostatic pressure of the liquid phase. Conversely, the lowest BOG rate is observed at the full tank condition (85%), where the pressure is minimized, comprising primarily hydrostatic pressure. Thus, the BOG rate is directly proportional to the pressure within the LNG tank.

When the Harbor Tug 2 x 2500HP vessel is operational, the BOG values exhibit a range of measurements: -0.428 m^3/min at full speed, -0.430 m^3/min at economic speed, -0.436 m^3/min at slow speed, -0.426 m^3/min at manoeuvring speed, and 0.427 m^3/min while towing. These results reveal an anomaly where LNG gas condensation occurs within the tank, attributed to a pressure drop that affects BOG formation rates. The thermal energy of the liquid LNG influences the gaseous phase, leading to condensation. This dynamic can fluctuate over time as atmospheric heat transfer into the LNG tank increases the thermal energy of the LNG gas, subsequently raising the BOG rate.

Acknowledgement

The authors gratefully acknowledge financial support from the Institut Teknologi Sepuluh Nopember for this work, under project scheme of the Publication Writing and IPR Incentive Program (PPHKI) 2024.

References

[1] K. Dong, G. Hochman, Y. Zhang, R. Sun, H. Li, and H. Liao, "CO2 emissions, economic and population growth, and renewable energy: Empirical evidence across regions," *Energy Economics*, vol. 75, pp. 180–192, Sep. 2018, doi: 10.1016/j.eneco.2018.08.017.

[2] F. Perez *et al.*, "Measurements of boil-off gas and stratification in cryogenic liquid nitrogen with implications for the storage and transport of liquefied natural gas," *Energy*, vol. 222, p. 119853, May 2021, doi: 10.1016/j.energy.2021.119853.

[3] S. Mokhatab, *Handbook of liquefied natural gas*. Amsterdam: Elsevier/Gulf Professional Publishing, 2014.

[4] J. L. Ferrín and L. J. Pérez-Pérez, "Numerical simulation of natural convection and boil-off in a small size pressurized LNG storage tank," *Computers & Chemical Engineering*, vol. 138, p. 106840, Jul. 2020, doi: 10.1016/j.compchemeng.2020.106840.

[5] A. Saleem, S. Farooq, I. A. Karimi, and R. Banerjee, "A CFD simulation study of boiling mechanism and BOG generation in a full-scale LNG storage tank," *Computers & Chemical Engineering*, vol. 115, pp. 112–120, Jul. 2018, doi: 10.1016/j.compchemeng.2018.04.003.

[6] E. N. Nugroho, *Desain LNG Storage Untuk Sistem Bahan Bakar Pada LNG Fueled Harbour Tug 2 x 2500 HP*.

IOP Conf. Series: Earth and Environmental Science **1423** (2024) 012041 doi:10.1088/1755-1315/1423/1/012041

Investigating Post-fire Critical Area of ship base on Numerical Fire Simulation

S Anggara[1] *, S Komariyah[1], R F Hariadi[1] and R Lesmana[1]

[1] Research and Development, Biro Klasifikasi Indonesia, Jakarta, Indonesia

*sony.anggara@bki.co.id

Abstract. Ship fires are recognized as significant maritime incidents that can lead to substantial losses, including loss of life. In events where fires severely compromise a vessel's structural integrity, the costs associated with comprehensive repairs can be exorbitant, placing considerable pressure on shipowners. Ensuring post-fire structural integrity is paramount. Therefore, it is essential to provide technical considerations that aid shipowners and classification societies in identifying critical areas of a ship's structure that require replacement. This study aims to investigate the post-fire critical areas of ship structures through numerical fire simulations. By leveraging the properties of steel, which can undergo microstructural changes when subjected to high temperatures and rapid cooling, critical structural areas can be identified based on fire-induced heat exposure. A specific case study involving a fire incident aboard a 7,500 GT bulk carrier, triggered by the failure of an auxiliary engine, is examined. The simulations utilize on-site data and actual post-fire conditions to draw robust conclusions. The insights derived from this research are expected to enhance the understanding of the impacts of fire on ship construction and facilitate more efficient and accurate post-fire inspections.

1. Introduction

Ship fires warrant special consideration due to their high frequency. According to available data [1], at least 37% of the 120 ship accidents over the past decade were due to fire incidents. This alarming statistic predominantly involves motor vessels, particularly ro-ro ferries, cargo ships, and bulk carriers. The high fire risks associated with ships stem from factors such as their cargo, machinery, and electrical systems. Ships that endure fire incidents often face total loss, or at a minimum, suffer significant structural damage. Hence, the post-fire treatment of ships is a critical discussion point, especially considering the substantial repair costs incurred by owners.

Shipowners must evaluate whether their vessels can remain operational post-accident from a financial standpoint, taking into account potential repair costs. An engineering analysis is necessary to assess the extent of required repairs while prioritizing safety. One such analysis includes determining whether the ship's structural strength can withstand operational loads as effectively as before or if a complete structural replacement is warranted due to compromised integrity. In the latter case, the ship may no longer be financially viable.

To illustrate the significance of this research, a case study involving a 115-meter-long bulk carrier is examined. The data detailing the event chronology is derived from incident reports and witness accounts compiled on-site. The fire occurred while the vessel was at dock, loading fertilizer cargo. Following the completion of approximately 5,000 MT of cargo, the fire alarm activated as flames erupted in the engine room due to an auxiliary engine malfunction. Consequently, the ship experienced a blackout. Three fire trucks and one tugboat were deployed to assist in extinguishing the flames. However, the fire spread to the accommodation deck and superstructure, resulting in significant damage to the aft section. The firefighting efforts lasted nearly six hours, utilizing seawater.

This study employs numerical fire simulations to gain a comprehensive understanding of the structural integrity post-fire, thereby aiding shipowners in determining which structural components necessitate replacement.

2. Methodology

Investigating material property changes at the micro-level requires understanding temperature variations experienced during a fire event. These temperature profiles can be generated through simulations using specialized engineering software. The Fire Dynamics Simulator (FDS) is an advanced tool capable of simulating diverse fire scenarios, including flame propagation from one fuel source to another.

While FDS is predominantly utilized in designing fire prevention and response systems in building structures, its application extends to maritime fire scenarios. To ensure effective operation, the BKI R&D team has conducted training sessions and knowledge transfer activities, facilitated by expert instructors who possess both academic and extensive practical experience with FDS in various fire safety projects. The stages of the ship fire simulation can be delineated into several phases [6]:

2.1 Modelling

Modelling is performed using external software, PyroSim. The model is constructed based on general arrangement drawings supplemented by field survey data. The comprehensive modelling includes the superstructure and deck house rooms. Each deck level—from the genset room on the operating deck to the main deck, upper deck, poop deck, "B" deck, "C" deck, bridge deck, wheelhouse, and top deck—is fully modelled, adhering to procedures outlined in reference [2]. Various components employed in the modelling process include [9]:

- **Obstructions:** Walls and slabs function as room separators, with walls defining vertical partitions and slabs representing decks or floors.
- **Holes:** These indicate access points such as doors, manholes, and hatch openings on decks, as well as ventilation features like windows and ducting. In this study, all doors and stairs are assumed to be open, allowing fire to spread unimpeded between rooms while oxygen entry maintains combustion.
- **Burners:** These components initiate the fire, with the genset serving as the primary heat source in this incident.
- **Reactions:** This element describes the chemical reactions that propagate fire, specifically in this case, methane.
- **Simulation Materials:** Identifying the materials used in obstructions is vital for determining fire propagation directions and temperature distributions. Field observations classify partition walls as multilayer wood/plywood, while external walls, decks, and engine casing/funnel walls are designated as steel.

2.2 Determining Input Parameters

Critical inputs for fire simulations include the Heat Release Rate (HRR) of the burner, indicating the amount of heat transferred between objects and influencing temperature distribution within the environment. The HRR for diesel fuel oil is reported to range from 42 to 46 MJ/kg [3].

The second input is time; due to computational limitations, simulations are conducted over

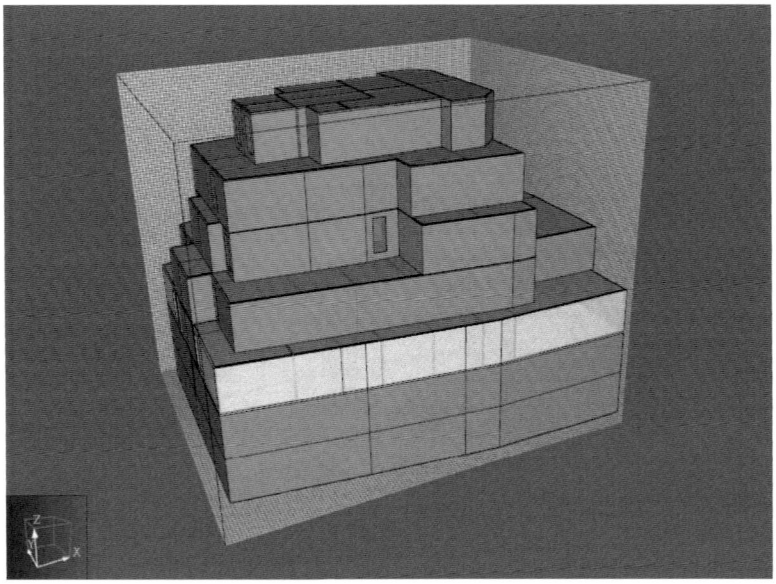

Figure 1. After hull and accommodation deck modelling in FDS

3,600 seconds. The first 20 seconds include the initial explosion, followed by fire propagation through partition walls, during which material temperatures are recorded.

2.3 Solver and Post-Processor

The simulations are conducted using FDS. Upon successful completion, temperature visualizations at various deck and wall positions can be analysed.

2.4 Material Testing

When steel is subjected to temperatures between 704°C (1,300°F) and 843°C (1,550°F) and subsequently cooled rapidly, certain microstructures transform into martensite—an incredibly hard yet brittle phase of steel [7]. If exposure temperatures remain below the transformation threshold, the mechanical properties of the steel can revert to pre-fire conditions after cooling, contingent on exposure duration and cooling rates [4]. Additionally, the degradation of steel's mechanical properties is influenced by the grade of steel used, with higher-strength steels experiencing greater deterioration than normal-strength steels.

To ascertain the mechanical and chemical properties of steel post-fire, several tests will be conducted, including Tensile Testing, Microstructure Analysis, Hardness Testing, and Impact Testing, with results to be presented in subsequent papers.

3. Result and Discussion

3.1 Simulation Result

The simulation output recorded the maximum temperature distribution at various locations (decks and partitions). The highest temperature was observed in the engine room, peaking at 1,000°C. Moving upwards within the structure, temperatures decreased, reaching a minimum of 296°C in the wheelhouse. Visualizations demonstrated that the fire extended to upper levels facilitated by the combustion of partition walls made from plywood. The fire subsequently spread through the after deck "B" door.

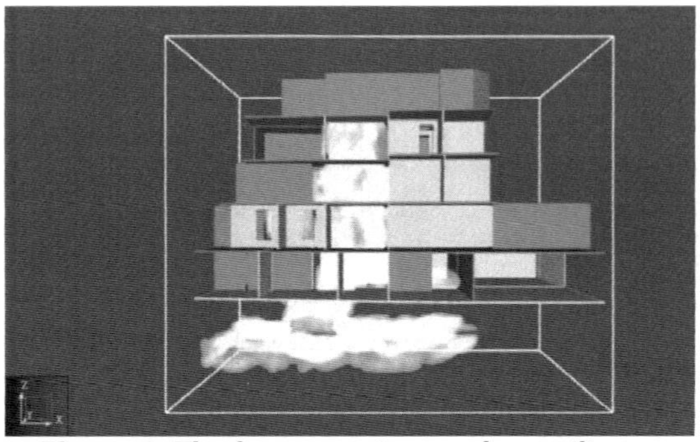

Figure 2. The first two minutes: the auxiliary engine ignites a fire that rapidly engulfs the engine room and spreads upward through the engine casing to upper accommodation levels.

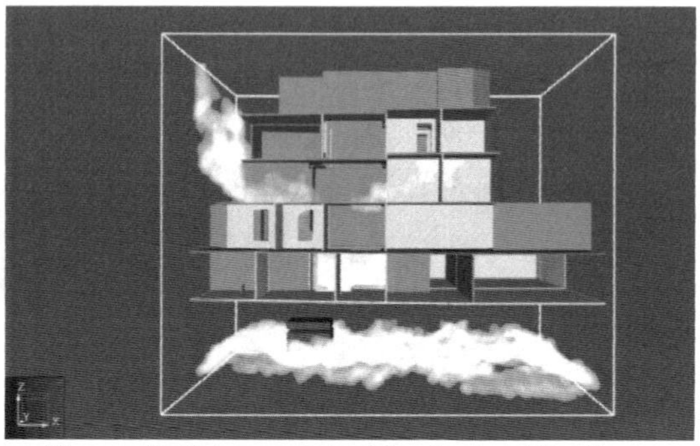

Figure 3. Between 2 to 15 minutes: The fire intensifies within the engine room while simultaneously spreading to higher accommodation spaces via openings such as windows and doors.

For analysis, samples were taken from areas experiencing significant deformation and visible fire damage. A total of nine test samples were collected from the locations marked in Figure 4. The temperatures experienced by sampled materials were recorded and presented in Table 1.

Table 1. Temperature experience by material selected for laboratory testing

Position	Temperature [^{0}C]
1	319
2	287
3	624
4	180
5	380
6	318
7	155
8	164
9	325

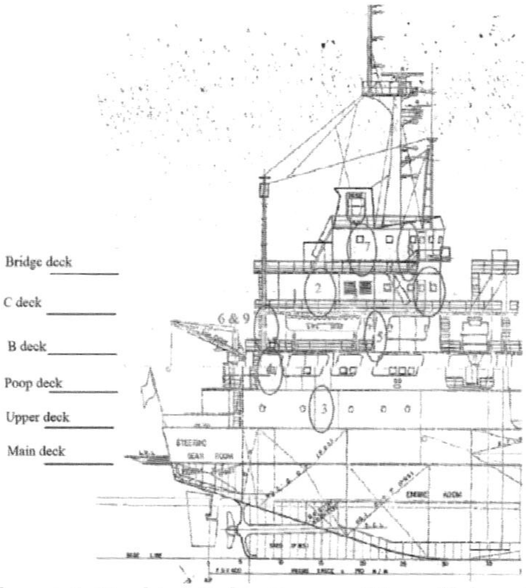

Figure 4. Marking indicating position of laboratory sample on ship

During the fire, it was discovered that the CO2 extinguishing system did not function adequately, necessitating the deployment of seawater by fire trucks and tugboats. Areas of the accommodation deck reachable by seawater were extinguished rapidly, whereas places like the engine room and poop deck required extensive time for complete extinguishment. To explore correlations with microstructural material changes, laboratory testing was conducted, revealing that structural components subjected to rapid cooling exhibited increased hardness compared to others. Sample 7 (engine casing) demonstrated a higher hardness value than other samples, albeit significantly below the martensitic hardness threshold, which begins at 56 HRC or 586 Brinell [5]. Consequently, no martensite formation was detected in any samples, whose hardness values remained below 586, likely due to temperatures during the fire being insufficient for phase change. Additional laboratory findings indicated that materials located near the fire source, yet not directly exposed to seawater for extinguishment, retained acceptable mechanical properties

Furthermore, utilizing seawater for firefighting resulted in corrosion of several structural components, as observed in field assessments. This corrosion raises concerns regarding potential structural integrity degradation over time, suggesting that these elements require replacement. Deformations were also identified in certain construction areas of the accommodation, presumably due to thermal expansion of the plates. The most significant deformation was noted in sample number 3. Deformed plates exhibited internal stress, leading to a reduction in material strength, as confirmed by decreases in the Rm value during tensile testing.

3.2 Recommendations for Future Protocols

The insights gained from this study prompt the formulation of detailed protocols for post-fire evaluations of maritime structures. Emphasizing the importance of material testing and specific criteria for assessing fire damage is paramount. Recommendations may include:

- **Detailed Inspection Protocols:** Establishing comprehensive guidelines for evaluating structural integrity and the potential for reusability in maritime contexts, inclusive of tensile and microstructural examinations.
- **Controlled Fire Scenarios**: Conducting advanced simulations and tests with controlled fire scenarios to refine predictive modelling for fire damage on various ship structures.
- **Corrosion Mitigation Practices:** Implementation of corrosion monitoring programs, particularly in areas impacted by seawater used during firefighting, to prevent long-term structural degradation.
- **Training and Knowledge Sharing**: Promoting workshops and training sessions for maritime professionals to familiarize them with post-fire evaluation techniques, ensuring they can effectively assess risks and maintain safety.
- **Collaboration with Classification Societies:** Encouraging cooperation between shipowners, classification societies, and research organizations to share findings and develop improved standards for fire risk management based on the latest research developments [10].

By adopting such measures, the maritime industry can enhance its operational resilience, mitigate eventual risks associated with fire incidents, and ensure the safety and structural integrity of vessels post-incident. The understanding of the effects of fire, as demonstrated in this case study, plays a pivotal role in developing smarter, safer, and more sustainable maritime practices going forward [8].

4. Conclusion

In summary, this study sheds light on the complexities surrounding post-fire structural integrity of steel ships. The findings underscore that while fire incidents pose a severe threat to maritime safety and cargo, not all fire exposures result in irrevocable damage to a vessel's structure. The combination of numerical fire simulations and laboratory testing allows for a comprehensive assessment of material characteristics post-fire.

Through meticulous examination of the steel's response to fire, particularly when subjected to elevated temperatures, we have established that temperatures exceeding 600°C can hinder the formation of martensitic structures, which are crucial for maintaining the mechanical integrity of steel components. The testing confirmed that rapid cooling processes may indeed enhance hardness; however, it should be noted that this increase in hardness does not guarantee material resilience.

Future research will be vital to deepen our understanding of the interplay between fire dynamics, material properties, and structural performance. This knowledge will equip maritime stakeholders with the tools necessary to enhance ship design, improve fire prevention systems, and ultimately to mitigate the repercussions of such catastrophic events. By prioritizing comprehensive post-fire analysis, the maritime industry can safeguard lives, assets, and the environment, reinforcing a culture of safety and accountability on the high seas. As we advance in these technical disciplines, we encourage collaboration between researchers, shipbuilders, and regulatory bodies to ensure best practices are adopted industry-wide.

References

[1] M. J. G. Perdana, I. P. Mulyatno and A. W. B. santosa, "Analisa Evakuasi Pada Kapal KM. Kirana IX 9168 GT Menggunakan Metode Agent Based Modelling Simulation Dalam Kondisi Kebakaran". Jurnal Teknik Perkapalan Universitas Diponegoro, Vol. 11, No. 2 April 2023

[2] Thunderhead Engineering, " PyroSim User Manual" , Version : 2022-1

[3] M. Bednarski, P. Orlinski and M. Wojs, "Evaluation of the heat release rate during the combustion process in the diesel engine chamber powered with fuel from renewable energy sources", Bulletin of the Polish Academy of Sciences, Technical Sciences 68(6):1333-1339, December 2020, DOI:10.24425/bpasts.2020.135394

[4] J. Wiley and Sons Ltd. "Microstructural characterization of materials". 2nd Edition, March 2008, DOI: 10.1002/9780470727133.ch2

[5] M. B. Eugine, " Brinell Hardness Test- determination of hardness of steel and aluminium", Journal of Materials Science and Engineering, Nov. 2012

[6] Höök, T., & Hjerpe, T. (2021). "Fire Dynamics for Fire Safety Engineering: A Numerical Approach." Fire Safety Science, Vol. 11, pp. 234-245.

[7] D. W. McGuire, "The Effects of Temperature on the Mechanical Properties of Structural Steel," Journal of Structural Engineering, 2019. DOI:10.1061/(ASCE)ST.1943-541X.0002196.

[8] International Maritime Organization (IMO), "Fire Safety in Ships: Guidelines and Best Practices." IMO Publications, 2020.

[9] B. G. G. B. Mackey, "Advanced Fire Simulation Techniques for Maritime Safety," Marine Technology and SNAME News, Vol. 60, No. 12, December 2020, pp. 45-54.

[10] N. J. M. Abdul Rahman, A. Abdullah, and H. Z. Mohd, "Impact of Fire on Structural Integrity of Marine Vessels," Journal of Marine Engineering and Technology, Vol. 22, No. 4, 2023, pp. 181-199, DOI:10.1080/20464177.2023.2235674

IOP Conf. Series: Earth and Environmental Science **1423** (2024) 012042 doi:10.1088/1755-1315/1423/1/012042

Steel Scrap Prediction from Indonesian Ship Recycling: Ship Type Point of View

W Mutmainnah[1*], F Fauzi[2], D R Mauliani[1], M A Kurniawan[1], K K Hendratna and Sunaryo[2]

[1] Biro Klasifikasi Indonesia, Jakarta 14320, Indonesia
[2] Department of Mechanical Engineering, Universitas Indonesia, Jakarta 16424, Indonesia

*E-mail: wanginingastuti@bki.co.id

Abstract. Hong Kong International Convention for the Safe and Environmentally Sound Recycling of Ships 2009 will entry into force on 26 Juni 2025, which is 24 month after Bangladesh and Liberia deposited their instrument to be a Contracted States to the Convention in 2023. As known widely, the output from the ship recycling activity is steel scrap that can be melted as a source of new steel. In Indonesia, the number of ship scrap is expected to be able to contribute to national steel industry. There are around 14.000 ships registered in Biro Klasifikasi Indonesia. More than 2000 ships have been operated for more than 25 years. The aim of this study is for predicting the weight of the steel scrap by calculating the Light Displacement Tonnage (LDT) of potential ships to be recycled from available data based on 11 ship types that has higher total GT. Total LDT estimation from ships more than 25 years is almost 2 million tons. This number would contribute positively to shipbuilding industry in Indonesia.

1. Introduction

The Hong Kong International Convention for the Safe and Environmentally Sound Recycling of Ships (HKC) was adopted in 2009 to ensure how a ship will be dismantled without risking a health and safety to human and environment. HKC covers requirements for ships that will be recycled and the Ship Recycling Facility (SRF) that will recycle the ship [1]. In June 2023, the HKC's requirement for entering into force has been met since Bangladesh and Liberia deposited their document to be a Contracting State and it is set to enter into force 24 months later. The main requirement for the ships that will be recycled is the Inventory of Hazardous Material (IHM). The IHM consists of Part I (hazardous material that is contained in ship structures and equipment), Part II (waste that is generated when the ship is in operation), and Part III (hazardous material that is stored in ship). The ship with more than 500GT that sail internationally should provide IHM Part I on board. While the IHM Part II and III should be provided when the ship will be sent to SRF. For the SRF, the main requirement is Document of Authorization to Conduct Ship Recycling (DASR). It is an authorization document from a competent authority that will be issued based on its Ship Recycling Facility Plan (SRFP).

In Indonesia, the ship recycling activity is regulated under Regulation No. 101 of 2014 on Hazardous Waste Management and Regulation of the Minister of Transportation No. 29 of 2014, that has been updated to No. 24 of 2022 on the Prevention of Marine Environmental Pollution (PM 24 of 2022) [2]. Government Regulation No. 101 of 2014 on the Treatment of Hazardous and

Toxic Wastes outlines the requirements for managing hazardous and toxic waste, including those generated from ship-recycling activities. However, it does not provide any guidelines on handling and treating the waste. Regulation of the Minister of Transportation No. 24 of 2022 on the Prevention of Marine Environmental Pollution provides more comprehensive requirements on ship recycling activities like HKC but with some additional requirements. While HKC only require IHM Part I should be on board the ships with more than 500GT that going internationally, the PM 24 of 2022 add the requirement for the domestic ships with 100GT and more to have IHM Part I, II, and III when the ship is in the end of its life [3].

Reported by NGO Ship Breaking Platform, the number of recycled ships since 2009 is 8078 ships [4]. In 2023, There were 444 ships that was dismantled around the world with total volume as 7.669.426 GT. The top three countries receiving ships for dismantling are Bangladesh, India, and Pakistan. Bangladesh has 170 ships to be dismantled (in total 3.232.593 GT), India 140 ships (in total 2.610.849 GT), and Pakistan 15 (in total 680.802 GT) [5]. According to their data, the average age of ships dismantled globally in 2023 is 34 years, with the oldest ship being 79 years old and the youngest being 14 years old. Of the 444 dismantled ships 11 were Indonesian-flagged ships with a total volume of 128.048 GT, scrapped in Indonesia it selves, India, and Bangladesh with the average age is 35 years old. Meanwhile, the total number of ships dismantled in Indonesia is 12 ships, with a total volume of 103.818 GT, originating from the flags of Indonesia, Seychelles, Singapore, and Taiwan.

There is no regulation in limiting the age of ships being operated. However, Hariyanto et al say that over 20 years old, ships are tending to be difficult to operate optimally [6]. According to data from DNV, more than 30% of recorded casualties between 2012 and 2021 occurred on vessels over 25 years old, with older general cargo and passenger vessels being the most involved in these casualties [7]. The older ship the more difficult for them to comply regulation especially for the decarbonisation era. In Indonesia, there are almost 14.000 active ships that is registered by Biro Klasifikasi Indonesia (BKI) where the oldest ship is 87 years old with average age is 14 years old [8]. As long as the ship is maintained well, the ship will be operated continuously as the ship owner will. However, the ship condition will meet some point that it need to be recycled.

The output from the recycled ship is steel scrap that can be used as material for a new steel. In India, circular economy has been practiced through its ship recycling industry since 1980s by converting waste to energy, waste to employment, and waste to wealth [9]. The materials generated by ship recycling activities are re-used, recycled, and refurbished. In average, about 85% of the ship material can be recovered by the ship recycling activity. Tola et al says that there is relationship between GDP and the number of ships dismantled [10]. When economic growth is high, the number of ships sent for demolition decreases, and vice versa. However, the demand for steel from downstream industries is a critical factor often overlooked in economic assessments. To see how this ship recycling activity in Indonesia affects to the Indonesian maritime industry, this paper is conducted by calculating the prediction of potential steel scrap from domestic ships with volume more than 100 GT based on PM 24 of 2022 Regulation for ship recycling.

2. Methodology

Methodology to conduct this research is shown in Figure 1 below. Research is started by conducting literature study and collecting data from BKI, NGO Ship Breaking Platform, and any other related documents. The literature study is conducted to understand the conditions under which a ship is considered to meet its End of Life (EoL) because it has no longer economical value thus it is considered for recycling. The study is resuming the recommended age for ships being recycled due to several reasons. The reasons why ship owners are considering for conducting recycling is learnt in this literature study.

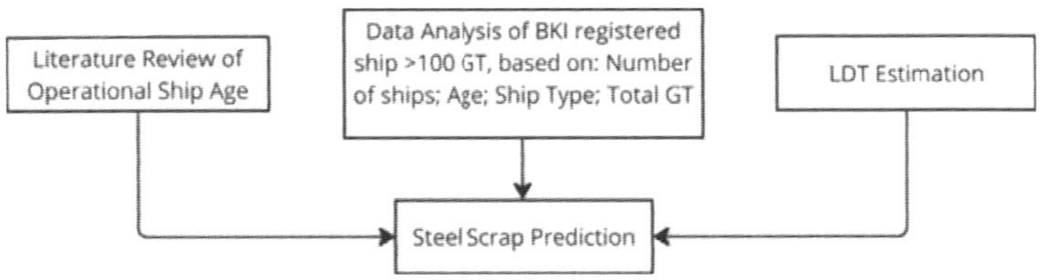

Figure 1. Research Flowchart

Simultaneously, the collected data is analysed. At first, data from BKI register for ships larger than 100 Gross Tonnage (GT) is processed based on age, ship type and its total GT to see the composition of Indonesian Ship. Subsequently, an analysis of Indonesian-Flagged ship data is conducted to create estimating equation of the Light Displacement Tonnage (LDT) for existing ships. The method for estimating LDT is utilizing linear regression from the available data for several ship types. The LDT data is crucial for understanding the potential contribution of ship recycling to the contribution of scrap steel from ships for steel industry. The estimated total LDT is then compared with the national steel needs in general and the specific needs for shipbuilding.

3. Indonesian Ship

There are 13.950 ships registered by BKI as per July 2023. Out of those ships, 13.155 ships with 100GT and more are required to comply PM 24 of 2022 by preparing IHM part I, II and III if the ships are intended for recycling. Figure 2 Shows number of ships and the total GT based on their age. It can be seen that the age distribution of ships registered by BKI with greater than 100 GT is most concentrated in the 10-15 year age range (30%) with a total volume of 6,171,365 GT (27%). However, it is specifically at the ages of 11 and 10 years (almost 9% of all ships and 7% from total GT for each age).

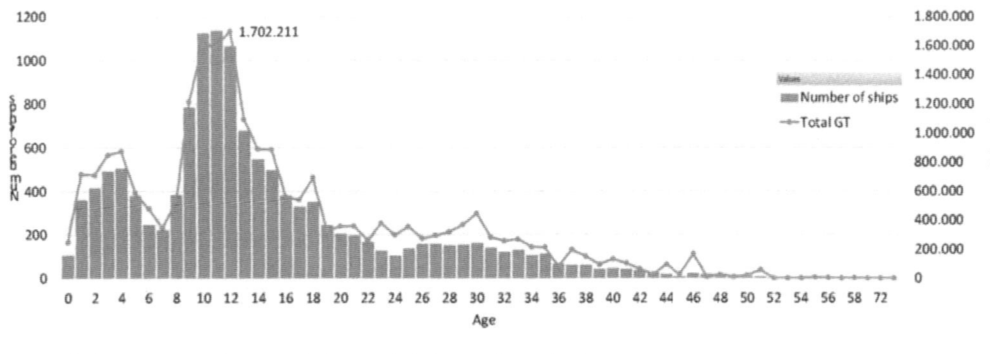

Figure 2. Number of ships and the total GT based on ship's age

Of all the ships registered by BKI, Table 1 shows the total GT and number of ships based on shipt type and its group of ship type. The ship type and its group classification is based on BKI Guidance for Classification and Construction (Pt.0 Vol B) that is issued in 2023 [11]. It can be seen that the largest number of ship type is Pontoon (31% of total number of ships and total volume is almost 11 million GT or 48% of total GT). The second largest number of ships is Tugs (37%), then followed by Barge (12%). However, if we see from total GT of all ships, the second largest volume is Barge (2.5 million GT or 11%), then followed by Oil Tanker (1.5 million GT or 7%).

Table 1. Total GT and number of ships based on ship type

Ship Type	Group of Ship Type	Total GT (%)	Number of Ships (%)
Pontoon	Pontoon and Barge	48%	31%
Barge	Pontoon and Barge	11%	12%
Oil Tanker	Oil Tanker	7%	3%
General Dry Cargo Ship	Dry cargo ship	6%	5%
Container ship	Container ship	5%	1%
RO-RO Passenger Ship	Passenger Ship	5%	3%
Tugs	Vessel for Special Operations	4%	37%
Bulk Carrier	Dry Cargo Ship	4%	0%
Passenger Ship	Passenger Ship	2%	1%
Liquefied Gas Carrier	Landing Craft	2%	0%
FSO	Floating Production Installation (FPI)	1%	0%
"X" Carrier	Dry cargo ship	1%	0%
FOI	Pontoon and Barge	1%	0%
Landing Craft	Landing Craft	1%	2%
Offshore Service Vessel	Vessel for Special Operations	1%	1%
FOLGT	Pontoon and Barge	1%	0%
Special Service Ship	Vessel for Special Operations	0%	0%
FPSO	Floating Production Installation (FPI)	0%	0%
Chemical Tanker	Chemical Tanker	0%	0%
Dredgers Ship	Pontoon and Barge	0%	0%
RO-RO ship	Passenger Ship	0%	0%
HSC	Pontoon and Barge	0%	0%
Tanker	Pontoon and Barge	0%	0%
Self Elevating Unit	Specific Offshore Unit	0%	0%
work ship	Vessel for Special Operations	0%	0%
Crew Boat	Offshore Service Vessel	0%	0%
Floating dock	Floating dock	0%	0%
Livestock Carrier	Liquefied Gas Carrier	0%	0%
Yacht	Yacht	0%	0%
Drilling vessel	Drilling Unit	0%	0%
CNG Carrier	Compressed Natural Gas Carrier	0%	0%
Fishing Vessel	Fishing Vessel	0%	0%

Table 2. Total GT from 11 largest GT ship type based on age range

Ship Type	0 yo	1-5 yo	6-10 yo	10-15 yo	16-20yo	21-25 yo	26-30 yo	31-35 yo	>35 yo	Total GT
Pontoon	1%	12%	9%	18%	6%	1%	1%	0%	0%	48%
Barge	0%	2%	4%	4%	1%	1%	0%	0%	0%	11%
Oil Tanker		0%	1%	1%	0%	1%	1%	1%	1%	7%
General Dry Cargo Ship		0%	0%	1%	1%	1%	1%	1%	1%	6%
Container ship		0%	1%	1%	0%	1%	1%	1%	0%	5%
RO-RO Passenger Ship		0%	1%	0%	1%	0%	1%	1%	1%	5%
Tugs	0%	1%	1%	1%	0%	0%	0%	0%	0%	4%
Bulk Carrier			0%	0%	1%	2%	1%		0%	4%
Passenger Ship		0%	0%	0%	0%	0%	1%	0%	0%	2%
Liquefied Gas Carrier			1%	0%		0%	0%	0%	0%	2%
FSO						0%	1%		0%	1%
Others	0%	0%	1%	1%	1%	0%	0%	1%	1%	6%
Total GT	**1%**	**17%**	**18%**	**27%**	**11%**	**7%**	**7%**	**5%**	**6%**	**1**

For more detail age range data, Table 2 shows 11 largest total GT based on ship types. From Table 2, it can be seen that the ships aged 0-20 years, the largest Total GT is found in Pontoons, followed by Barges. However, in older age ranges, the largest Total GT varies significantly. In the 21-25 year age range, Bulk Carriers dominate with a total of 435,938 GT. In the 26-30 year age range, the largest Total GT is found in Container ships, followed by General Dry Cargo Ships. In the

31-35 year age range and those over 35 years, the largest Total GT is found in RoRo Passenger Ships, followed by General Dry Cargo Ships and Oil Tankers, followed by General Dry Cargo Ships as well, respectively.

4. LDT Projection from Indonesian Ships

In scrap steel from ship transactions, the price is based on the Light Displacement Tonnage (LDT) of the vessel, which represents the weight of the empty ship that includes all its permanent equipment but without its ballast, cargo, crew, fuel, passengers, stores, nor water [5]. However, it typically includes the weight of permanent ballast and the water needed to operate steam machinery. To predict the amount of scrap steel from the contribution of old Indonesian ships that must meet the requirements of PM 24 of 2022, the LDT data of the ships needs to be processed. However, the available LDT data is limited, so the LDT values can be obtained using a regression approach.

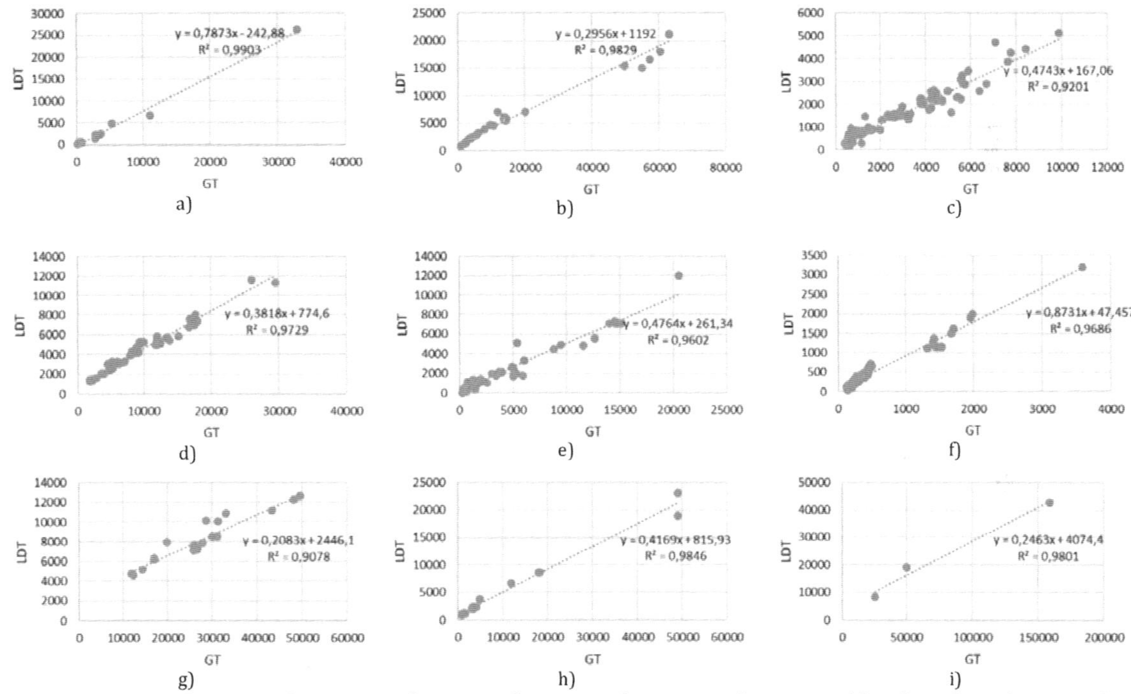

Figure 3. Scatter diagram of GT vs LDT for some ship type: a). Barge and Pontoon; b). Oil Tanker; c). General Dry Cargo; d). Container Ship; e). Passenger & Ro-Ro Passenger Ship; f). Tugs; g). Bulk Carrier; h). Liquefied

The available data for the Light Displacement Tonnage (LDT) of Indonesian-flagged ships amounts to 734 entries. Even though this data is not as extensive as the number of ships in the BKI register, the prediction of LDT based on the Gross Tonnage (GT) of other BKI-registered ships can be conducted by categorizing the ship types according to the eleven most common types listed in Table 2. From these eleven ship types with the largest GT, the LDT prediction can be grouped into seven categories due to the similarity between some ship types. The prediction calculations for LDT can be observed in the scatter plot in Figure 3 and the equations are listed in Table 3 below.

Table 3. LDT prediction for some ship type

Ship Type	LDT Prediction Equation	LDT Prediction (ton)
Barge and Pontoon	y=0,78773x-242,88	320.448
Oil Tanker	y=0,2956x+1192	168.807
General Dry Cargo	y=0,4743x+167,06	352.768
Container Ship	y=0,3818x+774,6	195.584
Passenger & Ro-Ro Passenger Ship	y=0,476x+261,34	410.022
Tugs	y=0,8731x+47,457	73.595
Bulk Carrier	y=0,2083x+2446,1	37.946
Liquefied Gas Carrier	y=0,4169x+815,93	26.808
FSO	y=0,2463x+4074,4	69.433
Others	y=0,3211x+654,98	169.686
Total		1.825.096

5. Discussion

HKC implementation is set to reduce the negative effect of ship dismantling activity through the safety of worker and its environment. Globally, ships that are being dismantled are vary in age, ship types, and conditions. According to a new report by Baltic and International Maritime Council (BIMCO), one of the largest international shipping associations in the world, the number of ships being recycled has been declined to its lowest level in twenty years. This declined number is caused by ships being kept operated longer, driven by market shocks, strong demand, and low order books of new ship [12]. Meanwhile, Mannan et al shows the main contributing factors for ship recycling are government policies, environmental and maritime regulation, and funding for capacity building [13]. Some reason why ship owner or operator decide their ships for recycling are the growth of the shipping industry, such as the existence of new technology and the capacity that affect to operational demands in the maritime sector. A requirement for double hull and/or bottom is also one of regulation reason for recycling consideration. Other conditions prompting ships to be recycled include old age, declining fuel efficiency, and rising maintenance and operational costs [14]. However, the exact recommended age for ships being recycled is difficult to be formulated. From DNV data, ships aged over 25 years has more incident [7]. For the decarbonising ship point of view, old ships need to be modified to comply emission requirement. Ships around 7-15 years old are expected to be able to comply GHG emission requirement if the ships are being retrofitted with multifuel engine. However, retrofitting ships more than 15 years is not preferable for ship owner in order to fulfil the emission requirement [15].

Generally, general cargo has average youngest age as at 14,2 years, while liquefied gas ship is at 22,9 years as the highest average age. Container ship average age is 16.3 years and chemical tanker is 16,4 years [16]. According to PFE Express, a logistic company, the average age of ship being recycled was 17.5 years old. Even though these recent years the average age is getting older, up to around 20 years old, ship recycling market is forecasted to be higher in 2024-2025. This prediction is caused by anticipating in arrival of new vessels [14]. To estimate the potential amount of steel scrap from Indonesian ships, an age of 25 years is used as the limit, considering it to be the highest age among the information collected, with the expectation of reducing the number of ship incidents as suggested by the research from DNV. Besides, there are almost 1200 ships is being build that are predicted to be done in 2024, as data BKI per July 2023, and it may be increased [8]. There will be possibility after the increased number of new ships, the old ship will be recycled. Even though the number of ships more than 25 years old is almost 2000 ships, there is a high possibility that in the next 2 years, the number of new ships can replace the old ship.

The total volume of ships registered by BKI that are over 25 years old is 19% of the total GT of all ships. The highest GT volume among these 25-year-old ships is for Passenger and Ro-Ro Passenger ships, followed by General Dry Cargo. From Table 2 above, it can be seen that the most common types of BKI ships, namely Barges and Pontoons, are operationally younger compared to General Dry Cargo ships or Passenger Ships and Ro-Ro Passenger Ships. The total LDT prediction values and for each ship type with the highest GT can be seen in Table 3. Total LDT prediction is almost 2 million tons. The highest LDT prediction is for Passenger Ships and Ro-Ro Passenger Ships, totalling 410,022 tons, followed by General Dry Cargo with 352,768 tons, then Barges and Pontoons with 320,448 tons. Although the GT of Barges and Pontoons accounts for only 10% of the total GT of BKI-registered ships over 25 years old, their LDT prediction is significant, comprising 18% of the total LDT prediction. Barges and Pontoons typically have quite heavy structures because they are designed to carry very large capacities and heavy cargo. The structural design to ensure the stability and strength of ships brings more weight of the structure. This results in the LDT being larger compared to the volume of the ship as measured in GT.

For a barge, the required steel estimation steel is approximately 1,300 tons. In East Java in 2020, 600 barge ships were needed, but only 10% are currently available. Therefore, it is estimated that around 700,000 tons of steel are needed to build these ships [17]. The availability of scrap steel from the estimated LDT of old Indonesian ships, which reaches nearly 2 million tons, will certainly help steel industry companies produce the required steel by melting down the scrap steel. The production of steel from scrap can also contribute to the reduction of carbon footprint, which is currently as a global issue. On a national level, the utilization of scrap steel from ships can increase the local content value (TKDN) of the steel products. The output of this ship recycling activity is positively enhance the shipbuilding industry chain. However, it needs supporting factors include government policies encouraging investment in new ship construction that can improve operational efficiency, and reduced maintenance and operational costs of old ships. This is related to the previous study conducted by Mannan et al that the south asian countries found opportunities from ship recycling activity, changing ship recycling status from liability become a commodity [18].

6. Conclusion

The implementation of the Hong Kong Convention (HKC) aims to mitigate the negative impacts of ship recycling activities by ensuring the safety of workers and the environment. Despite a significant decline in ship recycling to its lowest level in twenty years due to market conditions and extended ship service life, the need for recycling persists due to factors such as aging ships, declining fuel efficiency, and rising maintenance and operational costs. Ships aged over 25 years, which constitute 19% of the total Gross Tonnage (GT) of Indonesian ships, present a substantial potential for steel scrap, estimated at nearly 2 million tons. This scrap steel could greatly benefit the steel industry by reducing the carbon footprint and increasing the local content value (TKDN) of steel products.

The ship recycling not only supports the steel industry but also enhances the shipbuilding industry by promoting new ship construction, then it will improve operational efficiency, and reducing costs. Government policies encouraging investment in new ships can further support these efforts, turning ship recycling from a liability into a valuable commodity, as seen in South Asian countries.

References

[1] IMO, *Hong Kong International Convention for the Safe and Environmentally Sound Recycling of Ships,* 2009.

[2] S. Sunaryo , E. Djatmiko, S. Fariya, R. Kurt and S. Gunbeyaz, "A Gap Analysis of Ship-Recycling Practices in Indonesia," *Recycling,* vol. 6, p. 48, 2021.

[3] Ministry of Transportation Republic of Indonesia, *Regulation of the Ministry of Transportation No. 24 of 2022 on the Prevention of Marine Environmental Pollution,* 2022.

[4] NGO Ship Breaking Platform, "Ship Breaking: A Dirty and Dangerous Indystry," 01 06 2024. [Online]. Available: https://shipbreakingplatform.org/. [Accessed 01 06 2024].

[5] NGO Ship Breaking Platform, "The Toxic Tide: 2023 Shipbreaking Records," 01 02 2024. [Online]. Available: https://www.offthebeach.org/. [Accessed 01 06 2024].

[6] Hariyanto, I. N. Gutami, Z. Qonita, N. Shabrina, D. A. Wibowo and A. Muhtadi, "The Development of Ship Recycling Facility Cluster in Indonesia," in *The 7th International Conference on Marine Technology (SENTA 2022)*, Surabaya, 2022.

[7] Det Norske Veritas (DNV), "MARITIME IMPACT," Det Norske Veritas (DNV), 13 January 2022. [Online]. Available: https://www.dnv.com/expert-story/maritime-impact/Navigating-new-safety-challenges-after-a-decade-of-progress. [Accessed 1 June 2024].

[8] Biro Klasifikasi Indonesia, "Research Report on Ship Recycling in Indonesia," Jakarta, 2023.

[9] A. M. Hiremath and J. M. Vivek, "Circular Economy in Ship Recycling: An Indian Perspective," in *Fourth International Conference on Reuse and Recycling of Materials (ICRM) 2018*, Kottayam, Kerala, 2018.

[10] F. Tola, E. M. Mosconi, M. Marconi and M. Gianvincenzi, "Perspectives for the Development of a Circular Economy Model to Promote Ship Recycling Practices in the European Context: A Systemic Literature Review," *Sustainability,* vol. 15, no. 7, p. 5919, 2023.

[11] Biro Klasifikasi Indonesia, Guidance for Classification and Construction, Jakarta: BKI, 2023.

[12] M. Schuler, "Ship Recycling Plunges to Lowest Level in Decades," gCaptain, 04 April 2024. [Online]. Available: https://gcaptain.com/ship-recycling-plunges-to-lowest-level-in-decades/. [Accessed 01 May 2024].

[13] B. Mannan, M. J. Rizvi and Y. M. Dai, "Developing a MCDM-based Framework for Achieving a Circular Economy Through Ship Recycling," in *The International Conference on Marine Technology*, Dhaka, Bangladesh, 2022.

[14] PFE Express, "How Old is an Elderly Container Ship?," PFE Express, 2023. [Online]. Available: https://www.pfe.express/2023/12/how-old-is-an-elderly-container-ship/#:~:text=Typically%2C%20container%20ships%20are%20built,due%20to%20wear%20and%20tear.. [Accessed 01 05 2024].

[15] N. Graya, S. McDonagha, R. O'Sheaa, B. Smythc and J. D. Murphy, "Decarbonising ships, planes and trucks: An analysis of suitable low-carbon fuels for the maritime, aviation and haulage sectors," *Advances in Applied Energy,* vol. 1, p. 100008, 2021.

[16] T. Q. Pham, An assessment of development opportunities of ship recycling facilities based on the shipbuilding yards' infrastructures in Vietnam, Malmo: World Maritime University, 2019.

[17] T. K. Yunianto, "Prospek Industri Galangan Kapal untuk Mendongkrak Konsumsi Baja RI," KataData, 12 August 2020. [Online]. Available: https://katadata.co.id/berita/industri/5f33b61a3dd93/prospek-industri-galangan-kapal-untuk-mendongkrak-konsumsi-baja-ri. [Accessed 01 May 2024].

[18] B. Mannan, M. J. Rizvi and Y. M. Dai, "Ship recycling in developing economies of South Asia: Changing liability to a commodity," *Green Technologies and Sustainability,* vol. 2, no. 2, 2024.

IOP Conf. Series: Earth and Environmental Science **1423** (2024) 012043 doi:10.1088/1755-1315/1423/1/012043

Leveraging Bayesian Network to Improve the Marine Insurance's Condition Survey Form for Passenger Vessel

M Faishal[1*], R O S Gurning[1], A Santoso[1], and D H Waskito[2]

[1] Marine Engineering Department, Institut Teknologi Sepuluh Nopember, Surabaya, Indonesia
[2] Research Center for Transportation Technology, National Research and Innovation Agency, South Tangerang, Indonesia

*E-mail: m.faishal19@gmail.com

Abstract. To prevent the losses caused by accidents in passenger vessel, marine insurance is becoming one of the alternative solutions for the vessel's owner to retain their financial capability. One of the most essential phases in marine insurance is survey inspection, which utilises survey condition forms as guidance. However, the current form is not designed explicitly for particular vessel types, leading to inaccurate insurance assessment. The primary objective of this study is to identify and develop an adaptive and comprehensive condition survey form for passenger vessels in Indonesia by incorporating a qualitative and quantitative risk assessment, which will be the cornerstone of this study. The Bayesian Network (BN) approach was utilised to analyse several factors that lead to a sinking accident. Twenty-one accident data from Indonesian passenger vessels were utilised as base data for generating the conditional probability. The results indicated that vessel structural defect and unqualified crew are the two most significant factors contributing to the sinking accident's critical events, such as loss of stability and buoyancy. A practical improvement for the existing survey form has been proposed, which will benefit marine surveyor companies by enhancing the safety and accuracy of ship assessment.

1. Introduction

Indonesia relies heavily on maritime transportation to support its economy and connect its diverse regions [1]. Passenger ships are the lifeblood of this network, facilitating the movement of people and goods across vast distances. With over 300 ports and more than 1,200 passenger ships, Indonesia boasts one of the world's most active maritime transport systems [2]. However, this high volume of maritime traffic inherently introduces significant risks in terms of safety and financial loss [3].

Incidents involving passenger ships are common in Indonesia, with accidents causing injuries, loss of life, and significant financial burdens for ship owners and operators. In recent years, high-profile accidents like the sinking of the KM Sinar Bangun on Lake Toba in 2018 and the capsizing of the MV Lestari Maju in the same year highlighted the inherent risks of maritime travel in the country [4]. These incidents have underscored the need for robust safety protocols and adequate insurance coverage to protect passengers and ship owners from financial losses due to accidents and other unforeseen events.

Marine insurance is crucial in mitigating these risks, providing a safety net for ship owners against potential losses related to vessel damage, liability, and other unforeseen events [5]. In particular, Hull and Machinery (H&M) insurance covers physical damage to the ship, while Protection and Indemnity (P&I) insurance covers third-party liabilities, such as passenger injuries or environmental damage. However, accurate risk assessment is critical to determine appropriate coverage levels, where condition surveys come into play.

Survey forms are used as a pivotal instrument to evaluate the condition of vessels, guiding insurance underwriters in assessing risk and determining premiums. Condition surveys serve as a foundational tool for assessing the risks associated with maritime vessels, guiding insurance companies in setting premiums and terms of coverage. Unfortunately, many existing survey forms used in marine insurance are generic and not specifically designed for different types of ships. This lack of specificity can lead to inaccurate risk assessments, resulting in either underinsurance or overinsurance, which can create financial strain on ship owners or insurance companies, respectively [6].

Given Indonesia's unique maritime landscape, high frequency of trips, and diversity of ship types, a tailored approach to condition surveys is essential. This study addresses this gap by developing a comprehensive condition survey form explicitly designed for passenger vessels in Indonesia. By incorporating qualitative and quantitative assessments, the proposed survey form aims to generate a more accurate risk rating, providing a valuable tool for ship owners, insurers, and other stakeholders in the marine insurance industry. This comprehensive approach is expected to improve Indonesia's maritime transport system's overall safety and financial stability, reducing accidents and fostering a safer environment for passengers and crew.

Several risk assessment methods in the maritime sector can be carried out to obtain the factors that influence passenger ship accidents, one of which is the Bayesian Network (BN), which has been widely used in the maritime sector [7], [8]. Several authors have conducted BN to determine the factors that influence maritime accidents in sinking [9], grounding [10], collision [11], and piracy [12]. To improve and incorporate several factors, BN often has been integrated with HFACS to analyse human factors[13], [14], system theory [15], and FRAM [16] to improve the identification of the risk. Bayesian network (BN) is commonly used because of its reliability in analysing the relationship between factors through backward inference sensitivity analysis [17]. Therefore, the most influential factors can be further analysed as preventive measures.

The ability to identify the most significant factors in accident analysis can also be utilised in marine insurance survey forms, and based on the author's knowledge and literature reviews, there is no study on how to use BN on marine insurance survey forms. This study proposed to enhance the detail and accuracy of the marine insurance survey form by identifying the most significant factors contributing to ship sinking accidents in passenger vessels by employing the BN. After the most influential factors were considered, these factors were added to the current forms as additional items to be surveyed. The additional items would be related to specific items to be checked regarding the ship's machinery, structure, and even crew's certificate. Therefore, this study's novelty is that it will conduct the Bayesian network analysis of maritime accidents and incorporate the most influencing factor to generate a tailored approach to the marine insurance survey form.

2. Methods

2.1 Data Collection
To determine the most influencing factors of accidents in passenger vessels, the most critical step is defining the accident type related to the passenger vessel and how the survey form can improve the safety of the passenger vessel. Sinking accidents are caused mainly by the technical influence of the ship's condition, structure, machinery, or loading and unloading, which can cause the ship to lose moments of stability and buoyancy. These three factors are essential in marine insurance and can be comprehensively increased in detail. With the above considerations, this study uses the sinking accident investigation data of passenger vessels for Bayesian Network analysis.

There are 21 accident data of sinking accidents in Indonesia from 2007 to 2023 that were investigated by the NTSC [18]. Given the vast area of Indonesian waters and the high number of maritime accidents, the NTSC only investigates a few accidents (National Transportation Safety Committee of the Republic of Indonesia). The NTSC investigates ship accidents with more than 100 GT (Gross Tonnage) for passenger ships, ferries, RO-RO, and fishing vessels and ship accidents with more than 500 GT weight for cargo ships and tankers. Ship accidents must also include the following elements: casualties, damage or inoperability of ships and/or water facilities, and/or marine pollution.

2.2 Bayesian Network Theorem
Bayesian network, also known as a causal model, is a directed graphical model used to depict the conditional independence of a set of random variables [19]. BN has been widely used in several sectors other than maritime to analyse causal factors such as aviation [20], road transportation [21], and mining [22]. Using Bayesian networks requires the user to understand conditional probability adequately. For example, if a parent node A has a child node B connected by an arrow. Since event A has been known, the probability of event B can be calculated by:P(B | A). Meanwhile, when event B is known, the likelihood of event A can be calculated by Equation (1) and (2).

$$P(\mathrm{A}|\mathrm{B}) = \frac{P(A_i).P(B|A_i)}{P(B)} \qquad i = 1, 2, 3, 4, \dots, k \tag{1}$$

$$P(B_1, B_2, \rightarrow B_k) = \prod_{k=1}^{n} P(B_k\,|A_k) \tag{2}$$

Where P(Ai) is the prior probability of the hypothesis, P(B|Ai) is the conditional probability of B when Ai is determined, P(Ai|B) is the posterior probability as a result of bayesian analysis, and P(B) is the marginal probability of B.

2.3 Bayesian Network Modelling
Bayesian networks for sinking accidents are based on the cumulative integration of inputs from each sinking accident. Twelve nodes reflect the significant factors contributing to ship sinking accidents. Each of the 21 accidents was modelled, including the node connection. After a Bayesian network model based on the scenario of each accident, they were combined into one model that reflects all sinking accidents in Indonesia. Modelling and sensitivity calculations for the Bayesian network model were carried out using GENIE 4.1-academic software.

According to Figure 1, the BN model consists of several accident factor levels. The first is the ship's condition, which consists of the ship's age. The second level is the causal factors, which consist of "Engine Failure", "Weather", "Vessel structural defects", and human-related factors. The third level category is the triggering events, which comprises "Water Intake", "Overloading", and "Unstable Loading", which initiate a series of actions or consequences caused by causal factors at the second level and can influence the next level. The last level is the critical events that can cause the ship to sink: the "Loss of Stability" and "Loss of Buoyancy". Loss of stability refers to the ship's inability to recover to the upright position, while "Loss of Buoyancy" states the condition when the vessel cannot float properly. When one of the critical events occurs, the ship's probability of sinking will increase. The consequence node "Sinking" was added with two states, such as capsizing and tilting (heeling), since all the events investigated are sinking events with 100% automatic probability.

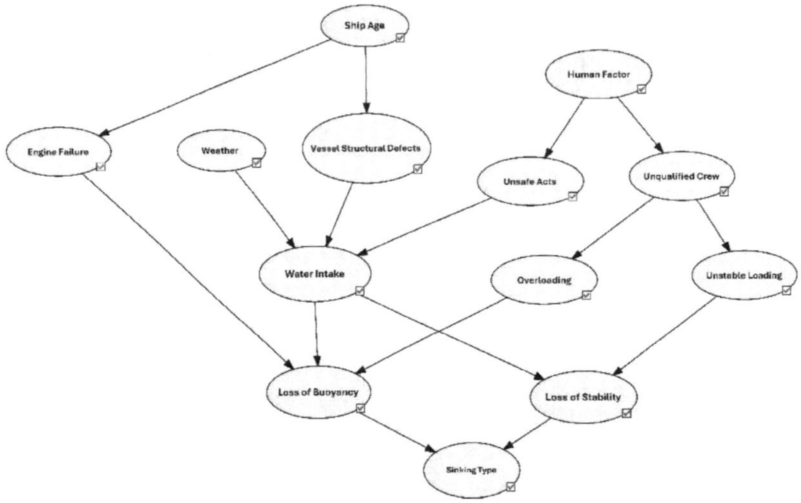

Figure 1. The proposed BN model that leads to sinking accidents for passenger vessels

Table 1. The occurrence and the prior probability of every factor

Factors (Nodes)	Occurrence (*f*)		Probability *P(f)*	
	Yes, Bad, Capsize	No, Good, Tilting	Yes, Bad, Capsize	No, Good, Tilting
Weather (Bad, Good)	9	12	42.86%	57.14%
Vessel Structural Defects	14	7	66.67%	33.33%
Engine Failure	6	15	28.57%	71.43%
Overloading	11	10	52.38%	47.62%
Unstable Loading	13	8	61.90%	38.10%
Water Intake	18	3	85.71%	14.29%
Loss of Stability	15	6	71.43%	28.57%
Loss of Buoyancy	15	6	71.43%	28.57%
Unsafe Acts	18	3	85.71%	14.29%
Unqualified Crew	9	12	42.86%	57.14%
Human Error	12	9	57.14%	42.86%
Sinking Type (Capsize, Tilting)	16	5	76.19%	42.86%

IOP Conf. Series: Earth and Environmental Science **1423** (2024) 012043 doi:10.1088/1755-1315/1423/1/012043

2.4 Prior and Conditional Probability

The prior probability value is obtained by dividing the occurrence of each factor in each accident (f) and the total accident number, 21. The occurrence is based on the report from the NTSC, and no assumption is included. For instance, if there is no evidence or statement in the report that the accident was caused by engine failure, it is categorised as "No". The detailed prior probabilities are shown in Table 1.

Table 2 shows how to calculate the conditional probability into the BN analysis for the "Loss of Stability" node. Of the 21 sinking cases, 13 were categorised as having unstable loading. Among these 13 cases, there were 10 cases where water intake caused the sinking. 9 out of 10 are led to the stability loss, which confirms the conditional probability of "Loss of stability" while "Unstable Loading" and "Water Intake" is 90%.

Table 2. Conditional Probability Table for Loss of Stability nodes

Relevant Nodes	**Conditional Cases**							
Unstable Loading	Yes (n=13)				No (n=8)			
Water Intake	Yes (n=10)		No (n=3)		Yes (n=8)		No (n=0)	
Loss of Stability	Yes (n=9)	No (n=1)	Yes (n=2)	No (n=1)	Yes (n=4)	No (n=4)	Yes (n=0)	No (n=0)
Probability	0.9	0.1	0.667	0.333	0.5	0.5	0.5	0.5

3. Results

3.1 Posterior Probability of Bayesian Network

Figure 2 depicts the posterior probabilities results from BN modelling after incorporating conditional probability. Regarding the sinking type, about 74.41% of ships were fully capsized and unable to be salvaged, which can be concluded that the probability to recover from sinking is negligible for passenger ships. From 21 sinking accidents, the probability of the ship experiencing the critical events of "Loss of stability" is 70.27%, while the "Loss of Buoyancy" has a probability of 74.27%. Those two critical events were heavily influenced by the triggering event of "Water Intake" with 85.1%, which shows that most of the sinking accidents commenced when the water infiltrated the ship structure, which can create stability disruption. The vessel structure defects also contribute with a 66.68% probability. Human factors also significantly contribute to sinking accidents, with the nodes "Unsafe acts" majorly with 85.71% probability.

3.2 Model Validation

Validation is an important part of the process since it provides confidence in the model's output and checks whether the model satisfies the natural occurrence compared to the real-case scenario. In order to validate the model that has been generated, three axioms should be accomplished, according to Pristrom [23].

Test of Axiom 1: The requirements for testing the model according to Axiom 1 involve checking whether the increase of parent nodes will also increase the probability of child nodes. In the BN for ship sinking accidents, the child nodes that will be checked are for "Loss of Buoyancy." For the loss of the buoyancy nodes, the parent nodes are "Engine Failure", Water Intake", and "Overloading". If a ship experiences a water intake into the compartment, the probability of buoyancy loss rises from 31.77% to 81.30%, as shown in Table 3. The results also indicate that by increasing the prior probability of each parent node, the posterior probability of their child nodes will increase. Therefore, the Bayesian network model satisfies the requirements of Axiom 1.

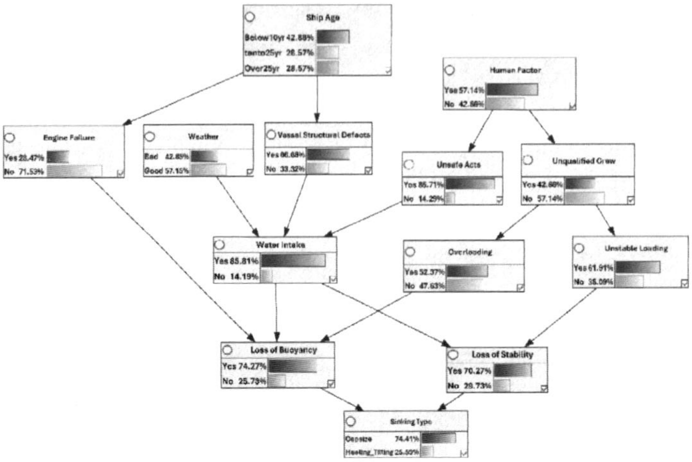

Figure 2. Posterior Probabilities results of contributing factors for sinking accidents

Table 3. Axiom 1 test of "Loss of Buoyancy"

Nodes	Condition	Probability of Error (%)	Posterior Probability of Loss of Buoyancy (%)
Engine Failure	Best	0	73.87
	Actual	28.47	74.27
	Worst	100	75.39
Water Intake	Best	0	31.77
	Actual	85.81	74.27
	Worst	100	81.30
Overloading	Best	0	61.96
	Actual	52.37	74.27
	Worst	100	81.80

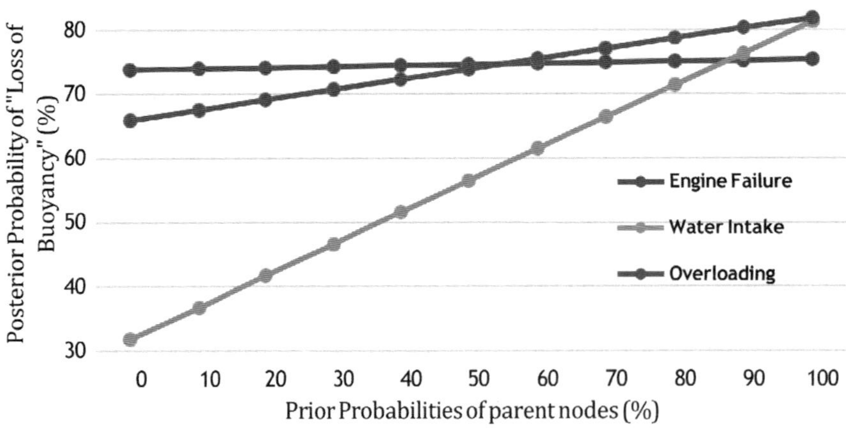

Figure 3. Results of Axiom 2 test of the BN model

Test of Axiom 2: Axiom 2 requires consistency in the probability change of the child nodes concerning the parent nodes. Figure 3 shows the change in the prior probability of "Engine Failure", "Water Intake", and "Overloading" towards the posterior probability of their child nodes, "Loss of Buoyancy". The probability of loss of buoyancy increases consistently without fluctuations in the graph. Therefore, the model has been found to fulfil the requirements of Axiom 2.

Test of Axiom 3: Axiom 3 states that the individual influence of each parent node on the child node should not be greater than the aggregate effect. For example, consider the child node "Water Intake," which consists of the parent nodes "Weather," "Vessel Structural Defects," "and "Unsafe Acts." If each parent node occurs independently, the probability of a sinking accident is 90.48%, 89.93%, and 88.2%, respectively. When all four parent nodes coincide, the probability of water intake is 100% greater than each parent node's. Similar calculations were performed for the other child nodes in the network, and all results satisfy Axiom 3.

3.3 Sensitivity Analysis

The main aim of the sensitivity analysis is to find the factors that influence the specific nodes the most. This study uses sensitivity analysis to improve the current marine insurance survey form and find the most sensitive factors contributing to sinking accidents involving passenger vessels. Firstly, the sensitivity was measured on two critical events, and the entropy reduction value denotes the sensitivity. Higher entropy affects data uncertainty more[24]. Table 4 shows the sensitivity for loss of buoyancy node, which shows that "Water Intake" is the most sensitive with 11.9%, followed by "Sinking Type" with 2.98%. Meanwhile, for the" Loss of Stability "node, the "Unstable Loading" is the most influencing factor, with 17.3%, followed by "Water Intake" with 4.47%.

In order to obtain the root causal factor that will be incorporated into the survey form, the sensitivity analysis was also analysed for "Water Intake" and "Unstable Loading" as the two most influencing factors of the two critical events of Loss of stability and Loss of Buoyancy. Figure 4 shows that vessel structural defects have the most entropy reduction, with 3.25% for the occurrence of water intake, and the unstable unqualified crew is the most impactful failure that contributes to the unstable loading of the ship with 6.24%, as shown in Figure 5.

Table 4. Sensitivity results of "Loss of Buoyancy" node

Loss of Buoyancy		**Loss of Stability**	
Node	Entropy Reduction (%)	Node	Entropy Reduction (%)
Water Intake	11.9	Unstable Loading	17.3
Sinking Type	2.98	Water Intake	4.47
Overloading	2.91	Unqualified Crew	1.47
Loss of Stability	0.974	Loss of Buoyancy	0.912
Unqualified Crew	0.63	Overloading	0.281
Unsafe Acts	0.444	Unsafe Acts	0.211
Vessel Structural Defect	0.391	Human Factor	0.136
Weather	0.18	Vessel Structural Defect	0.133
Human Factor	0.172	Sinking Type	0.0772
Unstable Loading	0.0497	Weather	0.0608
Ship Age	0.0367	Ship Age	0.00801
Engine Failure	0.0228	Engine Failure	0.000334

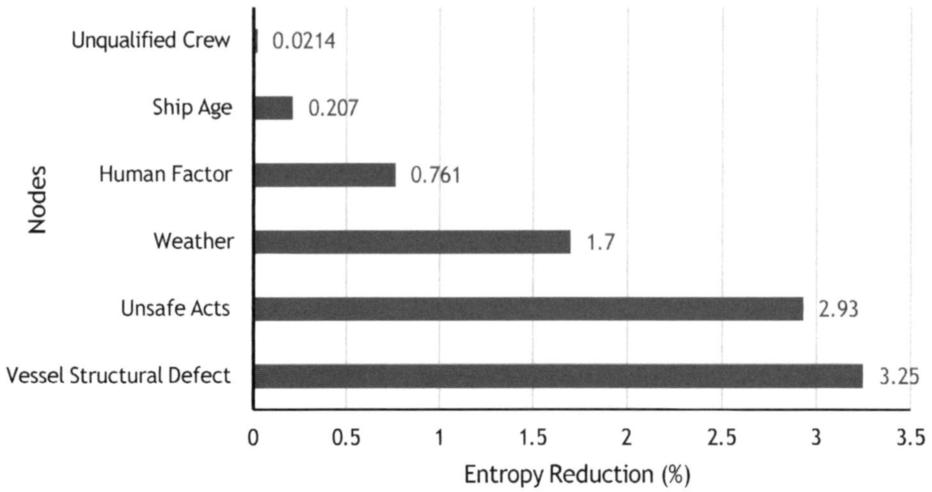

Figure 4. Sensitivity analysis of "Water Intake"

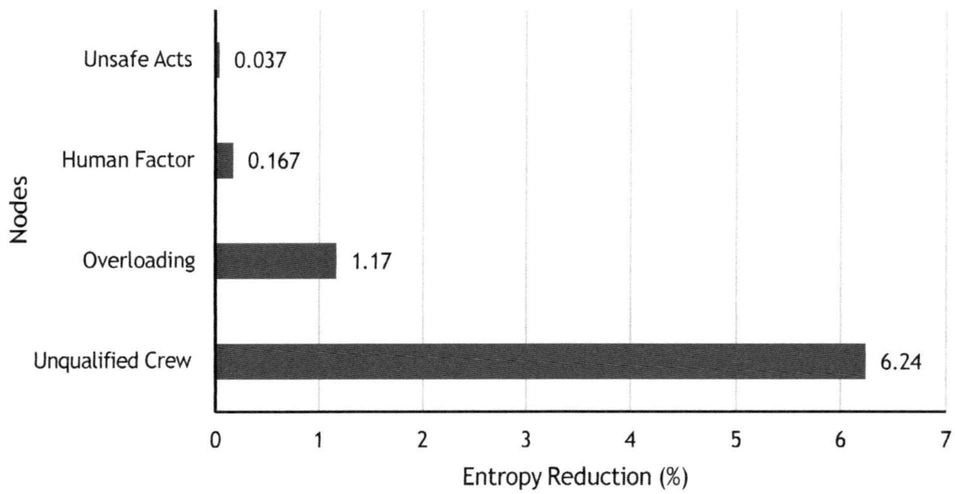

Figure 5. Sensitivity analysis of "Unstable Loading"

From the sensitivity analysis, it can be concluded that the two root causes of sinking accidents significantly contribute to these incidents: vessel structural defects, and unqualified crew. These findings are crucial for enhancing or revising the existing marine insurance survey explicitly tailored for passenger vessels.

3.4 Recommendation towards the existing survey form

Using P&I (Protection and Indemnity) survey forms for maritime insurance is not yet well-established in Indonesia. Instead, ship owners and insurers often rely on survey forms provided by P&I Clubs from abroad or online survey forms designed according to international standards. The existing marine insurance survey forms are based on several companies' H&M and P&I clubs. The baseline was from the IMO, Lloyds Register, DNV, and ABS. The form was adjusted depending on the needs and requirements of respective companies.

According to Table 6, the current survey item list is straightforward, not accommodating several items crucial for preventing accidents, such as sinking accidents on passenger vessels, whether ferry or Ro-Ro. For example, in the structural, there is no item for checking the ramp door and also the openings and holes in the double hull, where both items are very pivotal to prevent water leakage into the ship. As for crew assignment, the existing item list focuses on the number of crew members, but qualifications have not been included in the item list.

Table 6 also shows the recommendation according to the sensitivity analysis of BN, which was generated from the previous defects and failures in 21 accidents involving passenger vessels that were used for generating the BN model.

Table 5. The existing and recommendation item list for the marine insurance survey form

Component	Existing Survey item list	Additional
Structural	Side shell plating from sheer down to light water line and from aft to fore-ends	Stern Opening
	Deck and superstructure	Watertight Door
	Ballast tank	Ramp Door
		Void Tank
		Deck Opening
Crew assignment and qualifications	Manning	Check crews' certificate Check crews' attendance.

4. Discussion

This study uses Bayesian Network modelling to assess the influence of different factors contributing to ship sinking accidents for passenger vessels in Indonesia. The proposed model provides a comprehensive analysis of the sequential nature of these incidents. By adhering to the three axioms of Bayesian Network modelling, this framework demonstrates its potential to effectively model ship sinking accidents in the future. Moreover, the model is also substantial for obtaining the most significant factors in ship sinking accidents, which will be employed to improve the survey form for marine insurance.

The analysis of ship sinking accidents for passenger vessels reveals that "Unstable loading" is the most sensitive factors that contribute to the critical event "Loss of buoyancy", which is similar to the study from [25] highlighted through sensitivity analysis that the loss of stability due to cargo shifting increases the likelihood of ship sinking incidents. Furthermore, from the "Loss of stability" critical event, the most impacting factor is "Water Intake", which is similar to results from [9], which states that water intake is one of the most critical triggering events that lead to sinking accidents.

Human error is also essential in enhancing survey forms for marine insurance for sinking accidents. This study analysed two human error factors: "Unsafe acts" and "Unqualified crew". Unsafe acts significantly influence water intake, where the crew's negligence to close ramp door openings or doors causes water to enter the hull. Meanwhile, an unqualified crew impacts unstable loading, leading to capsizing or listing. The mismatch of the crew's position level to the certificate can affect the ability to calculate initial stability and manage cargo loading on the ship. These results are similar to previous studies [26], [27].

The results also show that the detailed level of the current marine insurance form can be enhanced by incorporating the most sensitive factors from BN analysis. The current form of marine insurance is considered to be a general checking practice. For instance, in terms of ship structure, the instructions for the surveyor only check the general condition instead of focusing on specific items. Several recommendations for improving or enhancing the component-level marine insurance survey form have been proposed. The additional items consist of more detailed ship components that must be addressed to prevent or assess damage from ship sinking accidents.

However, this study is also limited since it only addresses sinking accidents, which can be alleviated by incorporating other frequent maritime accidents such as fire, collision, and grounding should be incorporated into the BN. Moreover, it is crucial to incorporate accident data that the NTSC has not investigated to encompass a broader range of sea transportation modes in Indonesia. By including these uninvestigated incidents, the model can capture a more comprehensive representation of ship-sinking accidents, contributing to a more accurate risk analysis framework.

5. Conclusion

Given the unique risks in Indonesian maritime transportation, there is a clear need for customised condition survey forms designed explicitly for passenger vessels. These forms should include qualitative and quantitative assessments to understand the vessel's condition and associated risks thoroughly. This study aims to develop a condition survey form that addresses the specific risks of Indonesian passenger vessels, contributing to marine insurance knowledge and providing practical industry solutions. This study seeks to fill this gap by providing a specialised condition survey form incorporating insights from Bayesian Network (BN) analysis and addressing specific safety items. The results from BN indicated that three essential factors need to be emphasised in the marine insurance survey form. Based on the sensitivity analysis, vessel structural defects and unqualified crew are three factors recommended for the existing survey form. This study will be valuable to the marine insurance industry by providing a specialised condition survey form for Indonesian passenger vessels, addressing the unique risks that will occur in the foreseen future.

References

[1] Y. Rochwulaningsih, S. T. Sulistiyono, N. N. Masruroh, and N. N. Maulany, "Marine policy basis of Indonesia as a maritime state: The importance of integrated economy," *Mar. Policy*, vol. 108, no. March 2016, p. 103602, 2019, doi: 10.1016/j.marpol.2019.103602.

[2] Ministry of Transportation, "2020 Yearly Report," Jakarta, Indonesia, 2020.

[3] H. Li, K. Zhou, C. Zhang, M. Bashir, and Z. Yang, "Dynamic evolution of maritime accidents: Comparative analysis through data-driven Bayesian Networks," *Ocean Eng.*, vol. 303, no. February, p. 117736, 2024, doi: 10.1016/j.oceaneng.2024.117736.

[4] K. Heryandri, "The Importance of Ferry Ro-Ro Transportation in Indonesia and Its Contrary to The Lack of Attention on Ferry Ro-Ro Safety, Which Cause High Rate of Accidents and Fatalities," *Adv. Transp. Logist. Res.*, vol. 1, pp. 641–651, 2018.

[5] L. Johannsdottir, D. Cook, and G. M. Arruda, "Systemic risk of cruise ship incidents from an Arctic and insurance perspective," *Elementa*, vol. 9, no. 1, pp. 1–22, 2021, doi: 10.1525/elementa.2020.00009.

[6] K. X. Li, Y. Wang, and J. Min, "Quantitative analysis of materiality in marine insurance," *Marit. Policy Manag.*, vol. 36, no. 5, pp. 437–455, 2009, doi: 10.1080/03088830903187168.

[7] L. Fan, Z. Zhao, and J. Yin, "Research on dynamic influence mechanism of port state control and ship risk level," *Ocean Coast. Manag.*, vol. 250, no. May 2021, p. 107028, 2024, doi: 10.1016/j.ocecoaman.2024.107028.

[8] Ö . Ugˇurlu, S. Yıldız, S. Loughney, J. Wang, S. Kuntchulia, and I. Sharabidze, "Analyzing Collision, Grounding, and Sinking Accidents Occurring in the Black Sea Utilizing HFACS and Bayesian Networks," *Risk Anal.*, vol. 40, no. 12, pp. 2610–2638, 2020, doi: 10.1111/risa.13568.

[9] F. Ugˇurlu, S. Yıldız, M. Boran, Ö . Ugˇurlu, and J. Wang, "Analysis of fishing vessel accidents with Bayesian network and Chi-square methods," *Ocean Eng.*, vol. 198, no. August 2019, 2020, doi:

10.1016/j.oceaneng.2020.106956.
[10] S. T. Ung, "Navigation Risk estimation using a modified Bayesian Network modeling-a case study in Taiwan," *Reliab. Eng. Syst. Saf.*, vol. 213, 2021, doi: 10.1016/j.ress.2021.107777.
[11] K. X. Li, J. Yin, H. S. Bang, Z. Yang, and J. Wang, "Bayesian network with quantitative input for maritime risk analysis," *Transp. A Transp. Sci.*, vol. 10, no. 2, pp. 89–118, 2014, doi: 10.1080/18128602.2012.675527.
[12] M. Jiang and J. Lu, "The analysis of maritime piracy occurred in Southeast Asia by using Bayesian network," *Transp. Res. Part E*, vol. 139, no. 1, p. 101965, 2020, doi: 10.1016/j.tre.2020.101965.
[13] S. Adumene, M. Afenyo, V. Salehi, and P. William, "An adaptive model for human factors assessment in maritime operations," *Int. J. Ind. Ergon.*, vol. 89, no. December 2021, p. 103293, 2022, doi: 10.1016/j.ergon.2022.103293.
[14] O. Bayazit and M. Kaptan, "Dynamic risk analysis of allision in port areas using DBN based on HFACS-PV," *Ocean Eng.*, vol. 298, no. February, p. 117183, 2024, doi: 10.1016/j.oceaneng.2024.117183.
[15] S. Basnet, A. BahooToroody, J. Montewka, M. Chaal, and O. A. Valdez Banda, "Selecting cost-effective risk control option for advanced maritime operations; Integration of STPA-BN-Influence diagram," *Ocean Eng.*, vol. 280, no. April, p. 114631, 2023, doi: 10.1016/j.oceaneng.2023.114631.
[16] W. Qiao, X. Ma, Y. Liu, and W. Deng, "Resilience evaluation of maritime liquid cargo emergency response by integrating FRAM and a BN: A case study of a propylene leakage emergency scenario," *Ocean Eng.*, vol. 247, no. September 2021, p. 110584, 2022, doi: 10.1016/j.oceaneng.2022.110584.
[17] Y. Cao *et al.*, "Analysis of factors affecting the severity of marine accidents using a data-driven Bayesian network," *Ocean Eng.*, vol. 269, no. January, p. 113563, 2023, doi: 10.1016/j.oceaneng.2022.113563.
[18] National Transportation Safety Committee, "NTSC Transport Accident Investigation Statistics Book 2022," Jakarta, Indonesia, 2023.
[19] J. Pearl, *Probabilistic reasoning in intelligent systems: networks of plausible inference.* Morgan kaufmann, 1988.
[20] T. Zhou, J. Zhang, and D. Baasansuren, "A hybrid HFACS-BN model for analysis of mongolian aviation professionals' awareness of human factors related to aviation safety," *Sustain.*, vol. 10, no. 12, 2018, doi: 10.3390/su10124522.
[21] D. H. Waskito *et al.*, "Analysing the Impact of Human Error on the Severity of Truck Accidents through HFACS and Bayesian Network Models," *Safety*, vol. 10, no. 1, p. 8, 2024, doi: https://doi.org/10.3390/safety10010008.
[22] M. Mirzaei Aliabadi, H. Aghaei, O. Kalatpour, A. R. Soltanian, and A. Nikravesh, "Analysis of human and organisational factors that influence mining accidents based on Bayesian network," *Int. J. Occup. Saf. Ergon.*, vol. 26, no. 4, pp. 670–677, 2020, doi: 10.1080/10803548.2018.1455411.
[23] S. Pristrom, Z. Yang, J. Wang, and X. Yan, "A novel flexible model for piracy and robbery assessment of merchant ship operations," *Reliab. Eng. Syst. Saf.*, vol. 155, pp. 196–211, 2016, doi: 10.1016/j.ress.2016.07.001.
[24] M. Jiang, J. Lu, Z. Yang, J. Li, and M. Jiang, "Risk analysis of maritime accidents along the main route of the Maritime Silk Road : a Bayesian network approach Risk analysis of maritime accidents along the main route of the Maritime Silk Road : a Bayesian network approach," *Marit. Policy Manag.*, vol. 00, no. 00, pp. 1–18, 2020, doi: 10.1080/03088839.2020.1730010.
[25] O. Ugurlu, S. Yildiz, S. Loughney, J. Wang, S. Kuntchulia, and I. Sharabidze, "Analysing Collision, Grounding, and Sinking Accidents Occurring in the Black Sea Utilising HFACS and Bayesian Networks," 2020, [Online]. Available: http://researchonline.ljmu.ac.uk/id/eprint/8705/
[26] Ö . Ug˘urlu, S. Yıldız, S. Loughney, and J. Wang, "Modified human factor analysis and classification system for passenger vessel accidents (HFACS-PV)," *Ocean Eng.*, vol. 161, no. April, pp. 47–61, 2018, doi: 10.1016/j.oceaneng.2018.04.086.
[27] W. Qiao, Y. Liu, X. Ma, and Y. Liu, "A methodology to evaluate human factors contributed to maritime accident by mapping fuzzy FT into ANN based on HFACS," *Ocean Eng.*, vol. 197, no. September 2019, p. 106892, 2020, doi: 10.1016/j.oceaneng.2019.106892.

AUTHOR INDEX